# Vibronic Interactions and the Jahn-Teller Effect

# Progress in Theoretical Chemistry and Physics

## VOLUME 23

*Honorary Editors:*

Sir Harold W. Kroto *(Florida State University, Tallahassee, FL, U.S.A.)*
Pr Yves Chauvin *(Institut Français du Pétrole, Tours, France)*

*Editors-in-Chief:*

J. Maruani (formerly *Laboratoire de Chimie Physique, Paris, France)*
S. Wilson (formerly *Rutherford Appleton Laboratory, Oxfordshire, U.K.)*

*Editorial Board:*

V. Aquilanti *(Università di Perugia, Italy)*
E. Brändas *(University of Uppsala, Sweden)*
L. Cederbaum *(Physikalisch-Chemisches Institut, Heidelberg, Germany)*
G. Delgado-Barrio *(Instituto de Matemáticas y Física Fundamental, Madrid, Spain)*
E.K.U. Gross *(Freie Universität, Berlin, Germany)*
K. Hirao *(University of Tokyo, Japan)*
E. Kryachko *(Bogolyubov Institute for Theoretical Physics, Kiev, Ukraine)*
R. Lefebvre *(Université Pierre-et-Marie-Curie, Paris, France)*
R. Levine *(Hebrew University of Jerusalem, Israel)*
K. Lindenberg *(University of California at San Diego, CA, U.S.A.)*
R. McWeeny *(Università di Pisa, Italy)*
M.A.C. Nascimento *(Instituto de Química, Rio de Janeiro, Brazil)*
P. Piecuch *(Michigan State University, East Lansing, MI, U.S.A.)*
M. Quack *(ETH Zürich, Switzerland)*
S.D. Schwartz *(Yeshiva University, Bronx, NY, U.S.A.)*
A. Wang *(University of British Columbia, Vancouver, BC, Canada)*

*Former Editors and Editorial Board Members:*

| | |
|---|---|
| I. Prigogine (†) | H. Hubač (*) |
| J. Rychlewski (†) | M.P. Levy (*) |
| Y.G. Smeyers (†) | G.L. Malli (*) |
| R. Daudel (†) | P.G. Mezey (*) |
| M. Mateev (†) | N. Rahman (*) |
| W.N. Lipscomb (†) | S. Suhai (*) |
| H. Ågren (*) | O. Tapia (*) |
| D. Avnir (*) | P.R. Taylor (*) |
| J. Cioslowski (*) | R.G. Woolley (*) |
| W.F. van Gunsteren (*) | |

† deceased; * end of term

For previous volumes:
http://www.springer.com/series/6464

# Vibronic Interactions and the Jahn-Teller Effect

Theory and Applications

Edited by

**MIHAIL ATANASOV**
*Max-Planck Institute for Bioinorganic Chemistry, Stiftstr. 32-34, D-45470, Mülheim an der Ruhr, Germany*
*and*
*Bulgarian Academy of Sciences, Institute of General and Inorganic Chemistry, Sofia, Bulgaria*

**CLAUDE DAUL**
*Department of Chemistry, University of Fribourg, Fribourg, Switzerland*

*and*

**PHILIP L.W. TREGENNA-PIGGOTT**\*
*\*deceased*

*Editors*

Mihail Atanasov  
Institute of General and Inorganic Chemistry  
Bulgarian Academy of Sciences  
1113 Sofia  
Bulgaria  
atanasov@thch.uni-bonn.de

Claude Daul  
Department of Chemistry  
University of Fribourg  
Chemin du Musée 9  
CH-1700 Fribourg  
Switzerland  
claude.daul@unifr.ch

Philip L.W. Tregenna-Piggott*  
*deceased

ISSN 1567-7354  
ISBN 978-94-007-2383-2   e-ISBN 978-94-007-2384-9  
DOI 10.1007/978-94-007-2384-9  
Springer Dordrecht Heidelberg London New York

Library of Congress Control Number: 2011943349

© Springer Science+Business Media B.V. 2012  
No part of this work may be reproduced, stored in a retrieval system, or transmitted in any form or by any means, electronic, mechanical, photocopying, microfilming, recording or otherwise, without written permission from the Publisher, with the exception of any material supplied specifically for the purpose of being entered and executed on a computer system, for exclusive use by the purchaser of the work.

Printed on acid-free paper

Springer is part of Springer Science+Business Media (www.springer.com)

# PTCP Aim and Scope

## Progress in Theoretical Chemistry and Physics

*A series reporting advances in theoretical molecular and material sciences, including theoretical, mathematical and computational chemistry, physical chemistry and chemical physics and biophysics.*

## Aim and Scope

Science progresses by a symbiotic interaction between theory and experiment: theory is used to interpret experimental results and may suggest new experiments; experiment helps to test theoretical predictions and may lead to improved theories. Theoretical Chemistry (including Physical Chemistry and Chemical Physics) provides the conceptual and technical background and apparatus for the rationalisation of phenomena in the chemical sciences. It is, therefore, a wide ranging subject, reflecting the diversity of molecular and related species and processes arising in chemical systems. The book series *Progress in Theoretical Chemistry and Physics* aims to report advances in methods and applications in this extended domain. It will comprise monographs as well as collections of papers on particular themes, which may arise from proceedings of symposia or invited papers on specific topics as well as from initiatives from authors or translations.

The basic theories of physics – classical mechanics and electromagnetism, relativity theory, quantum mechanics, statistical mechanics, quantum electrodynamics – support the theoretical apparatus which is used in molecular sciences. Quantum mechanics plays a particular role in theoretical chemistry, providing the basis for the valence theories, which allow to interpret the structure of molecules, and for the spectroscopic models employed in the determination of structural information from spectral patterns. Indeed, Quantum Chemistry often appears synonymous with Theoretical Chemistry: it will, therefore, constitute a major part of this book series. However, the scope of the series will also include other areas of theoretical chemistry, such as mathematical chemistry (which involves the use of algebra and topology in the analysis of molecular structures and reactions); molecular mechanics, molecular dynamics and chemical thermodynamics, which play an important role in rationalizing the geometric and electronic structures of molecular assemblies and polymers, clusters and crystals; surface, interface, solvent and solid-stale effects; excited-state dynamics, reactive collisions, and chemical reactions.

Recent decades have seen the emergence of a novel approach to scientific research, based on the exploitation of fast electronic digital computers. Computation provides a method of investigation which transcends the traditional division between theory and experiment. Computer-assisted simulation and design may afford a solution to complex problems which would otherwise be intractable to theoretical analysis, and may also provide a viable alternative to difficult or costly laboratory experiments. Though stemming from Theoretical Chemistry, Computational Chemistry is a field of research in its own right, which can help to test theoretical predictions and may also suggest improved theories.

The field of theoretical molecular sciences ranges from fundamental physical questions relevant to the molecular concept, through the statics and dynamics of isolated molecules, aggregates and materials, molecular properties and interactions, and to the role of molecules in the biological sciences. Therefore, it involves the physical basis for geometric and electronic structure, states of aggregation, physical and chemical transformations, thermodynamic and kinetic properties, as well as unusual properties such as extreme flexibility or strong relativistic or quantum-field effects, extreme conditions such as intense radiation fields or interaction with the continuum, and the specificity of biochemical reactions.

Theoretical Chemistry has an applied branch – a part of molecular engineering, which involves the investigation of structure–property relationships aiming at the design, synthesis and application of molecules and materials endowed with specific functions, now in demand in such areas as molecular electronics, drug design and genetic engineering. Relevant properties include conductivity (normal, semi- and supra-), magnetism (ferro- and ferri-), optoelectronic effects (involving nonlinear response), photochromism and photoreactivity, radiation and thermal resistance, molecular recognition and information processing, biological and pharmaceutical activities, as well as properties favouring self-assembling mechanisms and combination properties needed in multifunctional systems.

Progress in Theoretical Chemistry and Physics is made at different rates in these various research fields. The aim of this book series is to provide timely and in-depth coverage of selected topics and broad-ranging yet detailed analysis of contemporary theories and their applications. The series will be of primary interest to those whose research is directly concerned with the development and application of theoretical approaches in the chemical sciences. It will provide up-to-date reports on theoretical methods for the chemist, thermodynamician or spectroscopist, the atomic, molecular or cluster physicist, and the biochemist or molecular biologist who wish to employ techniques developed in theoretical, mathematical and computational chemistry in their research programmes. It is also intended to provide the graduate student with a readily accessible documentation on various branches of theoretical chemistry, physical chemistry and chemical physics.

# Preface

Since the discovery of the Jahn-Teller effect in 1937 the concept of vibronic coupling has become a source of inspiration for many researchers – theoreticians and experimentalists, chemists and physicists. The Jahn-Teller theorem states that degenerate electronic states of non-linear molecules are energetically unstable with respect to nuclear displacements which lift the orbital degeneracy. It has been later shown that mixing of electronic states by vibrational modes (vibronic coupling) can be very strong indeed even if degeneracy is absent and when energy separation between the interacting states is quite large. Thus the vibronic coupling is the sole mechanism responsible for structural distortions and dynamics in molecules and solids, and is at the very heart of reactivity in chemistry both in thermochemical and photochemically induced reactions. Jahn-Teller intersections are now being recognized as prototype cases of conical intersections, where the nuclear and electronic motions are known to be inherently nonadiabatic in nature and electrons interchange freely between different potential energy surfaces. In the condensed phase, the significance of the Jahn-Teller effect is increasingly recognized, appreciated and applied to the origin of high-temperature superconductivity, ferroelectric phase transitions in perovskite crystals, superconductivity in fullerides, and of very large ("colossal") magnetoresistance in manganites. These and other related vibronic coupling problems are particularly challenging because the Jahn-Teller interaction competes with electronic correlation and relativistic (spin-orbit coupling) effects.

The present volume reports on the state-of-art in the field of the Jahn-Teller effect and vibronic interactions. Starting in the mid-1970s, Jahn-Teller symposia – biannual conferences gathered together chemists and physicist, theoreticians and experimentalist working in this largely interdisciplinary field. In this volume we report a major part of the contributions to this field presented at the XX International Conference on the Jahn-Teller effect held in Fribourg, 16th–20th August in Switzerland, 2010. Several contributions to the conference (K.A. Müller, V.Z. Polinger, D. Khomskii, etc.) are published elsewhere.

The book starts with an introductory chapter by Isaac Bersuker (Chap. 1) – a pioneer in the field – who provides a thorough and critical overview of the controversial issues related to the pseudo Jahn-Teller interactions in some cases of intersystem crossings. This chapter addresses the need of reinterpreting the experimental data in the Jahn-Teller system $Cu^{2+}$ doped MgO in light of the new ab-initio results and the improvement of the latter result by (still missing) multimode Jahn-Teller treatments etc.

Part I deals with the developments of the DFT treatments of the multimode Jahn-Teller problem (Chap. 2, M. Zlatar et al.) and discusses the extensive use of symmetry which allows a better understanding of both static and dynamic Jahn-Teller phenomena (contributions by B. Tsukerblat – Chap. 3 and M. Breza – Chap. 4, respectively). Part II presents the Jahn-Teller effect due to impurity centers in crystals, such as $Cu^{2+}$ doped MgO – a prototype of a dynamic Jahn-Teller effect thoroughly analysed in relation to temperature and field dependent EPR experiments (Chap. 5 by Riley et al.). The origin of the dynamic and static Jahn-Teller effect in the EPR spectra of transition metal impurities in insulators has been discussed in the light of ab-initio calculations by Garcia-Fernandez et al. (Chap. 6). The application of ultrasonic experiments to impurity centers in crystals with zinc-blende structure and tetrahedral coordination (Chap. 7 by V. Gudkov and I. Bersuker) offers the unique opportunity of evaluating empirically the linear Jahn-Teller coupling parameters. Vibronic coupling is manifested in Raman scattering in resonance with an electronic transition to a pseudo Jahn-Teller distorted excited state (Chap. 8 by I. Tehver et al.) and a theory of electronic transition from a non-degenerate to a two-fold degenerate state in an impurity is proposed by Hiznyakov et al. (Chap. 9). The many-electron multiplet theory as applied to oxygen vacancies in nanocrystalline $HfO_2$ and non-crystalline $SiO_2$ by Lucovsky at al. (Chap. 10) concludes this part.

Part III is dedicated to the manifestation of the Jahn-Teller effect in fullerenes, fullerides and related systems. The imaging of molecular orbitals of fullerene molecules on surface substrates by using scanning tunnelling microscopy and their splitting by static or dynamic Jahn-Teller effect is the subject of the very transparent and nicely visualized contribution by Dunn et al. (Chap. 11). The same group of authors analyses the Jahn-Teller effect in $p^3 \otimes h$ of $C_{60}^{3-}$ as found in the $T_c = 40$ K superconductor $A_3C_{60}$ (Chap. 12 by A. Lakin et al.), whereas Iwahara et al. (Chap. 13) focus on deriving vibronic coupling constants from experiment and (Muya et al., Chap. 14) deal with the theory and the investigation of pseudo-Jahn-Teller effect and chemical reactivity in the boron buckyball $B_{80}$.

Part IV examines the conical intersection and the interplay between Jahn-Teller coupling and spin-orbit coupling. Here Adhikari et al. (Chap. 15), using the $^2E'$ and $^2A'_1$ electronic states in a case study of the trimer of sodium, carry out a nuclear dynamics study employing diabatic ab-initio derived potential energy surfaces and non-adiabatic coupling terms proving the validity of the so-called "Curl condition". W. Ernst et al. (Chap. 16), present a thorough study that combines spectroscopy and theory and manifests the quenching of the $E \otimes \varepsilon$ Jahn-Teller coupling with increasing spin-orbit coupling across the series $K_3, Rb_3, Cs_3$. Along the same line, R. Fishman (Chap. 17) reviews the interaction between Jahn-Teller coupling due to the high-spin

Preface                                                                                                    ix

Fe(II) and spin-orbit coupling of the same ion in Fe(II)Fe(III) bimetallic oxalates as well as their relevance for the temperature- dependent magnetic properties.

Part V focuses on mixed valence compounds and their spectroscopic and magnetic properties that lead to exciting phenomena where Jahn-Teller coupling or pseudo Jahn-Teller coupling and electronic delocalization interfere. A vibronic pseudo Jahn-Teller coupling model for mixed valence dimers belonging to Robin and Day class II compounds is the subject of the contribution by Palii et al. (Chap. 18). This chapter elucidates the factors controlling the non-adiabatic Landau-Zener tunnelling between vibronic levels induced by a pulsed electric field. A mixed valence iron dimer $[L^1Fe_2(\mu\text{-OAc})_2](ClO_4)$ is investigated by Ostrovsky (Chap. 19) in the framework of the generalized vibronic model which takes into account both the local vibrations on the metal sites (Piepho-Krausz-Schatz model) and the molecular vibrations changing the intermetallic distance (suggested by Piepho).

Lucovsky et al. (Chap. 20) complete this part by presenting the many-electron multiplet theory as applied to the spectroscopic detection of hopping induced mixed valence Ti and Sc in $GdSc_{1-x}Ti_xO_3$.

The last Part II of the book deals with the cooperative Jahn-Teller effect and spin-coupling phenomena. A vibronic approach to the cooperative spin transitions in crystals based on cyano-bridged trigonal bipyramidal $M_2Fe_3$ complexes (M=Co,Os) subject of the report by Klokishner et al. (Chap. 21) allows one to determine the type and temperature of spin transitions in spin clusters originating from the changes of the spin in a single metal ion or caused by charge transfer between different ions in the cluster. Based on a model by Kamimura and Suwa (interplay between Jahn-Teller physics and Mott physics) the mechanism of the high-$T_c$ superconductivity and a new phase diagrams of high-$T_c$ oxo-cuprates has been proposed (Chap. 22, Ushio et al.). Completing this part are two chapters on a vibronic theory of ferroelectric transitions in hydrogen bonded phosphates (Chap. 23, Levin et al.) and on virtual phonon exchange influence on the magnetic properties of Jahn-Teller crystals with Jahn-Teller centers in a triply degenerate electronic states (Chap. 24, Kaplan et al.).

The editors hope that this volume will be of valuable use for researchers working in the field of Jahn-Teller coupling and vibronic coupling and in all closely related experimental and theoretical disciplines of physics and chemistry.

M. Atanasov
C.A. Daul
P.L.W. Tregenna-Piggott[*]

---

[*]It is with great sadness that we report the death of P.L.W. Tregenna-Piggott, the initiator of the Jahn-Teller Conference in 2010 and of the present volume.

# Acknowledgements

The Editors owe thanks to Prof. Dr. Frank Neese, Institute for Physical and Theoretical Chemistry, University of Bonn, and Max-Planck Institute for Bioinorganic Chemistry, Mülheim an der Ruhr, Germany, for generous support during the whole work on the edition of this book. Valuable help at the final stage of the edition of this Volume by Matija Zlatar, University of Belgrade, Belgrade, Serbia, is gratefully acknowledged.

# Contents

**1 Critical Review of Contributions to the Jahn-Teller Symposium JT2010 and Beyond** .......................................... 1
Isaac B. Bersuker

**Part I Jahn-Teller Effect and Vibronic Interactions: General Theory**

**2 Density Functional Theory Study of the Multimode Jahn-Teller Effect – Ground State Distortion of Benzene Cation** ..... 25
Matija Zlatar, Jean-Pierre Brog, Alain Tschannen,
Maja Gruden-Pavlović, and Claude Daul

**3 A Symmetry Adapted Approach to the Dynamic Jahn-Teller Problem** ...................................................... 39
Boris Tsukerblat, Andrew Palii, Juan Modesto Clemente-Juan,
and Eugenio Coronado

**4 Group-Theoretical Treatment of Pseudo-Jahn-Teller Systems** ........ 59
Martin Breza

**Part II Jahn-Teller Effect and Vibronic Interactions in Impurity Centers in Crystals**

**5 The Dynamic Jahn-Teller Effect in Cu(II)/MgO** ....................... 85
P.L.W. Tregenna-Piggott, C.J. Noble, and Mark J. Riley

**6 Dynamic and Static Jahn-Teller Effect in Impurities: Determination of the Tunneling Splitting** ............................................................... 105
Pablo Garcia-Fernandez, A. Trueba, M.T. Barriuso,
J.A. Aramburu, and Miguel Moreno

xiii

**7 Experimental Evaluation of the Jahn-Teller Effect Parameters by Means of Ultrasonic Measurements. Application to Impurity Centers in Crystals** .......................... 143
V.V. Gudkov and I.B. Bersuker

**8 Raman Scattering for Weakened Bonds in the Intermediate States of Impurity Centres** ........................ 163
Imbi Tehver, G. Benedek, V. Boltrushko, V. Hizhnyakov, and T. Vaikjärv

**9 Vibronic Transitions to a State with Jahn-Teller Effect: Contribution of Phonons** ................................................ 179
V. Hizhnyakov, K. Pae, and T. Vaikjärv

**10 Many-Electron Multiplet Theory Applied to O-Vacancies in (i) Nanocrystalline $HfO_2$ and (ii) Non-crystalline $SiO_2$ and Si Oxynitride Alloys** ................................................ 193
Gerry Lucovsky, Leonardo Miotti, and Karen Paz Bastos

**Part III  Fullerenes, Fullerides and Related Systems**

**11 $C_{60}$ Molecules on Surfaces: The Role of Jahn–Teller Effects and Surface Interactions** ........................................ 215
Janette L. Dunn, Ian D. Hands, and Colin A. Bates

**12 The Quadratic $p^3 \otimes h$ Jahn–Teller System as a Model for the $C_{60}^{3-}$ Anion** ........................................ 231
Andrew J. Lakin, Ian D. Hands, Colin A. Bates, and Janette L. Dunn

**13 Estimation of the Vibronic Coupling Constants of Fullerene Monoanion: Comparison Between Experimental and Simulated Results** ................................... 245
Naoya Iwahara, Tohru Sato, Kazuyoshi Tanaka, and Liviu F. Chibotaru

**14 Investigations of the Boron Buckyball $B_{80}$: Bonding Analysis and Chemical Reactivity** ...................................... 265
Jules Tshishimbi Muya, G. Gopakumar, Erwin Lijnen, Minh Tho Nguyen, and Arnout Ceulemans

Contents     xv

**Part IV    Conical Intersections and Interplay Between Jahn-Teller Coupling and Spin-Orbit Coupling: Theory and Manifestations in Magnetism and Spectroscopy**

**15   Adiabatic to Diabatic Transformation and Nuclear Dynamics on Diabatic Hamiltonian Constructed by Using *Ab Initio* Potential Energy Surfaces and Non-adiabatic Coupling Terms for Excited States of Sodium Trimer** ................. 281
Amit Kumar Paul, Somrita Ray, and Satrajit Adhikari

**16   Jahn–Teller Effect and Spin-Orbit Coupling in Heavy Alkali Trimers** ........................................................... 301
Andreas W. Hauser, Gerald Auböck, and Wolfgang E. Ernst

**17   Jahn-Teller Transitions in the Fe(II)Fe(III) Bimetallic Oxalates** ...... 317
R.S. Fishman

**Part V    Jahn-Teller Effect in Mixed Valence Systems**

**18   Coherent Spin Dependent Landau-Zener Tunneling in Mixed Valence Dimers** ................................................... 329
Andrew Palii, Boris Tsukerblat, Juan Modesto Clemente-Juan, and Eugenio Coronado

**19   Mixed Valence Iron Dimer in the Generalized Vibronic Model: Optical and Magnetic Properties** ............................... 351
Serghei M. Ostrovsky

**20   Spectroscopic Detection of Hopping Induced Mixed Valence of Ti and Sc in $GdSc_{1-x}Ti_xO_3$ for x Greater than Percolation Threshold of 0.16** ..................................... 361
Gerry Lucovsky, Leonardo Miotti, and Karen Paz Bastos

**Part VI    Cooperative Jahn-Teller Effect and Spin-Coupling Phenomena**

**21   Vibronic Approach to the Cooperative Spin Transitions in Crystals Based on Cyano-Bridged Pentanuclear $M_2Fe_3$ (M=Co, Os) Clusters** .................................................... 379
Serghei Ostrovsky, Andrew Palii, Sophia Klokishner, Michael Shatruk, Kristen Funck, Catalina Achim, Kim R. Dunbar, and Boris Tsukerblat

**22   On the Interplay of Jahn–Teller Physics and Mott Physics in the Mechanism of High $T_c$ Superconductivity** ....................... 397
H. Ushio, S. Matsuno, and H. Kamimura

**23 Vibronic Theory Approach to Ising Model Formalism and Pseudospin-Phonon Coupling in H-Bonded Materials** .......... 419

S.P. Dolin, Alexander A. Levin, and T.Yu. Mikhailova

**24 Virtual Phonon Exchange Influence on Magnetic Properties of Jahn-Teller Crystals: Triple Electronic Degeneracy** .... 429

Michael D. Kaplan and George O. Zimmerman

**Index** ................................................................ 441

# Contributors

**Catalina Achim** Department of Chemistry, Carnegie Mellon University, Pittsburgh, PA, USA

**Carolina Adamo** Cornell University, Ithaca, NY 14583, USA

**Satrajit Adhikari** Indian Association for the Cultivation of Science, Jadavpur, Kolkata 700032, India, pcsa@iacs.res.in

**J.A. Aramburu** Departamento de Ciencias de la Tierra y Física de la Materia Condensada, Universidad de Cantabria, Avda. de los Castros s/n. 39005 Santander, Spain, antonio.aramburu@unican.es

**Gerald Auböck** Ecole Polytechnique Fédérale de Lausanne, Institut des sciences et inǵenierie chimiques, CH-1015 Lausanne, Switzerland, gerald.aubock@epfl.ch

**M.T. Barriuso** Departamento de Física Moderna, Universidad de Cantabria, Avda. de los Castros s/n. 39005 Santander, Spain, barriust@unican.es

**Karen Paz Bastos** Department of Physics, North Carolina State University, Raleigh, NC 27695-8202, USA

**Colin A. Bates** School of Physics and Astronomy, University of Nottingham, Nottingham, UK

**G. Benedek** Donostia International Physics Center, San Sebastian, Spain

Department of Materials Science, University of Milano-Bicocca, Milan, Italy

**Isaac B. Bersuker** Institute for Theoretical Chemistry, The University of Texas at Austin, Austin, TX 78712, USA, bersuker@cm.utexas.edu

**V. Boltrushko** Institute of Physics, University of Tartu, Tartu, Estonia

**Martin Breza** Department of Physical Chemistry, Slovak Technical University, SK-81237 Bratislava, Slovakia, martin.breza@stuba.sk

**Jean-Pierre Brog** Department of Chemistry, University of Fribourg, Fribourg, Switzerland

**Arnout Ceulemans** Department of Chemistry and INPAC, Institute for Nanoscale Physics and Chemistry, Katholieke Universiteit Leuven, B-3001 Leuven, Belgium, Arnout.Ceulemans@chem.kuleuven.be

**Liviu F. Chibotaru** Division of Quantum and Physical Chemistry, University of Leuven, Celestijnenlaan 200F, B-3001 Leuven, Belgium, Liviu.Chibotaru@chem.kuleuven.be

**Juan Modesto Clemente-Juan** Instituto de Ciencia Molecular, Universidad de Valencia, Polígono de la Coma, s/n 46980 Paterna, Spain

Fundació General de la Universitat de València (FGUV), Valencia, Spain, juan.m.clemente@uv.es

**Eugenio Coronado** Instituto de Ciencia Molecular, Universidad de Valencia, Polígono de la Coma, s/n 46980 Paterna, Spain, eugenio.coronado@uv.es

**Claude Daul** Department of Chemistry, University of Fribourg, Fribourg, Switzerland, claude.daul@unifr.ch

**S.P. Dolin** NS Kurnakov Institute of General and Inorganic Chemistry, RAS, 119991 Leninskii prospect, 31, Moscow, Russia

**Kim R. Dunbar** Department of Chemistry, Texas A&M University, College Station, TX, USA, dunbar@mail.chem.tamu.edu

**Janette L. Dunn** School of Physics and Astronomy, University of Nottingham, Nottingham, UK, janette.dunn@nottingham.ac.uk

**Wolfgang E. Ernst** Institute of Experimental Physics, Graz University of Technology, Petersgasse 16, A-8010 Graz, Austria, wolfgang.ernst@tugraz.at

**Randy S. Fishman** Materials Science and Technology Division, Oak Ridge National Laboratory, Oak Ridge, TN 37831-6065, USA, fishmanrs@ornl.gov

**Kristen Funck** Department of Chemistry, Texas A&M University, College Station, TX, USA, kchambers@mail.chem.tamu.edu

**Pablo Garcia-Fernandez** Departamento de Ciencias de la Tierra y Física de la Materia Condensada, Universidad de Cantabria, Avda. de los Castros s/n. 39005 Santander, Spain, garciapa@unican.es

**G. Gopakumar** Max-Planck Institut für Kohlenforschung, Kaiser-Wilhelm-Platz 1, 45470 Mülheim an der Ruhr, Germany, gopakumar@kofo.mpg.de

**Maja Gruden-Pavlović** Faculty of Chemistry, University of Belgrade, Belgrade, Serbia, gmaja@chem.bg.ac.rs

**V.V. Gudkov** Ural Federal University, 19, Mira st., 620002 Ekaterinburg, Russia, gudkov@imp.uran.ru

Contributors

**Ian D. Hands** School of Physics and Astronomy, University of Nottingham, Nottingham, UK, ian.hands@nottingham.ac.uk

**Andreas W. Hauser** Institute of Experimental Physics, Graz University of Technology, Petersgasse 16, A-8010 Graz, Austria, andreas.w.hauser@gmail.com

**V. Hizhnyakov** Institute of Physics, University of Tartu, Tartu, Estonia, hizh@fi.tartu.ee

**Naoya Iwahara** Department of Molecular Engineering, Graduate School of Engineering, Kyoto University, Kyoto 615-8510, Japan, iwaharanaoya@t03.mbox.media.kyoto-u.ac.jp

**H. Kamimura** Research Institute for Science and Technology, Tokyo University of Science, 1-3 Kagurazaka, Shinjuku-ku, Tokyo 162-8601, Japan, kamimura@rs.kagu.tus.ac.jp

**Michael D. Kaplan** Department of Chemistry and Physics, Simmons College, 300 The Fenway, Boston, MA 02115, USA, michael.kaplan@simmons.edu

**Sophia Klokishner** Institute of Applied Physics, Academy of Sciences of Moldova, Kishinev, Moldova, klokishner@yahoo.com

**Andrew J. Lakin** School of Physics and Astronomy, University of Nottingham, Nottingham, UK, ppxal@nottingham.ac.uk

**Alexander A. Levin** NS Kurnakov Institute of General and Inorganic Chemistry, RAS, 119991 Leninskii prospect, 31, Moscow, Russia, Levin@igic.ras.ru

**Erwin Lijnen** Department of Chemistry and INPAC, Institute for Nanoscale Physics and Chemistry, Katholieke Universiteit Leuven, B-3001 Leuven, Belgium, Erwin.Lijnen@chem.kuleuven.be

**Gerry Lucovsky** Department of Physics, North Carolina State University, Raleigh, NC 27695-8202, USA, lucovsky@ncsu.edu

**S. Matsuno** Shimizu General Education Center, Tokai University, 3-20-1 Orido Shimizu-ku, Shizuoka 424-8610, Japan, smatsuno@scc.u-tokai.ac.jp

**T.Yu. Mikhailova** NS Kurnakov Institute of General and Inorganic Chemistry, RAS, 119991 Leninskii prospect, 31, Moscow, Russia

**Leonardo Miotti** Department of Physics, North Carolina State University, Raleigh, NC 27695-8202, USA

**Miguel Moreno** Departamento de Ciencias de la Tierra y Física de la Materia Condensada, Universidad de Cantabria, Avda. de los Castros s/n. 39005 Santander, Spain, morenom@unican.es

**Jules Tshishimbi Muya** Department of Chemistry and INPAC, Institute for Nanoscale Physics and Chemistry, Katholieke Universiteit Leuven, B-3001 Leuven, Belgium, Jules.Tshishimbi@chem.kuleuven.be

**Minh Tho Nguyen** Department of Chemistry and INPAC, Institute for Nanoscale Physics and Chemistry, Katholieke Universiteit Leuven, B-3001 Leuven, Belgium, Minh.Nguyen@chem.kuleuven.be

**C.J. Noble** Centre for Advanced Imaging, University of Queensland, St. Lucia, QLD 4072, Australia, c.noble@uq.edu.au

**Serghei M. Ostrovsky** Institute of Applied Physics of the Academy of Sciences of Moldova, Kishinev, Moldova, sm_ostrovsky@yahoo.com

**K. Pae** Institute of Physics, University of Tartu, Tartu, Estonia, kaja.pae@gmail.com

**Andrew Palii** Institute of Applied Physics, Academy of Sciences of Moldova, Academy Str. 5 Kishinev, Moldova, andrew.palii@uv.es

**Amit Kumar Paul** Indian Association for the Cultivation of Science, Jadavpur, Kolkata 700032, India, pcakp@iacs.res.in

**Somrita Ray** Indian Association for the Cultivation of Science, Jadavpur, Kolkata 700032, India, pcsr@iacs.res.in

**Mark J. Riley** School of Chemistry and Molecular Biosciences, University of Queensland, St. Lucia, QLD 4072, Australia, m.riley@uq.edu.au

**Tohru Sato** Department of Molecular Engineering, Graduate School of Engineering, Kyoto University, Kyoto 615-8510, Japan

Fukui Institute for Fundamental Chemistry, Kyoto University, Takano-Nishihiraki-cho 34-4, Sakyo-ku, Kyoto 606-8103, Japan, tsato@scl.kyoto-u.ac.jp

**Darrell G. Schlom** Cornell University, Ithaca, NY 14583, USA

**Michael Shatruk** Department of Chemistry and Biochemistry, Florida State University, Tallahassee, FL, USA, shatruk@chem.fsu.edu

**Kazuyoshi Tanaka** Department of Molecular Engineering, Graduate School of Engineering, Kyoto University, Kyoto 615-8510, Japan, ktanaka@moleng.kyoto-u.ac.jp

**Imbi Tehver** Institute of Physics, University of Tartu, Tartu, Estonia, tehver@fi.tartu.ee

**P.L.W. Tregenna-Piggott** Laboratory for Neutron Scattering, ETH Zürich & Paul Scherrer Institut, CH-5232 Villigen PSI, Switzerland

**A. Trueba** Departamento de Ciencias de la Tierra y Física de la Materia Condensada, Universidad de Cantabria, Avda. de los Castros s/n. 39005 Santander, Spain, truebaa@unican.es

**Alain Tschannen** Department of Chemistry, University of Fribourg, Fribourg, Switzerland

**Boris Tsukerblat** Chemistry Department, Ben-Gurion University of the Negev, Beer-Sheva 84105, Israel, tsuker@bgu.ac.il

**Hideki Ushio** Tokyo National College of Technology, 1220-2 Kunugida-chou, Hachioji 193-0997, Japan, ushio@tokyo-ct.ac.jp

**T. Vaikjärv** Institute of Physics, University of Tartu, Tartu, Estonia, taavi.vaikjarv@ut.ee

**George O. Zimmerman** Department of Physics, Boston University, 590 Commonwealth Ave., Boston, MA 02215, USA

**Matija Zlatar** Department of Chemistry, University of Fribourg, Fribourg, Switzerland, and Center for Chemistry, IHTM, University of Belgrade, Belgrade, Serbia, matijaz@chem.bg.ac.rs

# Chapter 1
# Critical Review of Contributions to the Jahn-Teller Symposium JT2010 and Beyond

**Isaac B. Bersuker**

**Abstract** A review of some important problems presented at the XX International Symposium on the Jahn-Teller effect (JTE) is given, outlining also the author's view on some controversial issues. It is shown that the presentation of the pseudo JTE (PJTE) as a second order perturbation-theory correction (and hence called "second order JTE") is misleading. The PJTE is a two-level (multilevel) problem that can only be reduced to a second order perturbation in very limited cases when there is no term crossover or when the crossing between the interaction states is avoided. This statement is illustrated by several examples demonstrating also the possibilities of the PJTE as a tool for molecular and solid state problem solving. The possible reason of the puzzling, drastically different results obtained for the magnitude of tunneling splitting in the impurity system $MgO : Cu^{2+}$ by experimental EPR measurements and theoretical ab initio calculations is discussed. More specifically, it is suggested that a reinterpretation of the experimental results is needed, while the calculations can be improved by taking into account the multimode nature of the problem. A brief discussion of the relation between orbital ordering and the cooperative JTE is given in view of a talk presented to the Symposium under an incorrect (wrong) title of "failure of the JTE physics". More elaborate is the controversy around the origin of ferroelectricity in perovskite type crystals, mostly $BaTiO_3$, discussed in a plenary talk to the Symposium. The vibronic PJT origin and order-disorder nature of the ferroelectric phase transitions in these crystals, predicted many years ago and presently fully confirmed experimentally, serves also as an indication of the predictive ability of the PJTE and should be

---

I.B. Bersuker (✉)
Institute for Theoretical Chemistry, The University of Texas at Austin, Austin, TX 78712, USA
e-mail: bersuker@cm.utexas.edu

M. Atanasov et al. (eds.), *Vibronic Interactions and the Jahn-Teller Effect:*
*Theory and Applications*, Progress in Theoretical Chemistry and Physics 23,
DOI 10.1007/978-94-007-2384-9_1, © Springer Science+Business Media B.V. 2012

recognized as such. Some historical notes were also made with regard of the JT origin of high-temperature superconductivity discussed in a plenary talk to the Symposium.

## 1.1 Introduction

This review paper resulted from the analysis of some up-to-date problems presented at the XX Symposium on the Jahn-Teller effect (JTE) and related issues, and prepared on request of the Organizing Committee. Obviously, only a few questions were reported in the talk to the Symposium and can be discussed in this paper of limited volume, and their choice is solely that of the author's.

In light of our experience in the field and of the presentations given at the Symposium, we found that one of the basic notions in the JTE theory, the pseudo JTE (PJTE), lacks both a precise definition and a unique interpretation in the current literature, leading to confusion and misunderstanding. Meanwhile, the PJTE has become one of the most important applications of the vibronic coupling theory which in many aspects surpasses the importance of the JTE itself. In this paper we tried to clarify this situation and demonstrate the importance of a proper understanding of the PJTE showing also several molecular and solid state examples from our papers illustrating this problem.

Another topic brought up at the Symposium is that one of tunneling splitting in JT systems, which was one of the first experimental demonstrations of the dynamical JTE predicted by the theory. A very puzzling situation emerged in the publications and in the contributions presented at the Symposium and related to the two-orders-of-magnitude difference between the experimentally (EPR) observed and ab initio calculated magnitude of the tunneling splitting in the impurity crystal $MgO : Cu^{2+}$ which is subject to the JTE problem $E \otimes e$. We show here our perceived inconsistencies in the interpretation of the experimental data and the possible improvement to the calculations that could solve this embarrassing problem.

There is already some history in the relation between the problems of orbital ordering (OO) in crystals with electronically degenerate centers and the cooperative JTE (CJTE) in such systems. A part of it was discussed at the previous Symposium on the JTE in Heidelberg. Orbital ordering that ignores the vibronic coupling (the JTE) is an easy way to try to handle the problems of JT crystals, but in most cases it is not the correct way. As the local vibronic coupling is much stronger than the orbital-orbital interaction between metal centers (which are not adjacent), there is not much sense to consider the latter ignoring the former. In this context, the talk presented to the Symposium under title of "failure of the JTE physics" looks provocative. A brief critical discussion of this subject is given in this review.

More attention is devoted to the problem of vibronic origin of ferroelectricity in perovskite type crystals, which demonstrates the prediction abilities of the JTE theory. Indeed, based on the vibronic theory (the PJTE), it was predicted that the fundamental property of ferroelectric phase transitions in so-called displacive ferroelectrics is triggered by the local vibronic coupling, being hence of order-disorder type, not displacive! This prediction was made when there was none experimental evidence that could confirm it, and therefore it was not well accepted at the time. Unfortunately, it remains obscure even now after full experimental confirmation of all basic predictions of the PJTE theory. There was a talk about this subject at the Symposium, and the present paper is to further defend this important contribution of the vibronic coupling theory and its prediction power.

Finally, the role of the JTE in the modern problem of high-temperature superconductivity (HTSC) presented at the Symposium by one of its two authors, K.A. Muller, is augmented here by some additional historical notes.

## 1.2 The Pseudo Jahn-Teller Effect: Definition, Understanding, and Misunderstanding

The theory of vibronic coupling defined as the coupling between electronic states and nuclear displacements has already a long history. Starting with the equilibrium configuration of the system in a non-degenerate electronic state and considering the nuclear displacements as a perturbation one can easily find that only totally symmetric vibrations contribute to the vibronic coupling because the matrix elements of lower symmetry displacements are zero by symmetry. This understanding has been dominating the vibronic coupling theory for many decades being widely used in molecular problems to explain vibrational implications in observable properties and, more elaborate, in solid state physics as electron-phonon interaction.

The situation became more complicated for some particular cases of molecular systems and local properties in crystals in degenerate electronic states after it was shown by Jahn and Teller [1] that in these cases the coupling to low-symmetry nuclear displacements is non-zero too, with the result that the starting high-symmetry configuration is no more in equilibrium with respect to the latter, it distorts spontaneously. This Jahn-Teller effect (JTE) has been developed forming now a whole trend which serves as a tool for problem solving when there is electronic degeneracy (see, e.g., in [2]). Obviously, a sufficiently small splitting of the electronic term that removes the degeneracy may not destroy completely the JTE (albeit modifying it), so the system remains unstable in this configuration and distorts spontaneously. This was shown by Opik and Pryce [3]. A small splitting at the point of crossing terms looks like a "pseudo crossing" (avoided crossing). Therefore the JTE in this situation is termed pseudo JTE (PJTE). Thus from the very beginning the PJTE was introduced as a two-level (multi-level) problem of a slightly destroyed degeneracy. If we denote the energy gap between the two states $|1>$ and $|2>$ by $\Delta$, we get for the condition of instability of the ground state $|1>$ in the direction of the low-symmetry displacements Q induced by the PJT coupling to the excited state $|2>$ the following inequality (H is the Hamiltonian) [2, 3]:

$$K = K_0 + K_v = K_0 - \frac{F_{12}^2}{\Delta} < 0 \tag{1.1}$$

where

$$F_{12} = \left\langle 1 \left| \left( \frac{\partial H}{\partial Q} \right)_0 \right| 2 \right\rangle \tag{1.2}$$

$$K_0 = \left\langle 1 \left| \left( \frac{\partial^2 H}{\partial Q^2} \right)_0 \right| 1 \right\rangle \tag{1.3}$$

and

$$K_v = -\frac{|F_{12}|^2}{\Delta} \tag{1.4}$$

is the PJTE contribution to the curvature of the lower state.

Hence the initial understanding of the PJTE introduced by Opik and Pryce [3] was that it may produce instability similar to the JTE one, provided the $\Delta$ value is sufficiently small. So the PJTE enlarged the applicability of the JTE adding to the particular cases of electronic degeneracy some more cases of "quasi-degeneracy" where the degeneracy is slightly removed. This understanding ("paradigm") changed when it was shown that the energy gap $\Delta$ in Eq. 1.4 is not a small parameter, it may be very large as the other parameters F and $K_0$ in this inequality vary significantly [2, 4]. By means of ab initio calculations it was shown that PJT instability may be induced by excited states with energy gaps of tenths of electronvolts (for instance, the instability of planar $NH_3$ is induced by the excited state at $\Delta \sim 14\,eV$ [4]). Moreover, further developments in this understanding resulted in the *theorem of instability* which states that any instability of any polyatomic system in a high-symmetry configuration in a non-degenerate electronic state is due to and only to the PJTE. In the presentation of Eqs. 1.1–1.4, it was shown that $K_0 > 0$ always, and hence instability may occur only due to the PJTE contribution $|K_v| > K_0$ [2, 4, 5] (see also Sect. 1.5 below).

If there are many excited states that contribute to the instability of the ground state, then Eq. 1.4 in the linear approximation with respect to the vibronic coupling changes to $K_v = \sum_n |F_{1n}|^2 / \Delta_{1n}$, so the total PJT contribution to the instability becomes similar to a second order perturbation theory correction, provided the $\Delta_{1n}$ values are sufficiently large. Second order perturbation corrections that lower the curvature of the ground state under low-symmetry nuclear displacements have been well known before, irrelevant to the PJTE [6, 7]. They were employed to explain the softening of some molecular systems in the direction of chemical reactions. But for small $\Delta_{1n}$ (in the sense of inequality (1.1)!) the perturbation theory becomes inapplicable and two-level (multi-level) problems should be formulated similar to the two-level case above.

The similarity to the second order perturbation formulas in some multilevel problems led to some controversies in terminology and interpretation of the PJTE in publications on this topic, including some presentations to this conference. An unacceptable statement is that the PJTE is a second order perturbation correction to the ground state often called *second order JTE*. This is not just a terminology issue. To begin with, it instills the impression that the PJTE is a small effect, smaller that the JTE, which, generally speaking, is wrong (see below). But the main failure of this definition is that it does not include the novel effects that are not possible in the JT presentation. In addition to the instability of the ground state which is similar to that of the JTE (but richer in its varieties) the PJTE may produce spin crossover that changes also the ground electronic state spin multiplicity, with far going consequences, in particular, in their magnetic and dielectric properties. There are no first and second order JTE, *both the JTE and the PJTE are induced by the first order vibronic coupling terms* in the Hamiltonian, but *are applied to different situations* and *produce different results* that may have some similar features.

The essential difference between the JTE and the PJTE is first of all in the different situations to which the vibronic coupling terms of the Hamiltonian are applied. While the JTE is applicable to limited classes of system with electronic degeneracy, the *PJTE is applicable to any polyatomic system without limitations*. In particular, the PJTE is present also in systems with the JTE, in which case there is no way to state that the PJTE is of second order (see below). The correct definition is that the PJTE is a nondegenerate two- or multi-level vibronic coupling problem (in case of more than two states some of them may be degenerate); in the particular case of only corrections to the ground state when there is no avoided crossing with the excited states (meaning the interaction between them is sufficiently small) the PJTE contribution can be estimated by means of the second order perturbation theory.

Let us illustrate the said above by some examples. In a recent publication on ab initio calculation of the $CO_3$ molecule [8] it was shown that involving the PJTE in a multilevel seven-state six-mode problem (that takes into account also the JTE in the excited states) we get the correct description of the adiabatic potentials energy surface (APES) with three equivalent global minima of $C_{2v}$ symmetry in addition to the more shallow minimum at the high-symmetry configuration $D_{3h}$ (Figs. 1.1 and 1.2). Note that starting with the latter configuration and involving second order perturbation theory we obtain just the curvature in the central minimum, but not the distorted configurations in the three global minima. This is also an example of the "hidden JTE" which we introduced earlier [9]; it emerges only when the excited

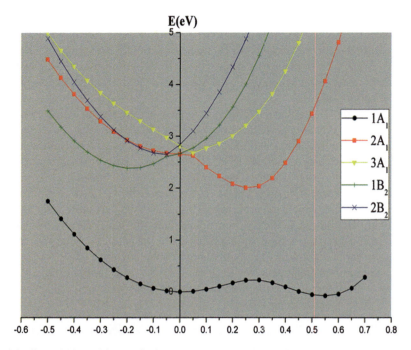

**Fig. 1.1** Ground 1A$_1$ and four excited states energy levels (2A$_1$ and 1B$_2$, 3A$_1$ and 2B$_2$ correspond to E$_1$ and E$_2$ in D$_{3h}$ symmetry, respectively) of CO$_3$ in the cross section of the APES along the interaction mode $q_\vartheta$ obtained by the MRCI + Q//SA-CASSCF (16/13) method of ab initio electronic structure calculations [8]

states (sometimes high in energy excited states) are involved in the problem via the JTE or PJTE.

Another example of the hidden JTE revealed by the PJTE is from a whole class of molecular systems and crystals with e$^2$ and t$^3$ electronic configurations. It was shown that in such systems a very strong PJTE between two excited electronic states in the high-symmetry configuration results in a spin crossover in which the lower component of the two PJT mixing states goes down and crosses the former ground state producing a global minimum with lower symmetry and changed spin multiplicity [10]. Figure 1.3 illustrates how these two minima are formed by the PJTE using the molecule Si$_3$ as an example [10]. Similar to such two molecular configurations, two crystalline configurations (phases) may be produced by the PJTE in crystals with e$^2$ or t$^3$ centers; Fig. 1.4 shows the results of calculations performed for LiCuO2 [11] in which the Cu$^{3+}$ centers have the electronic configuration e$^2$.

Finally, an example of the particular case when the PJTE can be taken into account by means of perturbation theory is illustrated in Table 1.1. It shows that when the two effects, JTE and PJTE, are present simultaneously the latter may be more important than the former. For a series of linear molecules in which there is no

1 Critical Review of Contributions to the Jahn-Teller Symposium JT2010 and Beyond

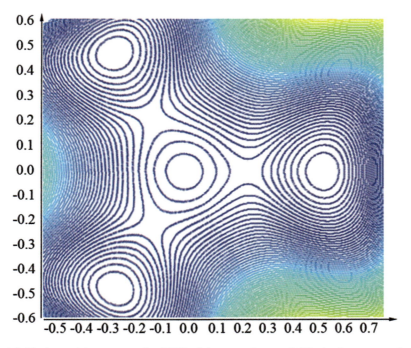

**Fig. 1.2** Equipotential curves on the APES of the ground state of $CO_3$ in the cross section of the $(q_\theta, q_\varepsilon)$ plane of the interaction $e$ mode obtained by ab initio calculations with the BCCD(T) method. In the central shallower minimum the molecule has the undistorted $D_{3h}$ configuration, whereas in the three equivalent global minima it is distorted to an acute triangular $C_{2v}$ geometry

ground state crossing or avoided crossing with excited states ab initio calculation of the Renner-Teller effect (RTE) and the PJTE taken into account as a second order perturbation effect were performed [12].

The potential energy Hamiltonian in this case is (this equation and the values of g in Table 1.1 are formally different from that in Ref. [12] as the constant g there was defined as $1/2$ of the present one):

$$\hat{U} = \frac{1}{2}(K_0 - p)(Q_x^2 + Q_y^2)\sigma_0 + \left(g - \frac{1}{2}p\right)[(Q_x^2 - Q_y^2)\sigma_z + 2Q_xQ_y\sigma_x] \quad (1.5)$$

where $p$ and $g$ are the PJTE and RTE contribution, respectively, $K_0$ is the primary force constant defined in Eq. 1.3, $\sigma_0$, $\sigma_z$, and $\sigma_x$ are Pauli matrices, and $Q_x$ and $Q_y$ are the bending coordinates. The first term in Eq. 1.5 describes the bending and the second term stands for the term splitting. Hence the PJTE represented by the constant $p$ influences both effects. From the results given in Table 1.1 we see that in all the cases under consideration the PJTE contribution $p$ to both the instability of the ground state and the RTE splitting is significantly larger than the pure RTE splitting constant $g$ [12]. For a JT ground state the Hamiltonian contains also linear

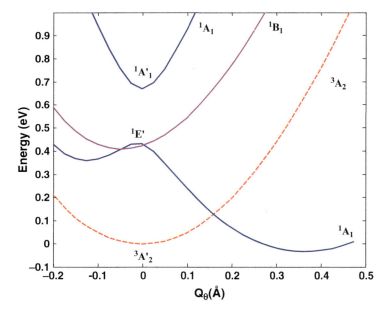

**Fig. 1.3** Cross section of the APES along the $Q_\theta$ coordinate for the states arising from the electronic e2 configuration of Si3. Its main features are (as predicted by the PJTE theory): a very weak JTE in the excited E state, a strong PJTE between the excited E and A states that produces the global minimum with a distorted configuration, and a second conical intersection along $Q_\theta$ (with two more, equivalent, in the full e space). The spin-triplet state is shown by *dashed line*

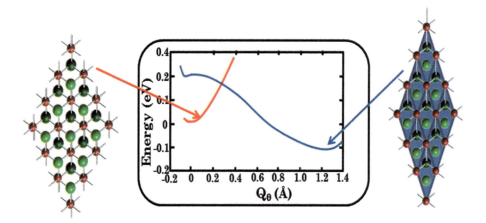

**Fig. 1.4** DFT calculation of the energies as a function of local distortions of the two configurations of the LiCuO$_2$ crystal, magnetic-undistorted (shown *left* in *red*) and nonmagnetic-distorted (*blue*), arising from the local electronic $e^2$ configuration of the Cu$^{3+}$ ions in the octahedral environment of the oxygen ions. The magnetic centers are ordered anti-ferromagnetically in the crystal, while in the lower-in-energy nonmagnetic phase the crystal is strongly tetragonally distorted. In the two lateral illustrations the two fragments of the layers of CuO$_6$ octahedrons in the corresponding two phases are shown with Cu in red, and O in light green above the plane and dark green beyond the plan [11]

# 1 Critical Review of Contributions to the Jahn-Teller Symposium JT2010 and Beyond

**Table 1.1** Numerical values (in $eV/\AA^2$) of the primary force constant $K_0$, vibronic coupling RT constant $g$, and PJT constant $p$ for some triatomic molecules obtained by direct ab initio calculation ($g$), and by fitting ab initio calculated energy level curves to Eq. 1.10 (for $K_0$ and $p$)

| Molecules/Constants | $K_0$ | g | p |
|---|---|---|---|
| HBF | 9.67 | 10.50 | 19.09 |
| HCO | 21.27 | 18.50 | 30.26 |
| HNN | 19.35 | 17.18 | 24.83 |
| $BH_2$ | 10.12 | 6.12 | 10.69 |
| $NH_2$ | 14.26 | 14.90 | 23.53 |
| $AlH_2$ | 6.91 | 3.58 | 13.33 |

vibronic coupling terms with the constant $F$:

$$\hat{U} = \frac{1}{2}(K_0 - p)(Q_\vartheta^2 + Q_\varepsilon^2)\sigma_0 + F(Q_\vartheta\sigma_z - Q_\varepsilon\sigma_x)$$
$$+ \left(g - \frac{1}{2}p\right)[(Q_\vartheta^2 - Q_\varepsilon^2)\sigma_z + 2Q_\vartheta Q_\varepsilon\sigma_x] \tag{1.6}$$

These (and many other) examples [2] demonstrate the richness of the PJTE as it is applicable to any polyatomic system without exceptions. In this sense it becomes more important than the JTE.

## 1.3 Tunneling Splitting in Systems with the JTE $E \otimes e$ Problem

Tunneling splitting in JT systems together with special features in EPR spectra have been in the first predictions of new observables that are due to the JTE. Presently there are many observations of these effects (see, e.g., in [2]). However, full theoretical and experimental elucidation of the tunneling splitting is still not achieved, and the controversy in evaluation of the magnitude of tunneling splitting in crystals with impurity centers reported to the Symposium is an example of the ongoing difficulties. The resolution of this issue seems to be most important as it is related to the basic features of the JTE in such systems.

Two papers, both of high-level performance, experimental EPR [13] and ab initio calculations [14], report essentially (drastically) different values of the magnitude of the tunneling splitting $3\Gamma$ in the system $MgO : Cu^{2+}$, namely, $3\Gamma \cong 4\,cm^{-1}$ in the experimental measurements and $3\Gamma \cong 234.79\,cm^{-1}$ in the calculations. The $Cu^{2+}$ ions in the octahedral environment of oxygen ions have a double degenerate electronic E term with an $E \otimes e$ JTE problem. The tunneling between the three equivalent minima along the bottom of the trough of the APES of this problem (Fig. 1.5) depends on the barrier height and width between them which, in turn, is determined by the contribution of the quadratic terms of the vibronic coupling [2].

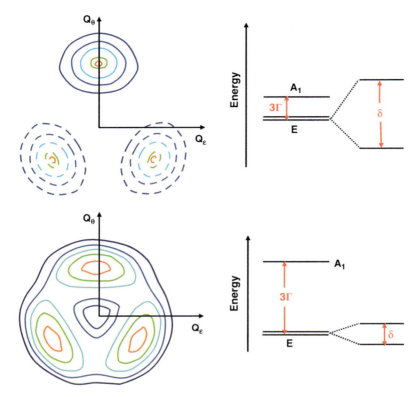

**Fig. 1.5** Illustration of the nuclear density and tunneling splitting in impurity centers with the $E \otimes e$ JT problem for the cases of static distortion (*upper part*) when the system is locked in one minimum by crystal imperfections, $\delta > 3\Gamma$, and dynamic distortion (*lower part*) when $\delta < 3\Gamma$. For $MgO : Cu^{2+}$ $\delta$ is assumed to be small, and hence the lower picture is applicable [13, 14]

The almost two orders of magnitude in the difference between the experimental and theoretical numbers cannot be attributed to just different approximations involved in the two papers, it should be more fundamental.

Our comments on this situation are as follows. The experimental work with EPR measurements at low temperatures seems to be performed at a high level accuracy and it is not questioned here, but its interpretation raises some concerns. Firstly, we noticed that the determination of the numerical value of $3\Gamma$ stays as an isolated procedure which can be easily separated from the main body of the experimental data. In other words, if we exclude the result $3\Gamma \cong 4\,cm^{-1}$, the interpretation of the spectrum remains intact.

Indeed, the vibronic reduction factor that determines the angular dependence of the spectrum depends mainly on the linear vibronic coupling constants, and hence it cannot serve as a source of the quadratic coupling constant. The value $3\Gamma \cong 4\,cm^{-1}$ vas derived from some additional (independent) absorption line positions that were attributed to transitions between the tunneling levels. Unfortunately, these line

1 Critical Review of Contributions to the Jahn-Teller Symposium JT2010 and Beyond 11

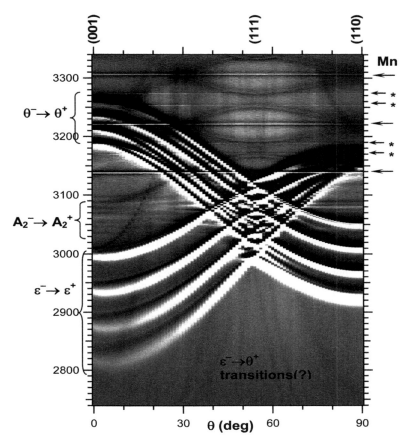

**Fig. 1.6** Angular dependence of the EPR spectrum of MgO : $Cu^{2+}$ from Ref. [13]. The authors assumed that the angular independent tiny (hardly seen) lines denoted as $A_2^- \to A_2^+$ represent the transition between the tunneling levels

positions, as shown in Fig. 1.6 (the authors [13] have no better pictures), are hardly seen in the spectrum, and to our understanding, their attribution to transitions between the tunneling levels seems to be unreliable. Our conclusion from this analysis is that the authors should try to repeat the experiments in better conditions in order to either confirm or retract this result.

On the other hand, the theoretical work on ab initio calculation of the full APES [14] including the barrier between the equivalent minima is performed on the highest level possible at present with one exception: they do not take into account that for impurity centers in crystals the JTE problem is a multimode problem [2, 15]. The multimode nature of the problem in such systems is due to the involvement of the next (to the first) coordination spheres in the distortions and its changes when the system transfers from one minimum to another by tunneling. In other words, the dynamical deformations within the first coordination sphere involve the next coordination spheres in a wave that rotates around the impurity center together with the JTE tunneling. Theoretical consideration of the multimode effects in impurity centers in a general form were published long ago [15], but at that time there were no computers to carry out numerical estimates. The multimode effect definitely increases the barrier between the multimode minima and hence lowers the value of the tunneling splitting. It is difficult to predict this effect in the case of $MgO : Cu^{2+}$ without numerical estimates, but it hardly reduces the $3\Gamma$ value by orders of magnitude.

## 1.4 Orbital Ordering Versus Cooperative JTE

There is already a while of discussion of the influence of vibronic coupling on orbital ordering which is well outlined in the review article [16]. The orbital ordering approximation (OOA) was introduced to simplify the treatment of crystals with electronic degenerate centers by assuming that the atomic orbitals of the degenerate states, if not coupled to the vibrations, can rotate freely (as in isolated atoms), and then the exchange interaction between them may results in their ordering along the lattice [17]. Assuming that the nuclei follow the electrons, one gets in this way the phase transitions to lower symmetry phases of the lattice. Obviously, this approach ignores the vibronic coupling, the local coupling of the electrons to the nuclear displacements (the JTE) or, more precisely, considers the electron-phonon interaction included in the orbital-orbital interaction constant in an averaged way. Thus in the very formulation of the OOA it is assumed implicitly that the orbital-orbital interaction between two JT centers is stronger than the local vibronic coupling, and hence the primary mechanism of phase transition in JT crystals is the OO, not the cooperative JTE.

This assumption is hardly acceptable for the majority of JT crystals for the simple reason that in such crystals the distance between the JT centers is relatively large and they are separated by ligands, hence the orbital interaction between them is an indirect superexchange interaction. The latter is known to be of the order of $10^2 \text{cm}^{-1}$, whereas the JTE stabilization energy may be of the order of $10^3-10^4 \text{cm}^{-1}$. For this reason in such crystals the orbital orientation in space is not free, and in any case the orbitals are "coated" by low-symmetry phonons, so the interaction between them is both electronic and elastic, just as described in the cooperative JTE (CJTE) [2, 18]. Among other difficulties of the OOA, it considers only metal-metal orbital interaction thus essentially excluding the nature of the ligands and the chemical bonding which were shown to play an important role in the phase transition and other properties of the JT crystals [2, 16]. In particular, if the ligands are positioned outside the line between the two interacting orbitals as, e.g., in $CsCuCl_3$, the direction of the ordered orbitals does not coincide with the direction of the chemical bonding, which leads to unphysical conclusions that the orbital-orbital interaction is either stronger than the chemical bonding, or it has nothing to do with the structure of the lattice.

Most discouraging for the OOA are some numerical estimates for specific crystals. If the structural phase transition in JT crystals is driven by the OO interaction, its transition temperature $T_{str}$ should be approximately the same as for the magnetic phase transition $T_{mag}$. Experimental data for $KCuF_3$ are [19]: $T_{str} = 800\,K$ and $T_{mag} = 38\,K$ ($T_{mag} = 22\,K$ for another modification). This is way out of any possibility to explain the origin of the structural phase transition in terms of orbital ordering without involving the CJTE. There are other difficulties in the attempts to apply the OOA to JT crystals [16] but, of course, there may be cases when this model is applicable, in particular, when the vibronic coupling of the JT centers is very weak, meaning the degenerate state electrons are almost free, like in metals.

In this situation with the OOA, a group of authors presented to the JT Symposium JT2010 a talk entitled: "Orbital order in $La_{0.5}Sr_{1.5}MnO_4$: failure of the standard Jahn-Teller physics" [20]. Under this provocative title, the authors show that in the crystal $La_{0.5}Sr_{1.5}MnO_4$ the oxygen environment of the $Mn^{3+}$ ions is in a tetragonally compressed octahedral symmetry for which, they state, the highest occupied atomic $(Mn^{3+})$ orbital in accordance with the JTE should be of the $d_{x2-y2}$ type, while the calculations and experiments show that in fact it is of $d_{z2}$ type, and this means that the JTE physics fails. Assuming that the factual part of this statement is accurate, its interpretation in relation to the JTE is unacceptable, in principle. To begin with, the JTE follows from the first principles, and as such it cannot fail provided it is employed correctly. In their statement they use the definition of the JTE in ideal systems given many years ago without taking into account the later developments, corrections, and extensions [2]. First, the system is far from being ideal in the sense of application of the simplest formulation of the JTE. Following the crystal structure of this system, there is very strong influence of the next coordination spheres via commonly shared oxygen atoms. In simplified formulation this leads to the multimode JTE. Then the quadratic terms of the vibronic coupling (not taken into account by the authors [20]) may be important. And the contribution of the PJTE may be significant too. Some of these effects were mentioned by the authors [20] and in another paper on this subject [21], but not all of them. In other words, the origin of the local electronic structure in this crystal should be sought for within the JTE theory in its modern formulation with modifications and extensions [2], and not as a failure of the "primitive" formulated JTE.

## 1.5 Vibronic Origin of Order-Disorder Phase Transitions in "Displacive" Ferroelectrics

The plenary lecture devoted to this topic at the conference reflected the ongoing intensive discussion of the origin of ferroelectricity in perovskite type crystals. Investigation of this interesting subject has already a long history. Till recently the overwhelming majority of papers devoted to this problem has been based on the, mostly phenomenological, ideas of Cochran [22], Anderson [23], and Ginsburg [24], that explain the spontaneous polarization of the crystal at the phase transition as resulting from the compensation of the local repulsion between the ions by the attractive long-range dipole-dipole interactions in the boundary optical phonon (odd) displacements of the sublattices ("displacive" phase transitions). This led to the conception of the soft mode that goes to zero at the point of phase transition, experimentally observed in many crystals; in the cases where the soft mode is not seen clearly, it was assumed to be dumped.

However, experimental evidence in increasing number of publications [25–39] indicates that the main idea that the lattice polarization is due to solely long-range forces may be wrong. Experiments performed mostly on $BaTiO_3$ and $KNbO_3$, including diffuse X-ray scattering [25], Raman spectra [26], optical refractive index [27], infrared reflectivity [28], X-ray absorption [29, 30], ESR with probing ions [31, 38], XAFS [32, 34, 35, 37], femtosecond resolution light scattering [33], NMR [36], elastic and dielectric measurements [39], etc., show convincingly (as a pattern) that the Ti or Nb ions are off-center displaced in all the phases, including the cubic paraelectric phase, and the phase transitions are essentially of order-disorder type. An important step toward the correct understanding and interpretation of the experimental data was made by the observation of the dynamics of the ions at the phase transition and estimation of its time-scale, achieved by means of EPR experiments with probing ions [31, 38] (see below).

Meanwhile a totally different approach to the problem based on the PJTE was suggested and published in 1966 [40] (with a more elaborate version in 1967 [41]; see the reviews [42, 43] and Sect. 8.3 in the book [2]) which explains directly the origin of the main features of all the above properties of ionic ferroelectric crystals. The theory developed based on this idea was termed "vibronic theory", sometimes called "two-band theory." In this approach it is shown that, in contrast to the ideas of displacive phase transitions, under certain conditions *there is no local repulsion in the off-center displacements* of some atoms in the unit cell. On the contrary, the local vibronic coupling between the ground and excited states of opposite parity, the PJTE, may lead to the spontaneous odd displacements forming local dipole moments, their cooperative interaction resulting in ferroelectric phase transitions of *order-disorder type* (in other language, mixing of two bands by boundary optical phonons leads to qualitatively the same result [42, 43]). This suggestion has been in severe dissonance with the dominant ideas of the time; there were not any experimental or other indirect indications of its feasibility (many scientists in this field hardly knew anything about the PJTE). This somewhat explains why the PJTE idea was not accepted and/or normally cited by other authors (the first paper on the vibronic theory prepared in 1964 was refused publication first by *Sov. Phys. JETP*, then by *Sov. Phys. Solid State*, and only *Phys. Lett.* agreed to publish the paper [40]); the papers that introduced this novel understanding of the problem are unfairly not normally cited till present, even after the main idea is confirmed experimentally.

The first experimental confirmation of the order-disorder nature of the phase transitions in $BaTiO_3$ by means of diffuse scattering of X-rays was published in 1968 [25], two years after the publication of 1966 [40] and one year after the

1 Critical Review of Contributions to the Jahn-Teller Symposium JT2010 and Beyond    15

1967 paper [41], but the authors [25] did not cite the papers [40, 41]; instead, they sent a letter of recognition to the author of the 1966 paper (see [44]). As a result, the experimental paper [25] is cited as the first one to discover the order-disorder character of the phase transitions in $BaTiO_3$, while the author of the factual first paper on this, in principle novel approach to the problem which predicted the observable new properties based on first principles is unfairly ignored.

The novel approach in its more modern formulation is based on the *theorem of instability* mentioned above. According to this theorem the primary force constant in Eq. 1.3 $K_0 > 0$ for all the symmetrized displacements Q of any polyatomic system in a high-symmetry nuclear configuration with a nondegenerate electronic state $|0>$ for which the first derivative $<0|(\partial H/\partial Q)_0|0> = 0$, and hence the instability of such systems may occur due to, and only to the PJTE contribution $K_v < 0$ (see in [2, 4, 5, 45]. The importance of this result is in the different nature of the two contributions, $K_0$ and $K_v$, to the curvature K of the APES. Presenting the wavefunctions of the crystal, for instance, in a tight-bonding approximation we can see that $K_0$ in Eq. 1.3 is a diagonal matrix element of second derivatives with regard to the odd displacements Q that includes local ("bare" elastic) and nonlocal long range interactions (via nonzero matrix elements of derivatives of mixed displacements in two cells), whereas $K_v$ in Eq. 1.4 depends on only off-diagonal matrix elements of first order derivatives with respect to the odd displacements. The matrix elements in $K_v$ are nonzero when the overlap between the wavefunction of the atoms of the two sublattices (oxygen and titanium in $BaTiO_3$) increases due to the nuclear displacements (enhancing their covalence bonding [2]). This means that the $K_v$ contribution and hence the instability is essentially of local origin (change in overlap of the atomic functions of non-near-neighbor atoms is negligible). The long-range interactions presented by $K_0$ are still most important as they play a key role in the realization of the inequality $|K_v| > K_0$. *The trigger mechanism of spontaneous polarization of the crystal is* thus *of local origin*, directly related to its atomic (electronic) structure, but the polarization depends also on the long-range interactions. In other words, the long-range interactions only, without the local instability, cannot produce the spontaneous polarization of the crystal. $K_0$ and $K_v$ depend also on the third ion in the perovskite (Ba in $BaTiO_3$) as it defines the size of the unit cell which influences the Ti-O interaction.

In case of $BaTiO_3$, in a two-level problem with an energy gap $\Delta$ the PJTE condition of instability for the titanium ions in the cubic position is given by the inequality (1.1) (with a similar condition in the band presentation), where the PJT vibronic coupling constant is that of mixing the occupied $p_\pi$ orbitals of the oxygen atoms with the unoccupied $d_\pi$ orbitals of titanium.

$$F = < 2p_\pi(O)|(\partial H/\partial Q)_0|3d_\pi(Ti) > \qquad (1.7)$$

The basic idea that the mixing of electronic states by the nuclear displacements leads to the instability of the lattice was developed in different approximations [42, 43, 46] and modified by other authors, in particular, by Kristoffel and Konsin [47] and Girshberg and Yacoby [48] (the latter prefer not to cite the works that

**Table 1.2** Phases, phase transitions, and disorder dimensionality in $BaTiO_3$ and $KNbO_3$ type crystals predicted by the PJT theory of ferroelectricity (1966) [40]

| PHASES: | Rhombo-hedral | Orthor-hombic | Tetragonal | Cubic |
|---|---|---|---|---|
| Direction of polarization | [111] | [011] | [001] | none |
| Dimensionality of disorder | 0 | 1 | 2 | 3 |
| Number of minima involved | – | Two | Four | Eight |
| Temperature of phase transition | $T_c(I) <$ | $T_c(II) <$ | $T_c(III)$ | |

introduced the original idea). Some authors just assume that there is an off-center displacement of the Ti ion in $BaTiO_3$, sometimes referring to lattice anharmonicity without explaining its origin (the earlier eight-minimum models are also of this kind). Such general assumptions are not related directly to the atomic and electronic structure of the crystal, they do not explain the microscopic origin of ferroelectricity (in fact, they assume what should be proved), in particular, why is this anharmonicity so specific for some perovskites only (e.g., why is $BaTiO_3$ ferroelectric, whereas $BaVO_3$ is not).

The vibronic theory explains the origin of the observable ferroelectric properties based on the condition of the PJTE instability that produces a special (anharmonic) form of the APES which is directly related to the atomic and electronic structure. For $BaTiO_3$ the main features of the APES were obtained already in the first publication on this topic [40]. It has eight equivalent minima along the eight trigonal directions of the Ti ions off-center displacements with 12 lowest barriers between each of the two near-neighbor minima (e.g., along [110] between the [111] and [11-1] minima), six higher barriers along [100] type directions and a maximum at the center of inversion.

By considering the motions along these APES in a quasi-classical approximation [49] the following picture of order-disorder phase transitions emerges (Table 1.2). At low temperatures the crystal rests in the fully ordered rhombohedral phase with the Ti ions in the lowest trigonal minima producing the spontaneous polarization along, e.g., the [111] axis; at the temperature of the first phase transition the lowest barrier between two near-neighbor minima (e.g., [111] and [11-1]) is overcome and the crystal in average becomes polarized along [110] with disorder along the direction that connects the two minima; at the next phase transition the next energy barrier is overcome, and the crystal distortions are averaged over four near-neighbor minima with a polarization along [100] and two perpendicular directions of disorder; finally at the next phase transition to the paraelectric phase the distortions are fully averaged over all the minima and the crystal is cubic in average with three-dimensional disorder. The phase transitions in this crystal are thus predicted by the theory to be in essence of order-disorder type, and this prediction is now fully confirmed experimentally. It is worthwhile to notice that in his paper on XAFS measurements in comparison with other methods Stern [37] describes the picture of phase transitions in $BaTiO_3$ (emerging from the experiments) *exactly* as it is given in the first paper on the vibronic theory in 1966 [40], with almost the same wording (!), but he does not cite this paper.

All the experimental results, mentioned above, are consistent with the PJT prediction of the local origin of the phenomenon and order-disorder nature of the phase transitions. However, there are some differences in the interpretation of the results of the experimental observations. In particular, some authors report the presence of a component of displacive origin in the phase transition (see, e.g., in [36–39]). Others see differences in the dynamics of the ferroelectric ions at the phase transition when measured with different methods (e.g., NMR [36] and XAFS [35] at the tetragonal-paraelectric phase transition in $BaTiO_3$). The differences in experimental observations of the same phenomena at the order-disorder phase transition can be attributed to their essential different time-scales, different "time of measurement" $\tau'$ in comparison with the time $\tau$ of the reorientational hoping of the metal ion between the minima of the APES, as well as to different space-scale (e.g., optical wave-length) measurements as compared with the unit cell dimensions [37–39].

With regard to the time-scale, the conclusions from the experimental data may be different for experiments with $\tau' > \tau$ and $\tau' < \tau$ because of the full or partial averages over the reorientation positions (APES minima) observed in the former case. In this respect evaluation of the time $\tau$ is most important. The first experiments in which the ion reorientation time was estimated were performed [31] by means of EPR with probing ions $Mn^{4+}$ substituting $Ti^{4+}$; the characteristic correlation time of the reorientational hoping between two near-neighbor [111] type minima was estimated to be $\tau \sim 10^{-9}$–$10^{-10}$ s. In a more recent paper [38] further details of this process were revealed in comparison with other experimental data. The origin of apparent differences in the interpretation of the NMR and EXAFS experiments with regard to the tetragonal-paraelectric phase transition in $BaTiO_3$, mentioned above, were explained as due to the essential difference in the time-scales of the two methods of measurement [37]: in NMR $\tau' \sim 10^{-8}$ s which is larger than $\tau$, whereas in EXAFS $\tau \sim 10^{-15}$ s which is smaller than $\tau$.

As for the displacive part of the phase transition which is seen in some of the experiments, it can be attributed to the space averages. Indeed, for instance, in the tetragonal phase, according to the vibronic theory, as described above, the system resonates between four minima situated around one of the tetragonal axis, so as in average the unit cell is a tetragonally distorted. Since in the tetragonal phase the lattice is ordered in this tetragonal direction (with disorder in two perpendicular directions), the crystal is tetragonally distorted too. This is seen as a displacive component in the paraelectric-tetragonal phase transition in experiments that perceive the averaged picture [38, 39]. For transitions between the tetragonal to orthorhombic and the latter to rhombohedral phases this effect is expected to be less pronounced because they are accompanied by changes in dynamics of only two minima. Note that the wavelength in the usually employed optical measurements of soft modes is much larger than the unit cell where the dynamics of the ions at the order-disorder phase transition takes place; the observed picture in such measurements is averaged over many unit cells, thus seen as a displacive component.

## 1.6 Remarks on the JTE in High-Temperature Superconductivity

The understanding that the JTE plays an essential role in the mechanism of high-temperature superconductivity (HTSC) was in the basic idea that led to this Nobel price discovery [50]. Recent developments and publications show that in essence this idea is correct, although the phenomenon as a whole is complicated and the JTE in it is "dressed" with so many other interactions (formation of JT polarons, density, stability and mobility of Cooper pairs, structural phase transitions, perfect diamagnetism, etc.) that in many cases it is not easy (not straightforward) to see this effect explicitly (see, e.g., some recent publications [51]). Therefore it is no wonder that there were (and are still) attempts to ignore the JT origin of HTSC trying to explain it by other (e.g., magnetic) interactions. On the primary atomic (electronic) level the JT interactions remain in the basic triggering mechanism of HTSC; the plenary talk presented to the Symposium by K.A. Muller demonstrated the latest convincing results on this topic [52].

In comments on this issue, I should like to straighten some "historical" facts. The first calculations on JT polarons, in general, were published before the discovery of HTSC [53] and are cited as an inspirational factor in this discovery [50]. The first specific calculations of the JT bipolaron in HTSC cuprates were performed about 20 years ago [54]. Figure 1.7 taken from that publication shows the scheme of the two $Cu^{2+}$ centers in the $CuO_2$ layer, each distorted by the PJT $(A_{1g} + B_{1g}) \otimes b_{1g}$ interaction, to form an anti-distortive pair, and the calculations [54] revealed the lowest energy arrangement. Although the latest publications on this topic use the same model of the bipolaron [55], they do not cite the paper [54] where it was considered long before them.

In development of the bipolaron model [56] and with more sophisticated calculations it was shown that the PJTE triggered by an one-electron oxidation of the $(CuO_2)^{2-}$ units in $La_{2-x}Sr_xCuO_4$ creates a JT multicenter polaron with 5–7 cooper centers (Fig. 1.8) and their interaction in the crystal produces polaronic stripes of JT distorted and anti-distortive ordered Cu centers shown in Fig. 1.9. These stripes form an ideal superstructure for HTSC and explain many of their properties [56]. The formation of stripes in HTSC crystals was confirmed experimentally [57]. Again, these interesting findings are not normally cited in many publications on this subject.

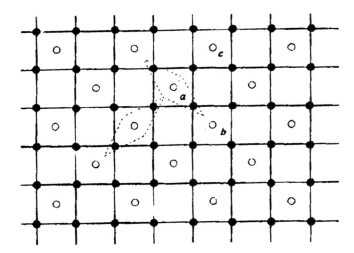

**Fig. 1.7** Schematic presentation of the bipolaron in the $CuO_2$ layer of a superconducting cuprate as an antidistortive ordered pair of PJT distorted $CuO_4$ squares (published about 20 years ago [54]). The idea is now used in recent theories involving the JTE [55]

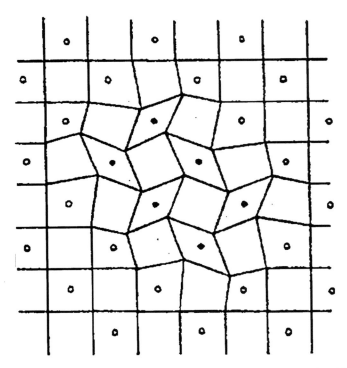

**Fig. 1.8** Extended polaron in the $CuO_2$ layer of a superconducting cuprate produced by the PJTE due to an additional electron hole (that emerged due to a corresponding impurity) shared by several 5–7) centers [56]

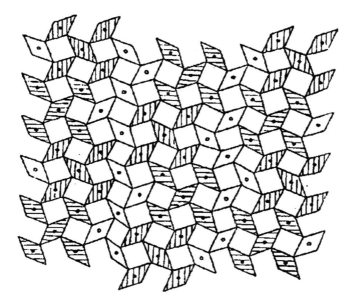

**Fig. 1.9** Polaronic stripes predicted based on the interaction of pairs of extended polarons shown in Fig. 1.8 [56]. Such stripes were observed experimentally [58]

## 1.7 Conclusions

In a brief discussion of several important problems of the JTE theory presented to the JT2010 International Symposium on the JTE the author's view on the subject and critical comments are presented. They are related to the correct definition and understanding the PJTE as a two-level or multi-level problem that, in general, cannot be reduced to a second order perturbation theory correction; to the controversy that emerged in the comparison of the experimental and theoretical evaluated magnitude of the tunneling splitting in some impurity crystals; the relation between orbital ordering and cooperative JTE; vibronic (PJT) origin of ferroelectricity in so-called displacive ferroelectrics of perovskite type crystals and the order-disorder nature of the phase transitions in these crystals, and some historical notes on the JT origin of high-temperature superconductivity.

**Acknowledgments** A part of this research was supported by Grant F -100 from the Welch Foundation.

## References

1. Jahn HA, Teller E (1937) Proc R Soc 161:220
2. Bersuker IB (2006) The Jahn-Teller effect. Cambridge University Press, Cambridge
3. Opik U, Pryce MHL (1957) Proc R Soc A 238:425

4. Bersuker IB, Gorinchoi NN, Polinger VZ (1984) Theor Chim Acta 66:161
5. Bersuker IB (1980) Nouv J Chim 4:139; Teor Eksp Khim 16:291; (1988) Pure Appl Chem 60(8):1167; Fiz Tverdogo Tela 30:1738
6. Bader RFW (1960) Mol Phys 3:137; (1962) Can J Chem 40:1164; Bader RFW, Bandrauk AD (1968) J Chem Phys 49:1666
7. Pearson RG (1976) Symmetry rules for chemical reactions. Orbital topology and elementary processes. Wiley, New York
8. Yang Liu, Bersuker IB, Wenli Zou, Boggs JE (2009) J Chem Theory Comp 5:2679
9. Bersuker IB (2009) In: Koppel H, Yarkony DR, Barentzen H (eds) The Jahn-Teller effect. Fundamentals and implications for physics and chemistry. Springer, Heidelberg, p 3
10. Garcia-Fernandez P, Bersuker IB, Boggs JE (2006) J Chem Phys 125:104102/11
11. Garcia-Fernandez P, Bersuker IB (2011) Phys Rev Lett 106:246406
12. Liu Y, Bersuker IB, Boggs JE (2010) Chem Phys 376:30
13. Riley M, Noble CJ, Tregenna-Piggott PLW (2009) J Chem Phys 130:104708
14. Garcia-Fernandez P, Trueba A, Barriuso MT, Aramburu JA, Moreno M (2010) Phys Rev Lett 104:035901
15. Polinger VZ, Bersuker GI (1979) Phys Stat Solidi (b) 95:153; (1979) Phys State Solidi (b) 95:403; (1980) Sov Phys Solid State 22:1485; Bersuker GI, Polinger VZ (1981) Solid State Commun 38:795; (1981) Sov Phys JETP 53:930
16. Polinger VZ (2009) In: Koppel H, Yarkony DR, Barentzen H (eds) The Jahn-Teller effect. Fundamentals and implications for physics and chemistry. Springer, Berlin/Heidelberg, pp 685
17. Kugel KI, Khomskii DI (1982) Sov Phys-Uspekhi 25:231
18. Kaplan MD (2009) In: Koppel H, Yarkony DR, Barentzen H (eds) The Jahn-Teller effect. Fundamentals and implications for physics and chemistry. Springer, Berlin/Heidelberg, p 653
19. Pavarini E, Koch E, Lichtenstein AI (2008) Phys Rev Lett 101:266405
20. Wu H, Chang CF, Schumann O, Hu Z, Cezar JC, Burnus T, Hollmann N, Brookes NB, Tanaka A, Braden M, Tjeng LH, Khomskii DI (2010) In: 20th international symposium on the Jahn-Teller effect, Abstracts, University of Fribourg, Fribourg, p 47
21. Sboychakov AO, Kugel KI, Rakhmanov AL, Khomskii DI (2011) Phys Rev B 83:205123
22. Cochran W (1959) Phys Rev Lett 3:412; (1961) Adv Phys 10:40
23. Anderson PW (1959) In: "Fizika Dielectrikov", AN SSSR, p 290
24. Ginsburg VL (1960) Fizika Tverdogo Tela 2:2031
25. Comes R, Lambert M, Guinner A (1968) Solid State Commun 6:715
26. Quitet AM, Lambert M, Guinier A (1973) Solid State Commun 12:1053; Comes R, Currat R, Denoyer F, Lambert M, Quittet M (1976) Ferroelectrics 12:3
27. Burns G, Dacol F (1981) Ferroelectrics 37:661
28. Gervais F (1984) Ferroelectrics 53:91
29. Ehses KH, Bock H, Fischer K (1981) Ferroelectrics 37:507
30. Itoh K, Zeng LZ, Nakamura E, Mishima N (1985) Ferroelectrics 63:29
31. Muller KA (1986) Helv Phys Acta 59:874; In: Bishop AR (1986) Nonlinearity in condensed matter, Springer/Heidelberg, p 234
32. Hanske-Petitpierre O, Yacoby Y, Mustre de Leon J, Stern EA, Rehr JJ (1992) Phys Rev B 44:6700
33. Dougherty TP, Wiederrecht GP, Nelson KA, Garrett MH, Jensen HP, Warde C (1992) Science 258:770
34. Sicron N, Ravel B, Yacoby Y, Stern EA, Dogan F, Rehr JJ (1994) Phys Rev B 50:13168
35. Ravel B, Stern EA, Vedrinskii RI, Kraisman V (1998) Ferroelectrics 206–207:407
36. Zalar B, Laguta VV, Blinc R (2003) Phys Rev Lett 90:037601
37. Stern E (2004) Phys Rev Lett 93:037601
38. Volkelm G, Muller KA (2007) Phys Rev B 76:094105
39. Bussman-Holder A, Beige H, Volkel (2009) Phys Rev B 79:184111
40. Bersuker IB (1996) Phys Lett 20:589
41. Bersuker IB, Vekhter BG (1967) Fizika Tverdogo Tela 9:2452
42. Bersuker IB, Vekhter BG (1978) Ferroelectrics 19:137

43. Bersuker IB (1995) Ferroelectrics 164:75
44. The letter states: "University of Paris...Orsay, January 23 [1969], ...*Dear Dr Bersuker, We just discovered your note "On the Origin of Ferroelectricity in Perovskite type Crystals" published in Physics Letters (1. April 1966). From a completely different approach we came to conclusions which are similar to yours. You will find enclosed our publications on BaTiO$_3$ and KNbO$_3$... Sincerely Yours, R. Comes*".
45. Bersuker IB, Polinger VZ (1989) Vibronic interactions in molecules and crystals. Springer, Berlin
46. Zenchenko VP, Vekhter BG, Bersuker IB (1982) Sov Phys JETP 55:943
47. Kristoffel NN, Konsin PI (1973) In: Titanat Baria Nauka, Moscow, p 11; Ferroelectrics 6:3; Konsin P, Kristoffel N (1999) Ferroelectrics 226:95
48. Girshberg YaG, Yacoby Y (1997) Solid State Commun 103:425; (1999) J Phys Condens Matter 11:9807; Girshberg YaG, Yacoby Y (2001) J Phys Condens Matter 13:8817
49. Bersuker IB, Vekhter BG (1969) Izvestia Acad Nauk SSSR, ser fiz 23:199
50. Bednorz JG, Muller KA (1986) Z Phys B 64:189; (1988) Revs Modern Phys 60:585
51. Muller KA (2007) J Phys Condens Matter 19:251002
52. Muller KA (2010) In: 20th international symposium on the Jahn-Teller effect, Abstracts, University of Fribourg, Fribourg, p 5
53. Höck K-H, Nickisch H, Thomas H (1983) Helv Phys Acta 56:237
54. Bersuker GI, Gorinchoy NN, Polinger VZ, Solonenko AO (1992) Superconductivity 5:1003
55. Kabanov VV, Mihailovic D (2000) J Supercond 13:950; Mihailovic D, Kabanov VV (2001) Phys Rev B 63:054505
56. Bersuker GI, Goodenough JB (1997) Physica C 274:267
57. Biancini A, Saimi NL, Rossetti T, Lanzera A, Perali A, Missori M, Oyangi H, Yamaguchi H, Nichibara Y, Ha DH (1996-I) Phys Rev B 54:12018

# Part I
# Jahn-Teller Effect and Vibronic Interactions: General Theory

# Chapter 2
# Density Functional Theory Study of the Multimode Jahn-Teller Effect – Ground State Distortion of Benzene Cation

**Matija Zlatar, Jean-Pierre Brog, Alain Tschannen, Maja Gruden-Pavlović, and Claude Daul**

**Abstract** The multideterminental-DFT approach performed to analyze Jahn-Teller (JT) active molecules is described. Extension of this method for the analysis of the adiabatic potential energy surfaces and the multimode JT effect is presented. Conceptually a simple model, based on the analogy between the JT distortion and reaction coordinates gives further information about microscopic origin of the JT effect. Within the harmonic approximation the JT distortion can be expressed as a linear combination of all totally symmetric normal modes in the low symmetry minimum energy conformation, which allows calculating the Intrinsic Distortion Path, IDP, exactly from the high symmetry nuclear configuration to the low symmetry energy minimum. It is possible to quantify the contribution of different normal modes to the distortion, their energy contribution to the total stabilization energy and how their contribution changes along the IDP. It is noteworthy that the results obtained by both multideterminental-DFT and IDP methods for different classes of JT active molecules are consistent and in agreement with available theoretical and experimental values. As an example, detailed description of the ground state distortion of benzene cation is given.

---

M. Zlatar (✉)
Department of Chemistry, University of Fribourg, Fribourg, Switzerland

Center for Chemistry, IHTM, University of Belgrade, Belgrade, Serbia
e-mail: matijaz@chem.bg.ac.rs

J.-P. Brog • A. Tschannen
Department of Chemistry, University of Fribourg, Fribourg, Switzerland

M. Gruden-Pavlović
Faculty of Chemistry, University of Belgrade, Belgrade, Serbia
e-mail: gmaja@chem.bg.ac.rs

C. Daul
Department of Chemistry, University of Fribourg, Fribourg, Switzerland
e-mail: claude.daul@unifr.ch

M. Atanasov et al. (eds.), *Vibronic Interactions and the Jahn-Teller Effect: Theory and Applications*, Progress in Theoretical Chemistry and Physics 23, DOI 10.1007/978-94-007-2384-9_2, © Springer Science+Business Media B.V. 2012

## 2.1 Introduction

In contempt of the great progress and development of various experimental techniques for studying the Jahn-Teller (JT) effect [1, 2], computational methods are necessary to understand the microscopic origin and to get deeper insight into the vibronic coupling effects. Traditional computational methods, can still be used even where non-adiabatic effects are important, if a perturbation approach is introduced to the Born-Oppenheimer approximation. This is possible if the adiabatic potential energy surface can be accurately determined, as in the case of Density Functional Theory (DFT). In this way all the standard concepts in theoretical chemistry are still useful for the elucidation and prediction of the properties of JT active molecules and many manifestations of the JT effect can be understood within the Born-Oppenheimer approximation. DFT is today widely used method in quantum chemistry and allows obtaining accurate results at low computational cost. Moreover, it helps to understand chemical origin of the effect under study. Contrary to some beliefs that DFT is not adequate for degenerate states [3,4], DFT can be applied to both degenerate and excited states, as formally proved by the reformulation of the original Hohenberg-Kohn theorems, i.e. constrained search method and finite temperature DFT [5]. Furthermore, Kohn-Sham (KS) DFT, as the most common practical way of using DFT, is based on the equations equivalent and fully compatible with the equations used in the wave-function based methods. However, precaution is necessary, conventional DFT cannot be used as a black-box, e.g. it needs to be extended to treat a multiplet problem [6–8]. KS-DFT in its present implementations is not able to derive correct energies in the case of orbital degeneracy [3, 9, 10]. On the other hand, multideterminental-DFT approach, developed by Daul et al. [11, 12], can be successfully applied for detailed analysis of the JT active molecules [10–20]. This method gives the JT parameters, Fig. 2.1, as well as corresponding geometries with good accuracy.

The set of parameters depicted on the Fig. 2.1 characterizes the potential energy surface of the JT active molecules. The distortion from a high symmetry (HS) nuclear arrangement, due to the JT effect, towards a lower symmetry (LS) energy minimum conformation, is a linear displacement on the $3N - 6$ potential energy surface. In the ideal case, the distortion would correspond to the movements of nuclei along one normal mode that belongs to a non-totally symmetric irreducible representation of the HS point group of a molecule. Nevertheless, this can be true only for simple molecules. In complex molecules, the JT distortion is a superposition of many different normal coordinates. The appraising of the influence of different normal modes on the JT effect is referred as a multimode problem. Recently, we have proposed to express the JT distortion as a linear combination of all totally symmetric normal modes in the LS minimum energy conformation [12]. Our approach to the multimode JT problem is an alternative to the method based on the transformations in the coordinate system which reduces multimode to one-mode problem [2, 21]. Recently, Bersuker et al. used the method of coordinate transformation of the HS symmetrized displacements of $CO_3$ to reduce

## 2 DFT and the Multimode Jahn-Teller Effect

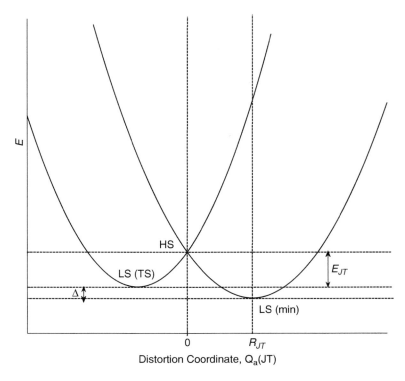

**Fig. 2.1** Qualitative cross section through the potential energy surface, along JT active distortion $Q_a$; Definition of the JT parameters—the JT stabilization energy, $E_{JT}$, the warping barrier, $\Delta$, the JT radius, $R_{JT}$

the two-mode to one-mode problem [22]. The normal coordinates in the LS are not identical with the normal coordinates in the HS, so the idea of using LS structure as a reference point is in contrast with usual treatment of the JT effect which starts from the HS configuration. Still, we can easily correlate the normal modes of the LS structure to the HS ones, using the method of Hug [23], thus having the connection with the common exploration of the JT effect. Our model is based on the symmetry rules developed by Bader [24–26] and Pearson [27, 28] for chemical reactions. Any displacement on the potential energy surface has to be totally symmetric and consequently a superposition of the totally symmetric normal coordinates. Within harmonic approximation the potential energy surface has a simple analytical form, hence with our model it is possible to directly separate contributions of different normal modes to the JT distortion, $\mathbf{R}_{JT}$, their energy contribution to the $E_{JT}$ and the forces at the HS point. This allows calculating the path of minimal energy, Intrinsic Distortion Path (IDP), exactly from the HS point to the LS energy minimum. Inspection of the IDP gives additional information about microscopic origin and mechanism of the distortion.

## 2.2 Computational Details

The DFT calculations reported in this work have been carried out using the Amsterdam Density Functional program package, ADF2009.01 [29–31]. The local density approximation (LDA) characterized by the Vosko- Willk-Nusair (VWN) [32] parametrization have been used for the geometry optimizations. An all electron Triple-zeta Slater-type orbitals (STO) plus one polarization function (TZP) basis set has been used for all atoms. All calculations were spin-unrestricted. Separation of orbital and geometrical symmetry, as used in the calculation of the energies of the HS nuclear configurations, is done using SYMROT subblock in the QUILD program, version 2009.01 [33], provided in the ADF2009.01 program package. Analytical harmonic frequencies were calculated [34, 35], and were analyzed with the aid of PyVib2 1.1 [36]. Vibrations are illustrated using the vibrational energy distribution representation [23]. The different colours indicate the direction of the displacement vector, while the volumes of the spheres are proportional to the contribution made by the individual nuclei to the energy of the vibrational mode.

## 2.3 Methodology

### 2.3.1 DFT for the Calculation of the Jahn-Teller Parameters

In order to get the JT parameters ($E_{JT}$, $\Delta$, $R_{JT}$), Fig. 2.1, it is necessary to know the energies and geometries of the HS and LS nuclear arrangements. For the LS structures, as they are in non-degenerate electronic states this is straightforward. A geometry optimization, constraining the structure to the LS point group, with proper orbital occupancy will yield the different LS geometries and energies, that correspond to the minimum and to the transition state on the potential energy surface.

Electronic structure of the HS point, on the other hand, must be represented with at least two Slater determinants, consequently, using a single determinant DFT is troublesome. In a non-empirical approach to calculate the JT distortion using DFT [11] it was proposed to use the average of configuration (AOC) type calculation to generate the electron density. This is a SCF calculation where the electrons of degenerate orbitals are distributed equally over the components of the degenerate irreps leading to a homogeneous distribution of electrons with partial occupation. In this way, the $A_1$ symmetry of the total density in the HS point group is retained. E.g. for $e^1$ configuration this will mean to place 0.5 electrons into each of the two $e$ orbitals. This calculation yields the geometry of the high symmetry species. Although, AOC calculation gives us geometry of a HS point, one needs to be cautious with using simply this energy. Electronic distribution with an electron (or a hole) evenly distributed between the degenerate orbitals leads to a lower energy than in the case of an integer orbital occupancy. Practical solution to the problem

2  DFT and the Multimode Jahn-Teller Effect

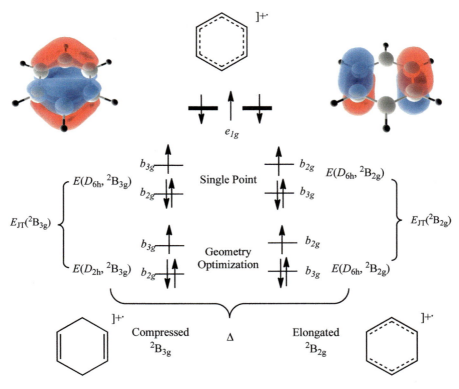

**Fig. 2.2** Multideterminental-DFT approach for the calculation of the JT parameters for the ground state distortion of $C_6H_6^+$

is to perform a single point calculation imposing the HS symmetry on the nuclear geometry and the LS symmetry of the electron density. This gives the energy of a Slater determinant with an integer electron occupation of the molecular orbitals. To obtain the energies of the degenerate states at HS point, it is necessary to evaluate the energies of all possible single determinants with integer electron occupations.

To summarize: the difference between the energy obtained by the single point calculations on the HS nuclear geometry with the LS electron density, and the energy obtained by geometry optimization of the LS structure with the same electron distribution is $E_{JT}$. $\Delta$ is the difference in energy between the two LS structures with different electron distributions. The JT radius, $R_{JT}$ is given by the length of the distortion vector ($\mathbf{R}_{JT}$) between the HS configuration ($\mathbf{R}_{HS}$) and the LS stationary points ($\mathbf{R}_{LS}$), where $\mathbf{u}$ is an unit vector.

$$\mathbf{R}_{JT} = \mathbf{R}_{HS} - \mathbf{R}_{LS} = R_{JT}\mathbf{u} \tag{2.1}$$

This computational procedure, for the particular example of the ground state JT effect in benzene cation, is outlined in the Fig. 2.2.

### 2.3.2 Analysis of the Multimode Jahn-Teller Effect

Geometry of the LS energy minimum conformation is chosen to be the origin of the configuration space, $\mathbf{R}_{LS} = \mathbf{0}$. Every point on the potential energy surface can be represented by a $3N$ dimensional vector $\mathbf{R}_X$ using mass-weighted generalized coordinates relative to the origin. Within the harmonic approximation, it is possible to express $\mathbf{R}_X$ as a linear combination of all totally symmetric normal coordinates ($N_{a_1}$) from the LS energy minimum conformation:

$$\mathbf{R}_X = \mathbb{Q}\mathbf{w}_X \tag{2.2}$$

$\mathbb{Q}$ is the $3N \times N_{a_1}$ matrix with a columns being mass-weighted totally symmetric normal coordinates, obtained by the DFT calculations in the LS minimum energy conformation. $\mathbf{w}_X$ is the $N_{a_1}$ dimensional vector containing the weighting factors, $w_{HSk}$, which can be easily obtained solving the linear problem, Eq. 2.2. $w_{HSk}$s represent the contribution of the displacements along the different totally symmetric normal coordinates to the $\mathbf{R}_X$. The energy of the nuclear configuration $\mathbf{R}_X$, $E_X$, relative to the energy of the origin, in harmonic approximation, is expressed as the sum of the energy contributions of all $N_{a_1}$ LS totally symmetric normal coordinates:

$$E_X = \sum_{k=1}^{N_{a_1}} E_{kX} = \frac{1}{2} \sum_{k=1}^{N_{a_1}} w_{Xk}^2 \mathbf{Q}_k^2 \lambda_k \tag{2.3}$$

where $\lambda_k$ are the eigenvalues of the Hessian from the DFT calculations in the LS minimum energy conformation. Using these considerations we can analyze the multimode JT problem by expressing the $\mathbf{R}_{JT}$ as a superposition of all the LS totally symmetric normal coordinates, and directly obtaining the energy contributions of all the normal modes to the total stabilization energy:

$$\mathbf{R}_{JT} = \mathbb{Q}\mathbf{w}_{kJT} \tag{2.4}$$

$$E_{JT} = \sum_{k=1}^{N_{a_1}} E_{kJT} = \frac{1}{2} \sum_{k=1}^{N_{a_1}} w_{kJT}^2 \mathbf{Q}_k^2 \lambda_k \tag{2.5}$$

The vector $\mathbf{R}_{JT} = \mathbf{R}_{HS}$ defines the straight path from the HS point to the LS minimum—direct path. Direct path is equivalent to the interaction mode of Bersuker et al. [2, 21, 22]. It contains essential information on the vibronic coupling at the HS point. Direct path is in general different from the minimal energy path from HS point on the potential energy surface to LS global minimum. The force along the normal mode $\mathbf{Q}_k$, $\mathbf{F}_{Xk}$, which drives the nuclei along that coordinate to the minimum, at any point $\mathbf{R}_X$ is defined as a derivative of the energy over the Cartesian coordinates. In the HS point this will lead information about the main driving force for the JT distortion from the HS to the LS. The total distortion force, $\mathbf{F}_{Xtot}$, is given as a vector sum of the individual forces. $\mathbf{F}_{Xtot}$ gives the direction from one to the another point on the adiabatic potential energy surface in a way of maximizing the decrease of energy.

$$\mathbf{F}_{\text{Xtot}} = \sum_{k=1}^{N_{a_1}} \mathbf{F}_{\text{Xk}} = \sum_{k=1}^{N_{a_1}} w_{\text{Xk}} \lambda_k \mathbb{M}^{1/2} \mathbf{Q}_k \qquad (2.6)$$

$\mathbb{M}$ is a diagonal $3N \times 3N$ matrix with atomic masses in triplicates as elements $(m_1, m_1, m_1, m_2, \ldots, m_N)$.

The simple analytical form of the potential energy surface in the harmonic approximation allows calculating the minimal energy path, IDP, exactly from the HS to the LS energy minimum. Along the IDP path the contributions of the different modes to the distortion will change, contrary to the direct path. At the HS only JT active modes contribute. After the first step, the symmetry is lowered and the other modes will mix in. Analysis of the IDP allows getting very detailed picture on the interaction between the deformation of the electron distribution and the displacements of the nuclei, not accessible through an experiment. We should make the remark that the $E_{\text{JT}}$ obtained with the IDP method is a complementary to the method based on the multideterminental-DFT described in previous section. For the IDP analysis one needs information about HS and LS geometries and LS normal modes, thus, avoiding the problems of evaluation of the energies of the degenerate states. Another important thing is that this model is completely theoretical, without additional fitting, all the vibrations that can contribute to the distortion are considered, and all the equations can be solved analytically in the harmonic approximation.

## 2.4 Results and Discussion

Using the multideterminental-DFT procedure in the conjunction with the IDP method we have analyzed a number of different JT active molecules, Table 2.1. The considered molecules, in HS nuclear configuration, have a doubly degenerate electronic ground state which is coupled with a doubly degenerate vibrations. Thus, they belong to the $E \otimes e$ JT systems. Apart from that, these systems differ in the nature of chemical bonding, symmetry of the distortion, range of the $E_{\text{JT}}$, $\Delta$ and $R_{\text{JT}}$, the number of atoms, and hence the number of different normal modes that need to be considered in the IDP analysis.

Without going into details, because some of these results have already been published [11, 12, 16, 18–20], we want to point out that the results obtained by multideterminental-DFT and IDP methods are consistent, and in a good agreement with experimental data and various theoretical methods, Table 2.1. In all cases we were able to rationalize which are the totally symmetric normal modes of the LS structure that contribute to the JT distortion at the HS point, their energy contribution to the $E_{\text{JT}}$, and how their contribution changes along the IDP, even in the case when we have to deal with 53 totally symmetric normal modes in

**Table 2.1** Results of multideterminental-DFT calculations and the IDP method performed to analyze the JT effect of selected compounds

| | Distortion | $E_{JT}$(DFT) | $E_{JT}$(IDP) | $E_{JT}$(exp) | $\Delta$ | $R_{JT}$ | $3N-6$ | $N_{a_1}$ |
|---|---|---|---|---|---|---|---|---|
| $Cu_3$ | $D_{3h} \longrightarrow C_{2v}$ | 0.53 [18] | 0.37 | 0.28–0.55 [2] | 0.12 | 1.12 | 3 | 2 |
| $VCl_4$ | $T_d \longrightarrow D_{2d}$ | 0.04 [11,12] | 0.01 | 0.03–0.08 [37–40] | 0.01 | 0.10 | 9 | 2 |
| $C_5H_5$ | $D_{5h} \longrightarrow C_{2v}$ | 1.25 [12] | 1.24 | 1.24 [41] | 0.00 | 0.17 | 24 | 9 |
| $CoCp_2$ | $D_{5h} \longrightarrow C_{2v}$ | 0.81 [16,18] | 0.80 | 0.15–1.05 [42] | 0.00 | 0.35 | 57 | 16 |
| $MnCp_2$ | $D_{5h} \longrightarrow C_{2v}$ | 0.71 [18] | 0.72 | $\sim 0.35$ [42,43] | 0.00 | 0.77 | 57 | 16 |
| $Cu(en)_3^{2+}$ | $D_3 \longrightarrow C_2$ | 2.12 [19] | 2.70 | $2.00 \pm 0.20$ [44] | 0.57 | 2.34 | 105 | 53 |
| $Cu(eg)_3^{2+}$ | $D_3 \longrightarrow C_2$ | 2.31 [19] | 2.50 | n.r. | 1.41 | 3.36 | 87 | 44 |

Energies are given in $10^3 cm^{-1}$, $R_{JT}$ in $(amu)^{1/2}$Å

$N$ is the number of atoms in a molecule, $N_{a_1}$ is the number of totally symmetrical vibrations in LS minimum

tris(ethylenediamine) Cu(II) [19]. In the beginning, the JT active modes, that are the basis of the non-totally symmetric irreps in HS point group, are dominating. If there are several appropriate vibrations, harder ones will be dominant in the first step, while the softer ones take over along the IDP. The contribution of the other modes is becoming more important for the direction of the relaxation with increasing deviation from the HS geometry. Their contribution to the stabilization energy however is minor.

As an example, results for benzene cation are presented in the following section, which allows the reader to understand the concepts behind both the multideterminental-DFT and IDP methods.

## 2.4.1 Multimode Jahn-Teller Effect in the Ground State of Benzene Cation

Benzene cation has a $^2E_{1g}$ ground electronic state in regular hexagonal nuclear configuration, $D_{6h}$ point group, with a three electrons (one hole) in the doubly degenerate orbital. Hence, benzene cation is JT active and prefers a conformation of lower symmetry. According to group theory, the distortion coordinate is $e_{2g}$ ($E_{1g} \times E_{1g} \subset A_{1g} + [A_{2g}] + E_{2g}$). Descent in symmetry goes to $D_{2h}$. In $D_{2h}$ electronic state splits into in $^2B_{2g}$ and $^2B_{3g}$. Qualitatively, direction of the distortion can be understood by looking to the frontier orbitals of the cation, Fig. 2.2. The direction of distortion is in the way of maximizing the bonding interactions and minimizing antibonding interactions in the doubly occupied molecular orbital originated from the $e_{1g}$ orbital in $D_{6h}$. The $^2B_{2g}$ state corresponds to the elongated (acute) geometry, with two C $-$ C bonds longer and four other shorter. According to our DFT calculations it is the global minimum on the potential energy surface. The $^2B_{3g}$ electronic state corresponds to the compressed (obtuse) geometry, with two shorter and four longer C $-$ C bonds. This structure is a transition state for the process of the pseudorotation along the lowest sheet of potential energy surface. JT effect in this

**Table 2.2** Results of the DFT calculations performed to analyze the JT effect in $C_6H_6{}^+$

| Occupation | State | Geometry | Energy |
|---|---|---|---|
| $e_{1g}^{0.75}e_{1g}^{0.75}$ | $^2E_{1g}$ | $D_{6h}$ | −70.9461 |
| $b_{3g}^2b_{2g}^1$ | $^2B_{2g}$ | $D_{6h}$ | −70.9122 |
| $b_{2g}^2b_{3g}^1$ | $^2B_{3g}$ | $D_{6h}$ | −70.9141 |
| $b_{3g}^2b_{2g}^1$ | $^2B_{2g}$ | $D_{2h}$ | −71.0212 |
| $b_{2g}^2b_{3g}^1$ | $^2B_{3g}$ | $D_{2h}$ | −71.0172 |
| $E_{JT}$ | $^2B_{2g}$ | | 879.2 |
| $E_{JT}$ | $^2B_{3g}$ | | 831.6 |
| $\Delta$ | | | 32.2 |
| $R_{JT}$ | $^2B_{2g}$ | | 0.09 |
| $R_{JT}$ | $^2B_{3g}$ | | 0.09 |
| $E_{JT}(IDP)$ | $^2B_{2g}$ | | 839.1 |
| $E_{JT}(IDP)$ | $^2B_{3g}$ | | 791.1 |

Energies (LDA) are given in eV, the JT parameters $E_{JT}$ and $\Delta$ are given in cm$^{-1}$, and $R_{JT}$ in (amu)$^{1/2}$Å

system is summarized in the Fig. 2.2, while the results of the DFT calculations are presented in the Table 2.2. According to our DFT calculations $E_{JT} = 880\,cm^{-1}$ for this system. The elongated structure is found to be more stable than the compressed one for $\Delta = 32\,cm^{-1}$. IDP method gives $E_{JT}$ in a good agreement with the DFT calculations, with the smaller values for around $40\,cm^{-1}$. $\Delta$ value from IDP method can be estimated from a difference between the two $E_{JT}$ for two electronic states and is $48\,cm^{-1}$, a slightly higher value than obtained with multideterminental-DFT. Small energy difference between the $^2B_{2g}$ and $^2B_{3g}$ states suggests that the second order JT effects and anharmonicity are small and that JT effect in benzene cation is dynamic.

Benzene cation has been thoroughly studied over the years [45–55]. The values of the JT parameters largely depend on the used method and model employed. Calculations based on the Hartree-Fock and MP2 methods, not surprisingly, overestimate the $E_{JT}$ [45]. Experimental methods and models that consider only one mode, on the other hand, are underestimating $E_{JT}$, e.g. experimental value of Lindner et al is $266\,cm^{-1}$ [46]. Recent calculations based on complete active space or DFT, including this study, are much more consistent and in general agreement that $E_{JT}$ falls in range between 700 and $1,000\,cm^{-1}$ [48, 53, 54]. Even more inconsistency is found for the value of $\Delta$, and alternately both states are reported to be a global minimum [47, 48]. This is due to the small value of $\Delta$ which is in the range of errors of the calculations. Experimentally $\Delta$ is estimated to be $8\,cm^{-1}$ [46] and the elongated form to be a minimum. Independently of the particular value, the $\Delta$ is rather small, smaller than zero point vibrational energy, and JT effect is dynamic. Köppel et al. [49–52] and Sardar et al. [55] did detailed dynamic study of the multistate multimode problem in benzene cation.

Benzene cation in $D_{6h}$ symmetry has four $e_{2g}$ and two $a_{1g}$ vibrations. After descent in symmetry to $D_{2h}$ they all become totally symmetric (one component of each pair in the case of the degenerate vibrations). Thus, in $D_{2h}$ symmetry there are

591 cm$^{-1}$ (C–C–C bend)   1166 cm$^{-1}$ (C–C–H bend)   1556 cm$^{-1}$ (C–C stretch)

**Fig. 2.3** Vibrational energy distribution representation of the three most important $a_{1g}$ vibrations in $D_{2h}$ symmetry of $C_6H_6^+$, corresponding to the three $e_{2g}$ JT active vibrations in $D_{6h}$ symmetry. The different *colours* indicate the direction of the displacement vector; the volume of the spheres is proportional to the contribution made by the individual nuclei to the energy of the vibrational mode

six totally symmetric normal modes, which can mix and all of them can contribute to the distortion. Out of six totally symmetric normal modes in $D_{2h}$, with our model we are able to identify the three most important vibrations contributing to the JT distortion: C–C–C bend (591 cm$^{-1}$), C–C–H bend (1,166 cm$^{-1}$) and C–C stretch (1,556 cm$^{-1}$). This is in agreement with previous studies [48, 53, 54]. These three vibrations are illustrated on the Fig. 2.3 using the vibrational energy distribution representation [23]. These three vibrations corresponds to $e_{2g}$ vibrations in $D_{6h}$ configuration of benzene cation, and contribute 99% to the $R_{JT}$ and 98% to the $E_{JT}$, Fig. 2.4.

IDP method gives further insight into the vibronic coupling in benzene cation. Figure 2.5 shows the energy differences between the IDP and the direct path and the changes of the forces of the different normal modes along the IDP. On the potential energy profile it is possible to distinguish two distinct regions. In the first region energy is changing faster. After 25% of the path, already c.a. 75% of the $E_{JT}$ is obtained. In this region all three dominant vibrations are contributing, with the hardest of the three, C–C stretch being most important. Contribution of the C–C stretch becomes minor in the second region. In the second region the change of the energy is small and potential energy surface is flat. In this region, the molecule, after achieving most of the stabilization energy due to the JT effect, relaxes towards the global minimum. The softest mode, C–C–C bend is the most important in this second region.

## 2.5 Conclusions

In this paper multideterminental-DFT method for the qualitative and quantitative analysis of the adiabatic potential energy surfaces of the JT active molecules is presented. It is shown how DFT can be successfully applied for the calculation of the JT parameters. In addition, the analysis of the multimode JT effect using the IDP method is shown. The essence of this model is to express the JT distortion as a linear combination of all totally symmetric normal modes in the LS minimum energy

## 2 DFT and the Multimode Jahn-Teller Effect

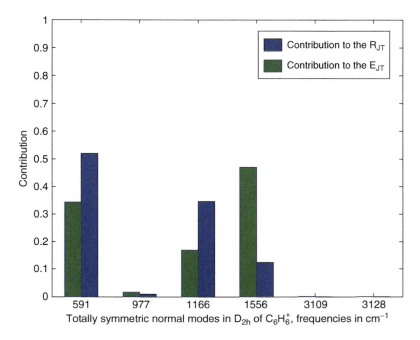

**Fig. 2.4** Contribution of the 6 $a_{1g}$ normal modes in $D_{2h}$ to the $E_{JT}$ (*green*) and to the $R_{JT}$ (*blue*) of $C_6H_6{}^+$

conformation. This is an alternative to the method that starts from the HS structure and reduces the multimode problem to a single interaction mode [2, 21, 22]. So far, IDP method has not been explored in the cases of the hidden JT effect, or where the influence of the pseudo JT effect is important, and combination of two different approaches may further develop this topic. The IDP analysis answers the questions which are the totally symmetric normal modes of the LS structure that contribute to the JT distortion at the HS point, how they contribute to the $E_{JT}$, and how their contributions change along the IDP. We want to point out that this reduction of the multimode problem to the IDP solves the structural part of the problem, but not the dynamical part, meaning the changes in vibrational frequencies by the multimode distortion.

The performance of the multideterminental-DFT and the IDP model has been evaluated for different JT active molecules. In all cases the results obtained by both methods are consistent and in the excellent agreement with experimental and theoretical values reported in the literature. The detailed description of the multimode JT distortion of the ground state of benzene cation is shown as an example of utility of both methods. Both schemes are fast and accurate and can be considered as reliable tools for the investigation of adiabatic potential energy surfaces of JT active molecules and for a better understanding of the JT effect.

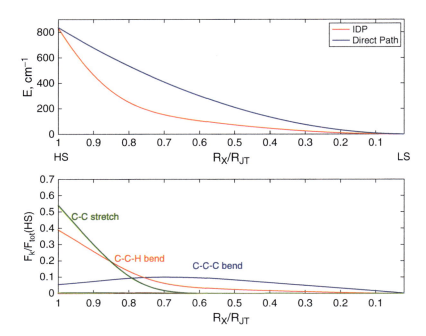

**Fig. 2.5** Difference between the direct Path and IDP (up) and changes of the forces (normalized to the total distortion force at HS point) of the 6 $a_{1g}$ normal modes in $D_{2h}$, along the IDP of $C_6H_6^+$

**Acknowledgements** This work was supported by the Swiss National Science Foundation and the Serbian Ministry of Science (Grant no. 172035). The authors would like to thank to Professor Carl-Wilhelm Schläpfer for the constant interest and advice.

# References

1. Jahn HA, Teller E (1937) Proc R Soc Lond Ser A 161:220
2. Bersuker IB (2006) The Jahn-Teller effect. Cambridge University Press, Cambridge
3. Bersuker IB (1997) J Comput Chem 18(2):260
4. Kaplan IG (2007) J Mol Struct 838:39
5. Parr RG, Yang W (1989) Density-functional theory of atoms and molecules. Oxford University Press, New York
6. Ziegler T, Rauk A, Baerends EJ (1977) Theor Chim Acta 43:261
7. Daul C (1994) Int J Quant Chem 52:867
8. Daul CA, Doclo KG, Stückl CA (1997) In: Chong DP (ed) Recent advances in density functional methods, Part II. World Scientific, Singapore, pp 61–113
9. Atanasov M, Daul C (2005) Chimia 59:504
10. Reinen D, Atanasov M, Massa W (2006) Z Anorg Allg Chem 632:1375
11. Bruyndonckx R, Daul C, Manoharan PT, Deiss E (1997) Inorg Chem 36:4251
12. Zlatar M, Schläpfer CW, Daul C (2009) In: Koeppel H, Yarkoni DR, Barentzen H (eds) The Jahn-Teller-effect fundamentals and implications for physics and chemistry. Springer series in chemical physics, vol 97. Springer, Berlin, pp 131–165

13. Kundu TK, Bruyndonckx R, Daul C, Manoharan PT (1999) Inorg Chem 38:3931
14. Atanasov M, Comba P (2007) J Mol Struct 38:157
15. Atanasov M, Comba P, Daul CA, Hauser A (2007) J Phys Chem A 38:9145
16. Zlatar M, Schläpfer CW, Fowe EP, Daul C (2009) Pure Appl Chem 81:1397–1411
17. Reinen D, Atanasov M, Köhler P, Babel D (2010) Coord Chem Rev 254:2703
18. Zlatar M, Gruden-Pavlović M, Schläpfer CW, Daul C (2010) J Mol Struct-Theochem 954:86
19. Gruden-Pavlović M, Zlatar M, Schläpfer CW, Daul C (2010) J Mol Struct-Theochem 954:80
20. Zlatar M, Gruden-Pavlović M, Schläpfer CW, Daul C (2010) Chimia 64:161
21. Bersuker IB, Polinger VZ (1989) Vibronic interactions in molecules and crystals. Springer, Berlin
22. Liu Y, Bersuker IB, Zou W, Boggs JE (2009) J Chem Theory Comput 5:2679
23. Hug W, Fedorovsky M (2008) Theor Chem Acc 119:113
24. Bader RFW (1960) Mol Phys 3:137
25. Bader RFW (1962) Can J Chem 40:1164
26. Bader RFW, Bandrauk AD (1968) J Chem Phys 49:1666
27. Pearson RG (1969) J Am Chem Soc 91(18):4947
28. Pearson RG (1976) Symmetry rules for chemical reactions. Willey, New York
29. Baerends EJ, Autschbach J, Bashford D, Bérces A, Bickelhaupt FM, Bo C, Boerrigter PM, Cavallo L, Chong DP, Deng L, Dickson RM, Ellis DE, van Faassen M, Fan L, Fischer TH, Guerra CF, Ghysels A, Giammona A, van Gisbergen S, Götz A, Groeneveld J, Gritsenko O, Grning M, Harris F, van den Hoek P, Jacob C, Jacobsen H, Jensen L, van Kessel G, Kootstra F, Krykunov MV, van Lenthe E, McCormack DA, Michalak A, Mitoraj M, Neugebauer J, Nicu VP, Noodleman L, Osinga VP, Patchkovskii S, Philipsen PHT, Post D, Pye CC, Ravenek W, Rodrguez JI, Ros P, Schipper PRT, Schreckenbach G, Seth M, Snijders JG, Sola M, Swart M, Swerhone D, te Velde G, Vernooijs P, Versluis L, Visscher L, Visser O, Wang F, Wesolowski TA, van Wezenbeek EM, Wiesenekker G, Wolff SK, Woo TK, Yakovlev AL, Ziegler T (2009) Adf2009.01. http://www.scm.com
30. Guerra CF, Snijders JG, te Velde G, Baerends EJ (1998) Theor Chem Acc 99:391
31. te Velde G, Bickelhaupt FM, van Gisbergen SJA, Guerra CF, Baerends EJ, Snijders JG, Ziegler T (2001) J Comput Chem 22:931
32. Vosko S, Wilk L, Nusair M (1980) Can J Phys 58:1200
33. Swart M, Bickelhaupt FM (2007) J Comput Chem 29:724
34. Bérces A, Dickson RM, Fan L, Jacobsen H, Swerhone D, Ziegler T (1997) Comput Phys Commun 100:247
35. Jacobsen H, Bérces A, Swerhone D, Ziegler T (1997) Comput Phys Commun 100:263
36. Fedorovsky M (2007) Pyvib2, a program for analyzing vibrational motion and vibrational spectra. http://pyvib2.sourceforge.net
37. Blankenship FA, Belford RL (1962) J Chem Phys 36:633
38. Morino Y, Uehara H (1966) J Chem Phys 45:4543
39. Johannesen RB, Candela GA, Tsang T (1968) J Chem Phys 48:5544
40. Ammeter JH, Zoller L, Bachmann J, Baltzer P, Bucher ER, Deiss E (1981) Helv Chim Acta 64:1063
41. Applegate BE, Bezant J, Miller TA (2001) J Chem Phys 114:4869
42. Bucher R (1977) Esr-untersuchungen an jahn-teller-aktiven sandwitchkomplexen. Ph.D. thesis, ETH Zürich
43. Ammeter JH, Oswald N, Bucher R (1975) Helv Chim Acta 58:671
44. Gamp E (1980) Esr-untersuchungen über den Jahn-Teller-effekt in oktaedrischen kupfer (ii)-komplexen mit trigonalen dreizähnigen liganden. Ph.D. thesis, ETH Zürich
45. Raghavachari K, Haddon RC, Miller TA, Bondybey VE (1983) J Chem Phys 79:1387
46. Lindner R, Müller-Dethlefs K, Wedum E, Haber K, Grant ER (1996) Science 271:1698
47. Muller-Dethlefs K, Peel JB (1999) J Chem Phys 111:10550
48. Applegate BE, Miller TA (2002) J Chem Phys 117:10654
49. Köppel H, Cederbaum LS, Domcke W (1988) J Chem Phys 89:2023
50. Döscher M, Köppel H, Szalay PG (2002) J Chem Phys 117:2645

51. Döscher M, Köppel H, Baldea I, Meyer HD, Szalay PG (2002) J Chem Phys 117:2657
52. Baldea I, Köppel H (2006) J Chem Phys 124:064101
53. Perebeinos V, Allen PB, Pederson M (2005) Phys Rev A 72:012501
54. Tokunaga K, Sato T, Tanaka K (2006) J Chem Phys 124:154303
55. Sardar S, Paul AK, Sharma R, Adhikari S (2009) J Chem Phys 130:144302

# Chapter 3
# A Symmetry Adapted Approach to the Dynamic Jahn-Teller Problem

**Boris Tsukerblat, Andrew Palii, Juan Modesto Clemente-Juan, and Eugenio Coronado**

**Abstract** In this article we present a symmetry-adapted approach aimed to the accurate solution of the dynamic Jahn-Teller (JT) problem. The algorithm for the solution of the eigen-problem takes full advantage of the point symmetry arguments. The system under consideration is supposed to consist of a set of electronic levels $\Gamma_1, \Gamma_2 \ldots \Gamma_n$ labeled by the irreducible representations (irreps) of the actual point group, mixed by the active JT and pseudo JT vibrational modes $\Gamma_1, \Gamma_2 \ldots \Gamma_f$ (vibrational irreps). The bosonic creation operators $b^+(\Gamma\gamma)$ are transformed as components $\gamma$ of the vibrational irrep $\Gamma$. The first excited vibrational states are obtained by the application of the operators $b^+(\Gamma\gamma)$ to the vacuum: $b^+(\Gamma\gamma)|n = 0, A_1\rangle = |n = 1, \Gamma\gamma\rangle$ and therefore they belong to the symmetry $\Gamma\gamma$. Then the operators $b^+(\Gamma\gamma)$ act on the set $|n = 1, \Gamma\gamma\rangle$ with the subsequent Clebsch-Gordan coupling of the resulting irreps. In this way one obtains the basis set $|n = 2, \Gamma'\gamma'\rangle$ with $\Gamma' \in \Gamma \otimes \Gamma$. In general, the Gram-Schmidt orthogonalization is required at each step of the procedure. Finally, the generated vibrational bases are coupled to the electronic ones to get the symmetry adapted basis in which the full matrix of the JT Hamiltonian is blocked according to the irreps of the point group. The approach is realized as a computer program that generates the blocks and evaluates

---

B. Tsukerblat (✉)
Chemistry Department, Ben-Gurion University of the Negev, Beer-Sheva 84105, Israel
e-mail: tsuker@bgu.ac.il

A. Palii
Institute of Applied Physics, Academy of Sciences of Moldova,
Academy Str. 5 Kishinev, Moldova
e-mail: andrew.palii@uv.es

J.M. Clemente-Juan • E. Coronado
Instituto de Ciencia Molecular, Universidad de Valencia, Polígono de la Coma,
s/n 46980 Paterna, Spain
e-mail: juan.m.clemente@uv.es; eugenio.coronado@uv.es

M. Atanasov et al. (eds.), *Vibronic Interactions and the Jahn-Teller Effect:*
*Theory and Applications*, Progress in Theoretical Chemistry and Physics 23,
DOI 10.1007/978-94-007-2384-9_3, © Springer Science+Business Media B.V. 2012

all required characteristics of the JT systems. The approach is illustrated by the simulation of the vibronic charge transfer (intervalence) optical bands in trimeric mixed valence clusters.

## 3.1 Introduction

JT and pseudo JT effects lead to a complicated dynamic problem that, in general, cannot be solved in an analytical way [1–3]. The difficulties are aggravated for the multimode systems or/and when the energy pattern contains several low lying levels mixed by the active JT and pseudo JT vibrations. This especially refers to the large scale mixed valence systems, like biologically important systems, iron-sulfur proteins [4–6], polynuclear reduced polyoxometalates (for example, Keggin and Wells-Dawson systems) in which several electrons are delocalized over metal network [7, 8], impurity metal ions in crystals [9–11], etc. In these cases the dimension of the truncated vibronic matrices to be diagonalized becomes very high due to sizeable electronic basis and multiple degeneracy of the excited vibrational levels. Convergence of the results is always questionably so that the truncation of the vibronic matrices can result in a dramatic lack of precision not only when the vibronic coupling is strong but also providing moderate or even relatively weak coupling.

For this reason many efforts have been applied toward elaboration of the approximate approaches among which the adiabatic approximation plays a key role. In many cases the adiabatic approximation provides a good qualitative insight on many features of the JT systems, for example, an approximate study of the broad electron-vibrational light absorption bands, asymmetric electronic distribution of the electronic density in complex mixed-valence systems (for example, in polynuclear polyoxometalates [8]), structural peculiarities of polyatomic molecules [2, 3], spin-crossover complexes [12] and crystals subjected to the structural phase transitions [13]. At the same time the applicability of the semiclassic adiabatic approximation is greatly restricted due to dynamic character of the vibronic coupling resulting in a discrete energy pattern of the hybrid electron-vibrational levels. Even in the favorable case of strong vibronic coupling the adiabatic approximation fails in the description of the physical characteristics of the JT systems related to the quantum structure of the levels, like spectroscopic phenomena, and in particular, wide optical bands for which transitions in the non-adiabatic anticrossing area of the potential surfaces play an important role in the adequate description of the band-shape [11].

Many efforts have been made to get an accurate solution of the dynamic problem (for a review see refs. [1–3, 11, 14–16]). A powerful approach is based on the symmetry (Lanczos algorithm) [9, 15, b] (see also references therein). Alternatively, a significant progress in solution of the dynamic problem has been achieved by the exploration of the Lie symmetries of the JT Hamiltonians which not only bring the beauty but really allow to essentially simplify the solution of the dynamic problem

# 3 A Symmetry Adapted Approach to the Dynamic Jahn-Teller Problem

[17, 18]. The well known examples are represented by the $E \otimes e$ Hamiltonians that possess O(2) symmetry in cubic, trigonal, pentagonal and hexagonal point groups. More complicated systems in which the electronic triplet in cubic systems is coupled to both type of active vibrations, $T_{1(2)}(e+t_2)$ belong to SO$_3$ Lie group. In a sophisticated case of $(A_{1g}+T_{1u}) \otimes (a_{1g}+t_{1u}+e_g+t_{2g})$ JTE in O$_h$ the Lie group O(4) have been found (for review see Ref. [2]). The last two cases (and the multimode problem in general) exhibit high symmetry only providing a special requirement to the interrelation between the two coupling constants and vibrational frequencies in the model of the linear vibronic coupling. This significantly reduces the area of practical applicability of the extremely elegant approaches based on the unitary symmetries. In fact, if the special requirements to the coupling constants so far mentioned are not fulfilled or/and the quadratic (or/and high order) vibronic terms are to be taken into account the unitary symmetries are reduced to the point ones. That is why the task of the full exploration of the advantages provided by the point symmetry of the JT and pseudo JT systems remains actual even nowadays when the computational abilities are strongly increased. The main goals of this article are to describe a symmetry-adapted approach aimed to an accurate solution of the dynamic problem and to illustrate this approach by the consideration of a relatively simple triangular mixed valence system.

## 3.2 General

The Hamiltonian of the JT (or in general, pseudo JT) system can be represented as:

$$H = H_e + \sum_i \hbar\omega_i \left( q_i^2 - \frac{\partial^2}{\partial q_i^2} \right) + \sum_i \upsilon_i \mathbf{O}_i q_i \tag{3.1}$$

Here $q_i$ are the dimensionless vibrational coordinates, $\omega_i$ are the vibrational frequencies so that the second term in Eq. 3.1 is the harmonic oscillator Hamiltonian. The symbol $i \equiv v\bar{\Gamma}\bar{\gamma}$ involves active vibrational irreps, $\bar{\Gamma}$, $\bar{\gamma}$, numerates the basis functions of the irrep $\bar{\Gamma}$ and the symbol $v$ is introduced in order to distinguish the repeated irreps $\bar{\Gamma}$. The electronic subsystem described by the Hamiltonian $H_e$ is supposed to consist of a set of the closely spaced electronic levels $\Gamma_1, \Gamma_2 \ldots \Gamma_t$ labeled by the irreps of the actual point group (the basis functions will be denoted by the symbols $\gamma_1, \gamma_2, \ldots \gamma_t$) of the system. Conventionally, this electronic Hamiltonian (and consequently the point group) is defined in the high symmetric nuclear configuration (all $q_i = 0$). The electronic levels that can be formed by the low lying crystal field states, spin-orbital interaction, etc. are mixed by the active JT and pseudo JT vibrational modes $\bar{\Gamma}_1, \bar{\Gamma}_2 \ldots \bar{\Gamma}_f$. The third term in Eq. 3.1 represents the linear (with respect to $q_i$) term of the vibronic interaction in which $\mathbf{O}_i$ are the vibronic matrices defined in the electronic basis restricted to the levels $\Gamma_1, \Gamma_2 \ldots \Gamma_n$ and $\upsilon_i$ are the dimensionless (in $\hbar\omega$ units) vibronic coupling parameters. In general, the approach suggested here in not limited to the case of the linear vibronic coupling. It is important to note that in general the matrices $\boldsymbol{O}_i$ are non-commuting that gives rise to a complicated multidimensional dynamic problem in which JT and pseudo JT interactions are interrelated.

The full JT Hamiltonian, Eq. 3.1, can be diagonalized in basis composed as the direct product of the electronic wave-functions $|\Gamma_f \gamma_f\rangle (f = 1, 2 \ldots t)$ and the states $|n_i\rangle (n_i = 0, 1, 2 \ldots)$ of the harmonic oscillators:

$$\left|\Gamma_f \gamma_f\right\rangle |n_1\rangle |n_2\rangle \ldots |n_k\rangle \equiv \left|\Gamma_f \gamma_f\right\rangle |n_1, n_2 \ldots n_k\rangle \tag{3.2}$$

Due to multiple degeneracy of the vibrational levels in multimode JT systems the size of the matrices of the full Hamiltonian proves to be rather large even if the basis, Eq. 3.2, is restricted to a relatively small number of the vibrational levels. On the other hand, while solving the dynamic JT problem the restriction of the basis leads to a significant lack of precision even providing moderate or even relatively weak JT coupling.

Let us introduce the creation and annihilation bosonic operators $b^+$ and $b$. The creation operator has the following property:

$$b^+ |n\rangle = \sqrt{n+1} |n+1\rangle \tag{3.3}$$

so that one can obtain:

$$\left(b^+\right)^n |0\rangle = \sqrt{n!} |n\rangle \tag{3.4}$$

By applying the creation operators to the vacuum state in a successive way

$$|n\rangle = \frac{1}{\sqrt{n!}} \left(b^+\right)^n |0\rangle \tag{3.5}$$

one can build the exited states $|n\rangle$.


## 3.3 Group-Theoretical Classification for a Two-Dimensional Oscillator

The vibrational functions in the full electron-vibrational basis in Eq. 3.2 do not possess definite symmetry properties with respect to the operations of the point symmetry group. To adapt them to the definite irreps of the point group let us consider as an example the case of a two-dimensional harmonic oscillator. For the sake of definiteness we will focus on the system with a trigonal symmetry $\mathbf{C}_{3v}$ (with $z$ along $C_3$ axis) that has three irreps: $A_1$ (basis: $z$), $A_2$ (basis: $L_z$). $E$(basis: $x, y$). The normal coordinates $q_x$ and $q_y$ of the two-dimensional harmonic oscillator form

# 3 A Symmetry Adapted Approach to the Dynamic Jahn-Teller Problem

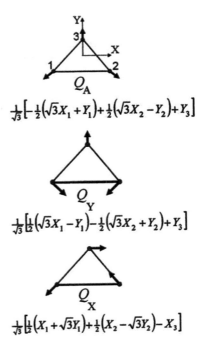

**Fig. 3.1** Vibrational coordinates of a triangular unit

basis $E$ and consequently the wave-functions can be labeled as $|n_x\rangle|n_y\rangle \equiv |n_x n_y\rangle$. In the case of a triangular molecule with the overall (including ligand surrounding) trigonal symmetry $\mathbf{C}_{3v}$ the explicit form of these vibrations is shown in Fig. 3.1 along with the molecular coordinate system.

The functions $|n_x n_y\rangle$ can be built by applying the creation operators as follows:

$$|n_x n_y\rangle = \frac{1}{\sqrt{n_x! n_y!}} \left(b_x^+\right)^{n_x} \left(b_y^+\right)^{n_y} |00\rangle \qquad (3.6)$$

Each level $n = n_x + n_y$ is $g = (n+1)$-fold degenerate, for example, the first excited level $n = 1$ is double degenerate ($n_x = 1$, $n_y = 0$ and $n_x = 0, n_y = 1$), etc. This shows that the dimension of the vibrational space is $p_N = \sum_{n=0}^{N}(n+1) = \frac{1}{2}(N+1)(N+2)$ where $N$ is the number of the vibrational levels included in the basis for the solution of the dynamic vibronic problem.

Each basis set with a given $n$, $|n_x n_y, n\rangle$ should be put in correspondence to a set $|l, m\rangle$, with $l$ being the quantum number of angular momentum. Value of $l$ for a definite $n$ is defined as $l = n/2$, so that $(2l+1) \equiv n+1$ - dimensional basis set corresponds to an irrep $D_p^{(l)}$ of $R_3$ with $p$ being the symbol of parity. Parity for the even $l$ should be defined as

$$p = (-1)^l = (-1)^{\frac{n}{2}},$$

while the parity is not assigned to the $D^{(l)}$ with half-integer $l$ (symbol $p$ is omitted). For several low lying levels one can obtain the following correspondence:

| $n$ | 0 | 1 | 2 | 3 | 4 | 5 | 6 | 7 | 8 |
|---|---|---|---|---|---|---|---|---|---|
| $D_p^{(l)}$ | $D_+^{(0)}$ | $D^{(1/2)}$ | $D_-^{(1)}$ | $D^{(3/2)}$ | $D_+^{(2)}$ | $D^{(5/2)}$ | $D_-^{(3)}$ | $D^{(7/2)}$ | $D_+^{(4)}$ |

At the next step the irreps $D^{(l)}$ of $R_3$ should be reduced in the actual point group $(C_{3v})$.

$$D_+^{(0)} \Rightarrow A_1$$

$$D^{(1/2)} \Rightarrow \bar{E}$$

$$D_-^{(1)} \Rightarrow A_1 + E$$

$$D^{(3/2)} \Rightarrow \bar{A}_1 + \bar{A}_2 + \bar{E}$$

$$D_+^{(2)} \Rightarrow A_1 + 2E$$

$$D^{(5/2)} \Rightarrow \bar{A}_1 + \bar{A}_2 + 2\bar{E}$$

$$D_-^{(3)} \Rightarrow 2A_1 + A_2 + 2E$$

$$D^{(7/2)} \Rightarrow \bar{A}_1 + \bar{A}_2 + 3\bar{E}$$

$$D_+^{(4)} \Rightarrow 2A_1 + A_2 + 3E \tag{3.7}$$

where $\bar{A}_1$, $\bar{A}_2$, $\bar{E}$ are the double-valued irreps of $C_{3v}$. Then, in this decomposition the double-valued irreps $\bar{A}_1$, $\bar{A}_2$, $\bar{E}$ (arising from odd $n$) should be replaced by the corresponding single-valued ones: $\bar{A}_1 \to A_1$, $\bar{A}_2 \to A_2$, $\bar{E} \to E$. Finally, the group – theoretical classification for the nine low lying vibrational states can be expressed as follows:

$$n = 0 \Rightarrow A_1,$$

$$n = 1 \Rightarrow E,$$

$$n = 2 \Rightarrow A_1 + E,$$

$$n = 3 \Rightarrow A_1 + A_2 + E,$$

$$n = 4 \Rightarrow A_1 + 2E,$$

$$n = 5 \Rightarrow A_1 + A_2 + 2E,$$

$$n = 6 \Rightarrow 2A_1 + A_2 + 2E,$$

$$n = 7 \Rightarrow A_1 + A_2 + 3E,$$

$$n = 8 \Rightarrow 2A_1 + A_2 + 3E, \tag{3.8}$$

3   A Symmetry Adapted Approach to the Dynamic Jahn-Teller Problem          45

This result allows to determine the irreps according to which the oscillator wave-functions are transformed without explicit calculation of this wave-functions and therefore to reduce the general time of calculation.

## 3.4   Symmetry Adapted Basis for a Two-Dimensional Harmonic Oscillator

The main issue of this section can be summarized as follows. The two functions $|01\rangle, |10\rangle$ form irrep $E$ (basis $x$ and $y$ in $C_{3v}$). The excited states behave like polynomials $x^{n_x} y^{n_y}$ (Eq. 3.6) and belong thus to the irreps. Creation operators $b_x^+$ and $b_y^+$ can be related to the irrep $E$, the wave-function of the ground state $|00\rangle$ (vacuum) belongs to $A_1$. The basis functions belonging to the definite irreps can be obtained for each $n$ by the application of the symmetry adapted polynomials constructed from operators $(b_x^+)^{n_x} (b_y^+)^{n_y}$ to the ground state $|00\rangle$. For example applying simply $b_x^+$ and $b_y^+$ one creates $E$ basis $|10\rangle \approx x$ and $|01\rangle \approx y$. Symmetry adapted operator polynomials (let say, of $\Gamma\gamma$ type, symbol $\gamma$ enumerates the basis functions) are applied to the ground state $|00\rangle$ thus creating the vibrational functions $|n, \Gamma\gamma\rangle$ with a given $n$ which belong to the irrep $\Gamma\gamma$.

Let us introduce the symmetry adapted polynomial operators $T_{\Gamma\gamma}^{(n)}$ constructed from operators $(b_x^+)^{n_x} (b_y^+)^{n_y}$ where the following notations are used: $n = n_x + n_y$, $\Gamma$ is one of the irreps $(A_1, A_2, E)$ corresponding to the set of the vibrational functions belonging the energy level $n$, symbol $\gamma$ enumerates the basis functions of the irrep $\Gamma$. The operators $T_{\Gamma\gamma}^{(n)}$ realize the transformation from the basis $|n_x n_y\rangle$ with a certain $n = n_x + n_y$ (a definite vibrational level) to the symmetry adapted basis $|n, \Gamma\gamma\rangle$:

$$|n, \Gamma\gamma\rangle = \sum_{\substack{n_x, n_y \\ (n_x + n_y = n)}} c(n_x n_y, \nu\, \Gamma\gamma) |n_x, n_y\rangle \tag{3.9}$$

where $\nu$ is an additional quantum number that is introduced in order to distinguish the states of the system in the case when the irrep $\Gamma$ occurs several times in the set $n$. By definition the operator $T_{\Gamma\gamma}^{(n)}$ acts on the vacuum state $|00\rangle \equiv |n_x = 0, n_y = 0\rangle$ and creates a basis function of the type $\Gamma\gamma$ belonging to a certain excited vibrational state $n(n = n_x + n_y)$:

$$\mathbf{T}_{\Gamma\gamma}^{(n)} |00\rangle = |n, \Gamma\gamma\rangle \tag{3.10}$$

For this reason we will refer $T_{\Gamma\gamma}^{(n)}$ to as multivibronic symmetry adapted creation operators which can be considered as the irreducible tensor operators of the type of $\Gamma\gamma$ in the actual point group ($\mathbf{C}_{3v}$ in the case under consideration). This allows one to build the multivibronic operators with the use of the well known technique for manipulation with the irreducible tensor operators in the point groups (see for example Ref. [19]).

Let us illustrate the approach by the evaluation of the multivibronic operators for several low lying vibrational levels and corresponding symmetry adapted basis functions. We will use a well developed technique [19] that allows to construct the symmetry adapted basis $\psi_{\Gamma\gamma} \equiv |\Gamma\gamma\rangle$ belonging to the irreps $\Gamma$ of a point group from the direct product $u_{\Gamma_1\gamma_1} v_{\Gamma_2\gamma_2} \equiv \langle \Gamma_1\gamma_1\Gamma_2\gamma_2\rangle$ of the basis functions $u_{\Gamma_1\gamma_1}$ and $v_{\Gamma_2\gamma_2}$ (bases of the irreps $\Gamma_1$ and $\Gamma_2$):

$$\psi_{\Gamma\gamma} = \sum_{\gamma_1\gamma_2} u_{\Gamma_1\gamma_1} v_{\Gamma_2\gamma_2} \langle \Gamma_1\gamma_1\Gamma_2\gamma_2 | \Gamma\gamma\rangle, \quad \Gamma \in \Gamma_1 \times \Gamma_2 \tag{3.11}$$

The coupling coefficients (Clebsch-Gordan coefficients) for all point groups are given by Koster et al. [20] where the circular (complex) basis $\psi^3_{-1} = (x - iy)$, $\psi^3_{+1} = -(x - iy)$ for the double degenerate irrep $E$ *is* used. The Clebsch-Gordan coefficients adapted to the real basis (as indicated) that are more convenient for our aids are the following:

| $A_2 \times E = E$ | | |
|---|---|---|
| $A_2 \times E$ | $u_2 v_x$ | $u_2 v_y$ |
| $\psi^3_x$ | $0$ | $-1$ |
| $\psi^3_y$ | $1$ | $0$ |

| $E \times E = A_1 + A_2 + E$ | | | | |
|---|---|---|---|---|
| $E \times E$ | $u_x v_x$ | $u_x v_y$ | $u_y v_x$ | $u_y v_y$ |
| $\psi_1$ | $1/\sqrt{2}$ | $0$ | $0$ | $1/\sqrt{2}$ |
| $\psi_2$ | $0$ | $1/\sqrt{2}$ | $-1/\sqrt{2}$ | $0$ |
| $\psi^3_x$ | $0$ | $-1/\sqrt{2}$ | $-1/\sqrt{2}$ | $0$ |
| $\psi^3_y$ | $-1/\sqrt{2}$ | $0$ | $0$ | $1/\sqrt{2}$ |

The notations of Koster et al. [20] are used with the obvious changes in order to adapt them to the real basis sets, we use also the Malliken notations for the irreps that are related to the Bethe notations as follows: $\Gamma_1 \leftrightarrow A_1, \Gamma_2 \leftrightarrow A_2, \Gamma_3 \leftrightarrow E$.

Creation operators $b_x^+$ and $b_y^+$ belong to the irrep $E$, the ground state $|00\rangle$ (vacuum) is full symmetric $(A_1)$ that means that in the case under consideration $T^{(1)}_{\Gamma\gamma} = b^+_{\Gamma\gamma}$.

Then for $n = 1$ one obtains:

$$T^{(1)}_{\Gamma\gamma} |00\rangle = |1, \Gamma\gamma\rangle \tag{3.12}$$

# 3 A Symmetry Adapted Approach to the Dynamic Jahn-Teller Problem 47

This leads to the following obvious result:

$$b_x^+ \left|n_x = 0, n_y = 0\right\rangle = \left|n_x = 1, n_y = 0\right\rangle \equiv |1, Ex\rangle$$
$$b_y^+ \left|n_x = 0, n_y = 0\right\rangle = \left|n_x = 0, n_y = 1\right\rangle \equiv |1, Ey\rangle$$

In general, according to the coupling scheme the operator $\mathbf{T}_{\Gamma\gamma}^{(2)}$ can be represented as a tensor product that is an alternative form of eq. (3.11):

$$\mathbf{T}_{\Gamma\gamma}^{(2)} = \left\{ b_{\Gamma_1}^+ \otimes b_{\Gamma_2}^+ \right\} \Gamma\gamma, \Gamma \in \Gamma_1 \times \Gamma_2 \tag{3.13}$$

The direct product $b_{E\gamma_1}^+ \otimes b_{E\gamma_2}^+$ is represented by the four terms that form the bases of the four-dimensional reducible representation of $C_{3v}$:

$$b_x^+ b_x^+, \; b_x^+ b_y^+, \; b_y^+ b_x^+, \; b_y^+ b_y^+ \tag{3.14}$$

This reducible representation can be decomposed into the irreducible ones accordingly to the direct product $E \times E = A_1 + A_2 + E$ by the use of the coupling scheme, Eq. 3.11 that gives the following result:

$$\mathbf{T}_{A_1}^{(2)} = \frac{1}{\sqrt{2}} \left( b_x^{+2} + b_y^{+2} \right)$$

$$\mathbf{T}_{Ex}^{(2)} = -b_x^+ b_y^+$$

$$\mathbf{T}_{Ey}^{(2)} = -\frac{1}{\sqrt{2}} \left( b_x^{+2} - b_y^{+2} \right)$$

$$\mathbf{T}_{A_2}^{(2)} = 0 \tag{3.15}$$

One can see that the operator $\mathbf{T}_{A_2}^{(2)}$ corresponding to the antisymmetric part $\{E \times E\}$ of the direct product $E \times E$ vanishes in the case of $n = 2$ so that the full dimension of the obtained basis is $g(n = 2) = 3$. Applying operators $\mathbf{T}_{\Gamma\gamma}^{(2)}$ to the vacuum state $\mathbf{T}_{\Gamma\gamma}^{(2)}|00\rangle = |2, \Gamma\gamma\rangle$ one obtains the three-dimensional vibrational basis for $n = 2$ with $\Gamma = A_1$, $E$ that corresponds to the result of the group-theoretical assignation (Sect. 3.3). The final results for the normalized states (that are also orthogonal) are the following:

$$|2, A_1\rangle = \frac{1}{\sqrt{2}} (|20\rangle + |02\rangle)$$

$$\left. \begin{array}{l} |2, Ex\rangle = |11\rangle \\[2mm] |2, Ey\rangle = -\frac{1}{\sqrt{2}} (|20\rangle - |02\rangle) \end{array} \right\} E \tag{3.16}$$

For the next excited level $n = 3$ one has to build the operators $T^{(3)}_{\Gamma\gamma}$ applying the subsequent step of the coupling procedure:

$$T^{(3)}_{\Gamma\gamma} = \left\{ \left\{ b^+_{\Gamma_1} \otimes b^+_{\Gamma_2} \right\}_{\Gamma_{12}} \otimes b^+_{\Gamma_3} \right\}_{\Gamma\gamma} \equiv \sum_{\gamma_{12}\gamma_3} \left\{ b^+_{\Gamma_1} \otimes b^+_{\Gamma_2} \right\}_{\Gamma_{12}\gamma_{12}} b^+_{\Gamma_3\gamma_3} \langle \Gamma_{12}\gamma_{12}\Gamma_3\gamma_3 | \Gamma\gamma \rangle$$

$$\Gamma_{12} \in \Gamma_1 \times \Gamma_2, \; \Gamma \in \Gamma_{12} \times \Gamma_3 \qquad\qquad (3.17)$$

The four-dimensional space for $n = 3$ can be split according to the irreps $A_1$ $A_2$ and $E$. Evaluation of the operators $T^{(3)}_{\Gamma\gamma}$ gives the following results:

$$T^{(3)}_{A_1} = \frac{1}{2} \left( \sqrt{3} b^{+2}_x b^+_y - b^{+3}_y \right)$$

$$T^{(3)}_{A_2} = \frac{1}{2} \left( b^{+3}_x - \sqrt{3} b^+_x b^{+2}_y \right)$$

$$T^{(3)}_{Ex} = \frac{1}{2} \left( \sqrt{3} b^{+3}_x - b^+_x b^{+2}_y \right)$$

$$T^{(3)}_{Ey} = \frac{1}{2} \left( \sqrt{3} b^{+2}_x b^+_y + b^{+3}_y \right) \qquad\qquad (3.18)$$

It should be noted that the application of the coupling scheme, Eq. 3.15, leads to the two identical operators of the $E$ type one of which is eliminated. Action of the operators $T^{(3)}_{\Gamma\gamma}$ to the vacuum state leads to the following symmetry adapted functions belonging to the $n = 3$ manifold:

$$|3,A_1\rangle = \frac{1}{2} \left( |03\rangle - \sqrt{3}|21\rangle \right),$$

$$|3,A_2\rangle = \frac{1}{2} \left( \sqrt{3}|12\rangle - |30\rangle \right),$$

$$\left.\begin{array}{l} |3,Ex\rangle = \frac{1}{2} \left( \sqrt{3}|30\rangle + |12\rangle \right) \\[2mm] |3,Ey\rangle = \frac{1}{2} \left( |21\rangle + \sqrt{3}|03\rangle \right) \end{array}\right\} E \qquad (3.19)$$

The basis set for $n = 4$ can be decomposed into three irreps of $C_{3v}$ : $A_1 + E + E$. Evaluation of the symmetry adapted basis functions by means of the procedure so far described gives the following expressions:

$$|4,A_1\rangle = \frac{1}{2\sqrt{2}} \left( \sqrt{3}|40\rangle + \sqrt{2}|22\rangle + \sqrt{3}|04\rangle \right)$$

$$\left.\begin{array}{l} |4,Ex\rangle = \frac{1}{\sqrt{10}} \left( |13\rangle - 3|31\rangle \right) \\[2mm] |4,Ey\rangle = \frac{1}{\sqrt{10}} \left( 2|04\rangle - \sqrt{6}|22\rangle \right) \end{array}\right\} E$$

$$|4,Ex\rangle = \frac{1}{\sqrt{10}}\left(|31\rangle - 3|13\rangle\right)$$
$$|4,Ey\rangle = \frac{1}{\sqrt{10}}\left(-2|40\rangle + \sqrt{6}|22\rangle\right)$$
$$\left.\right\}E$$

$$|4,Ex\rangle = -\frac{1}{\sqrt{2}}\left(|31\rangle + |13\rangle\right)$$
$$|4,Ey\rangle = \frac{1}{\sqrt{2}}\left(|04\rangle - |40\rangle\right)$$
$$\left.\right\}E \qquad (3.20)$$

One can see that the dimension $g = 7$ of the full basis in Eq. 3.20 exceeds the dimension $g(n=4) = 5$ and thus this set contains linearly dependent non-orthogonal functions. By comparing the basis in Eq. 3.20 with the result of the group-theoretical classification one can see that an excessive pair $E$ type functions is present in the basis given by Eq. 3.20.

The case of $n = 4$ illustrates what one can expect at the subsequent steps of the so far described approach, namely, the presence of several identical irreps resulting in an excessive dimension of the space, so that each step should be supplemented by the Gram-Schmidt procedure of orthogonalization within the repeated irreps. The Gram-Schmidt procedure takes an arbitrary basis and generates a new orthonormal one. It does this by sequentially processing the list of vectors and generating a vector perpendicular to the previous vectors in the list. As a result the obtained vectors are linearly independent. It should be noted that x-components of all $E$ bases are orthogonal to the y-components, and, of course, the orthogonality remains between different irreps and between the functions from the sets arising from different $n$. Due to this fact the Gram-Schmidt procedure of orthogonalization is to be applied only to the corresponding basis functions of the same irreps (let's say, to the $x$ type functions) with the same $n$.

In the case of $n = 4$ application of the Gram-Schmidt procedure gives the following orthonormal set of the symmetry adapted functions:

$$|4,A_1\rangle = \frac{1}{2\sqrt{2}}\left(\sqrt{3}|40\rangle + \sqrt{2}|22\rangle + \sqrt{3}|04\rangle\right)$$

$$|4,Ex\rangle = -\frac{1}{\sqrt{2}}\left(|13\rangle + |31\rangle\right)$$
$$|4,Ey\rangle = \frac{1}{\sqrt{2}}\left(|04\rangle - |40\rangle\right)$$
$$\left.\right\}E$$

$$|4,Ex\rangle = \frac{1}{\sqrt{2}}\left(-|13\rangle + |31\rangle\right)$$
$$|4,Ey\rangle = -\frac{1}{2\sqrt{2}}\left(|04\rangle - \sqrt{6}|22\rangle + |40\rangle\right)$$
$$\left.\right\}E \qquad (3.21)$$

The procedure can be extended to an arbitrary value of $n$ by the application of the multivibronic operator of the order $n$ to the vacuum state with the subsequent Gram-Schmidt procedure of orthogonalization at each step $n$. The multivibronic operator $\mathbf{T}_{\Gamma\gamma}^{(n)}$ can be built by extending the consequent coupling procedure to the order $n$:

$$\mathbf{T}_{\Gamma\gamma}^{(n)} = \left\{ \left\{ \left\{ b_{\Gamma_1}^+ \otimes b_{\Gamma_2}^+ \right\}_{\Gamma_{12}} \otimes b_{\Gamma_3}^+ \right\}_{\Gamma_{123}} \cdots \otimes b_{\Gamma_n}^+ \right\}_{\Gamma\gamma} \qquad (3.22)$$

$$\Gamma_{12} \in \Gamma_1 \times \Gamma_2, \ \Gamma_{123} \in \Gamma_1 \times \Gamma_2 \times \Gamma_3 \ldots, \Gamma \in \Gamma_{n-1} \times \Gamma_n$$

Finally, the symmetry adapted vibrational functions $|n, v\Gamma_v\gamma_v\rangle$ are to be coupled to the electronic ones $|\alpha\Gamma_e\gamma_e, SM\rangle$ that are eigen-functions of the Hamiltonian $H_e$ included in the JT/pseudo JT problem ($SM$ are the quantum numbers of the full spin and its projection, $\alpha$ is the additional quantum number that enumerate the repeated irreps $\Gamma_e$). To pass from the direct products of these two sets to the symmetry adapted electron-vibrational functions $|\alpha v, n, \Gamma\gamma\rangle$

$$|\alpha\Gamma_e\gamma_e, SM\rangle \otimes |n, v\Gamma_v\gamma_v\rangle \Rightarrow |\alpha v, n, \Gamma\gamma\rangle \qquad (3.23)$$

one should apply the standard coupling scheme:

$$|\alpha v, n, \Gamma\gamma\rangle = \sum_{\gamma_1\gamma_2} |\alpha\Gamma_e\gamma_e, SM\rangle |n, v\Gamma_v\gamma_v\rangle \langle\Gamma_e\gamma_e\Gamma_v\gamma_v| \Gamma\gamma\rangle, \quad \Gamma \in \Gamma_e \times \Gamma_v \quad (3.24)$$

This allows to achieve the final goal of the approach, namely to evaluate the basis in which the full matrix of the JT Hamiltonian is blocked according to the irreps of the point group. Finally, the approach is realized as an efficient computer program [21] that generates the blocks and evaluate required characteristics of the JT systems, like optical lines, thermodynamic characteristics, etc. The procedure generates basis assuming only point symmetry of the system without indication of the explicit form of the vibronic coupling, so it is not restricted by the linear terms of vibronic interactions and applicable also when the quadratic and high order terms are taken into account. Under this condition the approaches based on high symmetries lose their advantages because the actual symmetry is reduced to the point one.

## 3.5 Intervalence Optical Absorption in Mixed-Valence Trimeric Systems

In this section we will illustrate the approach by a relatively simple evaluation of the so-called intervalence optical absorption bands in the three-center triangular mixed-valence systems. These bands are associated with the photon induced electronic

**Fig. 3.2** Schematic representation of the structure of the trimeric complexes M$_3$O(RCO$_2$)$_6$L$_3$ (M – *filled circles*)

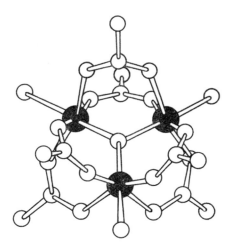

jumps between different sites (charge transfer bands) and their shapes and positions are closely related to the strength of the JT and pseudo JT interactions. In such systems one electron (or hole) is delocalized over three ions having in general non-compensated localized spins (spin cores). This leads to the spin-dependent delocalization of the extra electron that is called double exchange (see review article [22]). Many different types of the trimeric mixed-valence metal clusters are well known and studied. The most representative examples of mixed-valence trimers concern the tri-μ-oxo metal acetate complexes M$_3$O(RCO$_2$)$_6$L$_3$ where M$_3$ is either Fe$^{III}_2$Fe$^{II}$ or Mn$^{III}_2$Mn$^{II}$ and L = pyridine, H$_2$O (Fig. 3.2).

In this kind of mixed-valence systems the metal ions are connected by a central oxo ligand and by six peripheral carboxylate ligands. There are considerable experimental data for these complexes that indicate a thermally activated intramolecular electron hopping promoting a rapid valence delocalization at high temperatures but a valence trapping at low temperatures due to vibronic coupling.

Many organic compounds represent an alternative class of mixed valence systems that can be exemplified by an interesting compound C$_6$C$_3$(C$_6$Cl$_5$)$_3$. Depending on the oxidation state, this compound ranges from triradical to trianion forms for which one or two electrons are shared among three sites (Fig. 3.3).

Due to complexity of the vibronic problem in its general form the two main approximate vibronic models for mixed-valence systems have been proposed. The model formulated by Piepho, Krausz, and Schatz [23] and referred to as PKS model deals with the independent "breathing" displacements (that are assumed to be the normal coordinates) of the ions around the sites of the electron localization. Later on Piepho suggested a vibronic model [24] that takes into account also the multicenter vibrations that change the distances between the sites of localization and therefore lead to the electron-vibrational coupling due to modulation of the transfer integrals. Being very efficient and at the same time relatively simple, these models are able to describe the most important features of the phenomena related to mixed valency.

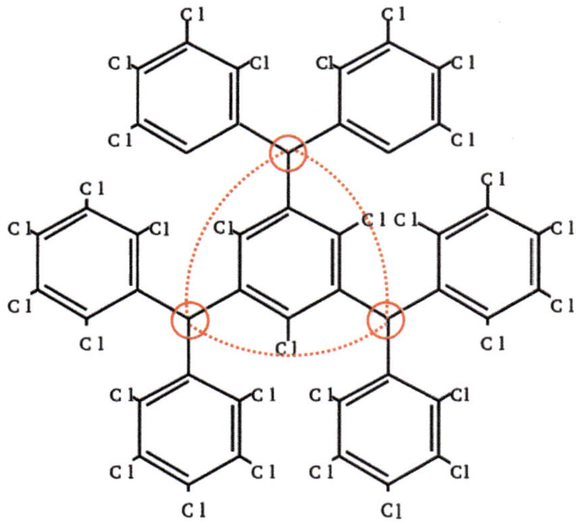

**Fig. 3.3** Organic compound $C_6C_3(C_6Cl_5)_3$ and illustration for the different oxidation states that range from triradical to trianion. The *circles* conventionally indicate three possible areas of the electron or hole localization

The energy pattern of a triangular unit containing one or two electrons delocalized over spinless ions consists of a singlet $A_1$ (or $A_2$) and a doublet $E$ separated by the gap $3|t|$ where $t$ is the hopping (transfer) integral (providing positive $t$ the low lying level is a singlet). One obtains $(A_{1(2)} + E) \otimes e$ PKS problem while intercenter vibrations give rise to the $(A_{1(2)} + E) \otimes (a_1 + e)$ problem (see review article [22] and Ref. [25]). Taking into account both types of the active vibrations one arrives at a relatively complicated three level-five mode combined JT/pseudo JT problem.

For illustration we consider a simple situation and assume that the PKS coupling dominates that is expected to be a reasonable approximation for the metal clusters while in organic compounds the intercenter vibrations are more important. Assuming that each center contains a non-degenerate orbital $\varphi_i(i = 1, 2, 3)$, Fig. 3.1, the symmetry adapted electronic basis can be represented as:

$$\psi_{A_1} = \frac{1}{\sqrt{3}}\left(\varphi_1 + \varphi_2 + \varphi_3\right),$$

$$\psi_{Ex} = \frac{1}{\sqrt{2}}\left(\varphi_1 - \varphi_2\right),$$

$$\psi_{Ey} = \frac{1}{\sqrt{6}}\left(\varphi_1 + \varphi_2 - 2\varphi_3\right) \tag{3.25}$$

The collective PKS coordinates can be expressed in terms of the "breathing" coordinates located at the sites as follows:

3 A Symmetry Adapted Approach to the Dynamic Jahn-Teller Problem

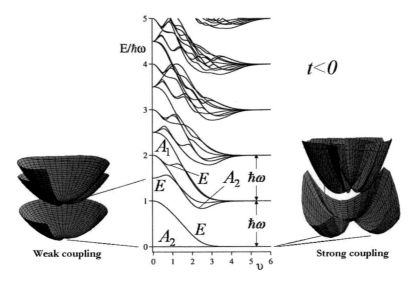

**Fig. 3.4** Example of the vibronic levels of the $(A_1 + E) \otimes e$ system vs PKS coupling parameter

$$\begin{pmatrix} q_{A_1} \\ q_X \\ q_Y \end{pmatrix} = \begin{pmatrix} 1/\sqrt{3} & 1/\sqrt{3} & 1/\sqrt{3} \\ 1/\sqrt{2} & -1/\sqrt{2} & 0 \\ 1/\sqrt{6} & 1/\sqrt{6} & -2/\sqrt{6} \end{pmatrix} \begin{pmatrix} q_1 \\ q_2 \\ q_3 \end{pmatrix} \quad (3.26)$$

Finally, the matrix of the adiabatic potential for the $(A_1 + E) \otimes e$ JT/pseudo JT problem in the basis of localized orbitals $\varphi_1, \varphi_2, \varphi_3$ can be written as:

$$U(q_x, q_y) = \frac{1}{2}(q_x^2 + q_y^2) + \begin{pmatrix} \upsilon\left(\frac{1}{\sqrt{2}}q_x + \frac{1}{\sqrt{6}}q_y\right) & t & t \\ t & \upsilon\left(-\frac{1}{\sqrt{2}}q_x + \frac{1}{\sqrt{6}}q_y\right) & t \\ t & t & -\frac{2}{\sqrt{6}}\upsilon q_y \end{pmatrix}$$
(3.27)

A representative vibronic spectrum containing symmetry labels is given in Fig. 3.4 for the case of $t < 0$. The total basis is restricted to $n = 48$ that gives $N = 3675$. The full basis is blocked into matrices with dimensions 625 ($A_1$), 600 ($A_2$) and 1225 (two identical matrices for $E$). At $\upsilon = 0$ one can find two interpenetrating system of the unperturbed harmonic oscillator levels shifted by the value $3|t|$ (see the adiabatic surfaces at the left side of Fig. 3.4). Providing strong vibronic coupling the levels are grouped again into an equidistant spectrum corresponding to the energy pattern in the deep parabolic minima of the lower sheet of the adiabatic potential (Fig. 3.4, right side).

Finally, the electric-dipole intervalence transitions are allowed and the non-vanishing matrices $d_x, d_y (\hat{d}_z = 0)$ of the dipole moment can be found as:

$$
\hat{d}_{Ex} = 
\begin{array}{ccc}
\psi_{A_1} & \psi_{Ex} & \psi_{Ey}
\end{array}
\begin{pmatrix}
0 & -eR/\sqrt{2} & 0 \\
-eR/\sqrt{2} & 0 & -eR/2 \\
0 & -eR/2 & 0
\end{pmatrix},
$$

$$
\hat{d}_{Ey} = 
\begin{array}{ccc}
\psi_{A_1} & \psi_{Ex} & \psi_{Ey}
\end{array}
\begin{pmatrix}
0 & 0 & -eR/\sqrt{2} \\
0 & -eR/2 & 0 \\
-eR/\sqrt{2} & 0 & eR/2
\end{pmatrix}, \tag{3.28}
$$

where $R$ is the distance between the center of the triangle and the apexes. The intensities (including population factors) of the individual vibronic lines for the intervalence light absorption are calculated as:

$$
\begin{aligned}
& D\left(\alpha v, n, \Gamma \to \alpha' v', n', \Gamma'\right) \\
& = \sum_{\gamma = x, y} \frac{\left(N_{\alpha v, n, \Gamma} - N_{\alpha' v', n', \Gamma'}\right)}{N} \left|\left\langle \alpha' v', n', \Gamma' \gamma \left| \hat{d}_{E'' \gamma''} \right| \alpha v, n, \Gamma \gamma \right\rangle\right|^2
\end{aligned} \tag{3.29}
$$

where $N_i = \exp(-E_i / k_B T)$ and $N = \sum_i N_i$. Some examples of the intervalence absorption bands are given in Fig. 3.5 for the cases of negative (Fig. 3.5a) and positive (Fig. 3.5b) transfer parameters. Occurrence of the structureless intervalence bands in mixed valence compounds reflects the effects of the dispersion of the active modes in crystals or/and the broadening of the discrete lines due to relaxation processes. For this reason the palisade of the discrete lines corresponding to the individual transitions as well as the enveloping curves are given in Fig. 3.5. The general character of the spectral distributions is closely related to the sign of the transfer integral that determines the interplay between the JT and pseudo JT effects. In fact, at $t > 0$ the ground electronic state $E$ exhibits JT effect while providing $t < 0$ the ground state is an orbital singlet and therefore the instability of the system occurs due to pseudo JT coupling if the last is strong enough (see right side of the Fig. 3.4). This affects the shapes of the bands that are always two-humped (or even more complicated) in the case of positive transfer and are bell-shaped providing weak or moderate negative transfer. More detailed description of the intervalence bands requires also the inclusion of intersite vibrations in the vibronic model. This will be done elsewhere along with the discussion of the experimental data.

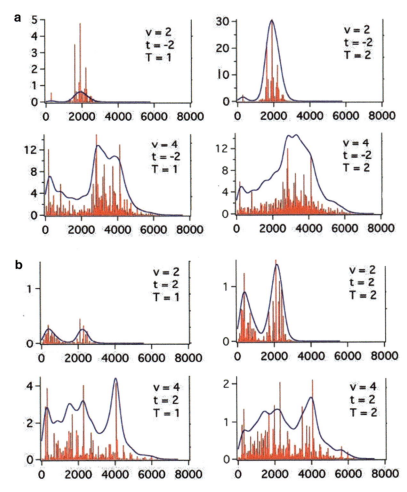

**Fig. 3.5** Shapes of the intervalence optical bands (arbitrary units for the intensities) for an one-electron triangular mixed-valence system in the PKS model providing negative (**a**) and positive (**b**) transfer integrals. All dimensionless parameters are in $\hbar\omega$ units ($k_B = 1$)

## 3.6 Concluding Remarks

In this article we have presented a powerful computational approach to the dynamic JT/pseudo JT systems. The approach is based on the receipt for the evaluation of the symmetry adapted vibronic basis. The algorithm for the solution of the eigenproblem takes full advantage of the point symmetry arguments and thus allows to reduce the vibronic matrices to full extent. The group-theoretical assignment of the vibronic states allows to predict the symmetry labels for the vibronic levels before evaluation. The approach is applicable to the arbitrary form of the vibronic coupling involving linear and high order terms. As a relatively simple illustrative example the intervalence optical absorption bands in mixed-valence triangular mixed valence systems within the PKS model are considered and briefly analyzed. The theoretical background is employed to elaborate an efficient program [21] that will be published elsewhere along with the detailed discussion of the trimeric and high nuclearity mixed valence clusters and complex polyoxometalates containing several electrons delocalized over a metal network.

**Acknowledgments** B.T. acknowledges financial support of the Israel Science Foundation (ISF, grant no. 168/09). A.P. thanks the Paul Scherrer Institute for financial support that made possible his participation in the Jahn-Teller Symposium. The financial support from STCU (project N 5062) and the Supreme Council on Science and Technological Development of Moldova is gratefully acknowledged. J.M.C.J. and E.C. thank Spanish MICINN (CSD2007-00010 CONSOLIDER-INGENIO in Molecular Nanoscience, MAT2007-61584, CTQ-2008-06720 and CTQ-2005-09385), Generalitat Valenciana (PROMETEO program), and the EU (MolSpinQIP project and ERC Advanced Grant SPINMOL) for the financial support. We thank Prof. V. Polinger for the discussion and Dr. O. Reu for his help in the artwork.

# References

1. Englman R (1972) The Jahn–Teller effect in molecules and crystals. Wiley, London
2. Bersuker IB, Polinger VZ (1989) Vibronic interactions in molecules and crystals. Springer, Berlin
3. Bersuker IB (2006) The Jahn-Teller effect. Cambridge University Press, Cambridge
4. Ding X-Q, Bominaar EL, Bill E, Winkler H, Trautwein AX, Drüeke S, Chaudhuri P, Weighardt K (1990) J Chem Phys 92:178; Gamelin DR, Bominaar EL, Kirk ML, Wieghardt K, Solomon EI (1996) J Am Chem Soc 118:8085
5. Marks AJ, Prassides K (1993) New J Chem 17:59; Marks AJ, Prassides K (1993) J Chem Phys 98:4805
6. (a) Borshch SA, Bominaar EL, Blondin G, Girerd G (1993) J Am Chem Soc 115:5155; (b) Bominaar EL, Borshch SA, Girerd JJ (1994) J Am Chem Soc 116:5362
7. Clemente-Juan JM, Coronado E (1999) Coord Chem Rev 361:193
8. (a) Borras-Almenar JJ, Clemente-Juan JM, Coronado E, Tsukerblat BS (1995) Chem Phys 195:1; (b) Borras-Almenar JJ, Clemente-Juan JM, Coronado E, Tsukerblat BS (1995) Chem Phys 195:17; (c) Borras-Almenar JJ, Clemente-Juan JM, Coronado E, Tsukerblat BS (1995) Chem Phys 195:29
9. (a) Polinger VZ, Boldirev SI (1986) Phys Stat Sol (b) 137:241; (b) Boldyrev SI, Polinger VZ, Bersuker IB (1981) Fiz Tverdogo Tela (Russ) 23:746
10. Sakamoto N, Muramatsu S (1978) Phys Rev B 17:868
11. Perlin YuE, Tsukerblat BS (1984) In: Perlin YuE, Wagner M (eds) The dynamical Jahn-Teller effect in localized systems, vol 7. Elsevier, Amsterdam, pp 251–346
12. Gütlich P, Hauser A, Spiering H (1994) Angew Chem Int Ed Engl 33:2024
13. Kaplan MD, Vekhter BG (1995) Cooperative phenomena in Jahn-Teller crystals Plenum, New York
14. Köppel H, Domcke W, Cederbaum LS (1984) Adv Chem Phys 57:59
15. Faraji S, Gindensperger E, Köppel H (2009) In: Köppel H, Yarkony DR, Barentzen (eds) The Jahn-Teller effect. Fundamentals and implications for physics and chemistry, Series of Chemical Physics, vol 97. Springer, Heidelberg, pp 239–276
16. Grosso G, Martinelli L, Parravicini GP (1995) Phys Rev B 51:13033

17. Pooler DR (1978) J Phys A 11:1045; Pooler DR (1980) J Phys A 13:1029
18. O'Brien MCM (1969) Phys Rev 187:329; O'Brien MCM (1971) J Phys C 4:2524
19. Tsukerblat B (2006) Group theory in chemistry and spectroscopy. Dover, Mineola/New York
20. Koster GF, Dimmok JO, Wheeler RG, Statz H (1963) Properties of the thirty-two point groups. MIT Press, Cambridge
21. Clemente-Juan JM, Palii A, Coronado E, Tsukerblat B (2011) J Comp Chem, submitted
22. Borras-Almenar JJ, Clemente-Juan JM, Coronado E, Palii AV, Tsukerblat BS (2001) In: Miller J, Drillon M (eds) Magnetoscience-from molecules to materials. Willey-VCH, New York, pp 155–210
23. (a) Wong KY, Schatz PN (1981) Prog Inorg Chem 28:369; (b) Piepho SB, Krausz ER, Shatz PN (1978) J Am Chem Soc 100:2996
24. (a) Piepho SB (1988) J Am Chem Soc 110:6319; (b) Piepho SB (1990) J Am Chem Soc 112:4197
25. Borrás-Almenar JJ, Coronado E, Ovstrosvsky SM, Palii AV, Tsukerblat BS (1999) Chem Phys 240:149

# Chapter 4
# Group-Theoretical Treatment of Pseudo-Jahn-Teller Systems

**Martin Breza**

**Abstract** According to Jahn-Teller (JT) theorem any nonlinear arrangement of atomic nuclei in electron degenerate state (except an accidental and Kramers degeneracy) is unstable. Pseudo-JT systems are treated as an analogy of JT theorem for pseudo-degenerate electron states. Stable nuclear arrangements of lower energy of such systems correspond to the minima of their potential energy surfaces (PES). In large systems the analytical description of their PES is too complicated and a group-theoretical treatment must be used to describe their stable structures. This may be based either on JT active coordinates – as in the epikernel principle – or on the electron states – as in the method of step-by-step descent in symmetry. This review explains the basic terms of group theory (especially of point groups of symmetry) and potential energy surfaces (extremal points and their characteristics) as well as the principles of group-theoretical methods predicting the extremal points of JT systems – the method of epikernel principle (based on JT active coordinates) and the method of step-by-step descent in symmetry (based on a consecutive split of the degenerate electron states). Despite these methods have been elaborated for the case of electron degeneracy, they are applicable to pseudo-degenerate electron states as well. The applications of both methods to pseudo-JT systems are presented on several examples and compared with the published results.

## 4.1 Introduction

The original complete formulation of Jahn-Teller (JT) theorem [1, 2] is too lengthy to be well known. Its generally used short formulation states that "any nonlinear

---

M. Breza (✉)
Department of Physical Chemistry, Slovak Technical University, SK-81237 Bratislava, Slovakia
e-mail: martin.breza@stuba.sk

M. Atanasov et al. (eds.), *Vibronic Interactions and the Jahn-Teller Effect:*
*Theory and Applications*, Progress in Theoretical Chemistry and Physics 23,
DOI 10.1007/978-94-007-2384-9_4, © Springer Science+Business Media B.V. 2012

arrangement of atomic nuclei in degenerate electron state is unstable (except Kramers degeneracy [3])." The authors supposed that the electron degeneracy is conditioned by the nuclear configuration symmetry. Nevertheless, the recent extended understanding of Jahn-Teller effect (JTE) includes both linear (Renner-Teller effect) and non-linear atomic configurations, both degenerate and pseudo-degenerate (pseudo-JTE) electron states as well as conical intersections, i.e. the crossings between potential energy surfaces corresponding to various electron states of the same multiplicity. The common aspect of these phenomena is the symmetry descent occurring through the coupling between different potential energy surfaces. The practical applications of this extended JTE involve stereochemistry, chemical activation, mechanisms of chemical reactions, spectroscopy, electron-conformational changes in biology, impurity physics, lattice formations, phase transitions, etc. [4].

The group-theoretical treatment of JTE is based on the original formulation of JT theorem and deals with the symmetry conditioned degenerate electron states. As a consequence, a stable nuclear configuration of lower symmetry in a nondegenerate electron state is formed. This nondegenerate state is obtained by splitting the parent degenerate electron state in a higher symmetry (parent) structure. Thus the problem of the symmetries of stable JT structures may be solved using the symmetries of distortion coordinates (most completely treated in the epikernel principle [5–8]) or of electron states (the method of step-by-step symmetry descent [8–12]).

The symmetry unconditioned conical intersection is understood as an accidental degeneracy and cannot be solved within the above mentioned group-theoretical treatment. Kramers twofold spin degeneracy demands the treatment based on double groups and only the solution based on the symmetries of electron states has been published yet [13]. The group-theoretical treatment of Renner-Teller effect leads to trivial results only (stable bent structures of $C_{2v}$ or $C_s$ symmetry groups). The aim of this study is to extend this treatment to the systems in pseudodegenerate electron states.

## 4.2 Point Groups of Symmetry

Symmetry properties of any object (such as a molecule) are described by its symmetry point group [14]. This group is a collection of its symmetry operations R (group elements) which must satisfy four group postulates:

(i) If any two operations, P and Q, are members of the group, then their combination P.Q = R must also be a member of the group

(ii) Under the rule of combination the associative law must hold, i.e.

$$(P \cdot Q) \cdot R = P \cdot (Q \cdot R) \tag{4.1}$$

# 4 Group-Theoretical Treatment of Pseudo-Jahn-Teller Systems

(iii) The group must contain the identity operation E which commutes with all other operations and leaves them unchanged. Thus

$$E \cdot R = R \cdot E = R \qquad (4.2)$$

(iv) Each member $R$ of the group must possess an inverse $R^{-1}$ which is also a member of the group. Thus

$$R^{-1} \cdot R = R \cdot R^{-1} = E \qquad (4.3)$$

There are several types of the symmetry operations (n and k are integral, for k = 1 the superscript is usually omitted) [14]:

  (i) Identity $E$
 (ii) Rotation $C_n^k$ through an angle $2\pi k/n$ about the n-fold rotation axis $C_n$ ($C_n$ with maximal n denotes a principal axis).
(iii) Inversion $i$ of all points through the unique centre of symmetry.
(iv) Reflection $\sigma$ of all points in a mirror plane $\sigma$.
 (v) Improper rotation-reflection operation $S_n^k$ is the rotation about the improper axis $S_n$ through an angle of $2\pi k/n$, combined with reflection k times in a plane normal to this axis.

The number of elements (operations) in a group is the order of the group (symbol h). A subgroup consists of a set of the elements within a group which, on their own, constitute a group. The order of the parent group (supergroup) is an integer multiple of the orders of each of its subgroups [14].

Each symmetry point group is represented by a Schönfliess symbol consisting of a capital letter and up to two suffixes [14]. Its properties are described by a character table (see Table 4.1 for $C_{4v}$, $C_{2v}$ and $C_s$ point groups). The table headline contains group operations $R$ (more exactly – the corresponding symmetry elements R). The first column contains the symbols of irreducible representations (IRs) describing the symmetry of molecular orbitals, electron states, symmetric coordinates, vibrations and so on (usually the electron states are denoted by capital letters and the remaining quantities by small letters). The effect of symmetry operations $R$ on IRs is described by their characters $\chi(R)$. One-dimensional (non-degenerate) IRs are denoted by symbols A or B. Degenerate (multidimensional) IRs are denoted by the symbols E (two-dimensional), T or F (three-dimensional), G or U (four-dimensional), H or V (five-dimensional) and so on (see their characters for the identity operation $E$). The first line under the headline of every table belongs to full-symmetric IR which is not changed by any symmetry operation (all characters equal to +1). Thus the full-symmetric vibration cannot change the point group of a molecule.

By removing $C_4$ rotations and $\sigma_d$ reflections in diagonal mirror planes from $C_{4v}$ group (dimension h = 8) we obtain its subgroup $C_{2v}$ of lower dimension (h = 4). The relations between their IRs may be obtained by the comparison of the characters in E, $C_2$ and $\sigma_v$ headed columns of $C_{4v}$ with the corresponding values for $C_{2v}$. We may

**Table 4.1** Character table of $C_{4v}$ and $C_{2v}$ point groups [14]

| $C_{4v}$ | E | $2\,C_4$ | $C_2$ | $2\sigma_v$ | $2\sigma_d$ |
|---|---|---|---|---|---|
| $A_1$ | +1 | +1 | +1 | +1 | +1 |
| $A_2$ | +1 | +1 | +1 | −1 | −1 |
| $B_1$ | +1 | −1 | +1 | +1 | −1 |
| $B_2$ | +1 | −1 | +1 | −1 | +1 |
| E | +2 | 0 | −2 | 0 | 0 |

| $C_{2v}$ | E | $C_2$ | $\sigma_v$ | $\sigma_v'$ |
|---|---|---|---|---|
| $A_1$ | +1 | +1 | +1 | +1 |
| $A_2$ | +1 | +1 | −1 | −1 |
| $B_1$ | +1 | −1 | +1 | −1 |
| $B_2$ | +1 | −1 | −1 | +1 |

| $C_s$ | E | $\sigma_h$ |
|---|---|---|
| $A'$ | +1 | +1 |
| $A''$ | +1 | −1 |

see that the double-degenerate E type IR of $C_{4v}$ is split into $B_1$ and $B_2$ type IRs of $C_{2v}$ as may be seen from the sum of their characters.

All characters of IR of $B_1$ type in the E, $C_2$ and $\sigma_v$ headed columns of $C_{4v}$ are positive. It means that these symmetry operations are conserved during any nuclear displacement described by this IR of $C_{4v}$ group. The subgroup formed in this way is called kernel or kernel subgroup as described by the formula

$$K(C_{4v}, b_1) = C_{2v} \tag{4.4}$$

For two-dimensional IR of E type we obtain

$$K(C_{4v}, e) = C_1 \tag{4.5}$$

For multidimensional IRs we may define epikernel groups as the intermediate groups between the parent and kernel groups, in our case

$$E(C_{4v}, e) = C_s(\sigma_v) \text{ or } C_s(\sigma_d) \tag{4.6}$$

i.e. $C_s$ group with preserved vertical or diagonal mirror planes.

There are often several epikernels corresponding to the same parent group and its multidimensional representation [5–8]. They may give rise to a chain of subgroups between the parent group and the kernel (lower and higher ranking epikernels). These epikernels can represent independent ways of symmetry lowering, leading to the kernel group along different paths. Kernels and epikernels of selected point groups and degenerate IRs are presented e.g. in [7, 8].

# 4 Group-Theoretical Treatment of Pseudo-Jahn-Teller Systems

Finally it must be mentioned that cyclic groups such as $C_n$, $S_n$, $C_{nh}$ and the tetrahedral groups $T$ and $T_h$ contain non-degenerate complex representations, which always occur in degenerate pairs with conjugate characters and hence form a reducible space of dimension two [7, 14]. Their kernel may be easily determined from its character set. However, since both irreducible components have complex conjugate transformational properties, it is impossible to find an epikernel subgroup, which leaves one component invariant, while transforming the other one.

## 4.3 Potential Energy Surface

The conception of potential energy surface (PES) is used in physics and chemistry for the description of structures, dynamics, spectroscopy and reactivity of compounds [8, 15]. It is generally connected with the adiabatic or Born-Oppenheimer approximation (APES). For any arrangement of N atomic nuclei it may be understood as the total energy $E_{tot}$ (i.e. the electronic energy with inter-nuclear repulsion contributions) function of its nuclear coordinates.

$$E_{tot} = f(Q_k, k = 1 \rightarrow n) \tag{4.7}$$

where the PES dimension $n = 3N - 6$ for nonlinear or $n = 3N - 5$ for linear arrangements. There are several PESs corresponding to the same molecular/ionic system in various charge, electron and spin states (usually denoted as ground and excited states). This treatment is based on the picture of the molecules/ions movement on a PES as well as on their transitions between various PESs depending on the system temperature and external fields.

PES extremal points fulfill the condition

$$\partial E_{tot} / \partial Q_k = 0 \tag{4.8}$$

for all $k = 1 \rightarrow n$. The stable structures of atomic nuclei correspond to PES minima whereas its first order saddle points correspond to transition states for the transition between neighboring minima. At low temperatures and if the energy barriers between PES minima are sufficiently high, a single structure may be observed. This is the case of a static JTE. Dynamic JTE corresponds to the situation when the structure is permanently changed along the pathways between several minima and as a result only the averaged structure of higher symmetry is observed.

The type of any PES extremal point may be determined by the corresponding energy Hessian (the matrix of cartesian second derivatives of the energy). All its eigenvalues must be positive for PES minima whereas its single negative eigenvalue corresponds to the first order saddle point. The group theory is able to predict the symmetry of these critical points of PES.

## 4.4 Jahn-Teller Active Coordinate

The term of JT active coordinate originates in the first order perturbation theory with the Taylor expansion of the perturbation operator being restricted to linear members [8, 15]. For the nuclear coordinate $Q_k$ we demand non-zero value of the first order perturbation matrix element

$$H_{ij}^{(1)} = \left\langle \Psi_i^0 | \partial H / \partial Q_k | \Psi_j^0 \right\rangle Q_k \neq 0 \tag{4.9}$$

in the space of non-perturbed wavefunctions $\Psi_i^0$. If we denote $\Gamma_i$, $\Gamma_k$ and $\Gamma_j$ the representations of $\Psi_i^0$, $Q_k$ and $\Psi_i^0$, respectively (operator $\boldsymbol{H}$ is full-symmetric), the integral

$$\left\langle \Psi_i^0 | \partial H / \partial Q_k | \Psi_j^0 \right\rangle \tag{4.10}$$

may be non-zero only if the direct product $\Gamma_{ikj}$ (reducible representation, in general)

$$\Gamma_{ikj} = \Gamma_i^* \otimes \Gamma_k \otimes \Gamma_j \tag{4.11}$$

contains full-symmetric IR within the point group of the unperturbed system (the asterisk denotes the complex conjugated value) [8, 14]. Alternatively $\Gamma_k$ must be contained in the direct product of the representations of both wavefunctions.

$$\Gamma_k \subset \Gamma_i \otimes \Gamma_j \tag{4.12}$$

If accounting for the hermicity of the $H_{ij}^{(1)}$ matrix element, the $\Gamma_k$ representation of JT active coordinate $Q_k$ must be contained in the symmetric direct product of the representations of the wavefunctions [8, 14].

$$\Gamma_k \subset [\Gamma \otimes \Gamma]^+ = \left[\Gamma^2\right]^+ \tag{4.13}$$

The characters of the representation $\Gamma_{ij} = \Gamma_i \otimes \Gamma_j$ are obtained by multiplying the corresponding characters of the contributing representations for the same symmetry operation $\boldsymbol{R}$ [8, 14]

$$\chi_{ij}(R) = \chi_i(R) \cdot \chi_j(R) \tag{4.14}$$

The characters $\chi_\Gamma^+$ and $\chi_\Gamma^-$ of the symmetric $[\Gamma^2]^+$ and antisymmetric $[\Gamma^2]^-$ direct product representation, respectively, for the symmetry operation $\boldsymbol{R}$ are defined as

$$\chi_\Gamma^\pm = \{[\chi_\Gamma(R)]^2 \pm \chi_\Gamma(R^2)\}/2 \tag{4.15}$$

where $\chi_\Gamma(R^2)$ is the character of the $R^2$ operation with $R^2 = E$ for $\boldsymbol{R} = \boldsymbol{E}, \boldsymbol{i}$ and $\sigma$ operations, whereas $\boldsymbol{R}^2 = \boldsymbol{C_{2n}}$ and $\boldsymbol{S_{2n}}$ for $\boldsymbol{R} = \boldsymbol{C_n}$ and $\boldsymbol{S_n}$, respectively [8, 14].

## 4.5 Epikernel Principle

The description of the symmetry characteristics of JT instabilities may be based on the symmetry of JT active coordinates described by their representations $\Lambda$ in the high-symmetric parent structure. Ceulemans et al. [5–8] formulated the epikernel principle as follows:

> *Extremum points on a JT PES prefer epikernels; they prefer maximal epikernels to the lower ranking ones. As a rule stable minima are to be found with the structures of maximal epikernel symmetry.*
>
> *Extremum points on a JT PES for an orbital doublet will coincide with epikernel configurations. If the distortion space conserves only one type of epikernel, minima and saddle points will be found on opposite sides of the same epikernel distortion. If the distortion space conserves two types of epikernels, minima and saddle points will be characterized by different epikernel symmetries.*

If the system happens to be in an epikernel configuration – in a non-degenerate state – there can be no symmetry lowering forces whatsover, driving the system out of this epikernel or kernel configuration into a surrounding kernel region. In contrast, kernel configurations can be subject to forces that are oriented towards epikernel configurations, since such symmetry increasing forces are totally symmetric under the point group operations of the kernel symmetry and therefore their existence does not violate the selection rule. As a result, non-degenerate epikernel states are indeed expected to be among the PES stationary points, and the more so, the higher their ranking [5–8, 14].

Since kernel $K(G, \Lambda)$ is a subgroup of epikernel $E(G, \Lambda)$, kernel extrema (if they exist) will be more numerous than epikernel extrema of a given type. In order to be stationary at all these equivalent points, the JT PES must be of considerable complexity. Only higher order term in the perturbation expansion (4.9) are able to generate non-symmetrical extrema. However – from a perturbational point of view – the dominance of higher order terms over the first- (and second-) order contributions is (extremely) unlikely. This rationalizes the epikernel principle as well.

## 4.6 The Method of Step-by-Step Descent in Symmetry

An alternative treatment to PES extrema of JT systems is based on their electron degeneracy removal with a symmetry decrease [8–12]. The method of step-by-step descent in symmetry supposes that the driving force of JT distortion is the (symmetry conditioned) electron degeneracy and its removal is connected with an energy decrease. During this process, some symmetry elements of the system are removed and a new symmetry group arises which is an immediate subgroup of the original (parent) group before the distortion. If the electron state (described by its IR) of the system in the immediate subgroup is non-degenerate (one-dimensional IR), the symmetry descent stops because there is no driving force anymore.

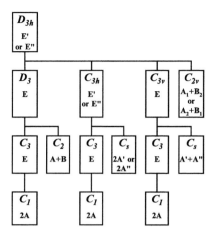

**Scheme 4.1** JT symmetry descent paths of $D_{3h}$ parent group and its subgroups (*upper lines in rectangles*) for two-dimensional IRs (*bottom lines in rectangles*)

Otherwise further symmetry elements are to be removed and the JT symmetry descent continues till a non-degenerate electron state is obtained. The relations between IRs describing the electron states within the same symmetry descent path are determined by group-subgroup relations. As several ways of symmetry descent (due to different symmetry elements removal) are possible (the parent group has several immediate subgroups), several symmetry descent paths may exist for the system in a degenerate electron state.

This problem is relatively simple in the case of double electron degeneracy where only the complete degeneracy removal is possible. Consequently, the JT stable groups may correspond to both PES minima and saddle points (and other PES extremal points as well). On the other hand, a partial degeneracy removal is possible in the systems with higher than double electron degeneracy. Thus the same symmetry group may be either JT stable or JT unstable for various chemical systems. JT unstable groups cannot correspond to PES minima but might correspond to other PES extrema types. Consequently, two types of PES saddle (or extremal) points may be distinguished – JT stable and JT unstable.

The possible JT symmetry descent paths for the most important point groups of symmetry have been published elsewhere [8] in the form of rectangles connected by lines which form the descent pathways (see Schemes 4.1–4.6). The symbols in each rectangle denote a point group (upper line) and the IR describing the electron state (bottom line). The paths connect the bottom side of rectangles at a given level with the upper side of the rectangles in the next level. The rectangles corresponding to JT stable groups and IRs are the end points of these paths (no path at the rectangle bottom side). In the JT unstable ones the path continues at the bottom side of their rectangle. The arrows at these lines indicate that the path continues in another scheme containing the rectangle with the same group and IR symbols.

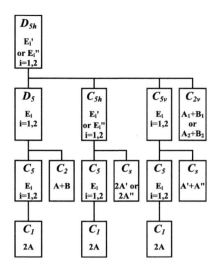

**Scheme 4.2** JT symmetry descent paths of $D_{5h}$ parent group and its subgroups (*upper lines in rectangles*) for two-dimensional IRs (*bottom lines in rectangles*)

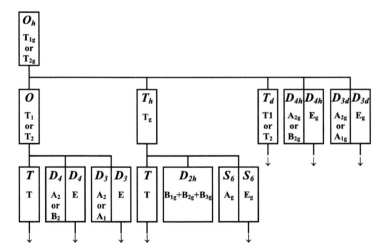

**Scheme 4.3** JT symmetry descent paths of $O_h$ parent group and its subgroups (*upper lines in rectangles*) for IRs (*bottom lines in rectangles*) $T_{1g}$ and $T_{2g}$. Analogous schemes may be obtained for ungerade IRs ($T_{1u}$, $T_{2u}$) replacing subscripts g by the u ones where appropriate. For continuation see Schemes 4.4 ($T_d$ and $T$ groups), 5 ($D_{4h}$ group) and 6 ($D_{3d}$ and $D_3$ groups) (adopted from Ref. [8])

## 4.7 Pseudo-Jahn-Teller Effect for Two-State Systems

The group-theoretical analysis of pseudo-JTE in this study is restricted to the simplest case of two-state systems but it might be very simply extended to more states. Each of the ground and excited electron state representations $\Gamma_0$ and $\Gamma_1$,

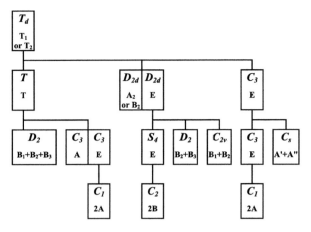

**Scheme 4.4** JT symmetry descent paths of $T_d$ parent group and its subgroups (*upper lines in rectangles*) for IRs (*bottom lines in rectangles*) $T_1$ and $T_2$

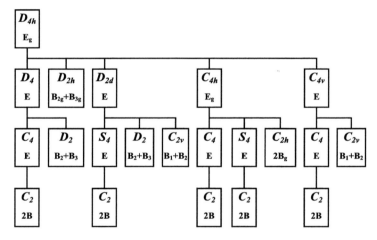

**Scheme 4.5** JT symmetry descent paths of $D_{4h}$ parent group and its subgroups (*upper lines in rectangles*) for IR $E_g$ (*bottom lines in rectangles*). An analogous scheme may be obtained for ungerade IR ($E_u$) replacing subscripts g by the u ones where appropriate

respectively, can be either one-dimensional (nondegenerate, N) or of a higher dimension (degenerate, D). The representations of JT active coordinates are given by the symmetric direct product $[(\Gamma_0 \oplus \Gamma_1)^2]^+$ (see Eq. 4.13). We distinguish the following five cases (the symbol A denotes the full-symmetric irreducible representation):

(i) both electron states are non-degenerate and of the same symmetry

$$\Gamma_0 = \Gamma_1 = N \tag{4.16}$$

4  Group-Theoretical Treatment of Pseudo-Jahn-Teller Systems

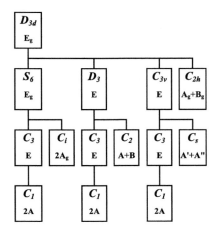

**Scheme 4.6** JT symmetry descent paths of $D_{3d}$ parent group and its subgroups (*upper lines in rectangles*) for two-dimensional IRs (*bottom lines in rectangles*). An analogous scheme may be obtained for ungerade IR ($E_u$) replacing subscripts g by the u ones where appropriate

Thus the representations of JT active coordinates are given by the relation

$$[(2N)]^+ = 3A \tag{4.17}$$

where the full-symmetric irreducible representation preserves the symmetry of the system. Therefore no JT distortion (and no symmetry descent) is possible.

(ii) both electron states are non-degenerate and of different symmetries

$$\Gamma_0 = N_1 \wedge \Gamma_1 = N_2 \wedge N_1 \neq N_2 \tag{4.18}$$

In this case we obtain for JT active coordinates

$$[(N_1 \otimes N_2)^2]^+ = 2A \oplus N_1 \otimes N_2 \tag{4.19}$$

where $N_1 \otimes N_2$ denotes the one-dimensional irreducible representation of lower symmetry which causes the symmetry descent of the system.

(iii) the ground electron state is non-degenerate and the excited one is degenerate

$$\Gamma_0 = N \wedge \Gamma_1 = D \tag{4.20}$$

Thus we obtain for JT active coordinates the relation

$$[(N \oplus D)^2]^+ = A \oplus N \otimes D \oplus [D^2]^+ \tag{4.21}$$

where both $N \otimes D$ and $[D^2]^+$ are multi-dimensional representations ($[D^2]^+$ being always reducible) which cause the symmetry descent of the system.

(iv) the ground electron state is degenerate and the excited one is non-degenerate

$$\Gamma_0 = D \quad \wedge \quad \Gamma_1 = N \tag{4.22}$$

We obtain the same result for JT active coordinates as in (iii)

$$[(D \oplus N)^2]^+ = [D^2]^+ \oplus D \otimes N \oplus A \tag{4.23}$$

The only difference is that in this case JTE in the degenerate ground electron states should be stronger than pseudo-JTE and in most cases we cannot distinguish between them.

(v) both electron states are degenerate

$$\Gamma_0 = D_1 \quad \wedge \quad \Gamma_1 = D_2 \tag{4.24}$$

JT active coordinates are as follows

$$[(D_1 \oplus D_2)^2]^+ = \left[D_1{}^2\right]^+ \oplus D_1 \otimes D_2 \oplus \left[D_2{}^2\right]^+ \tag{4.25}$$

where all three representations are reducible. Also in this case JTE in the degenerate ground electron state should be stronger than pseudo-JTE and in most cases we cannot distinguish between them.

Extremum points on pseudo-JT PES of the above systems may be obtained by epikernel principle described in Chap. 5 by using the representations of JT active coordinates obtained in the cases (ii)–(v).

An alternative treatment based on electron states (see Chap. 6) cannot be so exact as for JTE and the following approximations to the method of step-by-step symmetry descent are necessary to solve pseudo-JT problems:

A. Method of JTE in an excited degenerate electron state. We suppose an excitation of the system into the degenerate excited state and subsequent symmetry descent. Unfortunately, this treatment is unusable for the cases (ii) and (iv) with nondegenerate excited states.
B. Method of JTE in a supergroup with degenerate electron state which is split into the relevant pseudo-degenerate electron states. We suppose that our system belongs to a group on a symmetry descent path which starts at its supergroup describing the highest possible symmetry of the same system and that in this supergroup the system is in a degenerate electron state. The less-symmetric structures then belong to different symmetry descent paths. This treatment may meet problems for high-symmetric pseudo-JT structures.
C. Method of JTE in improved (more symmetric) compound in degenerate electron state. We suppose that a small "perturbation" has caused the symmetry decrease of a high-symmetric improved compound with a degenerate electron state into our less-symmetric real compound with pseudodegenerate electron states. In

the first step we solve the problem of JT symmetry descent of the improved compound in the degenerate electron state (i.e. without the "perturbation") and search for stable JT structures. In the next step we introduce the above mentioned "perturbation" into these JT stable structures which might cause their additional symmetry descent for the real compound. This method must be treated very carefully because of the definition of a small "perturbation".

## 4.8 Multimode Jahn-Teller and Pseudo-Jahn-Teller Systems

Zlatar et al. [16] investigated a multimode JTE at DFT level of theory (local density approximation using Vosko-Wilk-Nusair parametrization with TZP Slater-type orbitals). In the first step, they obtained a high-symmetry (HS) configuration $\mathbf{R}_{HS}$ using an average of configuration (AOC) geometry optimization with fractional orbital occupations. Stable low-symmetry (LS) structures $\mathbf{R}_{LS,min}$ with $E_{LS,min}$ energies corresponding to the PES minima have been obtained by geometry optimizations with the different (integer) LS electron distributions. The displacements of the atoms from their HS structure are defined as the vector

$$\mathbf{R}_{JT} = \mathbf{R}_{HS} - \mathbf{R}_{LS,min} \tag{4.26}$$

Single point calculations with fixed nuclear geometries $\mathbf{R}_{HS}$ and different LS electron distributions yield $E_{HS,LS,min}$ energies. Thus JT stabilization energies $E_{JT}$ are

$$E_{JT} = E_{HS,LS,min} - E_{LS,min} \tag{4.27}$$

Within a harmonic approximation the JT distortion is given as a linear combination of displacements along all totally symmetric normal coordinates in the LS conformation. The linear coefficients, or the weighting factors, $w_k$, define the contribution of each of these normal modes, $\mathbf{Q}_k$, to the distortion

$$\mathbf{R}_{JT} = \sum w_k \mathbf{Q}_k \tag{4.28}$$

Each of the totally symmetric normal modes contributes to the JT stabilization and $E_{JT}$ can be expressed as the sum of these energy contributions

$$E_{JT} = 0.5 \sum w_k^2 g_k \mathbf{Q}_k^2 = \sum w_k^2 E_k \tag{4.29}$$

where

$$g_k = (2\pi v_k)^2 \tag{4.30}$$

with $v_k$ being a frequency of the k-th normal mode.

VCl$_4$ belongs to the simplest JT molecules. In $T_d$ point group, a single d electron occupies e orbital. The electron ground state is $^2$E and the symmetric direct product yields

$$\left[E^2\right]^+ = A_1 \oplus E \qquad (4.31)$$

The degenerate electron state splits after the symmetry descent to $D_{2d}$ group into the non-degenerate $^2A_1$ and $^2B_1$ ones.

The results of the above DFT calculations [16, 17] have shown that within $3N - 6 = 9$ normal coordinates only one pair of e and one $a_1$ modes have non-zero linear vibronic coupling constant. Thus this can be the simplest case of the multimode problem, with possibly two JT active vibrations. The above mentioned analysis of the normal coordinates contributions indicates that the contribution of the e mode to the distortion is more than 99%, which is in agreement with usual consideration of VCl$_4$ system as an ideal, single mode problem.

Cyclopentadienyl radical is one of the most studied JT active molecules. The ground electron state of C$_5$H$_5$ in $D_{5h}$ symmetry is $^2E_1''$, with three electrons occupying the doubly degenerate orbital. The symmetric direct product

$$\left[\left(E_1''\right)^2\right]^+ = A_1' \oplus E_2' \qquad (4.32)$$

indicates that the distortion coordinate is of $e_2'$ symmetry. The descent in symmetry goes to $C_{2v}$ group and the degenerate electron state splits into the $^2A_2$ and $^2B_1$ ones. The results of the above mentioned DFT studies [16] have shown that the three most important $e_2'$ type vibrations contribute 90% to the JT distortion whereas two most important $e_1'$ type vibrations contribute around 10%. Their activity might be explained as the consequence of a pseudo-JT interaction of the $^2E_1''$ ground state with the $^2A_1''$ excited electron state as indicated by the symmetric direct product

$$[(E_1'' \oplus A_1'')^2]^+ = A_1' \oplus E_1' \oplus E_2' \qquad (4.33)$$

Both $e_1'$ and $e_2'$ vibrations belong to in-plane ring deformations, thus influencing the C–C bonding in similar way.

Cobaltocene has been subject of wide research, but only recently a detailed analysis of its JT distortion has been carried out [16, 18]. Its high symmetry conformation of $D_{5h}$ group with eclipsed rings (in $^2E_1''$ electron ground state) is more stable (by ca 161 cm$^{-1}$) than the $D_{5d}$ one with staggered rings (in $^2E_{1g}$ electron ground state), both having a single electron in a double degenerate orbital. In this study we will present only the results for the more stable eclipsed conformation. As implied by the symmetric direct product

$$\left[\left(E_1''\right)^2\right]^+ = A_1' \oplus E_2' \qquad (4.34)$$

the distortion coordinates are of $e_2'$ symmetry and the descent in symmetry goes to $C_{2v}$ group. The double degenerate electron state will split into the non-degenerate $^2A_2$ and $^2B_1$ ones in $C_{2v}$ symmetry group. There are six pairs of JT active $e_2'$ vibrations, four $a_1'$ and six pairs of $e_1'$ vibrations which are active in pseudo-JTE, most probably due to the interaction with the excited $^2A_1''$ electron state as indicated by the symmetric direct product

$$[(E_1'' \oplus A_1'')^2]^+ = A_1' \oplus E_1' \oplus E_2' \tag{4.35}$$

They become all totally symmetric in $C_{2v}$ symmetry group. The main contribution to the JT distortion arises from the four $e_2'$ type vibrations which contribute to about 95% of the total distortion vector, being predominantly located in the five-membered rings. The contribution of $e_1'$ types vibrations are significantly lower (ca 2%). This may be explained by different kind of these vibrations (the $e_1'$ ones are mainly skeletal deformations) unlike the above mentioned cyclo-$C_5H_5\cdot$ case.

Finally it may be concluded that in all the systems under study the pure JT contribution to the JT distortion dominates over the pseudo-JT one. This is in full agreement with the predictions in Chapter 7 concerning the cases (iv) and (v) for degenerate ground states.

## 4.9 Applications

### 4.9.1 Triphenylene Dianion

Hatanaka [19] has optimized the geometry of the triphenylene dianion (Fig. 4.1) at MP2/6-31G**//HF/6-31G** levels of theory. Its high-symmetric $D_{3h}$ structure should contain two electrons in double degenerate $e''$ orbitals $(\ldots(e'')^4(e'')^2$ electron

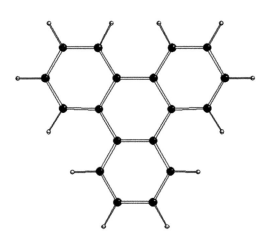

**Fig. 4.1** Structure of triphenylene

**Table 4.2** MP2/6-31G$^{**}$//HF/6-31G$^{**}$ symmetries, electron characteristics and relative energy ($\Delta$E) data of the optimized structures of triphenylene dianion [19]

| Symmetry | Electron configuration | Electron state | $\Delta$E (kcal/mol) |
|---|---|---|---|
| $D_{3h}(I)$ | $(e'')^2$ | $^1A_1', {}^1E'$ | ? |
| $D_{3h}(II)$ | $(e'')^2$ | $^3A_2'$ | 0.0 |
| $C_{2v}(I)$ | $(a_2)^2$ | $^1A_1$ | 2.41 |
| $C_{2v}(II)$ | $(b_1)^2$ | $^1A_1$ | 3.96 |
| $C_{2v}(III)$ | $(a_2)^1(b_1)^1$ | $^1B_2$ | 36.46 |
| $C_{2v}(IV)$ | $(a_2)^1(b_1)^1$ | $^3B_2$ | 13.54 |

configuration) which results either in $^1A_1'$ and $^1E'$ singlet electron states or in $^3A_2'$ triplet electron state as implied by the symmetric (4.36) and antisymmetric (4.37) direct products.

$$\left[(E'')^2\right]^+ = A_1' \oplus E' \tag{4.36}$$

$$\left[(E'')^2\right]^- = A_2' \tag{4.37}$$

The HF/6-31G$^{**}$ geometry optimization resulted into three different $C_{2v}$ structures in $^1A_1$ or $^1B_2$ electron states, a high-symmetric $D_{3h}$ structure in low-lying $^3A_2'$ electron state and a high-energy $C_{2v}$ structure in $^3B_2$ electron state (Table 4.2). Unfortunately, there is no information whether these structures correspond to minima or saddle points of the corresponding PES.

The existence of the $C_{2v}$ structures in singlet electron states may be explained by pseudo-JTE for $^1A_1'$ and $^1E'$ electron states in $D_{3h}$ symmetry

$$[(A_1' \oplus E')^2]^+ = 2A_1' \oplus 2E' \tag{4.38}$$

or by JTE in excited $^1E'$ state

$$[(E')^2]^+ = A_1' \oplus E' \tag{4.39}$$

both giving JT active coordinate of $e'$ symmetry. The stable $^3A_2'$ structure of $D_{3h}$ symmetry group cannot be involved into this pseudo-JT interaction and so the existence of $C_{2v}$ structure in $^3B_2$ electron state must be explained by splitting an energetically higher degenerate electron state.

$$D_{3h}(3E') \rightarrow C_{2v}\left(^3A_1\right) \text{ or } C_{2v}\left(^3B_2\right) \tag{4.40}$$

Epikernel principle predicts for both triplet and singlet cases the epikernel structures of $C_{2v}$ symmetry and the kernel ones of $C_s$ symmetry (not found).

**Fig. 4.2** Structure of cyclo-C$_3$H radical in $C_{2v}$ and $C_s$ symmetries

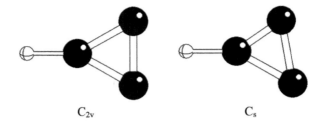

$$E(D_{3h}, e') = C_{2v} \tag{4.41}$$

$$K(D_{3h}, e') = C_s(\sigma_h) \tag{4.42}$$

The step-by-step symmetry descent treatment uses the method of JTE in excited double degenerate electron state. The symmetry descent paths (see Scheme 4.1)

$$D_{3h}\left(^{1,3}E'\right) \to C_{2v}\left(^{1,3}A_1\right) \text{ or } C_{2v}\left(^{1,3}B_2\right) \tag{4.43}$$

$$D_{3h}\left(^{1,3}E'\right) \to C_{3v}\left(^{1,3}E\right) \to C_s(^{1,3}A') \text{ or } C_s(^{1,3}A'') \tag{4.44}$$

explain the existence of both JT stable $C_{2v}$ and $C_s$ groups in singlet or triplet electron states in agreement with the calculations results.

### 4.9.2 Cyclo-C$_3$H Radical

High-level *ab initio* calculations [20–22] indicate that the planar $C_{2v}$ structure of cyclo-C$_3$H radical in $^2B_2$ ground electron state (Fig. 4.2) might be a transition state between two slightly distorted stable structures of $C_s$ symmetry in $^2A'$ electron state. This symmetry descent is ascribed to pseudo-JT interaction of the ground electron state with the first excited $^2A_1$ one in the planar structure. Using a complete active space self-consistent-field (CASSCF) calculations followed by a single and double configuration interaction mapping of the corresponding PES [22], the stable structure of $C_s$ symmetry has been found which is by only 27 cm$^{-1}$ lower in energy than the planar $C_{2v}$ structure. On the other hand, very recent MCSCF calculations [23] indicate that this symmetry decrease should be an artifact of the inadequate approximate solution of the Schrödinger equation which results in a wavefunction of the lower symmetry than those of the nuclear frame.

As indicated by the symmetric direct product for the above mentioned pseudo-JT interaction in $C_{2v}$ point group

$$[(B_2 \oplus A_1)^2]^+ = 2A_1 \oplus B_2 \tag{4.45}$$

we obtain the JT active coordinate of $b_2$ symmetry. According to the epikernel principle

$$K(C_{2v}, b_2) = C_s(\sigma'_v) \tag{4.46}$$

we obtain the $C_s$ kernel group in agreement with [22].

Within the step-by-step symmetry descent treatment we can use the method of JTE in a supergroup with degenerate electron state which is split into the relevant pseudo-degenerate electron states. If assuming the hypothetical $D_{3h}$ parent structure with the hydrogen atom in the centre of a carbon triangle in double degenerate $^2E'$ electron state, our structures may be explained by two descent paths (Scheme 4.1)

$$D_{3h}\left(^2E'\right) \to C_{2v}\left(^2A_1\right) \text{ or } C_{2v}\left(^2B_2\right) \tag{4.47}$$

$$D_{3h}\left(^2E'\right) \to C_{3v}\left(^2E\right) \to C_s\left(^2A'\right) \text{ or } C_s\left(^2A''\right) \tag{4.48}$$

Within this treatment, the JT stable $C_{2v}$ and $C_s$ structures are end points of two various symmetry descent paths of the hypothetical JT unstable parent structure of $D_{3h}$ symmetry. Nevertheless, it must be kept in mind that this high-symmetric cyclo-$C_3H$ structure is only a theoretical tool and not a real structure.

### 4.9.3 Cyclo-$C_5H_5{}^+$

Wörner and Merkt [24] have investigated the cyclopentadienyl cation and its fully deuterated isotopomer by pulsed-field-ionization zero-kinetic-energy (PFI-ZEKE) photoelectron spectroscopy and *ab initio* CASSCF(4,5)/cc-pVTZ calculations. It possesses a triplet ground electron state $^3A_2'$ of $D_{5h}$ equilibrium geometry (model A, see Fig. 4.3 and Table 4.3) and the first excited singlet state $^1E_2'$. The optimization of the geometry of the singlet state in $C_{2v}$ symmetry led to two different stationary points with a totally symmetric $^1A_1$ ground electron state (models B and C, see Table 4.3). By requiring the electron state symmetry to be $^1B_2$ the substantially less stable D structure has been obtained. The authors explain these structures by a very weak linear Jahn-Teller effect along the $e_1'$ modes (model D) and by an unusually strong pseudo-Jahn-Teller effect ($^1E_2' \oplus ^1E_1' \oplus ^1A_1'$ states) along the $e_2'$ modes (models B and C).

JT active coordinate in $D_{5h}$ geometry is implied by the symmetric direct product

$$[(E_2')^2]^+ = A_1' \oplus E_1' \tag{4.49}$$

According to the epikernel principle, we obtain the $C_s$ kernel and $C_{2v}$ epikernel groups

$$K\left(D_{5h}, e_1'\right) = C_s(\sigma_h) \tag{4.50}$$

# 4 Group-Theoretical Treatment of Pseudo-Jahn-Teller Systems

**Fig. 4.3** Structure of cyclo-$C_5H_5^+$

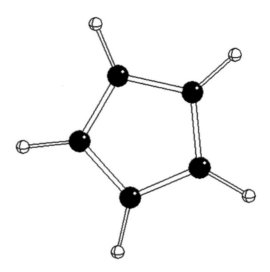

**Table 4.3** CASSCF(4,5)/cc-pVTZ bond lengths, d(C–C), and relative energies, $\Delta E$, of various conformations of cyclo-$C_5H_5^+$ [24]

| Model | Symmetry group | Electron state | $d(C_1-C_2/C_1-C_5)$ [Å] | $d(C_2-C_3/C_4-C_5)$ [Å] | $d(C_3-C_4)$ [Å] | $\Delta E$ [cm$^{-1}$] | |
|---|---|---|---|---|---|---|---|
| A | $D_{5h}$ | $^3A_2'$ | 1.381 | 1.381 | 1.381 | −5,580 | stable |
| B | $C_{2v}$ | $^1A_1$ | 1.443 | 1.349 | 1.534 | −3,700 | transition state |
| C | $C_{2v}$ | $^1A_1$ | 1.381 | 1.510 | 1.338 | −4,000 | stable |
| D | $C_{2v}$ | $^1B_2$ | 1.435 | 1.419 | 1.400 | −2,300 | stable |

$$E\left(D_{5h}, e_1'\right) = C_{2v} \tag{4.51}$$

which explain the existence of the D model $C_{2v}$ structure (the same kernel and epikernel groups may be obtained for $e_2'$ coordinate).

For pseudo-JTE we obtain JT active coordinates implied by the relation

$$\left[\left(A_1' \oplus E_1' \oplus E_2'\right)^2\right]^+ = 3A_1' \oplus 3E_1' \oplus 3E_2' \tag{4.52}$$

with JT active coordinates of $e_1'$ and $e_2'$ symmetries which explain the existence of all B-D model $C_{2v}$ structures optimized.

The method of step-by-step symmetry descent has for $E_1'$ and $E_2'$ electron states the descent path (see Scheme 4.2)

$$D_{5h}\left(^1E_i', i=1,2\right) \rightarrow C_{2v}\left(^1A_1\right) \text{ or } C_{2v}\left(^1B_1\right) \tag{4.53}$$

This means that the D structure cannot originate in the above mentioned degenerate electron states. Its existence is implied by $^1E_1''$ or $^1E_2''$ electron state and can be

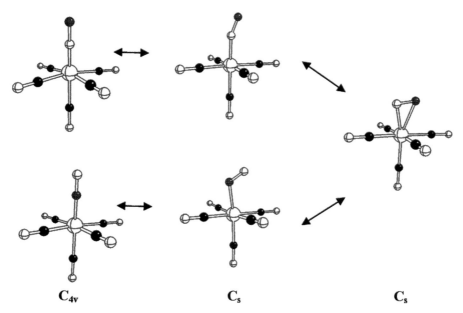

**Fig. 4.4** Intermediate structures for $[Fe(CN)_5NO]^{2-}$ photo-isomerization (Fe – large white, C – black, N – white, O – grey)

explained using the method of JTE in the (higher) excited electron state by the symmetry descent path

$$D_{5h}\left({}^1E_i'', i=1,2\right) \to C_{2v}\left({}^1A_2\right) \text{or} C_{2v}\left({}^1B_2\right) \tag{4.54}$$

An alternative explanation might be in an incorrect assignment of the $^1B_2$ electron state as $B_1$ and $B_2$ representations of $C_{2v}$ group differ only in the (opposite) signs of the characters of the $\sigma_v$ (in the molecular plane) and $\sigma_v'$ (perpendicular to the molecular plane) mirror planes.

### 4.9.4 $[Fe(CN)_5NO]^{2-}$ Photo-Isomerization

Atanasov and Schönherr [25] investigated the photoisomerization of $[Fe(CN)_5NO]^{2-}$ and $[Fe(CN)_5ON]^{2-}$ at DFT level of theory (the gradient corrected exchange functional of Becke and the correlation functional of Perdew in conjunction with the LDA parameterization of Vosko, Wilk and Nusair) using a triple zeta basis set (see Fig. 4.4). The structures of both nitrosyl and isonitrosyl compounds belonging to $C_{4v}$ symmetry group are in the non-degenerate $A_1$ ground electron state.

# 4 Group-Theoretical Treatment of Pseudo-Jahn-Teller Systems

**Table 4.4** Selected electron structure, relative energy ($\Delta E$) and geometry data of DFT optimized $[Fe(CN)_5NO]^{2-}$ and $[Fe(CN)_5ON]^{2-}$ complexes [25]

| Symmetry | Electron configuration | Electron state | $\Delta E$[eV] | Bond angle[°] Fe–N–O | Fe–O–N |
|---|---|---|---|---|---|
| $[Fe(CN)_5NO]^{2-}$ | | | | | |
| $C_{4v}$ | $\ldots(2b_2)^2$ | $^1A_1$ | 0.000 | 180.0 | 0.0 |
| | $\ldots(2b_2)^1(9e)^1$ | $^1E$ | ? | 180.0 | 0.0 |
| $C_s$ | $\ldots(11a')^1(23a'')^1$ | $^1A''$ | 1.397 | 140.8 | 24.1 |
| $C_{4v}$ | $\ldots(2b_2)^1(9e)^1$ | $^3E$ | 1.777 | 180.0 | 0.0 |
| $C_s$ | $\ldots(11a')^1(23a'')^1$ | $^3A''$ | 1.241 | 141.4 | 23.6 |
| $[Fe(CN)_5ON]^{2-}$ | | | | | |
| $C_{4v}$ | $\ldots(2b_2)^2$ | $^1A_1$ | 1.530 | 0.0 | 180.0 |
| | $\ldots(2b_2)^1(9e)^1$ | $^1E$ | ? | 0.0 | 180.0 |
| $C_s$ | $\ldots(23a')^1(11a'')^1$ | $^1A''$ | 2.000 | 26.9 | 138.1 |
| $C_{4v}$ | $\ldots(2b_2)^1(9e)^1$ | $^3E$ | 2.218 | 0.0 | 180.0 |
| $C_s$ | $\ldots(11a')^1(23a'')^1$ | $^3A''$ | 2.051 | 24.8 | 141.3 |

After the photo-excitation of the $C_{4v}$ structures into a double-degenerate E electron state the NO group is bent and the symmetry is lowered to $C_s$. In the next step these bent structures may be transformed into another $C_s$ structure with the sideways bonded NO ligand.

The results of DFT geometry optimizations confirm the higher energy of the isonitrosyl compound (Table 4.4). The energy of stable $C_{4v}$ structures of both compounds in the singlet $^1A_1$ electron state is lower than the corresponding $C_s$ one and thus it cannot be involved into pseudo-JT mechanism. Thus only excited electron states may be the source of the JT or pseudo-JT instability.

JT active coordinates for the double degenerate excited electron state as implied by the symmetric direct product in the $C_{4v}$ point group

$$\left[E^2\right]^+ = A_1 \oplus B_1 \oplus B_2 \tag{4.55}$$

may be of $b_1$ and $b_2$ symmetry. According to epikernel principle, the corresponding kernel groups

$$K\left(C_{4v}, b_1\right) = C_{2v}(\sigma_v) \tag{4.56}$$

$$K\left(C_{4v}, b_2\right) = C_{2v}(\sigma_d) \tag{4.57}$$

$$K(C_{4v}, b_1 \oplus b_2) = C_2 \tag{4.58}$$

are of $C_{2v}$ (with preserved vertical or diagonal mirror planes) or $C_2$ symmetries only and cannot explain the existence of the bent $C_s$ structures.

**Table 4.5** Splitting three-dimensional representations of $O_h$ group in $C_{4v}$ group

| Compound | $Fe(CN)_6]^{3-}$ | $[Fe(CN)_5NO]^{2-}$ $[Fe(CN)_5ON]^{2-}$ |
|---|---|---|
| Symmetry group | $O_h$ | $C_{4v}$ |
| Representations | $T_{1g}$ | $E \oplus A_2$ |
| | $T_{2g}, T_{1u}$ | $E \oplus A_1$ |
| | $T_{2u}$ | $E \oplus B_1$ |

Only the pseudo-JT interaction of the double-degenerate electron state with any of non-degenerate higher excited states produces JT active coordinate of e type as indicated by the symmetric direct product

$$[(N \oplus E)^2]^+ = 2A_1 \oplus E \oplus B_1 \oplus B_2 \qquad (4.59)$$

with $N = A_1$, $A_2$, $B_1$ or $B_2$. According to the epikernel principle

$$K(C_{4v}, e) = C_1 \qquad (4.60)$$

$$E(C_{4v}, e) = C_s(\sigma_v) \text{ or } C_s(\sigma_d) \qquad (4.61)$$

we may obtain the desired epikernels of $C_s$ symmetry with preserved vertical or diagonal mirror planes.

The step-by-step symmetry descent treatment uses the method of JTE in improved (more symmetric) compound in a degenerate electron state. In our case we suppose only small differences between a high-symmetric $[Fe(CN)_6]^{3-}$ complex ($O_h$ symmetry group) and our $[Fe(CN)_5NO]^{2-}$ or $[Fe(CN)_5ON]^{2-}$ complexes. Thus the replacement of $CN^-$ by NO is a small perturbation which causes the splitting of a triple degenerate electron states in $O_h$ structures into the double degenerate ones in $C_{4v}$ structures as may be seen in Table 4.5.

The JT symmetry descent paths for $[Fe(CN)_6]^{3-}$ complex (see Schemes 4.3–4.6)

$$O_h \left( T_{1j} \text{ or } T_{2j}, j = g, u \right) \rightarrow D_{4h} \left( E_j, j = g, u \right) \rightarrow C_{4v}(E) \qquad (4.62)$$

$$O_h \left( T_{1j} \text{ or } T_{2j}, j = g, u \right) \rightarrow D_{3d} \left( E_j, j = g, u \right) \rightarrow C_{3v}(E) \rightarrow C_s \left( A' \right) \text{ or } C_s \left( A'' \right) \qquad (4.63)$$

explain the existence of both JT unstable $C_{4v}$ and JT stable $C_s$ structures. The subsequent replacement of $CN^-$ by NO or ON in a mirror plane of these structures does not change their symmetry group. Thus we have explained the existence of $C_{4v}$ and $C_s$ symmetry groups of $[Fe(CN)_5NO]^{2-}$ and $[Fe(CN)_5ON]^{2-}$ complexes in agreement with the calculations as a consequence of pseudo-JTE.

## 4.10 Conclusions

This study deals with group-theoretical analysis of pseudo-JT systems. The results of two alternative treatments based on JT active coordinates and on electron states are compared in predicting the symmetries of the structures caused by pseudo-JTE. The methods of epikernel principle as well as the method of step-by-step symmetry descent have been developed for JT systems with symmetry conditioned electron degeneracy. They produce the same results for low-dimensional groups but may differ for higher symmetry groups. This study represents the extension of both methods to pseudo-JT systems.

Both methods have their advantages and shortages. The method of epikernel principle seems to be incomplete due to its restriction to the first order perturbation theory and linear expansion of the perturbation potential. Another problem is the applicability to the groups with complex characters (problematic epikernels for $T_h$, $T$, $C_{nh}$, $C_n$ and $S_n$ groups with n > 2). There is an unresolved problem how to distinguish JT active symmetric coordinates and the antisymmetric ones of the same representation such as the b type one in the general case

$$[(A \oplus B)^2]^+ = 2A \oplus B \tag{4.64}$$

$$[(A \oplus B)^2]^- = B \tag{4.65}$$

Nevertheless, the epikernel principle seems to be more suitable for pseudo-JT problems than the concept of step-by-step symmetry descent.

The concept of step-by-step symmetry descent is based exclusively on group theory and does not account for any approximations used in solving the Schrödinger equation. On the other hand, it does not explain the mechanisms which are responsible for JT and pseudo-JT distortions. Moreover, the approximations used for pseudo-degenerate electron states reduce its completeness – the method of JTE in excited electron states cannot produce complete results. The remaining methods seem to be too sophisticated and must be treated very carefully.

Finally it may be concluded that the above mentioned treatments should be used jointly in dealing with JT or pseudo-JT problems. It must be kept in mind that the group theory provides only the symmetry group and not the coordinates of stable JT or pseudo-JT systems as does an analytical solution based on perturbation theory which is, however, usable for small systems only. On the other hand, only the group-theoretical treatment is able to bring reliable and usable structure predictions for large JT or pseudo-JT systems. Further theoretical studies in this field are still desirable.

**Acknowledgments** Slovak Grant Agency VEGA (Project No. 1/0127/09) is acknowledged for financial support.

# References

1. Jahn HA, Teller E (1937) Proc R Soc Lond A 161:220
2. Jahn HA (1938) Proc R Soc Lond A 164:117
3. The Kramers degeneracy theorem states that the energy levels of systems with an odd number of electrons remain at least doubly degenerate in the presence of purely electric fields (i.e. no magnetic fields). Kramers HA (1930) Kon Acad Wet Amsterdam 33:959
4. Bersuker IB (2006) The Jahn-Teller effect. Cambridge University Press, London
5. Ceulemans A, Beyens D, Vanquickenborne LG (1984) J Am Chem Soc 106:5824
6. Ceulemans A (1987) J Chem Phys 87:5374
7. Ceulemans A, Vanquickenborne LG (1989) Struct Bond 71:125
8. Breza M (2009) In: Koeppel H, Yarkoni DR, Barentzen H (eds) The Jahn-Teller-Effect. Fundamentals and implications for physics and chemistry, Springer Series in Chemical Physics, vol 97. Springer, Berlin, p 51
9. Pelikán P, Breza M (1985) Chem Zvesti 39:255
10. Breza M (1990) Acta Crystallogr B 46:573
11. Breza M (2002) J Mol Struct Theochem 618:165
12. Breza M (2003) Chem Phys 291:207
13. Breza (1991) Chem Pap 45:473
14. Salthouse JA, Ware MJ (1972) Point group character tables and related data. Cambridge University Press, London
15. Boča R, Breza M, Pelikán P (1989) Struct Bond 71:57
16. Zlatar M, Schläpfer C-W, Daul C (2009) In: Koeppel H, Yarkoni DR, Barentzen H (2009) The Jahn-Teller-effect. Fundamentals and implications for physics and chemistry, Springer Series in Chemical Physics, vol 97, Springer, Berlin/Heidelberg, pp 131–165
17. Bruyndockx R, Daul C, Manoharan PT, Deiss E (1997) Inorg Chem 36:4251
18. Zlatar M, Schläpfer C-W, Fowe EP, Daul C (2009) Pure Appl Chem 81:1397
19. Hatanaka M (2009) J Mol Struct (Theochem) 915:69
20. Jiang Q, Rittby CML, Graham WRM (1993) J Chem Phys 99:3194
21. Yamamoto S, Saito S (1994) J Chem Phys 101:5484
22. Yamagishi H, Taiko H, Shimogawara S, Murakami A, Noro T, Tanaka K (1996) Chem Phys Lett 250:165
23. Halvick P (2007) Chem Phys 340:79
24. Wörner HJ, Merkt F (2007) J Chem Phys 127:034303
25. Atanasov M, Schönherr (2002) J Mol Struct (Theochem) 592:79

# Part II
# Jahn-Teller Effect and Vibronic Interactions in Impurity Centers in Crystals

# Chapter 5
# The Dynamic Jahn-Teller Effect in Cu(II)/MgO

**P.L.W. Tregenna-Piggott, C.J. Noble, and Mark J. Riley**

**Abstract** A true dynamic Jahn-Teller effect in the solid state has proven to be quite elusive, as pure compounds suffer from cooperative effects, while doped systems are susceptible to small crystal imperfections that lock the system into static distortions. Cu(II) doped into the cubic host MgO represents a rare example of such a dynamic Jahn-Teller system and has been the subject of numerous experimental and theoretical studies. Recently we have presented high resolution low temperature Electron Paramagnetic Resonance (EPR) spectra of Cu(II)/MgO as a function of the applied field direction. These spectra indicate that at temperatures as low as 1.8 K the Cu(II) centre is in a degenerate vibronic state of E symmetry that is delocalized over the ground potential energy surface, indicating a true dynamic Jahn-Teller effect. The experiments also show us that this system has a potential energy surface with three equivalent minima, each at three equivalent tetragonally elongated geometries, separated by low barriers. Relaxation from the anisotropic E type spectrum to an isotropic spectrum occurs at temperatures above 6 K. The observation of the dynamic Jahn-Teller effect in this system is due to small barrier heights between the minima and the random crystal strain, which is small when compared to the tunneling splitting. We examine the limitations of the experiment in being able to determine these quantities separately and suggest future experiments that may shed further light on this fascinating system.

---

P.L.W. Tregenna-Piggott (deceased)
Laboratory for Neutron Scattering, ETH Zürich & Paul Scherrer Institut,
CH-5232 Villigen PSI, Switzerland

C.J. Noble
Centre for Advanced Imaging, University of Queensland, St. Lucia, QLD 4072, Australia
e-mail: c.noble@uq.edu.au

M.J. Riley (✉)
School of Chemistry and Molecular Biosciences, University of Queensland, St. Lucia,
QLD 4072, Australia
e-mail: m.riley@uq.edu.au

M. Atanasov et al. (eds.), *Vibronic Interactions and the Jahn-Teller Effect:*
*Theory and Applications*, Progress in Theoretical Chemistry and Physics 23,
DOI 10.1007/978-94-007-2384-9_5, © Springer Science+Business Media B.V. 2012

## 5.1 Introduction

The Jahn-Teller effect is the manifestation of a rule from symmetry arguments that a non-linear molecule in a degenerate electronic state will distort to remove the degeneracy [1, 2]. The coupling of close lying electronic states to vibrational distortions in this manner is a breakdown of the Born-Oppenheimer approximation, which lies at the heart of the quantum description of molecules. Without external perturbations, the Jahn-Teller distortion will be dynamic, the molecule will have a time averaged undistorted geometry and the electronic degeneracy is replaced by a vibronic (vibrational-electronic) degeneracy. Six coordinate copper(II) complexes have a doubly degenerate E electronic ground state at an octahedral geometry and can be considered as classic examples of Jahn-Teller molecules.

In the $E \otimes e$ Jahn-Teller effect the E electronic state couples to a doubly degenerate e vibration and the resulting potential energy surface as a function of the two components $(Q_\theta, Q_\varepsilon)$ of the vibrational coordinates gives rise to the famous "Mexican hat" surface for linear vibronic coupling. When higher order effects are included, the "warped Mexican hat" results with three equi-energetic minima at points in $(Q_\theta, Q_\varepsilon)$ space that correspond to the geometries where the symmetry is lowered from octahedral to tetragonal. The position of these minima depend on the relative signs of the coupling constants [2] and for Cu(II) compounds they are usually found at a positive distortion along the $Q_\theta$ vibrational coordinate (and equivalent directions), which corresponds to an tetragonally elongated octahedron [3].

In the solid state, one usually finds the symmetry of copper(II) complexes lower than octahedral. The barriers between the minima of a warped Mexican hat potential surface are such that the system is easily trapped within a single minimum, which often results in cooperative effects in pure copper compounds. Even in the absence of cooperative effects when Cu(II) is doped into high symmetry host crystals, a true dynamic Jahn-Teller effect in the solid state is quite rare. Small inhomogeneities, or strain, in the crystal lattice will tend to trap the system in a static distortion of a single well at low temperature. The ability of the strain to localize such systems depends on the size of this strain relative to the tunneling frequency between the equivalent minima. Historically, the Cu(II)/MgO system has been a very important system as one of the few examples of a dynamic Jahn-Teller effect in the solid state and was instrumental in prompting many important early theoretical studies on the Jahn-Teller effect.

Recently, we have re-examined the low temperature EPR spectrum at higher angular resolution, and have resolved many of the outstanding problems of this system. The system continues to inspire a number of theoretical studies [4]. In this paper we discuss the Cu(II)/MgO EPR spectrum and outline what conclusions can and cannot be determined experimentally. Finally, we indicate some directions for future experiments that may prove fruitful.

## 5.2 Electron Paramagnetic Resonance (EPR) Spectroscopy

The EPR spectrum of Cu(II)/MgO was first measured by Orton et al. [5] who noted that an isotropic spectrum at 4.2 K was replaced by an anisotropic spectrum at 1.2 K. Coffman [6] studied the angular variation of the low temperature spectrum and found that it was not the superposition of the EPR spectra of three statically distorted species. Instead the spectrum of a single species was observed exhibiting the cubic anisotropy expected for a $^2E(\Gamma_8)$ state with full cubic symmetry. This was interpreted as indicating tunneling between the three tetragonally distorted configurations, as described earlier by Bersuker [7], that results in the cubic anisotropy being reduced by a factor of two. Ham [8] then generalized this result in terms of a "vibronic reduction factor" that takes the limiting value of 1/2 in the strong Jahn-Teller coupling limit. Following this, the low temperature angular variation and line-shapes were explained by Reynolds et al. [9] who used the concept of random crystal strain together with the "three-state" model of Ham [10] to explain the strain broadened line-shape of the spectra.

### 5.2.1 Description of the EPR Spectrum with Reference to Previous Studies

The EPR spectrum of Cu(II)/MgO at X-band (ca. 9.8 GHz) and a temperature of 1.8 K is shown in Fig. 5.1 as an image plot, allowing the main features of the spectra to be visualized. The figure shows the spectra plotted vertically as a function of the magnetic field direction as it is rotated from the H$||$(001) direction ($\theta = 0°$) to the H$||$(110) direction ($\theta = 90°$), passing through the H$||$(111) direction at $\theta = 54.7°$.

In the H$||$(001) direction ($\theta = 0°$), the spectrum shows two sets of four hyperfine lines with strain broadened line-shapes. The expected derivative line-shape of the low field lines are distorted to be more positive (blue), while that of the high field lines to be predominately more negative (red) at this ($\theta = 0°$) and all other angles. In addition, the broadening of the outer hyperfine lines is greater than the innermost lines in both the high and low field sets. For rotations away from the (001) and (110) directions, the high field hyperfine set is composed of more than four lines, particularly evident at $\theta \sim 20°$ and $\theta \sim 80°$. These extra lines were previously assumed to be due to sample misalignment [9]. At H$||$(110) the high field set become so close together that the individual hyperfine transitions can no longer be resolved. Again, this was not understood, because the hyperfine parameters found from the analysis of the rest of the spectrum predicted the high field hyperfine lines to be well separated in this direction [9]. The spectra for the magnetic field approaching the (111) direction show an increase in the apparent intensity, while the line-width decreases. The spectra become complex, clearly showing more than two sets of four hyperfine lines. The main four positive low field and four negative

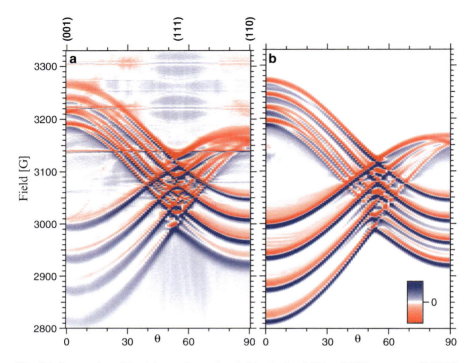

**Fig. 5.1** Image plot of the (**a**) experimental and (**b**) calculated X-band EPR spectrum (9.39759 GHz) at 1.8 K. The spectra are plotted against the field angle θ which takes the magnetic field direction from H||(001) to H||(110)

high field sets appear to have avoided crossings at H||(111). This complex behavior has not previously been observed; it has been reported that the spectrum coalesces to four hyperfine lines for H||(111) [9]. As the temperature is raised to 6 K, this anisotropic spectrum in Fig. 5.1a is replaced by a spectrum that does not show an angular dependence and can be analyzed in terms of the single isotropic $g-$ and $A$-values.

The new analysis [11] results in the simulation shown in Fig. 5.1b and accounts for the three main features observed in the spectra that were not clearly understood previously [9]:

1. The unresolved hyperfine lines at H||(110)
2. The > four hyperfine lines at particular values of θ and
3. The complexity of the spectrum at H||(111)

The key to understanding these features is to use numerical methods to simulate the spectrum rather than perturbation formulae which are based on assumptions such as $|A_1| > |A_2|$ which does not hold in this case. The first unexplained feature can be reproduced in the simulations with the correct choice of $A_1$ and $A_2$ which results in $|A_1| << |A_2|$. With these new hyperfine parameters there are particular field orientations when the quadrupole coupling, parameterized by $P_2$, has a dominate effect in inducing $\Delta M_I \neq 0$ type transitions, correctly accounting for the second unexplained feature. Finally, the third unexplained feature above is found to be due to the distribution of strain values being centered about a mean zero value, rather than a distribution about a non-zero mean value. At $H||(111)$ where the effects of random strain are far less, the spectra calculated numerically are far more complex than that given by perturbation formula.

### 5.2.2 Simulation of the EPR Spectrum

The numerical simulation of the EPR spectrum shown in Fig. 5.1b was made using the "eigenfields" method and details have been given [11]. Other methods, such as determining the energy levels by matrix diagonalization and stepping through the field to match the resonant frequency, could equally well be used. However, it should be realized that there are a large number of possibly transitions to search. Using 24 basis states (three vibronic, two electronic spin and four nuclear spin) there are 144 EPR type transitions, all of which must be examined for possible intensity.

The EPR spectrum can be modeled in terms of a cubic spin Hamiltonian appropriate to a $^2E(\Gamma_8)$ state operating on the lowest vibronic levels of a E⊗e Jahn-Teller system using the parameters given in Table 5.1. The parameter q is the "Ham" vibronic reduction parameter, which we fix as $q = 1/2$. We find that the first excited singlet is of $A_2$ symmetry, indicating that the $CuO_6$ center has the expected Jahn-Teller potential energy surface with three minima at three equivalent tetragonally elongated distorted octahedral geometries. Small random crystal strains have a dominant influence on the spectra and we find that the spectra can be reproduced with a Gaussian distribution of the random strain centered at zero, with a spread given by a half width at half height value of $2.0 \text{cm}^{-1}$. Our analysis differs from previous workers' in that both a much larger cubic anisotropy in the hyperfine values, as well as a nuclear quadrupole term ($P_2 = +11.0 \times 10^{-4} \text{cm}^{-1}$) are required to account for the hyperfine structure. The very weak transitions within the lowest excited singlet are observed directly, giving an estimate of the tunneling splitting as $\sim 4 \text{cm}^{-1}$. We conclude that the Cu(II)/MgO system can be described as an almost purely dynamic Jahn-Teller case, with most spectral features accounted for using a single isolated $\Gamma_8(^2E)$ vibronic state.

## 5.3 Symmetry Aspects of the $^2E(\Gamma_8)$ System

### 5.3.1 The Spin Hamiltonian for a $^2E(\Gamma_8)$ System

The $^2E$ state in $O_h$ symmetry consists of doubly degenerate spin and doubly degenerate orbital components, of which only the orbital components are coupled

**Table 5.1** Spin Hamiltonian parameters for Cu(II) in MgO

| T/K | Strain[a] | $\Delta_2/\text{cm}^{-1}$ | $g_1$ | $g_2$ | $A_1\ (\times 10^{-4}\ \text{cm}^{-1})$ | $A_2\ (\times 10^{-4}\ \text{cm}^{-1})$ | $P_2\ (\times 10^{-4}\ \text{cm}^{-1})$ |
|---|---|---|---|---|---|---|---|
| 77[9] | – | – | 2.1933 | – | $\pm 18.7$ | – | – |
| 1.3[9] | $\bar{\delta}/3\Gamma = 0.12$ | – | 2.1924 | 0.216 | $\pm 37.8$ | $\pm 22$ | – |
| 6.0 | – | – | 2.193 | – | $\pm 18.5$ | – | – |
| 1.8 | $\bar{\delta} = 0$ | $\Delta_2 = 4$ | 2.190 | 0.220 | $-19.0$ | $-84.0$ | $+11.0$ |
|  | $\delta_\Delta = 2\text{cm}^{-1}$ | | | | | | |

[a]The strain parameters are quantified by a Gaussian distribution with the mean value, $\bar{\delta}$, and half-width $\delta_\Delta$ see text, $\Delta_2$ is denoted by $3\Gamma$ in [9]

to non-totally symmetric vibrations by the Jahn-Teller effect. In the octahedral double point group [12] the four spin-orbit components of the $^2E$ state transform as $\Gamma_6(S = 1/2) \times \Gamma_3(E) = \Gamma_8$. Since $S = 3/2$ forms a basis for the irreducible representation $\Gamma_8$, it is possible to formulate the $^2E$ spin Hamiltonian in terms of an $S' = 3/2$ pseudo-spin basis. This has been done, for example, by Bleaney [13, 14] where it was noted that the spin operators, transforming as $\Gamma_4$, appear twice in the $\Gamma_8 \times \Gamma_8$ direct product, indicating that two g-values are in general required to describe the Zeeman effect in this cubic situation. To give rise to a linear Zeeman splitting, the non-standard spin Hamiltonian, including hyperfine interactions, is of the form

$$H = g_a\mu_B \sum_i S_i H_i + g_b\mu_B \sum_i S_i^3 H_i + A_a \sum_i S_i I_i + A_b \sum_i S_i^3 I_i \qquad (5.1)$$

For $\Gamma_8$ systems based on orbitally non-degenerate electronic states, the linear terms in Eq. 5.1 usually suffice, with the symmetry allowed $S_i^3 H_i$ terms, if needed, being very small [15].While this type of formalism could be applied to a $\Gamma_8$ state originating from a $^2E$ system, it became clear that it was more convenient to use a basis where spin and orbital parts are kept separate, in what we will call the $^2E$ basis [10].

A more natural basis for a $\Gamma_8$ state based on an orbital doublet $^2E$ is to use the orbital operators $U_\theta$ and $U_\varepsilon$ defined as

$$U_{A1} = \begin{pmatrix} 1 & 0 \\ 0 & 1 \end{pmatrix} \quad U_\theta = \begin{pmatrix} -1 & 0 \\ 0 & 1 \end{pmatrix} \quad U_\varepsilon = \begin{pmatrix} 0 & 1 \\ 1 & 0 \end{pmatrix} \qquad (5.2)$$

For an isolated $\Gamma_8$ state, the spin Hamiltonian will be of the form

$$H = G_{A1}U_{A1} + G_\theta U_\theta + G_\varepsilon U_\varepsilon \qquad (5.3)$$

where $G_{A1}$, $G_\theta$ and $G_\varepsilon$ are functions of the $S = 1/2$ spin operators, transforming as the $A_1$, $E_\theta$ and $E_\varepsilon$ irreducible representations, respectively, in the octahedral point group.

$$G_{A1} = g_1\mu_B\mathbf{H.S} + A_1\mathbf{I.S}$$

$$G_\theta = 1/2g_2\mu_B(3H_zS_z - \mathbf{H.S}) + 1/2A_2(3I_zS_z - \mathbf{I.S}) + 1/2P_2(3I_zI_z - \mathbf{I.I}) + S_\theta$$

$$G_\varepsilon = 1/2\sqrt{3}g_2\mu_B(H_xS_x - H_yS_y) + 1/2\sqrt{3}A_2(I_xS_x - I_yS_y)$$

$$+ 1/2\sqrt{3}P_2(I_xI_x - I_yI_y) + S_\varepsilon \tag{5.4}$$

## 5.3.2 The Relationship Between the $^2E$ and $S' = 3/2$ Bases

The spin Hamiltonians given in Eqs. 5.1 and 5.2–5.4 are mathematically equivalent and the parameters $g_a, g_b, A_a, A_b$ of the former can be related to the $g_1, g_2, A_1, A_2$ of the latter by evaluating the $4 \times 4$ matrices and comparing matrix elements explicitly.

One finds

$$g_a = 7/6g_1 - 13/12g_2 \quad g_b = -2/3g_1 + 1/3g_2 \tag{5.5}$$

with analogous expressions for $A_a, A_b$.

In the $^2E$ basis the isotropic part of Zeeman terms is bigger than the anisotropic part, i.e. $g_1 = 2.19 > g_2 = 0.22$ (Table 5.1); compared to the $S' = 3/2$ basis, where one finds $g_a = 2.32$, $g_b = -1.39$. Looking at the hyperfine terms, the opposite is true: $A_1 = -19 \times 10^{-4}\text{cm}^{-1}$, $A_2 = -84 \times 10^{-4}\text{cm}^{-1}$; while $A_a = -69 \times 10^{-4}\text{cm}^{-1}$, $A_b = -15 \times 10^{-4}\text{cm}-1$. The fact that $|A_1| > |A_2|$ does not hold has implications for perturbation formulae that assume this. At first glance in terms of these hyperfine values it may seem more appropriate to use the $S' = 3/2$ basis. However Ham has shown [8, 10] that the Jahn-Teller effect reduces the non-totally symmetric orbital parts of the spin Hamiltonian. This vibronic reduction cannot be easily implemented in the $S' = 3/2$ basis (5.1) where the spin and orbital parts are mixed.

## 5.3.3 Ham Reduction Factors

The Hamiltonian given in Eq. 5.3 operates on pure electronic states $\psi_\theta, \psi_\varepsilon$. We wish to study the vibronic states $\Psi_\theta, \Psi_\varepsilon$, and the matrix elements of $U_{A1}, U_\theta, U_\varepsilon$ in the vibronic basis are

$$<\Psi_\theta|U_{A1}|\Psi_\theta> = <\Psi_\varepsilon|U_{A1}|\Psi_\varepsilon> = 1 \quad <\Psi_\theta|U_{A1}|\Psi_\varepsilon> = <\Psi_\varepsilon|U_{A1}|\Psi_\theta> = 0$$

$$-<\Psi_\theta|U_\theta|\Psi_\theta> = <\Psi_\varepsilon|U_\theta|\Psi_\varepsilon> = q \quad <\Psi_\theta|U_\theta|\Psi_\varepsilon> = <\Psi_\varepsilon|U_\theta|\Psi_\theta> = 0$$

$$<\Psi_\theta|U_\varepsilon|\Psi_\theta> = <\Psi_\varepsilon|U_\varepsilon|\Psi_\varepsilon> = 0 \quad <\Psi_\theta|U_\varepsilon|\Psi_\varepsilon> = <\Psi_\varepsilon|U_\varepsilon|\Psi_\theta> = q$$

$$\tag{5.6}$$

where q is a real number $<1$ which represents a vibrational overlap. Comparing the matrix elements in (5.6) and (5.2), one finds that the spin Hamiltonian appropriate for vibronic states is then:

$$H = G_{A1}U_{A1} + q(G_\theta U_\theta + G_\varepsilon U_\varepsilon) \tag{5.7}$$

The $G_{A1}$, $G_\theta$ and $G_\varepsilon$ functions in Eq.(5.7) remain the same as those given in Eq.5.4. It can be seen that while (5.7) operates in the vibronic basis, the net effect of the vibrational parts of the vibronic functions is to reduce the electronic operators of $G_\theta$ and $G_\varepsilon$ symmetry by the factor q. This factor is known as the vibronic reduction or Ham factor [8, 10]. The values of q must be calculated numerically, but for strong Jahn-Teller coupling, q approaches the value of $1/2$.

## 5.4 Perturbation Expressions

The strain broadened nature of the spectra requires an EPR simulation to consider a large number of strain values. For that reason, the original work of this system used perturbation formula to develop simple analytic expressions for the calculation of the resonant fields [9]. It is instructive to examine these expressions and compare them to the numerically exact calculations. In addition, it allows a better understanding of which terms are important in determining the ground vibronic state. As many terms have similar magnitudes, care must be taken in using the perturbation formulae. Figure 5.2 shows the results of using two such formulae [8, 9, 13].

$$h\nu = (g_1\mu_B H + A_1 m) \pm q\,(g_2\mu_B H + A_2 M_I)\{1 - 3u\}^{1/2} \tag{5.8}$$

Here $u = \xi^2\eta^2 + \eta^2\zeta^2 + \zeta^2\xi^2$, with $\xi$, $\eta$, $\zeta$ being the directional cosines of the magnetic field and $M_I = -3/2, -1/2, +1/2, +3/2$, is the nuclear spin quantum number.

$$h\nu = g_1\mu_B H \pm qg_2\mu_B H\{1 - 3u\}^{1/2}\cos(\omega - \phi_s) + M_I A$$
$$A^2 = A_1^2 \pm 2qA_1A_2\{1 - 3u\}^{1/2}\cos(\omega - \phi_s)$$
$$+ 1/2q^2A_2^2[1 + \{1 - 3u\}^{1/2}\cos(\omega - 2\phi_s)]$$
$$\tan\omega = \sqrt{3}(\xi^2 - \eta^2)/(3\zeta^2 - 1) \tag{5.9}$$

Here the nuclear spins are not quantized in the direction of the strain, but along directions defined by both the strain direction, $\phi_s$, and the angle $\omega$ given in Eq. 5.9 (see also equations (21.74b) in Ref. [13]).

In both cases it is assumed that the strain determines the orbital nature of the ground state. Within each Kramers doublet thus determined, the Zeeman interaction

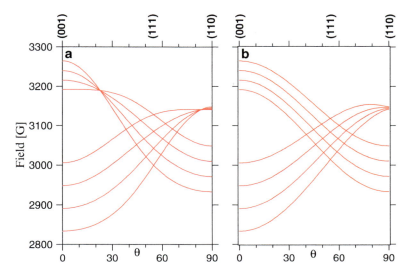

**Fig. 5.2** Comparison of the resonant field positions with perturbation expressions (**a**) Eq. 5.8 and (**b**) Eq. 5.9. The parameters given in Table 5.1 have been used

quantizes the spin direction (but does not mix the two Kramers doublets). In the formula given by Eq. 5.8 shown in Fig. 5.2a, the nuclear spin quantum numbers are assumed diagonal within each spin component (i.e. they do not mix), whereas the formula given by Abragam and Bleaney [13] and shown in Fig. 5.2b uses an effective field for the quantization of the nuclear spin [16]. This includes the off-diagonal hyperfine terms within each Kramers doublet. In Fig. 5.2b the four hyperfine lines are calculated not to cross, as seen experimentally.

The $M_I$ values are good quantum numbers for $H \| (001)$. When the hyperfine lines are predicted to cross in Fig. 5.2a at $\theta \sim 22°$ and $82°$, the nuclear spin states are undergoing strong mixing. The nuclear quadrupole term has a large effect at these particular orientations and causes many more than the four $\Delta M_I = 0$ type transitions to have intensity. Neither of perturbation formulae in Fig. 5.2 can reproduce these spectral features, which are observed experimentally and found in the exact eigenfields calculation (Fig. 5.1b).

## 5.5 The Crucial Role of Strain in the Spectra

The strain of a $^2E$ Jahn-Teller system can be given in terms of the tetragonal and orthorhombic components $S_\theta$ and $S_\varepsilon$ in Eqs. 5.4 and 5.7

$$H_{ST} = q(S_\theta U_\theta + S_\varepsilon U_\varepsilon) \quad (5.10)$$

where, again, the reduction factor q reduces the magnitude of the parameters from the values they would have if they were acting on an electronic basis only. The strain can be equivalently given in terms of a magnitude, $\delta_s$, and direction, $\phi_s$: $S_\theta = \delta_s \cos \phi_s$, $S_\varepsilon = \delta_s \sin \phi_s$.

The effect of a small strain perturbation on the vibronic wave-functions of an $E \otimes e$ potential energy surface can be very large. They are very easily localized within the separate minima, implying that very small random imperfections of the crystal will have a large effect on the experiment. We model this by assuming that the random crystal strain has a distribution that is equally likely for all values of $\phi_s$ and has a Gaussian distribution in the radial direction. We find that only a small spread of values, $\delta_\Delta$ (full width at half height) $= 2\,\mathrm{cm}^{-1}$ about a *zero* mean value, is enough to account for the experimental observations.

The strain within the lowest vibronic E state explains most of the strain broadening, in particular the "powder-like" appearance of the single crystal spectra. It requires coupling with higher-lying vibronic singlet states to account for the *relative* broadening within each set of hyperfine lines and this can be used to determine the symmetry of the excited singlet and hence the position of the minima on the potential surface. We examine these points in turn below.

### 5.5.1 Large Strain and the Static Jahn-Teller Effect

Figure 5.3 shows the effect of introducing a large strain into the $^2E$ spin Hamiltonian. We neglect the hyperfine lines for simplicity and Fig. 5.3a give the resonant field positions with zero strain and shows the effect of q in reducing the cubic anisotropy. The solid lines gives the X-band resonant field positions for $q = 1$ and the short dashed lines for $q = 1/2$ calculated for $g_1 = 2.19$, $g_2 = 0.22$. The long dashed line is the position of the singlet state (A) seen here to be isotropic and independent of angle with zero strain. We have labeled the two "branches" appearing in the spectrum as $\theta$ and $\varepsilon$. Note that these only refer to the actual $\theta$ and $\varepsilon$ Kramers doublets at $H\|(001)$, and these states are mixed at other angles, however they serve as convenient labels for future reference. When the strain is large, then one of the two components ($\theta$ or $\varepsilon$) mixes with the singlet state as shown in Figure 5.3b, c, depending on whether the excited singlet is of $A_2$ or $A_1$ symmetry, respectively.

For an $E \otimes e$ potential energy surface with three equivalent minima at tetragonally elongated (compressed) geometries, the excited singlet will be of $A_2(A_1)$ symmetry [2]. For a singlet state of $A_2$ symmetry, the $A_2$ (dashed) and $\varepsilon$ branch (short dash) will mix and "repel" each other, moving in the directions shown by the arrows in Fig. 5.3b. At $H\|(001)$, one resonant field ($\varepsilon$ branch) moves lower and the $A_2$ branch moves higher to become equal to the resonant fields of the $\theta$ branch. These resonant field positions correspond to $g_\| > g_\perp$ as expected for a tetragonally elongated complex. For a tetragonally compressed complex, the lowest singlet is

5 The Dynamic Jahn-Teller Effect in Cu(II)/MgO

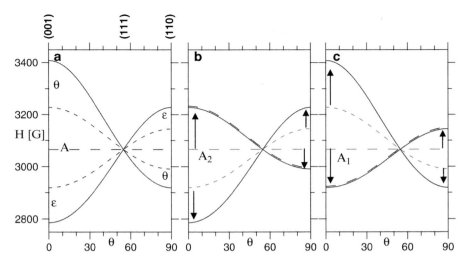

**Fig. 5.3** The resonant field positions (neglecting the hyperfine coupling). (**a**) The dynamic case, zero strain. The cubic anisotropy with vibronic reduction ($q = 1/2$; *dotted line*), and no vibronic reduction (*solid line*). The *dashed line* is the excited vibronic singlet. (**b**) The static case for an $A_2$ singlet lowest. The strain mixes the $\varepsilon$ and $A_2$ branches. (**c**) The static case for $A_1$ singlet lowest. The strain mixes the $\theta$ and $A_1$ branches. In (**b**) and (**c**) in the limit of large strain (*solid lines*), the spectra are the superposition of three axial spectra. The *arrows* indicate the shifts for increasing strain

of $A_1$ symmetry and mixes with the $\theta$ branch, the states moving in the directions indicated by the arrows in Fig. 5.3c. The resonant fields of the $\theta$ branch move higher and those of the $A_1$ branch move lower to become equal with the $\varepsilon$ branch, corresponding to $g_\| < g_\perp$.

The relationship between the tetragonal $g$-values and the cubic $g$-values is given by

$$g_\| = g_1 + g_2 \quad g_\perp = g_1 - 1/2\, g_2 \quad A_2 \text{ singlet (elongated)}$$
$$g_\| = g_1 - g_2 \quad g_\perp = g_1 + 1/2\, g_2 \quad A_1 \text{ singlet (compressed)} \qquad (5.11)$$

The experimental spectrum for rotation from $H\|(001)$ to $H\|(110)$, as shown in Fig. 5.1a, is symmetrical about a mean position when averaged over the hyperfine values similar to the cubic case as shown in Fig. 5.3a. This is very different for the spectrum of three superimposed tetragonal species as shown in Fig. 5.3b, c, and demonstrates that we are dealing with small values of the strain. In addition, the experimental spectrum only show the two branches expected for a single species with $^2E$ cubic anisotropy for an arbitrary field direction, rather than the superposition of spectra of three tetragonal species which would be seen if this were a case of large strain and a static Jahn-Teller effect.

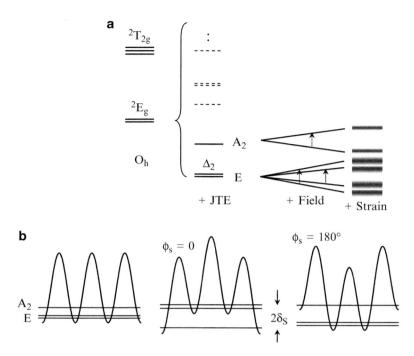

**Fig. 5.4** Hierarchy of energy levels with (**a**) small strain and (**b**) large strain that corresponds to a tetragonal elongation ($\phi_S = 0$) and a compression ($\phi_S = 180°$)

## 5.5.2 The Vibronic Energy Levels in the Presence of Strain

It may be appropriate at this point to discuss the vibronic energy levels associated with cases of small and large strain. Figure 5.4a shows the hierarchy of energy levels for a Cu(II) complex in $O_h$ symmetry. On going from left to right, one starts with the pure electronic states at $O_h$ symmetry. If the vibrational part of the wavefunction is included, then action of the Jahn-Teller effect is to give a ladder of vibronic levels (Fig. 5.4a center). Finally when the spin part of the wavefunction is included and a magnetic field applied, each of the vibronic Kramers doublets are split to give the EPR transitions indicated. The resonant fields of these transitions are very orientation dependent except for that of the singlet state which is isotropic and falls between the two resonant field positions of the E vibronic state as shown in Figs. 5.3a and 5.4a. A distribution of small random strains will then "smear" these levels and mix the vibronic functions and this will be discussed further below.

The Mexican hat Jahn-Teller potential energy surface results in the three minima shown in Fig. 5.4b left. Solving for the vibronic levels within this surface, one calculates a vibronic E state lowest in energy with a close lying singlet $A_2$ to higher energy. The energy separation, $\Delta_2$, between the E and $A_2$ states (sometimes called "$3\Gamma$") [9] is a tunneling splitting and should not be confused with the splitting of the $^2E_g$ state by a tetragonal distortion from $O_h$ symmetry. Considering the static distortion at one of the three minima of the Jahn-Teller surface, the splitting of the $^2E_g$ state corresponds to the separation of the two sheets of the Jahn-Teller surface $\sim 4E_{JT}$, which is much larger that the separation of the lowest vibronic levels. There are many other vibronic levels besides the lowest E-A$_2$ states, as indicated by the dashed lines in Fig. 5.4a but they will not concern us in discussing the EPR spectrum.

The case of large strain is shown in Fig. 5.4b. For a tetragonal elongation ($\phi_S = 0$), one minimum is lowered with respect to the other two by an energy $2\delta_S$. Similarly for a strain corresponding to a tetragonal compression ($\phi_S = 180°$) the levels are again separated by $2\delta_S$, but this time two equal minima are below the higher third minimum. The vibronic levels also follow this energy ordering of the minima, but the two vibronic functions that are localized in the two minima with the same energy, are non-degenerate and their energy separation represents the tunneling splitting between the double minima. Consider the lowest minimum (Fig. 5.4b center) corresponding to a tetragonal elongation, one would expect the electronic part of the localized vibronic function to have an $x^2 - y^2$ type ground state. The vibronic levels with the orthogonal $z^2$ type electronic function occur at $\sim 4E_{JT}$ to higher energy and are associated with the upper sheet of the $E \otimes e$ Jahn-Teller surface. The spectacular action of the strain shown in Figs. 5.3 and 5.4, refers to the lowest vibronic levels on the lower-energy JT surface solely. Figure 5.3b shows the action of the strain for a $x^2 - y^2$ type ground state (A$_2$ vibronic singlet) and Fig. 5.3c for a $z^2$ type ground state (A$_1$ vibronic singlet).

### 5.5.3 Strain Broadening, or "Powder-Like" Single Crystal Spectra

We assume that the strain is randomly oriented in $S_\theta$, $S_\varepsilon$ space such that $\phi_s$ can assume any value in the range $0 \leq \phi_s \leq 2\pi$ with equal probability. Lets now examine the effect on the resonant field positions as a function of $\phi_s$ for a small magnitude of the strain ($\delta_s = 0.05\text{cm}^{-1}$) at particular field orientations using numerical methods. Figure 5.5a and b show this dependence on $\phi_s$ for H$||$(001) and H$||$(110) respectively. As well as the $\theta$ and $\varepsilon$ branches, the resonant field positions for the excited singlet state are also shown in Fig. 5.5a. This isotropic signal shows no dependence on $\phi_s$.

There are several points of interest in Fig. 5.5. First, the resonant field positions in both orientations are at the same position for $\phi_s = 0$ and $\phi_s = \pi$. That is, in this dynamic Jahn-Teller case, the spectra are almost identical for a small strain, independent on whether the strain is in the direction of a tetragonal elongation or a compression. Compare this to the very different spectra expected for a static Jahn-Teller effect with large strain (Fig. 5.3). The only difference is that the energy ordering of the states is reversed, meaning that there will be very small intensity

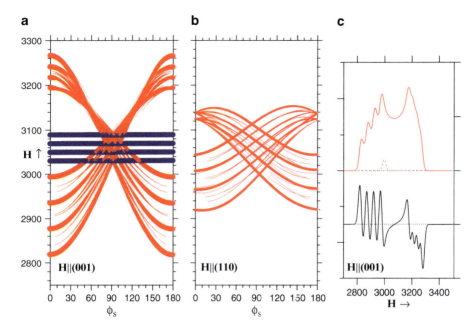

**Fig. 5.5** The calculated resonant field positions as a function of the strain angle $\phi_s$ for (**a**) $H||(001)$ and (**b**) $H||(110)$. The same parameters from Table 5.1 have been used. (**c**) The summation over $\phi_s$ with Gaussian line-shapes (dotted) for $H||(001)$ in absorption (*top*) and the derivative (*bottom*)

effects due to population differences. The energy of the θ state will be lowest for $\phi_s = 0$ (i.e. the Cu(II) ground state will have an electron hole in the higher $\varepsilon(x^2 - y^2)$ orbital) and the energy of the ε state will be lowest for $\phi_s = \pi$ (Cu(II) ground state $\theta(z^2)$ orbital).

The second point to note is that the resonant fields vary continuously as $\phi_s$ varies between 0 and $\pi$ such that the θ branch at $\phi_s = 0$ becomes the ε branch at $\phi_s = \pi$, and vice versa. If one averages over all these values, then one may at first expect the spectra to be a continuum between the extreme values as the individual transitions for particular values of $\phi_s$ overlap. What actually happens is shown schematically in Fig. 5.5c for $H||(001)$, where the Gaussian line-shapes for each of the resonant field positions in Fig. 5.5a are summed over the $0 \leq \phi_s \leq 2\pi$ values. The derivative of the summed absorption spectrum (Fig. 5.5c bottom) then has the distorted shape more often associated with powder spectra that are averaged over all molecular orientations (or equivalently: averaged over all field directions). However the "powder-like" shape of the spectrum shown at the bottom of Fig. 5.5c is due to the strain broadening of a single field direction $H||(001)$ of an oriented crystal. The "powder-like" description of the spectra is given in inverted commas as, unlike a real powder spectrum, the spectra is still very anisotropic, showing a distinct angular dependence of the field direction. However, for a particular field orientation, the spectrum is distorted in a manner which resembles a "powder-like" lineshape as shown in Fig. 5.5c.

Interestingly, the peak positions in the average spectrum correspond to the calculated extreme resonant field positions for $\phi_s = 0$ or $\pi$. This is where the curves of the resonant field positions as a function of $\phi_s$ have zero slope. This is not the case for all field orientations. For the H$||$(110) direction shown in Fig. 5.5b, the highest field hyperfine lines are calculated to occur at $\phi_s \sim 30°$, $150°$; not $\phi_s = 0°$, $180°$.

### 5.5.4 The Sign of the Warping

The variation of the broadening of hyperfine lines of the low temperature spectra arise due to the different sensitivity of the hyperfine lines to the strain broadening. As shown previously for large strain in Fig. 5.3b, an excited singlet of $A_2$ symmetry will mix with the $\varepsilon$ branch. This will result in a broadening of the $\varepsilon$ branch; which are the low field lines for H$||$(001) and the high field lines for H$||$(110). This is what is observed experimentally (Fig. 5.1a) and this then confirms that the excited singlet state has $A_2$ symmetry and the underlying E $\otimes$ e potential energy surface has the minima at tetragonally elongated rather than compressed geometries.

As the magnitude, $\delta_s$, of the strain increases then there becomes a difference in the resonant field positions for $\phi_s = 0$ and $\phi_s = \pi$, ie the plots in Fig. 5.5 are then not quite symmetrical. (Such plots are given in Fig. 9 of reference [11]) For an $A_2$ excited singlet, the low field lines at H$||$(001) and the high field lines at H$||$(110) are broadened, but further, the outermost hyperfine lines are broadened faster than the innermost ones.

### 5.5.5 Decoupling the Complications in the (111) Direction

Figure 5.6 shows two calculations using the "exact" eigenfields method of calculating all possible resonant field positions and intensities as a function of the field direction $\theta$. In both cases, the parameter values of Table 5.1 have been used and the two calculations differ only in that in Fig. 5.6a, no strain has been included, while in Fig. 5.6b, a small non-zero value of the strain ($\delta_s = 0.05\text{cm}^{-1}$, $\phi_s = 0°$) has been used. Paradoxically, a very much more complex spectrum is seen for the high symmetry case, while the inclusion of a low symmetry term greatly simplifies the spectrum, particularly when H$||$(111). As anticipated by Mary O'Brien (as given in a footnote in Ham [8]) the situation can be very complicated as even when the anisotropic terms due to $g_2$, $A_2$, $P_2$ become zero for H$||$(111), off-diagonal elements neglected in the perturbation formula become important.

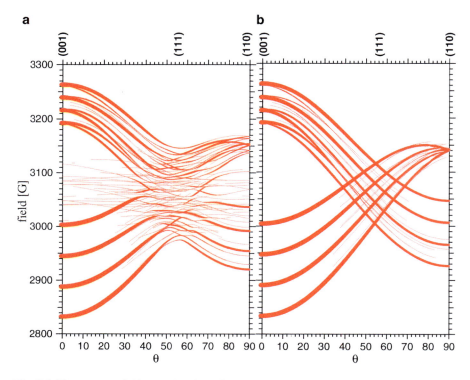

**Fig. 5.6** The resonant fields as a function of the angle θ rotated from (001) to (110) using the parameters given in Table 5.1. The line thickness indicates the transition intensity. (**a**) $\delta_s = 0$ and (**b**) $\delta_s = 0.05\,\text{cm}^{-1}$, $\phi_s = 0°$

## 5.6 Outstanding Questions – Future Experiments

### 5.6.1 Quantifying the Tunneling Splitting/Strain

The parameters given in Table 5.1 found from fitting the EPR spectra do not directly yield the coupling constants and barrier heights of the Jahn-Teller potential energy surface. We know that we are in a region of a strong Jahn-Teller coupling as q is close to 1/2. If the energy of the excited singlet state (or tunneling splitting) were known, then the barrier heights between the three equivalent minima could be determined. However, as we have seen, the main features of the EPR spectrum can be reproduced with an isolated $^2E(\Gamma_8)$ vibronic ground state. It is only in the details of the broadening that one can see the influence of the first excited singlet vibronic state. The slight broadening of the low field set of lines at H||(001) and the high field set at H||(110) is due to the strain coupling with the lowest singlet vibronic state that must have $A_2$ symmetry. However, the energy of this lowest singlet ($\Delta_2$) cannot be directly determined as it is the quantity $\delta_s/\Delta_2$ which is important in coupling

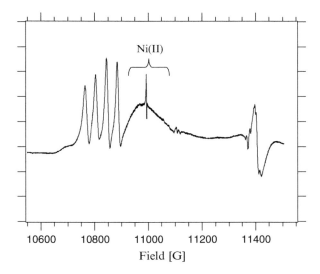

**Fig. 5.7** The Q-band spectrum (34.0401 GHz) of Cu(II)/MgO at 1.8 K with H∥(110). The sharp and broader underlying features indicated are due to Ni(II) ions [20]

the $A_2$ and E states. The direct determination of $\Delta_2$ from observing the expected isotropic EPR spectrum of the $A_2$ state by thermal population is problematic as the rapid relaxation with increasing temperature renders the whole spectrum isotropic at ∼4 K. Very weak signals that may be due to the population of the $A_2$ state have been seen, which would imply a value of $\Delta_2 \sim 4\,\mathrm{cm}^{-1}$ [11]. But extra experimental confirmation is desirable as a much larger value ($\Delta_2 \sim 235\,\mathrm{cm}^{-1}$) has been obtained from CASPT2 cluster calculations [4, 17].

We have recently attempted [18] to observe the fluorescence from the $^2T_{2g}$ excited state of Cu(II)/MgO as has previously been seen for Cu(II) doped into cubic and near cubic fluoride hosts [19]. It was hoped that the $\Delta_2$ splitting may be resolvable in the zero phonon lines in the fluorescence spectrum, but unfortunately it was found that our samples have a previously unknown very small amount of Ni(II) impurity. This impurity prevents Cu(II) fluorescence as the lowest energy $^3T_{2g}$ Ni(II) states fall below the $^2T_{2g}$ Cu(II) states.

The Ni(II) impurity is not seen in the X-band EPR spectrum (Fig. 5.1), but appears in the Q-band spectrum at high microwave power, as a very distinctive single sharp signal on a broader background signal as shown in Fig. 5.7. This spectrum has been previously observed [20] and described as a sharp $|-1>\rightarrow|+1>$ two photon transition on top of the broader $|-1>\rightarrow|0>$ and $|0>\rightarrow|+1>$ single photon transitions. The strain results in a distribution of the zero field splitting D values that separates the $|0>$ and $|\pm1>$ states. The energy of the $|-1>\rightarrow|+1>$ transition is unaffected by this "D-strain". The width of the spectrum $\Delta H$ is related to the width of the D distribution by $\Delta D = \mu_B \Delta H$. In the absence of vibronic reduction, the strain in MgO for a non-Jahn-Teller system will split the $e_g(x^2 - y^2, z^2)$ d-orbitals by the amount $\Delta_e = 2S_\theta$. This tetragonal splitting, $\Delta_e$, will give rise to a zero field splitting D value can be obtained from perturbation theory:

$$D \sim \xi^2 \Delta_e / \Delta_o^2 \qquad (5.12)$$

Here $\xi$ is the spin–orbit coupling constant and $\Delta_o$ is the octahedral ligand field splitting, which for this system take the values of 625 and 8300cm$^{-1}$ respectively. As a rough approximation the $\Delta H \sim 200$ G width of the Ni(II) $|0> \rightarrow |\pm 1>$ transitions observed in the Q-band spectra (Fig. 5.7) correspond to a spread in the strain values $\delta_\Delta \sim \mu_B \Delta H \Delta_o^2 / 2\xi^2 = 0.8 \, \text{cm}^{-1}$. This rough estimate is of the correct order of magnitude to the $\delta_\Delta \sim 2\text{cm}^{-1}$ inferred from modeling the X-band spectra [11]. While this very approximate approach requires a more complete simulation of the Ni(II) spectrum, this way of determining the strain distribution is appealing as it gives an estimate of the strain on the actual crystal being used for the Cu(II)/MgO study. This then precludes any sample dependence of the crystal strain that may be caused by different preparation techniques.

### 5.6.2 Other Systems

The obvious other dynamic Jahn-Teller system that may require further investigation is Cu(II) doped into CaO [21]. This was first studied at the time when Cu(II)/MgO was being studied, and was interpreted with the same formalism [9,22]. Cu(II)/CaO distinguishes itself as one of the very few examples of a Cu(II) Jahn-Teller system where the minima of the potential energy surface are thought to be at a compressed, not elongated, geometry. While the spectra was able to be understood satisfactorily using the three state model. There remain a number of questions. The isotropic hyperfine value at low temperature $|A_1| = 31.2 \times 10^{-4} \text{cm}^{-1}$ is different to the value at high temperature $|A_1| = 21.8 \times 10^{-4} \text{ cm}^{-1}$, whereas they would be expected to be the same as is seen here for Cu(II)/MgO. In addition, it has been reported [23] that $g_1$ and $A_1$ values show temperature dependence over the 4–77 K range.

## 5.7 Conclusions

We have shown that the Cu(II) centre in MgO is characterized by a Jahn-Teller ground state potential energy surface that has three equivalent minima at elongated octahedral geometries. The low symmetry perturbation to this $E \otimes e$ surface is small, and can be modeled by a distribution of strains about a zero mean value. This results in an EPR spectrum characteristic of a vibronic E ground state of a true dynamic Jahn-Teller system.

**Acknowledgments** We would like to acknowledge the many years friendship and collaboration with Phillip Tregenna-Piggott, you will be sadly missed.

# References

1. Jahn HA, Teller E (1937) Proc R Soc (Lond)A 161:220
2. Bersuker IB (2006) The Jahn-Teller effect. Cambridge University Press, Cambridge
3. Riley MJ (2001) Topics Curr Chem 214:57
4. Garcia-Fernandez P, Sousa C, Aramburu JA, Barriuso MT, Moreno M (2005) Phys Rev B 72:155107
5. Orton JW, Auzins P, Griffiths JHE, Wertz JE (1961) Proc Phys Soc Lond 78:554
6. Coffman RE (1968) J Chem Phys 48:609
7. Bersuker IB (1963) Soviet Phys JETP 17:836
8. Ham FS (1968) Phys Rev 166:307
9. Reynolds RW, Boatner LA, Abraham MM, Chen Y (1974) Phys Rev B 10:3802
10. Ham FS (1972) In: Geschwind S (ed) Electron paramagnetic resonance. Plenum, New York, p 1
11. Riley MJ, Noble CJ, Tregenna-Piggott PLW (2009) J Chem Phys 130:104708
12. Koster GF, Dimmock JO, Wheeler RG, Statz H (1963) Properties of the thirty-two point groups. MIT Press, Cambridge
13. Abragam A, Bleaney B (1986) Electron paramagnetic resonance of transition ions. Dover, New York
14. Bleaney B (1959) Proc Phys Soc (Lond) A73:939
15. Ham FS, Ludwig GW, Watkins GD, Woodbury HH (1960) Phys Rev Lett 5:468
16. Weil JA, Anderson JH (1961) J Chem Phys 35:1410
17. Garcia-Fernandez P, Trueba A, Barriuso MT, Aramburu JA, Moreno M (2010) Phys Rev Lett 104:035901
18. Riley MJ, Hall J, Krausz ER (unpublished 2010)
19. Dubicki L, Krausz ER, Riley MJ (1989) J Am Chem Soc 111:3452; Dubicki L, Riley MJ, Krausz ER (1994) J Chem Phys 101:1930
20. Orton JW, Auzins P, Wertz JE (1960) Phys Rev Lett 4:128
21. Tregenna-Piggott PLW (2003) Adv Quantum Chem 44:461
22. Coffman RE, Lyle DL, Mattison DR (1968) J Phys Chem 72:1992
23. Low W, Suss JT (1963) Phys Lett 7:310

# Chapter 6
# Dynamic and Static Jahn-Teller Effect in Impurities: Determination of the Tunneling Splitting

**Pablo Garcia-Fernandez, A. Trueba, M.T. Barriuso, J.A. Aramburu, and Miguel Moreno**

**Abstract** In this paper we review the concepts of dynamic and static Jahn-Teller (JT) effects and study their influence in the electron paramagnetic resonance (EPR) spectra of transition metal impurities in wide gap insulators. We show that the key quantities involved in this problem are the tunneling splitting, usually denoted $3\Gamma$, and the splitting, $\delta$, between $\sim 3z^2 - r^2$ and $\sim x^2 - y^2$ energy levels due to the random strain field which is present in every real crystal. It is pointed out that in the $E \otimes e$ JT problem the kinetic energy of nuclei involved in the ground state plays a key role for understanding the actual value of $3\Gamma$ and thus the existence of dynamic JT effect. The results of ab initio calculations on a variety of JT systems show that $3\Gamma$ values span a much larger range than previously suggested. In particular, we find that $3\Gamma = 235\,\text{cm}^{-1}$ for $MgO:Cu^{2+}$ while $3\Gamma = 10^{-4}\,\text{cm}^{-1}$ for $KCl:Ag^{2+}$. We also show that the dynamic JT effect can only appear for such large values of $3\Gamma$ as those found in $MgO:Cu^{2+}$ since usual strain fields lead to $\delta$ values of the order of $10\,\text{cm}^{-1}$ that would otherwise localize the system in only one of the JT wells. The present results explain satisfactorily why the JT effect for $Cu^{2+}$ and $Ag^{2+}$ impurities in MgO is dynamic while static for $Ag^{2+}$-doped CaO and alkali chlorides. The origin of such a difference is discussed in detail.

---

P. Garcia-Fernandez (✉) • A. Trueba • J.A. Aramburu • M. Moreno
Departamento de Ciencias de la Tierra y Física de la Materia Condensada, Universidad de Cantabria, Avda. de los Castros s/n. 39005 Santander, Spain
e-mail: garciapa@unican.es; truebaa@unican.es; antonio.aramburu@unican.es; morenom@unican.es

M.T. Barriuso
Departamento de Física Moderna, Universidad de Cantabria,
Avda. de los Castros s/n. 39005 Santander, Spain
e-mail: barriust@unican.es

M. Atanasov et al. (eds.), *Vibronic Interactions and the Jahn-Teller Effect: Theory and Applications*, Progress in Theoretical Chemistry and Physics 23, DOI 10.1007/978-94-007-2384-9_6, © Springer Science+Business Media B.V. 2012

## 6.1 Introduction

Structural instabilities play a key role for understanding the properties displayed by molecules and solids [1–3]. Indeed the geometrical structure exhibited by a compound at low temperatures is often not the one expected on simple grounds as it can be unstable with respect to symmetry lowering distortions. For instance, while $KMF_3$ compounds (M = Mg, Zn, Ni) are all cubic, a similar lattice like $KMnF_3$ is not cubic but tetragonal at low temperatures [4].

The Jahn-Teller (JT) effect belongs to the domain of structural instabilities. It implies that if a non-linear compound in a given electronic state is assumed to display a geometrical structure involving orbital degeneracy such a structure is unstable with respect to symmetry reducing distortions [1]. Despite instabilities driven by the JT effect are thus restricted to systems with *perfect* orbital degeneracy it has attracted a great deal of research due to the *different* experimentally observed phenomena [5, 6]. In particular the JT effect has been widely explored in the case of substitutional impurities in cubic insulating lattices. It is well known that while impurities like $Mn^{2+}$ or $Ni^{2+}$ in $KMgF_3$ display a cubic symmetry this is not the case for $d^9$ impurities ($Cu^{2+}$, $Ag^{2+}$, $Ni^+$) in cubic fluoroperovskites [5, 7–11]. In fact, in the last systems the electronic ground state would be $^2E(t_{2g}{}^6e_g{}^3)$ if the local symmetry is strictly octahedral. Nevertheless, the local geometry observed experimentally at low temperatures is tetragonal as a result of the so called static JT effect leading to an orbitally singlet ground state. Electron Paramagnetic Resonance (EPR) measurements carried out on $d^9$ and low spin $d^7$ impurities in cubic halide lattices, not perturbed by any close defect, unambiguously prove the existence of a static JT effect [1, 6–15]. However, a quite different behaviour is observed in a few cases where the EPR angular pattern at low temperatures surprisingly reveals the existence of a *local cubic* symmetry. This different manifestation of the JT effect observed in systems like $Cu^{2+}$- or $Ag^{2+}$-doped cubic MgO is denoted dynamic [16–19]. In these cases the measured angular pattern looks similar to that expected for a pure $^2E(t_{2g}{}^6e_g{}^3)$ electronic state although there are some subtle differences between both situations as discussed in Sect. 6.2.

A crucial issue in the realm of the $E \otimes e$ JT effect is thus to understand what the actual microscopic reasons for observing either a dynamic or a static JT effect through EPR spectra are. It was early pointed out by Ham [20] that the dynamic or static character exhibited by $d^9$ and low spin $d^7$ ($t_{2g}{}^6e_g{}^1$ ground state configuration) impurities in cubic lattices strongly depend on the actual value of the so-called tunneling splitting, $3\Gamma$. Despite the relevance of that quantity, firstly introduced by Bersuker [21], its actual value is far from being established from available experimental information for the explored $E \otimes e$ JT systems. A parallel situation is often encountered when looking at other key quantities like the stabilization energy, $E_{JT}$, or the barrier, B, among equivalent distorted configurations. As a relevant example, the values of $E_{JT}$ estimated from the analysis of *available* experimental data for $Cu^{2+}$-doped MgO or CaO lattices go from 350 to $16,000\,cm^{-1}$ [18, 22–24]. Seeking to determine what the actual values of $E_{JT}$, B and $3\Gamma$ quantities are, a good step forward has been obtained in recent times from *ab initio* calculations [25–30]. Accordingly, the trend displayed by the calculated $3\Gamma$ values for a variety of $E \otimes e$ JT systems [30] sheds light on the conditions for observing experimentally a dynamic behavior.

The present article is aimed at clarifying some main issues concerning the JT effect displayed by impurities in cubic insulating lattices. It is worth noting that active electrons from a transition metal impurity, M, in insulators are essentially *localized* in the $MX_N$ complex formed with the N nearest anions [10]. The existence of electronic localization, which is the fingerprint of any insulating material [31, 32], thus makes possible to understand most of problems of impurities considering *only* the $MX_N$ complex itself and a buffer involving some neighbor atoms.

The present work is arranged as follows. In Sect. 6.2 the differences between EPR spectra corresponding to static and dynamic JT effect are shortly recalled. For the sake of clarity the main characteristics of EPR spectra expected for a pure $^2E$ electronic state are also briefly exposed in that section. The theoretical basis of the JT effect for impurities in a *perfect* crystal is reviewed in Sect. 6.3, while Sect. 6.4 deals with the effects of unavoidable random strains present in every *real crystal*. Finally, in Sect. 6.5 we present numerical values for the main quantities involved in JT problems for a number of $d^7$ and $d^9$ transition-metal impurities in insulators, including the value of the tunneling splitting.

## 6.2 Static and Dynamic $E \otimes e$ Jahn-Teller Effect: Differences in EPR Spectra

In the case of impurities in cubic insulating lattices the existence of a dynamic JT effect at low temperatures has been proved essentially through magnetic resonance techniques [16–19, 33]. For this reason we shall briefly expose the main differences between EPR spectra characterizing either a static or a dynamic JT effect. For simplicity we shall only consider the dependence of the gyromagnetic factor with respect to the applied magnetic field, **H**.

### 6.2.1 Static Jahn-Teller Effect

Let us consider a substitutional $d^9$ or $d^7$ impurity in a cubic lattice like NaCl or $KMgF_3$ (Fig. 6.1). When a static JT effect takes place EPR spectra at low temperatures show the *simultaneous* existence of three different tetragonal centres which, in general, are not magnetically equivalent (Fig. 6.2). The principal $C_4$ axis of each centre is one of the three [100], [010] or [001] axes of the cube. It should be noted that the measured concentration of three centers, if no external strain is

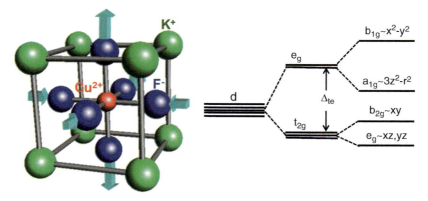

**Fig. 6.1** Illustration of the complex formed in KMgF$_3$ when doped with Cu$^{2+}$ ions (*left*) and the splitting undergone by the d-levels of the impurity from the free-ion to an axially elongated tetragonal complex passing through a octahedral configuration (*right*)

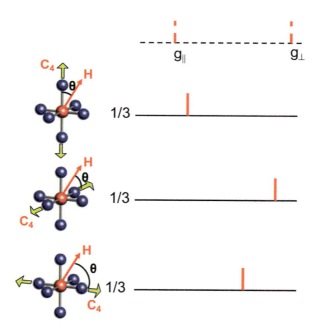

**Fig. 6.2** Illustration of the EPR signal of a S = 1/2 system presenting a static JT effect. At the top the position of the EPR signal when the magnetic field, **H**, is parallel or perpendicular to the C$_4$ axis is shown. Below are depicted the contributions coming from the three centers (whose C$_4$ axis is one of the three principal axes of the octahedron) for a given orientation of **H**. Due to the random character of internal strains the number of centres with a given C$_4$ axis is one third of the total number of impurities

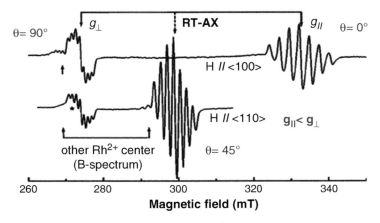

**Fig. 6.3** Experimental EPR spectrum of the NaCl:Rh$^{2+}$ center (Taken from [13]) presenting a static JT effect

applied, is the same. The effective spin Hamiltonian, $H_S$, for such centres with $S = 1/2$ is thus that corresponding to a tetragonal complex

$$H_S = \beta\{g_\| H_Z S_Z + g_\perp (H_X S_X + H_Y S_Y)\} \quad (6.1)$$

Here the meaning of symbols is standard and thus OZ means the $C_4$ axis and $\beta$ is the Bohr magneton. According to (6.1) the gyromagnetic factor, $g$, only depends on the angle, $\Omega$, between $C_4$ and **H**

$$[g(\Omega)]^2 = g_\|^2 \cos^2\Omega + g_\perp^2 \sin^2\Omega \quad (6.2)$$

and thus $g_\|$ and $g_\perp$ are the *extreme* values of the observed gyromagnetic factor.

An example of static JT effect is shown on Fig. 6.3 for the NaCl:Rh$^{2+}$ system when the charge compensation is far from the RhCl$_6^{4-}$ complex [12]. If **H**$\|$ [001], one of the three centres has $g = g_\|$ while in the other two $g = g_\perp$. Under an *elongated* tetragonal distortion the orbital degeneracy in $e_g$ is removed and the unpaired electron is placed in $|\theta\rangle \equiv |3z^2 - r^2\rangle$. Thanks to the admixture of this level with $yz$, $xz$ levels coming from $t_{2g}$ via the operators $l_x$ and $l_y$ involved in the spin-orbit coupling the differences between $g_\|$ and $g_\perp$ come out. If, only for simplicity, we discard in this analysis the covalency then $g_\|$ and $g_\perp$ are given, in second order perturbations, by [5]

$$g_\| = g_0 \quad ; \quad g_\perp = g_0 + 6\xi/\Delta_{te} \quad (6.3)$$

Here $\xi$ means the spin-orbit coefficient of the impurity, $\Delta_{te} \equiv 10Dq$ is the separation between $e_g$ and $t_{2g}$ orbitals in cubic symmetry and $g_0 = 2.0023$. In the case of a $d^9$ impurity in a tetragonally elongated geometry (or a $d^7$ impurity in compressed geometry) the unpaired electron resides in the $|\varepsilon\rangle \equiv |x^2 - y^2\rangle$ orbital and then $g_\parallel$ and $g_\perp$ are given by

$$g_\parallel = g_0 + 8\xi/\Delta_{te} \quad ; \quad g_\perp = g_0 + 2\xi/\Delta_{te} \tag{6.4}$$

Thus in both cases the separation between *extreme* values, $|g_\parallel - g_\perp|$, is the same.

Although $g_\parallel$ and $g_\perp$ in Fig. 6.3 are determined varying **H** on a single crystal both quantities can also be derived through the magnetic resonance of a policrystal. Let $\nu_\mu$ the microwave frequency used in the EPR experiment and $H_\parallel$ and $H_\perp$ the fields given by the resonance condition

$$h\nu_\mu = \beta g_\parallel H_\parallel = \beta g_\perp H_\perp \tag{6.5}$$

If, for instance, $g_\parallel > g_\perp$, according to (6.3) and (6.5) the absorbed power, $I(H)$, will be null when $H < H_\parallel$ while microwave absorption will take place when $H_\parallel \leq H \leq H_\perp$. Therefore, $I(H)$ would present two steps at $H = H_\parallel$ and $H = H_\perp$. As in EPR the derivative $dI(H)/dH$ is actually recorded two *peaks* at $H = H_\parallel$ and $H = H_\perp$ are thus observed in a powder spectrum. By this reason, the bands associated to both peaks are markedly *asymmetric* at variance with what is normally found in EPR spectra of single crystals.

### 6.2.2 EPR Spectrum of a Purely Electronic $^2E$ State

For the clarity of the present analysis let us now consider the characteristics of EPR spectra expected when the local symmetry is strictly cubic and thus the $|\theta\rangle$ and $|\varepsilon\rangle$ orbitals are degenerate. This issue was firstly investigated by Abragam and Pryce [5,34]. As all matrix elements of the orbital angular momentum and spin orbit operator are null within the $\{|\varepsilon\rangle, |\theta\rangle\}$ basis the differences between $g$ and $g_0$ arise again through the admixture with $t_{2g}$ levels via spin-orbit coupling. The corresponding effective Hamiltonian operating in the $\{|\varepsilon\rangle, |\theta\rangle\}$ basis is

$$H_S = \beta g_1 \boldsymbol{SH}\hat{I} + \frac{g_2\beta}{2}\left\{(3H_Z S_Z - \boldsymbol{HS})\widehat{U}_\theta + \sqrt{3}(H_X S_X - H_Y S_Y)\widehat{U}_\varepsilon\right\} \tag{6.6}$$

where $\hat{I}$ is the unit operator, while

$$\widehat{U}_\theta = \frac{1}{2}\left\{L_Z^2 - \frac{L^2}{3}\right\} \quad ; \quad \widehat{U}_\varepsilon = \frac{1}{2\sqrt{3}}\left\{L_X^2 - L_Y^2\right\} \tag{6.7}$$

The $\hat{U}_\theta$ and $\hat{U}_{\varepsilon r}$ operators are represented in the $\{|\varepsilon>, |\theta>\}$ basis by the Pauli matrices

$$U_\theta = \begin{bmatrix} 1 & 0 \\ 0 & -1 \end{bmatrix} \quad ; \quad U_\varepsilon = \begin{bmatrix} 0 & 1 \\ 1 & 0 \end{bmatrix} \tag{6.8}$$

As $\hat{U}_\theta \sim 3z^2 - r^2$ and $\hat{U}_\varepsilon \sim x^2 - y^2$ then $H_S$ in (6.6) is invariant when we rotate the axes and the magnetic field under the operations of the cubic group. The values of $g_1$ and $g_2$ are given by

$$g_1 = g_0 + 4\xi/\Delta_{te} \quad ; \quad g_2 = 4\xi/\Delta_{te} \tag{6.9}$$

Let $\mathbf{H}\|OZ'$. Writing $S_Z = S_{Z'}\zeta$, $S_X = S_{Z'}\xi$ and $S_Y = S_{Z'}\eta$ (where $\zeta^2 + \xi^2 + \eta^2 = 1$) and keeping only the terms in $S_{Z'}$ in (6.6) the two resonances emerging from the two Kramers doublets involved in the $^2E$ state are characterized by

$$g(\zeta,\xi,\eta) = g_1 \pm g_2\sqrt{1 - 3\{\zeta^2\xi^2 + \zeta^2\eta^2 + \eta^2\xi^2\}} \tag{6.10}$$

Therefore, measuring $g(\zeta, \xi, \eta)$ one can determine the two quantities $g_1$ and $g_2$, which by virtue of (6.9) should fulfill the relation

$$\frac{g_2}{g_1 - g_0} = 1 \tag{6.11}$$

According to the expression (6.10), which reflects the underlying cubic symmetry, the extreme values of resonance frequencies are characterized by $g_H$ and $g_L$ measured when $\mathbf{H}$ is parallel to any principal axis

$$g_L = g_0 + 8\xi/\Delta_{te} \quad ; \quad g_H = g_0 \tag{6.12}$$

### 6.2.3 EPR Characteristics of an $E \otimes e$ Dynamic Jahn-Teller Effect

In systems like $Cu^{2+}$- or $Ag^{2+}$-doped MgO several authors have explored the main characteristics of EPR spectra associated with a dynamic JT effect [16–19]. These spectra though measured in single crystals exhibit *asymmetric* line shapes which are similar in appearance to those observed in powder spectra [18–20]. Despite this similarity the EPR spectra of systems displaying a dynamic JT effect are not isotropic but the dependence on the magnetic field, $\mathbf{H}$, of the lines follow the *pattern*

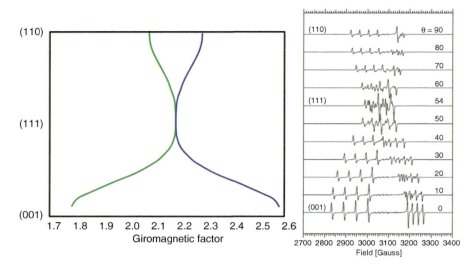

**Fig. 6.4** Illustration of the variation of the EPR signal with the observation angle of a system with $S = 1/2$ presenting a dynamic JT effect. Theoretically predicted curve (Eq. 6.10) (*left*) and experimental MgO:Cu$^{2+}$ spectrum (Taken from Ref. [18]) (*right*). Note that in the experimental spectrum the lines are further split by the hyperfine coupling with the spin of the Cu$^{2+}$ nucleus

described by Eq. 6.10. However, although the experimental dependence (Fig. 6.4) of gyromagnetic factors on ($\zeta$, $\xi$, $\eta$) do exhibit the cubic symmetry reflected in Eq. 6.10 the $g_1$ and $g_2$ quantities involved in Eq. 6.10 and defined in Eq. 6.9 have to be replaced by $g_1^*$ and $g_2^*$, where

$$g_1^* = g_1 \quad ; \quad g_2^* = q g_2 \tag{6.13}$$

Looking at experimental $g_1^*$ and $g_2^*$ values the $q$ factor is simply derived from

$$\frac{g_2^*}{g_1^* - g_0} = q \tag{6.14}$$

The experimental $g_1^* = 2.1924$ and $g_2^* = 0.108$ values measured for MgO:Cu$^{2+}$ [18, 19] thus give a factor $q$ close to 1/2. The existence of this *reduction* factor on $g_2$ stresses that the relation (6.11) is not fulfilled and thus the observed EPR signal cannot be associated with a pure $^2E$ electronic state [35]. How the existence of cubic symmetry and this reduction are both compatible with the JT effect are analyzed in next sections.

It is worth noting now that if $g_2^* = g_2/2$ it implies that the extreme values of the gyromagnetic factor, $g_H$ and $g_L$, are now given by

$$g_H = g_0 + 6\xi/\Delta_{te} \quad ; \quad g_L = g_0 + 2\xi/\Delta_{te} \tag{6.15}$$

Thus the separation between extreme values, $g_H - g_L$, is *smaller* than that corresponding to a static JT effect.

## 6.3 The Jahn-Teller Effect on Impurities in Perfect Crystals

### 6.3.1 The Origin of the Instability: A Simple View

Let us consider a $d^7$ impurity which is initially located in a cubic site and displays an octahedral coordination (Fig. 6.1). If, in this situation, $R_{oct}$ means the metal ligand distance let us now move the four equatorial ligands inwards a distance $a \ll R_{oct}$ while the two axial ligands are shifted outwards a distance $2a$ (Fig. 6.5). Accordingly, the axial ($R_{ax} = R_{oct} + 2a$) and equatorial ($R_{eq} = R_{oct} - a$) metal-ligand distances are now different. Under this *small* $Q_\theta$ distortion of ligands the degeneracy in the $e_g$ level is broken. If the octahedron is elongated ($Q_\theta > 0$) then the energy of the $3z^2 - r^2$ level, where the unpaired electron should be located, decreases (Fig. 6.1). This reduction of electronic energy induced by the distortion is the source of the JT instability [1]. A similar decrease of energy due to the tetragonal distortion should also happen when there are three electrons in the $e_g$ shell (such as in the ground state of $d^9$ impurities, like $Cu^{2+}$ or $Ag^{2+}$) but obviously not in the ground state configuration $(3z^2 - r^2)^1 (x^2 - y^2)^1$ of $Ni^{2+}$. In the latter case the phenomenon of *disproportionation*, triggered by the pseudo JT effect may play a similar role to the JT in $d^7$ or $d^9$ impurities [36].

Another simple view on the JT instability is reached just considering the forces on ligands due to *changes* of electronic density [37]. Let us consider a $d^7$ impurity in an artificial $(3z^2 - r^2)^{0.5}(x^2 - y^2)^{0.5}$ electronic configuration as a starting point. This configuration, which is fully similar to that for the ground state of $Ni^{2+}$, leads to an electronic density which has cubic symmetry and thus exerts the same force on all ligands. If at a given metal-ligand distance, $R_{oct}$, the total force on a

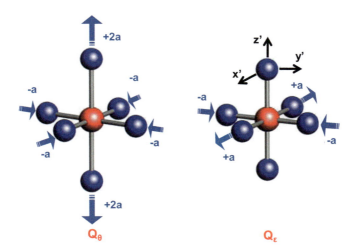

**Fig. 6.5** Illustration of the $e_g$-normal modes for an octahedral transition metal complex involved in the $E \otimes e$ JT effect. The local $\{x', y', z'\}$ coordinates describe the shift of the ligand with respect to the equilibrium position corresponding to octahedral symmetry

ligand is null for the $(3z^2 - r^2)^{0.5}(x^2 - y^2)^{0.5}$ configuration and then we move to a realistic $(3z^2 - r^2)^1$ configuration there is a variation of the electronic density, $\Delta n$, given by

$$\Delta n = \frac{1}{2}\{n(3z^2 - r^2) - n(x^2 - y^2)\} \tag{6.16}$$

Here, for instance, $n(3z^2 - r^2)$ is just the electronic density associated with the $3z^2 - r^2$ orbital. Obviously the *differential* density, $\Delta n$, pushes outwards the two ligands lying along the OZ axis while it pulls inwards the four ligands of the equatorial plane thus generating a $Q_\theta$ distortion [37]. Obviously, the force arising from $\Delta n$ can also modify the position of further ions from the impurity. However, the theory of elasticity tell us that if an impurity, placed in a three dimensional crystal, moves outwards the closest ions, lying at a distance R, by a quantity, $u$, then those lying at 2R would undergo a displacement $\sim u/4$. Therefore, if in a lattice with NaCl structure the substitutional impurity leads to a ligand relaxation of $\sim 5\%$ the relaxation of second neighbours along $<100>$ directions would be practically negligible.

For simplicity, in the analysis carried out in Sects. 6.3 and 6.4 we shall consider that in the distortion coordinates, $Q_\theta$ and $Q_\varepsilon$, only displacements of ligands are involved (Appendix 1).

## 6.3.2 Description of the $E \otimes e$ Linear Jahn-Teller Effect

### 6.3.2.1 Adiabatic Hamiltonian and Wavefunctions

As a starting point, let us consider the $d^7$ impurity seeing a local $O_h$ symmetry in the $(3z^2 - r^2)^{0.5}(x^2 - y^2)^{0.5}$ configuration. The corresponding metal-ligand distance, R, at equilibrium is $R_{oct}$ and the adiabatic Hamiltonian, $H(\mathbf{r}, R_{oct})$, exhibits cubic symmetry when the electronic coordinates are transformed under the cubic group. Let us now switch on the actual $e_g^1$ configuration and explore the *changes* on electronic levels due to *small* displacements involving the $Q_\theta$ and $Q_\varepsilon$ coordinates of the $e_g$ vibrational mode. This coupling is represented by a linear term $H_{LV} = V_\theta(\mathbf{r})Q_\theta + V_\varepsilon(\mathbf{r})Q_\varepsilon$ where, for instance, $V_\theta(\mathbf{r})$ and $Q_\theta$ both transform like $3z^2 - r^2$. Thus, this linear vibronic coupling term keeps cubic symmetry provided operations are carried out on *both* electronic and nuclear coordinates.

Although $H_{LV}$ is the responsible for having adiabatic minima without a local cubic symmetry it is necessary to consider the elastic term associated to the $e_g$ mode for reaching an equilibrium situation. In our notation this term is controlled by the $V_{2a}$ constant. Moreover, it is also crucial to take into account the anharmonicity of the $e_g$ mode (inevitably present in any real molecular vibration) in the description in order to have barriers among equivalent minima [10, 25, 26]. This is controlled by the elastic constant $V_{3a}$. Accordingly the adiabatic Hamiltonian, $H(\mathbf{r}, Q_\theta, Q_\varepsilon)$, around $Q_\theta = Q_\varepsilon = 0$ is of the form

$$
\begin{aligned}
H(\mathbf{r}, Q_\theta, Q_\varepsilon) = H(\mathbf{r}, R_{oct}) + \{V_\theta(\mathbf{r})Q_\theta + V_\varepsilon(\mathbf{r})Q_\varepsilon\} \\
+ V_{2a}\{Q_\theta^2 + Q_\varepsilon^2\} + V_{3a}\{Q_\theta^3 - 3Q_\theta Q_\varepsilon^2\}
\end{aligned}
\tag{6.17}
$$

Here all terms keep the cubic symmetry. Of course, the $V_{2a}$ constant in Eq. 6.17 is related to the angular vibration frequency, $\omega_e$, of the $e_g$ mode by the relation

$$
2V_{2a} = M_X \omega_e^2
\tag{6.18}
$$

where $M_X$ means the ligand mass. Working within *the frozen* $\{x^2 - y^2, 3z^2 - r^2\}$ electronic basis, $V_\theta(\mathbf{r})$ and $V_\varepsilon(\mathbf{r})$ are represented by $V_{1e}U_\theta$ and $V_{1e}U_\varepsilon$, respectively, where the $U_\theta$ and $U_\varepsilon$ matrices are defined in Eq. 6.8. Using now the polar coordinates $\rho$ and $\varphi$ defined by

$$
Q_\theta = \rho \cos\varphi; \quad Q_\varepsilon = \rho \sin\varphi
\tag{6.19}
$$

we can look into the solutions to the JT adiabatic Hamiltonian, $H_{JT}^A$, given by

$$
H_{JT}^A = V_{1e}\rho\{U_\theta \cos\varphi + U_\varepsilon \sin\varphi\} + V_{2a}\rho^2 + V_{3a}\rho^3 \cos3\varphi
\tag{6.20}
$$

Typically $|V_{1e}|$ is around $1\,\mathrm{eV/\mathring{A}}$ while $V_{2a}$ is usually $\sim 5\,\mathrm{eV/\mathring{A}}^2$ and $V_{3a} \sim 1\,\mathrm{eV/\mathring{A}}^3$ [10, 25, 26]. If, in a first step, we ignore the anharmonic term the minima, corresponding to the *lowest* solution, $E_-(\rho, \varphi)$, and the associated JT energy are given by

$$
\rho = \rho_0 = \frac{V_{1e}}{2V_{2a}} \quad ; \quad E_{JT} = \frac{1}{2}|V_{1e}|\rho_0
\tag{6.21}
$$

Therefore, at this level of approximation the distortions described by the circle $\rho = \rho_0$ are all equivalent and thus $E_-(\rho_0, \varphi)$ is independent on $\varphi$. The expected cubic symmetry is however restored when the anharmonic term in Eq. 6.20 is taken into account. In such a case $E_-(\rho, \varphi)$ is given by

$$
E_-(\rho, \varphi) \approx -(1/2)V_{1e}\rho_0 + V_{3a}\rho_0^3 \cos3\varphi
\tag{6.22}
$$

and then $E_-(\rho, \varphi)$ is warped as shown in Fig. 6.6. If $V_{3a} < 0$ there are only three equivalent minima at $\varphi = 0, \pm 2\pi/3$ and the barrier between two of them, B, is thus given by

$$
B \approx 2|V_{3a}|\rho_0^3
\tag{6.23}
$$

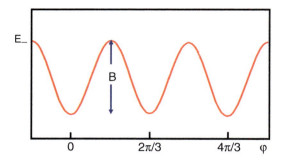

**Fig. 6.6** Cross-section of the energy surface of a E⊗e JT problem varying the polar angle $\varphi$ while keeping the value of the radial coordinate fixed to that of the global minimum of energy ($\rho = \rho_0$)

Although there are other factors which contribute to the barrier it has been pointed out that the anharmonicity of the $e_g$ mode usually plays a dominant role [26]. Moreover, if the $MX_N$ complex is elastically decoupled from the rest of the lattice then $V_{3a} < 0$ and the adiabatic minima correspond to an elongated conformation [25]. The phenomenon of elastic decoupling happens in cases like $Cu^{2+}$- or $Rh^{2+}$-doped NaCl and also for $KMgF_3:M^{3+}$ (M = Fe, Cr) where a host cation is replaced by an impurity with a higher nominal charge [10, 25].

Within the adiabatic approximation the electronic wavefunction, $\phi_-(\mathbf{r}, \varphi)$, associated with $E_-(\rho, \varphi)$ obtained from the diagonalization of Eq. 6.20 is given by [1]

$$\phi_-(\mathbf{r}, \varphi) = \cos\frac{\varphi}{2} \left|3z^2 - r^2\right\rangle - \sin\frac{\varphi}{2} \left|x^2 - y^2\right\rangle \quad (6.24)$$

It should be noted that this electronic wavefunction *changes sign* when $\varphi$ is replaced by $\varphi + 2\pi$ [38, 39]. This change in the phase of the electronic wavefunction phase is usually called geometric phase [40, 41] or *Berry phase* [42, 43].

### 6.3.2.2 Motion of Nuclei: Vibronic Levels with a Large Barrier

Let us now consider the motion of nuclei. In general, this requires adding the kinetic energy of nuclei in Eq. 6.20 and solving the Hamiltonian

$$H_{JT} = \frac{\hbar^2}{2M_X} \left\{ \frac{\partial^2}{\partial \rho^2} + \frac{1}{\rho}\frac{\partial}{\partial \rho} + \frac{1}{\rho^2}\frac{\partial^2}{\partial \varphi^2} \right\} + V_{1e}\rho\{U_\theta \cos\varphi + U_\varepsilon \sin\varphi\}$$

$$+ V_{2a}\rho^2 + V_{3a}\rho^3 \quad (6.25)$$

6 Dynamic and Static Jahn-Teller Effect in Impurities: Determination of the Tunneling    117

If the two solutions, $E_-(\rho_0, \varphi)$ and $E_+(\rho_0, \varphi)$, of the adiabatic Hamiltonian (Eq. 6.20) are well separated then $E_-(\rho_0, \varphi)$ is, in a first approximation, the effective potential energy for the motion of nuclei corresponding to the lowest solution. As $E_-(\rho_0, \varphi)$ exhibits a parabolic behavior around the $\varphi = 0, \pm 2\pi/3$ minima (Fig. 6.6) then, when warping is big enough, several harmonic oscillator-like levels can be accommodated in each well [21, 38].

Let us now consider the *lowest* level for each one of the three wells. When the barrier is infinite, the energy of the three levels will be exactly the same and thus the ground state is triply degenerate. However, this accidental degeneracy is partially removed when the barrier between two equivalent wells becomes finite giving rise to a doublet E state and a singlet [21, 38, 39]. Seeking to clarify this relevant matter we can write the total wavefunction, $\Psi(r, \rho, \varphi)$, describing the ground and first vibronic excited states, as follows

$$\Psi(r, \rho, \varphi) \approx \phi_-(r, \phi)\sigma(\rho - \rho_0)\chi(\varphi) \tag{6.26}$$

Here $\phi_-(r, \varphi)$ is the electronic wavefunction defined in Eq. 6.24 while $\sigma(\rho - \rho_0)$ and $\chi(\varphi)$ are vibrational wavefunctions associated with $\rho$ and $\varphi$ coordinates respectively. The $\sigma(\rho - \rho_0)$ function is peaked at $\rho = \rho_0$ [38]. As in $\Psi(r, \rho, \varphi)$ only orbital variables are involved then

$$\Psi(r, \rho, \varphi + 2\pi) = \Psi(r, \rho, \varphi) \tag{6.27}$$

Therefore, taking into account the Berry phase in $\phi_-$ (Eq. 6.24) the last condition implies that [36, 37]

$$\chi(\varphi + 2\pi) = -\chi(\varphi) \tag{6.28}$$

This last equation involves a problem: we either assume that $\chi$ is a $2\pi$ periodic function which is bivaluated or we take $\chi$ as a monovaluated function with a $4\pi$ period. In this work we will use this last convention both for wavefunctions where the Berry phase is present (Eq. 6.28) as for the ones where it is absent, in which $\chi$ is a monovaluated $2\pi$ periodic function. Let us now discuss the form of $\chi(\varphi)$ when the barrier is big enough. If $\chi_n(\varphi - \varphi_n)$ means the vibrational ground state corresponding to the isolated $n$-well we can try to express $\chi(\varphi)$ as a linear combination of $\chi_n(\varphi - \varphi_n)$ thus implying the existence of tunneling along the different wells. However, as pointed out by O'Brien [38], in order to fulfill (6.28) and keep $\chi$ monovaluated, we can consider, in principle, six (instead of three) equivalent wells placed at $\varphi_n = 0, 2\pi/3, 4\pi/3, 2\pi, 8\pi/3$ and $10\pi/3$. Writing now $\chi(\varphi)$ as a linear combination of *localized* $\chi_n(\varphi - \varphi_n)$ wavefunctions we obtain

$$\chi(\varphi) = \sum_{i=1}^{6} \alpha_i \chi_i(\varphi - \varphi_i) \tag{6.29}$$

Now we evaluate the Hamiltonian (Eq. 6.25) using the wavefunction given in Eq. 6.26. After some simple algebra we find

$$E = \langle \Psi | H_{JT} | \Psi \rangle = E_0 - \frac{\hbar^2}{2M_X} \left\langle \sigma \left| \frac{1}{\rho^2} \right| \sigma \right\rangle \left[ \left\langle \phi_- \frac{\partial^2 \phi_-}{\partial \varphi^2} \right\rangle + \left\langle \chi \frac{\partial^2 \chi}{\partial \varphi^2} \right\rangle \right]$$
$$+ V_{3a} \langle \sigma | \rho^3 | \sigma \rangle \langle \chi | \cos 3\varphi | \chi \rangle \tag{6.30}$$

Where the second and third terms come, respectively, from the angular-depending parts of the kinetic and potential energies while the first term is

$$E_0 = \langle \sigma | \frac{\hbar^2}{2M_X} \left( \frac{\partial^2}{\partial \rho^2} + \frac{1}{\rho} \frac{\partial}{\partial \rho} \right) + V_{2a} \rho^2 - V_{1e} |\rho| \, | \sigma \rangle \tag{6.31}$$

Substituting Eqs. 6.24 and 6.29 into Eq. 6.30 we obtain,

$$E = E_0 - \frac{\hbar^2}{2M_X} \left\langle \sigma \left| \frac{1}{\rho^2} \right| \sigma \right\rangle \left[ -\frac{1}{4} + \sum_{i,j}^6 \alpha_i \alpha_j \left\langle \chi_i \frac{\partial^2 \chi_j}{\partial \varphi^2} \right\rangle \right]$$
$$+ V_{3a} \langle \sigma | \rho^3 | \sigma \rangle \sum_{i,j}^6 \alpha_i \alpha_j \langle \chi_i | \cos 3\varphi | \chi_j \rangle \tag{6.32}$$

If the functions $\chi_i$ are localized inside the wells we can, approximately, consider that they only interact with the immediate neighbouring wells and that all other interactions are zero. Taking into account that the $\chi_n$ and $\sigma$ functions are gaussian-like we can write the relevant integrals as

$$\left\langle \chi_i \frac{\partial^2 \chi_i}{\partial \varphi^2} \right\rangle = I_0^k \tag{6.33}$$

$$\left\langle \chi_i \frac{\partial^2 \chi_{i\pm1}}{\partial \varphi^2} \right\rangle \approx \beta_k S \tag{6.34}$$

$$\langle \chi_i | \cos 3\varphi | \chi_i \rangle = I_0^p \tag{6.35}$$

$$\langle \chi_i | \cos 3\varphi | \chi_{i\pm1} \rangle \approx \beta_p S \tag{6.36}$$

$$\left\langle \sigma \left| \frac{1}{\rho^2} \right| \sigma \right\rangle \approx \rho_0^{-2} \tag{6.37}$$

$$\langle \sigma | \rho^3 | \sigma \rangle \approx \rho_0^3 \tag{6.38}$$

where $I_0^k$ and $I_0^p$ are positive integrals associated to the zero-point energy and $\beta_k$ and $\beta_p$ are positive constants associated to the interaction between wells through the kinetic and potential energy operators, respectively. $S = \langle \chi_i | \chi_{i\pm1} \rangle$ is the overlap between the $\chi_n$ wavefunctions in neighbouring wells.

## 6 Dynamic and Static Jahn-Teller Effect in Impurities: Determination of the Tunneling 119

Let us now focus on the value of $\alpha_i$. Using Eq. 6.29 it turns out that the condition of presence of the Berry phase (Eq. 6.28) simply requires that

$$\alpha_i = -\alpha_{i+3} \tag{6.39}$$

It is important to note that this condition is a direct consequence of the presence of the JT effect in the system. For the sake of comparison it is also useful to consider the nuclear dynamics of systems where similar distortions to those associated to the JT effect appear but whose origin is not the orbital degeneracy leading to the JT effect. This situation arises, for instance, in the ground state of the $CuF_3$ molecule [36] whose geometry is an obtuse triangle, similar to that of triangular molecules like $Na_3$ that present a $E \otimes e$ JT effect [26]. However, the distortion in this case arises due to the $(A + E) \otimes e$ pseudo JT effect and, as a consequence, the Berry phase is absent [36, 43]. Similarly, the first excited doublet of $Na_3$ displays a distorted geometry but the Berry phase is not present [44]. In the realm of doped insulators there are impurities like $Fe^+$ or $Cu^+$ in cubic lattices like $SrCl_2$ which move *off-centre* along $<001>$ type directions although its ground state has *no* orbital degeneracy [10, 45–47]. Such instability, obviously not due to a JT effect, leads to the existence of six equivalent wells where the Berry phase is again absent thus implying

$$\chi(\varphi + 2\pi) = \chi(\varphi) \tag{6.40}$$

and, as a consequence

$$\alpha_i = \alpha_{i+3} \tag{6.41}$$

The form of $\chi(\varphi)$ for the vibrational singlet and doublet can be well derived from representations of the $C_6$ group which change sign under a $\pi$ rotation if the Berry phase is present. So, as shown in Appendix 2 and illustrated in Fig. 6.7, $\chi(\varphi)$ for the case of the singlet and doublet vibronic cases can be respectively written as

$$\chi_A(\varphi) = \frac{1}{\sqrt{6(1+S)}} \left( \chi_1(\varphi) - \chi_2 \left( \varphi - \frac{2\pi}{3} \right) + \chi_3 \left( \varphi - \frac{4\pi}{3} \right) \right.$$
$$\left. - \chi_4(\varphi - 2\pi) + \chi_5 \left( \varphi - \frac{8\pi}{3} \right) - \chi_6 \left( \varphi - \frac{10\pi}{3} \right) \right) \tag{6.42}$$

$$\chi_{E\theta}(\varphi) = \frac{1}{\sqrt{12(1+S)}} \left( 2\chi_1(\varphi) + \chi_2 \left( \varphi - \frac{2\pi}{3} \right) - \chi_3 \left( \varphi - \frac{4\pi}{3} \right) \right.$$
$$\left. - 2\chi_4(\varphi - 2\pi) - \chi_5 \left( \varphi - \frac{8\pi}{3} \right) + \chi_6 \left( \varphi - \frac{10\pi}{3} \right) \right) \tag{6.43}$$

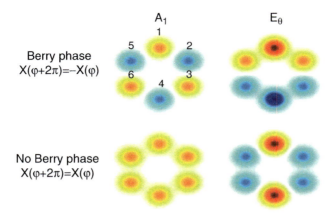

**Fig. 6.7** Illustration of the E and A vibrational wavefunctions localized at the bottom of the Mexican hat for the JT (with Berry phase) and non-JT (without Berry phase) cases. The plots are performed in polar coordinates with the angle varying between 0 and $4\pi$ so that the wavefunction is monovaluated (see text)

The notation $\chi_A(\varphi)$ or $\chi_{E\theta}(\varphi)$ simply means that such wavefunction is *part* of the total wavefunction, $\Psi_A(r, \rho, \varphi)$ or $\Psi_{E\theta}(r, \rho, \varphi)$, of the vibronic singlet or doublet, respectively. The form of these wavefunctions for the case where there are six equivalent distortions albeit no JT effect is given in Appendix 2 and illustrated in Fig. 6.7. Evaluating Eq. 6.32 for A and E$\theta$ for both the JT and non-JT cases we find

$$E_A^{JT} = E_0 - \frac{\hbar^2}{2M_X\rho_0^2}\left[-\frac{1}{4} + \frac{6I_0^k - 12\beta_k S}{6(1+S)}\right] + V_{3a}\rho_0^3 \frac{6I_0^p - 12\beta_p S}{6(1+S)} \quad (6.44)$$

$$E_{E\theta}^{JT} = E_0 - \frac{\hbar^2}{2M_X\rho_0^2}\left[-\frac{1}{4} + \frac{12I_0^k + 12\beta_k S}{12(1+S)}\right] + V_{3a}\rho_0^3 \frac{12I_0^p + 12\beta_p S}{12(1+S)} \quad (6.45)$$

$$E_A^{NJT} = E_0 - \frac{\hbar^2}{2M_X\rho_0^2} \frac{6I_0^k + 12\beta_k S}{6(1+S)} + V_{3a}\rho_0^3 \frac{6I_0^p + 12\beta_p S}{6(1+S)} \quad (6.46)$$

$$E_{E\theta}^{NJT} = E_0 - \frac{\hbar^2}{2M_X\rho_0^2} \frac{12I_0^k - 12\beta_k S}{12(1+S)} + V_{3a}\rho_0^3 \frac{12I_0^p - 12\beta_p S}{12(1+S)} \quad (6.47)$$

Observing the previous expressions it is clear that if the barrier is very high and the $\chi_n$ functions are complete localized (implying a null overlap, $S = 0$) the singlet and doublet states are degenerate both for the JT and non-JT cases. On the other hand, we can observe that the overlap-depending terms have *opposite signs* depending on whether they come from kinetic or potential energies and also whether it is the JT or non-JT case. In the JT case the kinetic energy contribution favors a doublet ground state and an excited singlet state while the potential energy favors the opposite behavior. In the non-JT case this pattern is reversed and the kinetic energy favors a singlet ground state and an excited doublet. Experimentally and theoretically [1, 16–19, 38, 39] it is found that the doublet is always lower in JT cases and the singlet is lower in non-JT cases. Thus, *the dominant contribution for the tunneling splitting must be the kinetic energy*. We will see below that these trends can be easily theoretically proven for the fully dynamic case. The sign of the overlap-depending terms can be simply understood if $\chi(\varphi)$ is taken as a molecular orbital built out of atomic orbitals $\chi_n$ (Eq. 6.29). Then if the sign of $\chi$ changes when passing from $\varphi = \varphi_n$ to $\varphi = \varphi_{n\pm 1}$ the combination of $\chi_i$ and $\chi_{i\pm 1}$ orbitals is antibonding-like and the overlap term will have a negative sign. This fact, combined with the negative sign in front of the kinetic energy term in Eqs. 6.44–6.47 will lead to an increase in the energy of the state. On the other hand, if the sign remains unchanged the orbitals are bonding-like and the energy of the state decreases. Observing Fig. 6.7 it is clear that in the JT case $\chi_A(\varphi)$ is a strongly antibonding orbital while $\chi_{E\theta}(\varphi)$ is predominantly bonding. On the other hand, in the non-JT case $\chi_A(\varphi)$ is clearly bonding while $\chi_{E\theta}(\varphi)$ is largely antibonding. This allows us to qualitatively understand the ordering of the lowest vibronic states in the JT and non-JT cases.

Before leaving this section it is worth noting that the *instantaneous* geometry displayed by nuclei at a given $\varphi$ value is, in general, orthorhombic. However, it does not imply that the ground state is a singlet due to the vibronic character of the *total* wavefunction. Indeed it displays a cubic symmetry when we look at the same time at the electronic wavefunction and the *accompanying* deformation of nuclei.

### 6.3.2.3 Vibronic Levels with No Barrier

When there is no barrier, $V_{3a}$ involved in Eqs. 6.17 and 6.25 is null. If the total wavefunction, $\Psi(r, \rho, \varphi)$, is well represented by Eq. 6.26 then the effective potential energy describing the motion of nuclei is independent on the $\varphi$ coordinate. Accordingly the $\chi(\varphi)$ wavefunction can be of the form

$$\chi(\varphi) = ce^{-im\varphi} \tag{6.48}$$

implying a free rotation along the whole circle characterized by $\rho = \rho_0$. In this limit case the kinetic energy, $E_m$, associated with this motion is simply given by

$$E_m = \frac{\hbar^2}{2M_X\rho_0^2}m^2 \tag{6.49}$$

It should be noted now that due to the boundary condition given by Eq. 6.28 only *half integers* values are allowed for $m$. This relevant fact thus precludes a singlet ground state associated with the $m = 0$ value and where that kinetic energy is null which is the expected result when the JT effect is absent. By contrast, the vibronic ground state including the Berry phase of the JT effect, corresponds to $m = \pm 1/2$ and is again a doublet. It is worth noting however that in this limit case the vibronic first excited state corresponds to $m = \pm 3/2$ and it is not a singlet. This accidental degeneracy is removed when the barrier is not exactly zero [38].

Let us now write the form of two vibronic wavefunctions transforming like $|\theta > \equiv 3z^2 - r^2$ and $|\varepsilon > \equiv x^2 - y^2$ when no warping is present. In such a case $\Psi_{E\theta}(r, \rho, \varphi)$ and $\Psi_{E\varepsilon}(r, \rho, \varphi)$ are simply given by

$$\Psi_{E\theta}(r,\rho,\varphi) = \sigma(\rho - \rho_0)\phi_-(r,\varphi)\frac{1}{\sqrt{\pi}}cos\frac{\varphi}{2}$$

$$\Psi_{E\varepsilon}(r,\rho,\varphi) = \sigma(\rho - \rho_0)\phi_-(r,\varphi)\frac{1}{\sqrt{\pi}}sin\frac{\varphi}{2} \qquad (6.50)$$

where the electronic wavefunction $\phi_-(r, \varphi)$ is defined in Eq. 6.24.

### 6.3.2.4 EPR Spectra of a Vibronic Doublet: Reduction Factor

The analysis carried out up to now on the $E \otimes e$ JT effect in perfect cubic crystals indicates that the ground state is always a vibronic E doublet. This means that the vibronic wavefunctions $|\theta_V\rangle \equiv \Psi_{E\theta}(r,\rho,\varphi)$ and $|\varepsilon_V\rangle \equiv \Psi_{E\varepsilon}(r,\rho,\varphi)$ transform like the purely electronic $|\theta > \equiv 3z^2 - r^2$ and $|\varepsilon > \equiv x^2 - y^2$ wavefunctions. Accordingly operators like $\hat{U}_\theta$ and $\hat{U}_\varepsilon$ (defined in Eq. 6.7) are represented in the *new* $\{|\theta_V\rangle, |\varepsilon_V\rangle\}$ basis by the matrices $q\boldsymbol{U}_\theta$ and $q\boldsymbol{U}_\varepsilon$. The reduction factor [20, 35], $q$, is thus obtained calculating for instance $\langle \theta_V |\hat{U}_\theta|\theta_V\rangle$ and comparing the result with $\langle \theta|\hat{U}_\theta|\theta\rangle = -1$. As $\hat{U}_\theta$ acts on the electronic wavefunction $\phi_-(r, \varphi)$, given by Eq. 6.24, it can easily be seen that

$$\left\langle \phi_-(r,\varphi)\left|\hat{U}_\theta\right|\phi_-(r,\varphi)\right\rangle = \cos\varphi \qquad (6.51)$$

and thus $q$ is just given by [19]

$$q = \int_o^{2\pi} cos\varphi[\chi_{E\theta}(\varphi)]^2 d\varphi \qquad (6.52)$$

As shown in Appendix 2, in the case of high warping the probability of finding the system in the well around $\varphi = 0$ is 4/6 while that for the well centered at $\varphi = 2\pi/3$ or $\varphi = 4\pi/3$ (where $\cos\varphi = -1/2$) is only 1/6. For this reason $q = 1/2$. It should

6 Dynamic and Static Jahn-Teller Effect in Impurities: Determination of the Tunneling    123

be noted that the *same* $q = 1/2$ value is obtained when there is no warping at all and $\chi_{E\theta}(\varphi)$ is just proportional to $cos\frac{\varphi}{2}$ as shown in Eq. 6.50.

The value $q = 1/2$ has been derived assuming that only the electronic wavefunction of the lowest solution, $\phi_-(r, \varphi)$, is involved in the total wavefunction, $\Psi(r, \rho, \varphi)$, such as it shown in Eq. 6.26. When the admixture with the electronic wavefunction related to the highest adiabatic solution $E_+(\rho_0, \varphi)$ is taken into account then $q$ is given by [20]

$$q = \frac{1}{2}\left\{1 + \left(\frac{\hbar\omega_E}{4E_{JT}}\right)^2\right\} \tag{6.53}$$

This correction to the value $q = 1/2$ thus plays a minor role if $E_{JT} \geq \hbar\omega_E$.

The present analysis thus sheds light on the origin of EPR spectra observed for systems like $Cu^{2+}$- or $Ag^{2+}$-doped MgO [16–19]. As it was already pointed out in Sect. 6.2, the angular dependence of the EPR signal looks similar to that expected for a pure $^2E$ electronic state but with a smaller spectral extension due to the reduction factor $q$. In accord with the present discussion the existence of this reduction factor is the signature of vibronic levels and thus of a dynamic $E \otimes e$ JT effect in $Cu^{2+}$- or $Ag^{2+}$-doped MgO [16–19].

Despite this fact we do not understand through the present discussion why the EPR spectra of systems like $MgO : Cu^{2+}$ exhibit asymmetric lines similar to those observed in powder spectra [18, 19]. Furthermore, the analysis performed in Sect. 6.3 indicates that the ground state for impurities like $Cu^{2+}$, $Ag^{2+}$, $Rh^{2+}$ or $Ni^+$ in *perfect* cubic crystals and octahedral coordination should always be a vibronic doublet. Accordingly, the corresponding EPR spectra at low temperatures would exhibit the dynamic behavior discussed in this section. However, as emphasized in the Introduction, the EPR spectra explored up to now on $d^9$ and $d^7$ impurities in cubic halides [6–15] show the existence of a static JT effect.

## 6.4 The Jahn-Teller Effect on Impurities in Real Crystals

The key role played by the unavoidable random strains present in any real crystal for properly understanding all experimental data associated with substitutional impurities like $Cu^{2+}$, $Ag^{2+}$, $Rh^{2+}$ or $Sc^{2+}$ in cubic lattices was stressed by Ham [20]. Such *internal* strains are the result of defects like dislocations or undesirable impurities which are present in every real crystal at low temperatures. As it has been pointed out the major part of dislocations formed in melt-grown crystals arises from the release of thermal stresses. Moreover, in crystals grown by solution crystal growth is known to be strongly favoured on the steps and kinks which appear in the intersection of screw dislocation with a surface [48–50].

It should be noted that dislocations are defects which are never in thermodynamic equilibrium [50]. Indeed if in a dislocation line we consider a length $b \sim 2$–$3$ Å the corresponding elastic energy is of the order of $1$ eV. Similarly the replacement of an atom of the pure lattice by an impurity often involves an *increase*, $\Delta U$, of the internal energy associated with the local distortion. However, if close to the melting point, the impurity is able to move *quickly* [10, 51] along the different lattice sites the corresponding increase of configurational entropy favours its incorporation into the lattice. Nevertheless, once the temperature is well below the melting point, the impurity can, in practice, be *locked* in a given lattice site and thus it is only able to vibrate around that *fixed* position. Therefore, undesirable impurities are found in crystals at $T = 0$ K even if $\Delta U > 0$ and thus its presence cannot again be understood in terms of thermodynamic equilibrium.

## 6.4.1 Evidence of Random Strains in Real Crystals

This section is addressed to gain information on random strains from the analysis of the so-called inhomogeneous broadening of optical bands associated with impurities in insulating materials [52]. The equilibrium geometry for the ground and excited states of impurities in insulators are usually different. In particular, for transition-metal impurities such a difference comes out when the excited state energy depends on 10Dq, a quantity extremely sensitive to the actual value of the impurity–ligand distance [10]. This phenomenon is behind of bandwidths in the range $0.1$–$1$ eV observed in absorption and emission spectra at room temperature [52, 53]. This broadening is called homogeneous provided all the impurities are strictly equivalent. In such a case, when we illuminate with a sharp laser on a small part of the absorption band we should obtain the whole emission spectrum. Along this view, when the equilibrium geometry of an excited state coincides with that for the ground state we expect to see a sharp line whose width is just determined by the lifetime if all impurities are *equivalent*. The last situation happens for the $^2E\left(t_{2g}{}^3\right)$ excited state of $Cr^{3+}$ impurities in oxide lattices which belongs to the *same* configuration as the ground state $^4A_2(t_{2g}{}^3)$ and thus its energy is independent on 10Dq. However, this is not the case of the excited state $^4T_2(t_{2g}{}^2e_g)$ whose energy is just equal to 10Dq [54].

Therefore, if the measured lifetime for the $^2E(t_{2g}{}^3) \rightarrow^4 A_2(t_{2g}{}^3)$ sharp transition of $Al_2O_3{:}Cr^{3+}$ at $T = 4.2$ K is $\tau = 4$ ms [55] one would expect an emission bandwidth, W, of the order of $\hbar/\tau$ while the experimental value is $\sim 1$ cm$^{-1}$ and thus *nine orders* of magnitude higher. This remarkable difference has been pointed out to reflect that not all $Cr^{3+}$ impurities in $Al_2O_3$ are exactly equivalent because they are subject to different internal strains [52, 55]. Therefore, the observed broadening arises from that *inhomogeneity*. This inhomogeneous broadening is revealed when shining with a sharp laser $(\Delta \nu \ll 1$ cm$^{-1})$ on the $^4A_2(t_{2g}{}^3) \rightarrow^2 E(t_{2g}{}^3)$ absorption band. In such a case only the $Cr^{3+}$ centres under a given internal strain are excited

6 Dynamic and Static Jahn-Teller Effect in Impurities: Determination of the Tunneling 125

and thus the corresponding emission is narrowed [48, 51]. Let us call $E_{E \to A}$ the energy of the $^2E(t_{2g}{}^3) \to {}^4 A_2(t_{2g}{}^3)$ transition of $Al_2O_3{:}Cr^{3+}$ which is very slightly dependent upon the average metal-ligand distance, R. Indeed from experiments of luminiscence under hydrostatic pressure it has been measured $dE_{E \to A}/dP = -7\,cm^{-1}/GPa$ [56] from which it can be derived $dE_{E \to A}/dR = 29\,cm^{-1}/pm$ using the bulk modulus of $Al_2O_3$, $B_o = 254\,GPa$ [57]. Therefore, calling $\bar{e}$ to an average isotropic strain

$$\bar{e} = \frac{1}{3} \left( \overline{e_{xx}} + \overline{e_{yy}} + \overline{e_{zz}} \right) \tag{6.54}$$

then the bandwidth, W, caused by the different internal strains felt by the involved $Cr^{3+}$ impurities can be estimated through the expression

$$W \approx \frac{dE_{E \to A}}{dR} R_a \bar{e} \tag{6.55}$$

Here $R_a = 1.97$ Å just means the equilibrium $Cr^{3+}$-$O^{2-}$ distance at ambient pressure [58]. Thus the experimental figure $W \sim 1\,cm^{-1}$ [55] implies $\bar{e} \sim 5.10^{-4}$ characterizing the internal strains present in the studied $Al_2O_3{:}Cr^{3+}$ sample.

## 6.4.2 Influence of Random Strains on the EPR Spectrum of a Vibronic Doublet

Let us again consider a $d^9$ or $d^7$ impurity located in a cubic site with octahedral coordination. The existence of non-isotropic internal strains around the impurity can split the degeneracy associated with an E state. In a cubic crystal the five components of the non-isotropic strain tensor transform like $E + T_2$. Let us now consider the two components, $e_{\theta\theta} \sim 3z^2 - r^2$ and $e_{\varepsilon\varepsilon} \sim \sqrt{3}(x^2 - y^2)$, which belong to E. In terms of cartesian components of the strain tensor, $e_{\theta\theta}$ and $e_{\varepsilon\varepsilon}$ are written as

$$e_{\theta\theta} = e_{zz} - \frac{1}{2}(e_{xx} + e_{yy}) \quad ; \quad e_{\varepsilon\varepsilon} = \frac{\sqrt{3}}{2}(e_{xx} - e_{yy}) \tag{6.56}$$

Obviously the same splitting on a doublet should be produced by a $e_{zz}$ strain along a [001] axis than by a $e_{xx}$ strain along an *equivalent* [100] axis. This means that the electron-strain coupling should preserve cubic symmetry when we *formally* act on both electronic and strain variables. Thus that coupling, $H_{ES}$, can be written as

$$H_{ES} = e_{\theta\theta}V_\theta^s(\boldsymbol{r}) + e_{\varepsilon\varepsilon}V_\varepsilon^s(\boldsymbol{r}) \tag{6.57}$$

where again $V_\theta^s(\boldsymbol{r}) \sim 3z^2 - r^2$ and $V_\varepsilon^s(\boldsymbol{r}) \sim \sqrt{3}(x^2 - y^2)$.

Let us now explore what happens when the vibronic singlet is *far apart* from the vibronic doublet. In such a case we have to consider the effects due to $H_{ES}$ only *within* the vibronic doublet described by $|\theta_V\rangle \equiv \Psi_{E\theta}(r,\ \rho,\varphi)$ and $|\varepsilon_V\rangle \equiv \Psi_{E\varepsilon}(r,\rho,\varphi)$ wavefunctions. It should be noted that as $T_2 \not\subset E \otimes E$ only the E strains produce a splitting in the vibronic doublet. Moreover, according to a generalization of the Wigner-Eckart theorem for finite groups, as E appears only once in the decomposition of $E \otimes E$ then the matrix elements of $V_\theta^s(r)$ and $V_\varepsilon^s(r)$ in the $\{|\theta_V\rangle,\ |\varepsilon_V\rangle\}$ basis are simply proportional to those of $V_\theta(r)$ and $V_\varepsilon(r)$ operators appearing in the description of the JT coupling in Eq. 6.17 [54]. Therefore, the corresponding matrices $\mathbf{V}_\theta^*$ and $\mathbf{V}_\varepsilon^*$ can be written as

$$\mathbf{V}_\theta^* = qV_s\mathbf{U}_\theta \quad ; \quad \mathbf{V}_\varepsilon^* = qV_s\mathbf{U}_\varepsilon \tag{6.58}$$

where, for instance

$$V_s = \langle\theta\,|V_\theta^s(r)|\,\theta\rangle \tag{6.59}$$

Let us now consider a JT impurity which, for simplicity, is subject *only* to a $e_{\theta\theta} > 0$ strain. If we assume that the effects induced by such strain on $|\theta\rangle \equiv 3z^2 - r^2$ and $|\varepsilon\rangle \equiv x^2 - y^2$ levels mainly come from the distortion on the ligand sphere the constant $V_s$ in Eq. 6.59 can be related to the JT constant, $V_{1e}$, defined in Eq. 6.20. Indeed in that case the energy increase undergone by the $|\theta\rangle$ level, given by $V_s e_{\theta\theta}$, should coincide with $V_{1e}Q_\theta$ where

$$Q_\theta = \frac{2}{\sqrt{3}}e_{\theta\theta}R_{oct} \tag{6.60}$$

Therefore [20],

$$V_s = \frac{2}{\sqrt{3}}V_{1e}R_{oct} \tag{6.61}$$

Accordingly, if $|V_{1e}| \sim 1\,\mathrm{eV/\mathring{A}}$ [25, 26] and take $|e_{\theta\theta}| \sim 5.10^{-4}$ then the splitting, $\delta = 2V_s e_{\theta\theta}$, induced on $|\theta\rangle$ and $|\varepsilon\rangle$ levels would be $\sim 10\,\mathrm{cm}^{-1}$. Bearing in mind this figure let us now explore the form of EPR spectra coming from the vibronic doublet under the influence of an internal strain described by $e_{\theta\theta}$ and $e_{\varepsilon\varepsilon}$. Adding the electron-strain coupling term to Eq. 6.6 and working now in the $\{|\theta_V\rangle,\ |\varepsilon_V\rangle\}$ basis it is obtained

$$H_S = \beta g_1 \mathbf{SH} + \frac{qg_2\beta}{2}\left\{(3H_ZS_Z - \mathbf{HS})U_\theta + \sqrt{3}(H_XS_X - H_YS_Y)U_\varepsilon\right\}$$
$$+ qV_s[e_{\theta\theta}\mathbf{U}_\theta + e_{\varepsilon\varepsilon}\mathbf{U}_\varepsilon] \tag{6.62}$$

As $\beta H g_0 \sim 0.3\,\mathrm{cm}^{-1}$ for $H \sim 3000\,\mathrm{G}$ this fact just means that the strain term is the dominant one in Eq. 6.62 leading to an splitting of the vibronic doublet. Thus,

# 6 Dynamic and Static Jahn-Teller Effect in Impurities: Determination of the Tunneling   127

after diagonalizing the last term in (6.62) for a *given set* of $e_{\theta\theta}$ and $e_{\varepsilon\varepsilon}$ *within* the $\{|\theta_V\rangle, |\varepsilon_V\rangle\}$ basis the new eigenfunctions, $\{|\theta_V^*\rangle, |\varepsilon_V^*\rangle\}$ are given by

$$|\theta_V^*\rangle = cos\frac{\varphi^*}{2}|\theta_V\rangle + sin\frac{\varphi^*}{2}|\varepsilon_V\rangle \quad ; |\varepsilon_V^*\rangle = -sin\frac{\varphi^*}{2}|\theta_V\rangle + cos\frac{\varphi^*}{2}|\varepsilon_V\rangle \quad (6.63)$$

where $\varphi^*$ is defined from $e_{\theta\theta}$ and $e_{\varepsilon\varepsilon}$ as follows

$$e_{\theta\theta} = \varepsilon cos\varphi^* \quad ; \quad e_{\varepsilon\varepsilon} = \varepsilon sin\varphi^* \quad (6.64)$$

Therefore, according to Eq. 6.63 even if the two components of the doublet are no longer degenerate due to the action of an internal strain, wavefunctions like $|\theta_V^*\rangle$ or $|\varepsilon_V^*\rangle$ spread over the three wells and are *not yet localized* on one of them.

Expressing now Eq. 6.62 in the new $\{|\theta_V^*\rangle, |\varepsilon_V^*\rangle\}$ basis it is found

$$H_S = \beta g_1 \mathbf{SH} + \frac{qg_2\beta}{2}\left\{(3H_ZS_Z - \mathbf{HS})U_\theta^* + \sqrt{3}(H_XS_X - H_YS_Y)U_\varepsilon^*\right\} + q\varepsilon V_s U_\theta$$

$$U_\theta^* = \begin{bmatrix} cos\phi^* & -sin\phi^* \\ -sin\phi^* & -cos\phi^* \end{bmatrix} \quad ; \quad U_\varepsilon^* = \begin{bmatrix} sin\phi^* & cos\phi^* \\ -cos\phi^* & -sin\phi^* \end{bmatrix} \quad (6.65)$$

Thus, according to Eq. 6.65 if $\mathbf{H} = 0$ the splitting in the vibronic doublet is equal to $\delta^* = 2q\varepsilon V_s$. If now the system is, for instance, in the *singlet* $|\theta_V^*\rangle$ state let us now look into the influence of a magnetic field $\mathbf{H} = H(\zeta, \xi, \eta)\| \ OZ'$. Writing $S_Z = S_{Z'}\zeta$, $S_X = S_{Z'}\xi$ and $S_Y = S_{Z'}\eta$ and retaining only the terms in $S_{Z'}$ in (6.52) the spin Hamiltonian for the $|\theta_V^*\rangle$ state can simply be written as

$$H_S = \beta g_1 H S_{Z'} + \frac{qg_2\beta}{2}H S_{Z'}\left\{(3\zeta^2 - 1)cos\varphi^* + \sqrt{3}(\xi^2 - \eta^2)sin\varphi^*\right\} \quad (6.66)$$

Therefore, in accord to the orthorhombic local symmetry induced on the impurity by a general internal strain (involving both $e_{\theta\theta}$ and $e_{\varepsilon\varepsilon}$ components) the angular pattern displayed by the gyromagnetic factor is given by

$$g = g_1 + \frac{qg_2}{2}\left\{(3\zeta^2 - 1)cos\varphi^* + \sqrt{3}(\xi^2 - \eta^2)sin\varphi^*\right\} \quad (6.67)$$

Bearing in mind that internal strains are *randomly* distributed along the crystal and that in a typical EPR experiment the concentration of JT impurities can be about $10^{18}\,cm^{-3}$ then, for a given orientation of the magnetic field, all possible resonances between the two extreme values are observed. For instance, if $\mathbf{H}\|$ [001] the lowest resonance appears at $g_L = g_0 + (2\lambda/\Delta)$ and the highest one at $g_H = g_0 + (6\lambda/\Delta)$ if we take $q = 1/2$. These values coincide with those given in Eq. 6.15. Therefore, if there is microwave absorption in the range $g_L \leq g \leq g_H$ but not when $g \leq g_L$ or $g \geq g_H$ then peaks at $g = g_L$ and $g = g_H$ will be observed in the EPR spectrum where the *derivative* of the absorbed intensity, $dI(H)/dH$, is actually recorded. For this reason also, the lines associated with such peaks are asymmetric. This band shape is thus rather similar to that found when looking at EPR of powder samples [19–21]. However, in the present case $g_L$ and $g_H$ do depend on the orientation of the magnetic field while the EPR spectra of powder samples are isotropic.

Let us now look into the *extreme values* of the gyromagnetic factor, $g_{extr}$, corresponding to a general orientation of $\mathbf{H} = H(\zeta, \xi, \eta)$, which determine the *peaks* observed in a normal EPR experiment. Thus, according to Eq. 6.67 such extreme values are determined by the condition

$$tan\varphi^* = \frac{\sqrt{3}(\xi^2 - \eta^2)}{(3\zeta^2 - 1)} \tag{6.68}$$

and thus the values of $g_{extr}$ when $\mathbf{H}$ is parallel to $(\zeta, \xi, \eta)$ are given by

$$g_{extr}(\zeta, \xi, \eta) = g_1 \pm q g_2 \sqrt{1 - 3\{\zeta^2\xi^2 + \zeta^2\eta^2 + \eta^2\xi^2\}} \tag{6.69}$$

Therefore, taking into account Eq. 6.10 and Sects. 6.2.3 and 6.3.2.4 the angular dependence described by Eq. 6.69 is exactly the same expected for a vibronic doublet in a perfect cubic crystal. The highest and lowest $g$-values are thus given by Eq. 6.15. This important result, firstly pointed out by Ham [20,33], thus clarifies why a signal displaying a cubic angular pattern can be observed although the two levels of the vibronic doublet are not, in general, degenerate as a result of internal strains present in any real crystal. Moreover, the present analysis strongly suggests that the observation of a dynamic EPR spectrum at low temperatures requires that the separation, $3\Gamma$, between the excited vibronic singlet and the ground state is big enough when compared to the splitting, $\delta^*$, caused by internal strains on the vibronic doublet.

### 6.4.3 Admixture of the Excited Singlet State Due to Strains on the EPR Spectrum of a Vibronic Doublet

Let us explore how the admixture with the excited singlet state favours the localization in one well. If we start in the domain where warping effects are important then, taking into account Eq. 6.26, when $\phi_-$ and $\chi$ have been substituted by their values (Appendix 2) $|\theta_V\rangle$ can be shortly be expressed as

$$\Psi_{E\theta}(\mathbf{r}, \rho, \varphi) = |\theta_V\rangle = \frac{1}{\sqrt{6}}\{2\Phi_Z - [\Phi_X + \Phi_Y]\} \tag{6.70}$$

# 6 Dynamic and Static Jahn-Teller Effect in Impurities: Determination of the Tunneling 129

where, for instance

$$\Phi_Z = |3z^2 - r^2\rangle \chi_1(\varphi)\sigma(\rho - \rho_0) \quad ; \quad \Phi_X = |3x^2 - r^2\rangle \chi_2\left(\varphi - \frac{2\pi}{3}\right)\sigma(\rho - \rho_0)$$
$$(6.71)$$

In the same vein the wavefunction associated with the singlet $A_1$ can be written

$$\Psi_A(\mathbf{r}, \rho, \varphi) = |A_V\rangle = \frac{1}{\sqrt{3}}\{\Phi_Z + \Phi_X + \Phi_Y\} \tag{6.72}$$

It should be noted that a random strain described by $e_{\theta\theta}$ and $e_{\varepsilon\varepsilon}$ components is *not coupled* in first order to the singlet A state because $E \not\subset A \times A$. However that strain induces a mixing between the vibronic doublet and the singlet A state. Let us now consider the symmetry-allowed admixture of $|A_V\rangle$ in $|\theta_V\rangle$ induced by a *small* internal $e_{\theta\theta}$ strain and controlled by the parameter $\gamma$. The resulting wavefunction $|\tilde{\theta}_V\rangle$ will be of this form

$$|\tilde{\theta}_V\rangle \cong |\theta_V\rangle + \frac{\gamma}{\sqrt{6}}|A_V\rangle = \frac{1}{\sqrt{6}}\{(2+\gamma)\Phi_Z - (1-\gamma)\Phi_X - (1-\gamma)\Phi_Y\} \tag{6.73}$$

Thus, once the interaction between the vibronic doublet with the singlet is switched on we are increasing the weight of the Z well ($\varphi = 0$). This is thus a first step towards the localization of the system in one of the wells which is the characteristic of the static JT effect. Along this line the gyromagnetic factor associated with the perturbed $|\tilde{\theta}_V\rangle$ wavefunction is $g = g_0 + (2\xi/\Delta_{te})[1 - 2\gamma]$. Therefore, this g-factor is smaller than the lowest one, $g_L = g_0 + 2\xi/\Delta_{te}$, given in Eq. 6.15. This implies that the spectral range is increasing with respect to the value $g_H - g_L$ obtained for an *isolated* doublet E.

According to this reasoning, $\gamma$ in Eq. 6.73 will be of the order of $\delta^*/(3\Gamma)$ and thus an increase of warping favours a reduction of the tunneling splitting, $3\Gamma$, and consequently a transition from dynamic to static regime. When warping is big enough tunneling is destroyed by internal strains making that the three wells at $\varphi_n = 0$, $2\pi/3$, $4\pi/3$ are no longer equivalent. In such a case the electronic wavefunction corresponding to the Z-well ($\varphi = 0$) would be $|\theta\rangle \equiv 3z^2 - r^2$ and the corresponding energy equal to $V_s e_{\theta\theta}$. Similarly, the energy associated with X ($\varphi = 2\pi/3$) and Y ($\varphi = 4\pi/3$) wells are given by $V_s e_{\theta\theta}(X)$ and $V_s e_{\theta\theta}(Y)$, respectively, where the expression of $e_{\theta\theta}(X)$ and $e_{\theta\theta}(Y)$ in terms of $e_{\theta\theta}$ and $e_{\varepsilon\varepsilon}$ is just given by

$$e_{\theta\theta}(X) = -\frac{1}{2}e_{\theta\theta} + \frac{\sqrt{3}}{2}e_{\varepsilon\varepsilon} \quad ; \quad e_{\theta\theta}(Y) = -\frac{1}{2}e_{\theta\theta} - \frac{\sqrt{3}}{2}e_{\varepsilon\varepsilon} \tag{6.74}$$

Thus, if on a given lattice site $V_s e_{\theta\theta} < V_s e_{\theta\theta}(X) < V_s e_{\theta\theta}(Y)$ then *only* the Z-well ($\varphi = 0$) will be populated at T $= 0$K. However, due to the random character of internal strains the number of sites where the Z-well ($\varphi = 0$) has the smallest energy will be the same as those for the X and Y wells.

When the temperature increases the system can reach higher vibrational states of a given well and thus jumps more easily to the other wells with a jump frequency, $v_J$. This motion is helped by the thermostat formed by the phonons of the lattice (where the JT impurity is placed) and can lead to an isotropic EPR spectrum provided the motional narrowing condition $hv_J > (g_{\parallel} - g_{\perp})\beta H$ is fulfilled. It should be noted that this dynamical behavior is *not coherent* as it is driven by the *collisions* with phonons. Therefore, this situation is quite different from the dynamic JT effect discussed in Sect. 6.3 where the total wavefunction involves a *coherent* tunneling among the three Z ($\varphi = 0$), X ($\varphi = 2\pi/3$) and Y ($\varphi = 4\pi/3$) wells. Nevertheless, it should be remarked that this coherent pattern is observed *only* at temperatures around 2 K or below because at T = 6 K an isotropic EPR spectrum characterized by $g = g_1$ due to motional narrowing is already observed in cases like $MgO : Cu^{2+}$ [16–19].

### 6.4.4 Comment on the Estimation of the Tunneling Splitting, $3\Gamma$, from Experimental Data

From the analysis carried out in Sect. 6.4.3 it emerges that the existence of a dynamic JT effect in real crystals is favoured by the condition

$$\delta^* = 2q\varepsilon V_s < 3\Gamma \tag{6.75}$$

Despite the relevance of this condition, firstly put forward by Ham [20], the estimated values of the tunneling splitting, $3\Gamma$, from available experimental data raise doubts. For instance, Boatner et al. [17] *assumed* that the isotropic EPR spectrum already detected at T = 4.2 K for $MgO:Ag^{2+}$ is the spectrum coming from the thermal population of the excited vibronic singlet and not from the motional narrowing of the spectrum corresponding to the vibronic doublet. In agreement with this assumption, Boatner et al. [17] give $3\Gamma = 4.8\,cm^{-1}$ for $MgO:Ag^{2+}$. A similar figure ($3\Gamma = 3\,cm^{-1}$) was derived from an *interpretation* of Raman spectra carried out on $CaO:Cu^{2+}$ [22]. In general, the accepted values of the tunneling splitting, $3\Gamma$, for $Cu^{2+}$- and $Ag^{2+}$-doped oxide lattices are currently lying in the $3–15\,cm^{-1}$ range [16, 17, 19, 20, 22]. For instance, from the behaviour of small lines observed around the zero-phonon line in $MgO:Fe^{2+}$ a value $3\Gamma = 14\,cm^{-1}$ corresponding to the excited $^5E_g$ state was assumed by Hjortsberg et al. [59].

Nevertheless, it is hard to accept that the actual $3\Gamma$ values for $Cu^{2+}$-doped MgO and CaO and in $MgO:Ag^{2+}$ are in the $3–10\,cm^{-1}$ range. In fact, if $\delta^*$ is estimated to be around $5\,cm^{-1}$ it would be comparable to the accepted $3\Gamma$ values and thus it is certainly not sure that the condition given by Eq. 6.75 be fulfilled. Furthermore,

6 Dynamic and Static Jahn-Teller Effect in Impurities: Determination of the Tunneling 131

analyzing the strain effects on EPR spectra of $MgO:Cu^{2+}$ Reynolds et al. [18] concluded that the ratio $\delta^*/3\Gamma$ would be around 0.10. Then, if $3\Gamma = 4\,cm^{-1}$ it would imply a value of the splitting $\delta^*$ resulting from the internal strain which is one order of magnitude smaller than the figure obtained for $Al_2O_3$ from experimental data on ruby [55].

As pointed out in the introduction, the use of *ab initio* calculations has recently been of help to understand how $3\Gamma$ varies on passing from $MgO:Cu^{2+}$ or $MgO:Ag^{2+}$ (where the JT effect is dynamic) to $Rh^{2+}$, $Cu^{2+}$ and $Ag^{2+}$ impurities in cubic halides which in all studied cases display a static JT effect [30]. Such results are discussed in some detail in the next section.

## 6.5 Ab Initio Calculations of $3\Gamma$ in $d^7$ and $d^9$ Impurities in Insulators

The objective of this section is to determine with the help of *ab initio* calculations the value of $3\Gamma$ for various $d^9$ and $d^7$ impurities in insulating materials that span the range from the dynamic JT effect to a fully static one. According to the analysis carried out in the foregoing Sect. 6.4 such calculations are undertaken in order to *rationalize* the experimental results obtained on a variety of JT systems. The calculations are carried out in two steps. In a first step, an accurate representation of the energy surfaces involved in the JT problem is obtained from adiabatic calculations of the electronic structure of the impurity. In a second step, the parameters in a JT model are fitted to the numerical energy surface and the vibronic states and wavefunctions are obtained from solving the eigenvalue problem associated to the full Hamiltonian. The tunneling splitting is then obtained from the difference of energies between the two lowest vibronic levels. In the next section we detail how these calculations are performed.

### 6.5.1 Computational Details

In order to carry out ab initio calculations of transition-metal impurities in wide-gap insulators it is very important to consider first that in these systems the active electrons are strongly localized on the impurity and its first neighbours, the ligands, that form a $MX_6$ complex [10, 25–29]. Thus, the properties of the impurity center can be usually accurately obtained just by considering quantum mechanically the complex and, also, a few surrounding shells of ions from the crystal to simulate the short-range interactions of the complex with the embedding crystal. We have used 19-ion clusters of $MX_6A_6X_6$-type to simulate the presence of a transition metal impurity, M, inside a crystal of AX-type with

NaCl structure. These clusters are embedded with *ab initio* Model Potentials (AIMPs) that are full-ion pseudopotentials obtained by Hartree-Fock calculations to accurately simulate the short-range interaction potential between the cluster and the surrounding lattice [60]. The long-range interaction of the cluster with the rest of the ionic solid is obtained by placing a large set of point charges that accurately reproduce the electrostatic potential the crystal exerts over the complex.

In order to obtain the ground and excited JT energy sheets we use the multireference Complete Active Space Self Consistent Field (CASSCF) method [61]. The active space consists of the $e_g$ orbitals with strong $nd(M)$ character that contain the unpaired electron(s) and the empty $a_{1g}$ orbital with $(n + 1)s(M)$ character that, in many cases, lies at low-energies above $e_g$. While the CASSCF method provides a very good qualitative description of the wavefunction, including most of the static electron correlation, in order to obtain high accuracy it is necessary to correct these energies to obtain the dynamic electron correlation. We perform this task by applying second-order-perturbation theory (PT2) to the CASSCF obtained energies to obtain the final CASPT2 energies [62]. The orbital wavefunctions are expressed using gaussian basis-sets of very high quality [63] and the equilibrium geometries of the minimum and transition states in the Mexican hat were obtained by a full optimization of the complex in $D_{4h}$ symmetry. All calculations above were carried out using the MOLCAS ab initio package [62]. In these first principles calculations no adjustable parameter have been employed.

In a second step, we parameterize the JT surfaces obtained in the previous calculation by using a second-order JT Hamiltonian

$$
\begin{aligned}
H_{JT} = {} & \frac{\hbar^2}{2M_X}\left\{\frac{\partial^2}{\partial Q_\theta^2} + \frac{\partial^2}{\partial Q_\varepsilon^2}\right\} + V_{2a}\rho^2 \\
& + U_\theta\left(V_{1e}Q_\theta + V_{2e}\left(Q_\theta^2 - Q_\varepsilon^2\right)\right) + U_\varepsilon(V_{1e}Q_\varepsilon + 2V_{2e}Q_\theta Q_\varepsilon)
\end{aligned} \tag{6.76}
$$

This Hamiltonian is then evaluated by functions of type

$$
\Psi_{abc}(\boldsymbol{r}, Q_\theta, Q_\varepsilon) \approx \phi_a(\boldsymbol{r})\chi_{\theta b}(Q_\theta)\chi_{\varepsilon c}(Q_\varepsilon) \tag{6.77}
$$

where $\phi_a$ is the electronic function that can take the values $|3z^2 - r^2\rangle$ or $|x^2 - y^2\rangle$ while $\chi_{\theta b}$ and $\chi_{\varepsilon c}$ are harmonic oscillator functions along $Q_\theta$ and $Q_\varepsilon$ whose dispersion can be used as a variational parameter to optimize the basis-set. We evaluated Eq. 6.76 with the basis-set described by Eq. 6.77, where the used number of harmonic oscillator functions per dimension is 50 [30]. Accordingly, the number of basis functions employed in any calculation is equal to 5000. Diagonalization of the $5000 \times 5000$ matrix allowed us to obtain the energy of the lowest vibronic states converged to a precision $10^{-6}\,\mathrm{cm}^{-1}$.

# 6 Dynamic and Static Jahn-Teller Effect in Impurities: Determination of the Tunneling    133

**Table 6.1** CASPT2 ab initio results characterizing the Jahn-Teller effect in several $d^7$ and $d^9$ impurities in wide-gap insulators, including the Jahn-Teller deformation $(\rho_0)$, the Jahn-Teller energy $(E_{JT})$ and the barrier between wells (B)

| System | $\rho_0(\text{Å})$ | $E_{JT}(cm^{-1})$ | $B(cm^{-1})$ | $3\Gamma(cm^{-1})$ | $\delta/3\Gamma$ |
|---|---|---|---|---|---|
| MgO:Cu$^{2+}$ | 0.0875 | 445 | −11 | 234.79 | 0.09 |
| CaO:Cu$^{2+}$ | 0.1288 | 567 | −25 | 130.53 | 0.15 |
| SrO:Cu$^{2+}$ | 0.2110 | 914 | −71 | 41.06 | 0.49 |
| MgO:Ag$^{2+}$ | 0.1025 | 685 | −16 | 200.31 | 0.10 |
| CaO:Ag$^{2+}$ | 0.2250 | 977 | −136 | 29.43 | 0.68 |
| NaCl:Ag$^{2+}$ | 0.3256 | 1,623 | 599 | 0.49 | 40.82 |
| KCl:Ag$^{2+}$ | 0.5196 | 2,363 | 1024 | $5.10^{-4}$ | $4.10^4$ |
| NaCl : Rh$^{2+}$ | 0.3200 | 1,832 | 511 | 0.50 | 40.00 |

A positive (negative) value of the barrier just means that the equilibrium conformation at adiabatic minima corresponds to an elongated (compressed) octahedron. The obtained tunneling splitting and a typical value for the $\delta/3\Gamma$ ratio (taking $\delta = 20\,cm^{-1}$) is also given

## 6.5.2 Results

The main results of adiabatic calculations, including distortions in the elongated and compressed $D_{4h}$ geometries, JT energies and barriers for a wide number of systems, are presented in Table 6.1. As it can be seen the JT distortions and stabilization energies vary from a large value $(\rho_0 \sim 0.52\text{Å})$ in soft host lattices like KCl to a small one in stiff lattices like MgO $(\rho_0 \sim 0.10 \text{ Å})$. The barrier, B, between the JT wells is seen to strongly increase with the distortion which is in agreement with the fact that the main origin of the barrier is cubic elastic anharmonicity [26].

Due to the lack of structural experimental data it is difficult to directly compare most of these results with experimental values. Perhaps, the experimentally better studied system is MgO:Cu$^{2+}$ that has been explored using Extended X-ray Absorption Fine Structure (EXAFS) by several groups. In the experiments of Hildebrand et al. [64], Asakura et al. [65] the geometry of the complex appeared to be *cubic* with a metal-ligand distance of around 2.11 Å. This result is fully compatible with the average metal-ligand distance found in our calculations $(2R_{eq} + R_{ax})/3 = 2.14$ Å [25, 30] which in turn is coincident within 1.5% with the figure calculated by Pascual et al. [28]. It should be noted that a quite similar value has been measured by EXAFS for the impurity-ligand distance of *divalent* Ni$^{2+}$ and Co$^{2+}$ impurities in MgO [66]. From EXAFS experiments carried out on MgO:Cu$^{2+}$ Dick et al. [67] have suggested that the JT distortion is huge, with two metal-ligand distances as different as $R_{ax} = 2.55$ Å and $R_{eq} = 2.18$ Å, thus implying $\rho_0 = 0.43$ Å a figure certainly higher than $\rho_0 = 0.0875$ Å given in Table 6.1. The conclusion reached by Dick et al. [67] is however not consistent with the existence of a dynamic JT effect in MgO:Cu$^{2+}$ as it prevents to observe the system *locked* in a *particular* distorted configuration. A further discussion on this matter is given later.

The reliability of the results obtained through present high level ab initio calculations is supported by the comparison with spectroscopic parameters experimentally measured. For example, the calculated optical spectra of KCl:Ag$^{2+}$ and NaCl:Rh$^{2+}$

are in a good agreement with experimental data [27, 29] while the main trends exhibited by the hyperfine and superhyperfine tensors of the last system [12–14] are also well understood through the *ab initio* calculations [29, 68].

Along this line the $q$ reduction factor (given in Eq. 6.53), which is a direct proof of the presence of dynamic JT effect, has been experimentally measured in $MgO:Cu^{2+}$. Using Eq. 6.53 and the data presented in Table 6.1 we have obtained the value $q = 0.54$ that is certainly close to the experimental value $q = 0.56$ [18, 19].

As regards the JT stabilization energy of $MgO:Cu^{2+}$, although there is no direct experimental measurement of $E_{JT}$ there are arguments supporting that our calculated value $E_{JT} = 445\,cm^{-1}$ (Table 6.1) is not unreasonable. Indeed such a value is not far from the figure $E_{JT} = 495\,cm^{-1}$ derived by Pascual et al. [28] using also CASPT2 calculations. Along this line these authors obtain a value $\rho_0 = 0.109$ Å which is not far from that reported in Table 6.1, thus supporting a value of $|R_{ax} - R_{eq}| < 0.1$ Å at the adiabatic minima. It is worth noting that the calculations by Pascual et al. reproduce well the position of two metal-to-ligand charge-transfer transitions observed experimentally for MgO samples containing only about 35 ppm of $Cu^{2+}$ [28]. At the same time these authors report the energy separation between the ground state and the crystal field state denoted as $^2T_{2g}$ using cubic notation. Such an excitation is calculated to be equal to $10,000\,cm^{-1}$ which is close to the maximum of the dominant band observed (at $\sim 11,500\,cm^{-1}$) through optical absorption for $Cu_xMg_{1-x}O$ mixed crystals [23, 69]. According to the small distortion calculated at the adiabatic minima the splitting in the excited $^2T_{2g}$ state is found to be only of $700\,cm^{-1}$ and thus much smaller than the experimental bandwidth at room temperature ($\sim 4,000\,cm^{-1}$). Moreover, the optical absorption data obtained on $Cu_xMg_{1-x}O$ mixed crystals indicate that there is absorption in the $4,500$–$6,000\,cm^{-1}$ region whose shape depends on the copper concentration [23, 69]. The origin of this optical band was suggested to arise from the $a_{1g}(3z^2 - r^2) \rightarrow b_{1g}(x^2 - y^2)$ electronic transition of the $CuO_6^{10-}$ center [23]. In solid state chemistry this centre is designated by $[CuO_{6/6}]^0$ pointing out that the $Mg^{2+} \rightarrow Cu^{2+}$ substitution does not lead to any charged species.

Since the energy of the $a_{1g}(3z^2 - r^2) \rightarrow b_{1g}(x^2 - y^2)$ transition is equal to $4E_{JT}$ the referred interpretation of the absorption in the $4,500$–$6,000\,cm^{-1}$ range implies $E_{JT} \cong 1,300\,cm^{-1}$ which is thus *three times bigger* than the calculated values giving $E_{JT} \leq 500\,cm^{-1}$ [25, 28, 30]. For this reason Pascual et al. [28] proposed that absorption in the $4,500$–$6,000\,cm^{-1}$ region for $Cu_xMg_{1-x}O$ mixed crystals ($2.5\% < x < 7.5\%$) is not due to isolated $CuO_6^{10-}$ monomers actually uncoupled to any close defect. Accordingly such an absorption can arise from a $Cu^{2+}$ perturbed by a nearby defect, the formation of aggregates or even the stabilization of a part of copper in valences such as $Cu^0$ or $Cu^+$ [28]. It is worth noting that Cordischi et al. have proved by EPR the formation of axial $Cu^{2+}$ centres in MgO even at low copper concentration [70]. Such signal has been ascribed to the $Cu^{2+}$ impurity either associated with some defect or placed close to the surface. Concerning the existence of aggregates Raizman et al. have proved by EPR the formation of $Cu^{2+}$ pairs in $CaO:Cu^{2+}$ [71].

6 Dynamic and Static Jahn-Teller Effect in Impurities: Determination of the Tunneling    135

It should be noticed now that a value of $E_{JT}$ for $MgO:Cu^{2+}$ derived from experimental data has been reported by Höchli et al. [24]. From the temperature dependence of the spin-lattice relaxation time, $T_1$, inferred from the experimental bandwidth, these authors obtain $E_{JT} \cong 370 \, cm^{-1}$. This result thus supports that the actual value of $E_{JT}$ for $MgO : Cu^{2+}$ can be in the neighborhood of $500 \, cm^{-1}$ and not around $15,000 \, cm^{-1}$ as it was assumed by Reynolds et al. [18].

In Table 6.1 are also displayed the calculated values of $3\Gamma$ for $MgO:Cu^{2+}$ and other JT systems. We can observe that values span a very large range of over 6 orders of magnitude from $234 \, cm^{-1}$ in $MgO:Cu^{2+}$ to $5 \cdot 10^{-4} \, cm^{-1}$ in $KCl:Ag^{2+}$. While some of these values are very small others are much larger than those in the range $1$–$15 \, cm^{-1}$ assumed in several works [16, 17, 19, 20, 22]. The value of $3\Gamma$ is clearly connected to the distortion, the larger the distortion is the smaller the tunneling splitting gets. This trend can be explained due to the direct link between the tunneling splitting and the overlap of the vibrational functions localized in the JT wells (Eqs. 6.44–6.47).

It is interesting to see how in $MgO:Cu^{2+}$ there is almost no barrier so that the origin of the tunneling splitting is completely due to the kinetic energy of the vibronic ground state. To be more specific, the value of that kinetic energy $E\left(m = \pm\frac{1}{2}\right) = \frac{\hbar^2}{8M_X\rho_0^2}$ in the absence of barrier (6.49) is found to be about *three times* bigger than the calculated barrier for $MgO : Cu^{2+}$ in Table 6.1. This fact thus implies that in $MgO : Cu^{2+}$ the system undergoes a nearly free rotation in the $\rho = \rho_0$ circle. Along this line, using Eq. 6.49 and the value of $\rho_0$ found in Table 6.1 we obtain that for $MgO : Cu^{2+}$ the tunneling splitting is $272 \, cm^{-1}$ which compares well with the full calculation, $234 \, cm^{-1}$.

It should be noted that the existence of a dynamic JT effect in $MgO:Cu^{2+}$ makes difficult the detection of any particular distortion in EXAFS experiments. According to the present discussion all distortions characterized by $\{\rho = \rho_0, \varphi\}$ are in practice equally probable. As X-rays take an instantaneous picture of the distortion in a given complex we could in principle observe all the distortions corresponding to the different $\varphi$ values if the impurity concentration is $\sim 10^{18} \, cm^{-3}$. However, in practice only the *envelope* is seen and thus the EXAFS spectrum is similar to that expected for a $Cu^{2+}$ impurity under perfect cubic symmetry. This reasoning is thus consistent with the experimental findings by Hildebrand and Martin [64].

In the case of $Ag^{2+}$ and $Rh^{2+}$ impurities in ACl lattices (A $=$ Na, K) the results embodied in Table 6.1 indicate that $V_{3a} < 0$, the barrier B is higher than $500 \, cm^{-1}$ and thus the adiabatic equilibrium geometry should correspond to an elongated octahedron. This conclusion is in perfect agreement with the experimental g-tensor of these systems displaying a *static* JT effect. It should be noticed that the $MCl_6^{4-}$ complex (M $=$ Ag, Rh) formed in ACl lattices (A $=$ Na, K) by the *divalent* impurity is elastically *decoupled* from the rest of the lattice involving monovalent cations [10, 25]. Accordingly, if a ligand is moved from its equilibrium position by a quantity equal to $z'$ (Fig. 6.5) the anharmonic energy of the M–Cl bond is higher when the distance is reduced ($z' < 0$) than when $z' > 0$, and thus $V_{3a}$ is expected to be negative. This situation no longer holds in cases like $MgO:Cu^{2+}$ where the impurity replaces another divalent cation. In that case it can happen that the energy increase in the $Cu^{2+}$-$O^{2-}$-$Mg^{2+}$ fragment is smaller when $z' < 0$ than when $z' > 0$ [25]. Although the calculations carried out on $MgO:Cu^{2+}$ support that B is certainly small $(< 25\,cm^{-1})$ we cannot rely on the sign of B despite it is found to be negative in Refs. [25, 28, 30]. Along this line, EPR experiments carried out on $CaO:Ag^{2+}$ [17] reveal the existence of a static JT effect consistent with a calculated $3\Gamma$ value only equal to $29\,cm^{-1}$. However, such experiments support that the equilibrium configuration corresponds at an elongated rather than to a compressed octahedron.

### 6.5.3 Effect of the Strain

As seen in Sect. 6.4.3 the dynamic or static JT character depends strongly on the ratio between tunneling splitting $3\Gamma$ and the splitting, $\delta^*$, due to random strains. In particular, the dynamic JT effect appears when $3\Gamma >> \delta^*$. According to our previous estimations $\delta^* \sim 5\,cm^{-1}$ although this value is obviously sample dependent. With this $\delta^*$ value it is clear that a dynamic JT effect would never be observed for a tunneling splitting of only a few $cm^{-1}$. On the other hand, the values presented in Table 6.1, show clearly why EPR measurements in $MgO:Cu^{2+}$ are always consistent with a dynamic JT effect [19, 20] while those for $NaCl:Rh^{2+}$ or $KCl:Ag^{2+}$ are always static [12, 13, 15]. In fact, taking into account that $3\Gamma$ is calculated to be $234\,cm^{-1}$ for $MgO:Cu^{2+}$ if the ratio $\delta^*/3\Gamma$ is equal to $\sim 0.10$ then $\delta^*$ would be around $20\,cm^{-1}$ for the sample measured by Reynolds et al. [18]. It is worth noting that a dynamic JT effect has also been found in the low temperature EPR spectra of $MgO:Ag^{2+}$ and $CaO:Cu^{2+}$ [16–20]. This finding is fully supported by the calculations of Table 6.1 giving $3\Gamma$ values higher than $100\,cm^{-1}$ in both cases. By contrast, a strong reduction of the calculated $3\Gamma$ quantity is obtained on passing to $CaO:Ag^{2+}$ $(3\Gamma = 29.4\,cm^{-1})$ or $SrO:Cu^{2+}$ $(3\Gamma = 41.06\,cm^{-1})$. In both cases the EPR spectra essentially display a static JT behavior [17, 72].

A pertinent question at this moment is what the $\delta/3\Gamma$ ratio is at the transition between dynamic to static JT effects. This question can be answered by calculating what the probability of finding the system in a particular well is when a tetragonal strain is applied (Fig. 6.8). In order to perform this calculation we took the parameters used to calculate the tunneling splitting in $MgO:Cu^{2+}$ and added a tetragonal strain to the JT Hamiltonian (Eq. 6.76),

$$H_{JT+S} = H_{JT} + \delta U_\theta \qquad (6.78)$$

In order to calculate the localization we integrated the nuclear density in three radial sectors whose limits contain the transition state that separates the JT wells. The probability of finding the distortion in each of the three wells as a function of the splitting, $\delta$, induced by the tetragonal strain is shown in Fig. 6.9. It should be remarked that for $\delta/2 = 25\,cm^{-1}$ almost 80% of the probability of finding the system is localized on one well. Therefore this result indicates that in order to

6 Dynamic and Static Jahn-Teller Effect in Impurities: Determination of the Tunneling    137

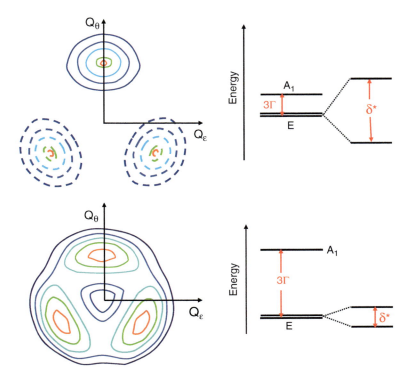

**Fig. 6.8** Illustration of the nuclear density and vibronic levels for the cases of the static (*up*) and dynamic (*down*) JT effects. In the static case the system is captured in one minimum (*solid line*) due to the effects of the local strain field producing a separation, $\delta$, between $3z^2 - r^2$ and $x^2 - y^2$ electronic levels and a splitting $\delta^* = q\delta$ on the vibronic doublet. In the dynamic case, when the condition $3\Gamma \gg \delta^*$ is fulfilled, the nuclear density has an almost equal contribution from each well. The position of the first vibronic singlet state (lying at an energy $3\Gamma$ above the vibronic doublet) is also shown in the figure. This singlet state belongs to $A_1$ when its wavefunction is described by Eqs. 6.71 and 6.72 and to $A_2$ when $3z^2 - r^2$ is replaced by $x^2 - y^2$ [20]

have a dynamic JT effect the ratio $\delta/3\Gamma$ should be smaller than $\sim 0.3$ and thus $\delta^*/3\Gamma \lesssim \sim 0.15$. This plot explains the reason why very few dynamic systems have been found in nature: the tunneling splitting has to be much larger than the random strains which are usually of the order of $10\,\text{cm}^{-1}$ and this can only be achieved in systems that present a very small JT distortion.

As emphasized in Sect. 6.4.2 the EPR spectra of *real* systems displaying a dynamic JT effect exhibit a broadening due to random strains which produce asymmetric bands. It should be noted that this characteristic is significantly modified when we consider systems displaying a clear static JT effect. Let us take as a guide the case of a $Cu^{2+}$ impurity in a cubic lattice like NaCl whose equilibrium geometry corresponds to an elongated octahedron [6]. All sites where the local $C_4$ axis of the $Cu^{2+}$ impurity is parallel to OZ verify $e_{\theta\theta} < e_{\theta\theta}(X) < e_{\theta\theta}(Y)$. Nevertheless, among the centers with $C_4 \| OZ$ not all of them are subject to the same $e_{\theta\theta}$ strain.

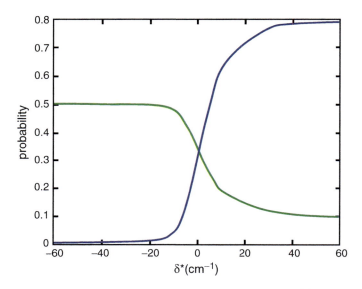

**Fig. 6.9** Localization of the ground vibrational wavefunction of MgO:Cu$^{2+}$ in a given well for a varying tetragonal strain which induces a splitting $\delta^*$, between $3z^2 - r^2$ and $x^2 - y^2$ electronic levels. The probability of finding the system in the well in the direction of the strain is represented in blue while that corresponding to one of the other two wells is plotted in *green* (Note that although the probability is equal to 1/3 for $\delta^* = 0$ it raises up to $\sim 80\%$ for $\delta^* \sim 25\,\text{cm}^{-1}$ while is practically equal to zero for $\delta^* = -25\,\text{cm}^{-1}$)

This fact thus induces changes on the $\Delta_{te} \equiv 10Dq$ quantity and the associated $g_{||} - g_0$ and $g_\perp - g_0$ quantities. Nevertheless, if $\Delta_{te} \sim 10^4\,\text{cm}^{-1}$ and $\delta \sim 10\,\text{cm}^{-1}$ we expect changes on $g_{||}$, $\Delta g_{||}$, which are only around $4.10^{-4}$ if $g_{||} - g_0 \sim 0.4$. Therefore, if we work in the X band ($\nu = 9,000\,\text{MHz}$) random strains induce a broadening of EPR lines $\beta H \Delta g_{||} \sim 2\,\text{MHz}$. As the bandwidth measured in EPR at low temperatures is usually in the range 10–100 MHz this means that the observed broadening is not mainly due to random strains. In fact, as it is well known such broadening arises mainly from superhyperfine interactions not resolved in EPR spectra.

## 6.6 Conclusions

In summary, we have reviewed the origin of the dynamic and static JT effects and shown their experimental features in the EPR spectra of transition-metal impurities in insulators. Moreover, we performed ab initio calculations to show the value of the tunneling splitting that controls the dynamics of a JT system and compared it with the strains. We found that realistic values for the tunneling splitting in JT systems are much larger than the ones previously expected. Such $3\Gamma$ values which are higher

# 6 Dynamic and Static Jahn-Teller Effect in Impurities: Determination of the Tunneling

than $\sim 100\,\mathrm{cm}^{-1}$ are necessary to overcome the effects of random strains naturally present in every crystal and that favour the localization of the system in one of the three JT wells as is characteristic of a static JT effect.

In order to reach $3\Gamma$ values comparable or higher than $100\,\mathrm{cm}^{-1}$ the results reported in this work support that this situation is helped when the condition

$$E\left(m = \pm\frac{1}{2}\right) = \frac{\hbar^2}{8M_X\rho_0^2} \geq 2V_{3a}\rho_0^3 \tag{6.79}$$

is fulfilled. This necessary condition means in practice that the distortion at adiabatic minima should verify $\rho_0 < 0.15$ Å for observing a dynamic JT effect at low temperatures. This figure implies $|R_{ax} - R_{eq}| < 0.13$ Å, a relatively small JT distortion which is difficult to find for host lattices much softer than MgO.

**Acknowledgements** The authors want to thank to Dr. Mark J. Riley for his relevant comments concerning MgO:Cu$^{2+}$ and Dr. Leona C. Nistor for her useful information on crystal growth. The support by the Spanish Ministerio de Ciencia y Tecnología under Project FIS2009–07083 is acknowledged.

## Appendix 1

The normal coordinates $Q_\theta$ and $Q_\varepsilon$ for a MX$_N$ complex can be expressed as

$$\begin{aligned}
Q_\theta &= \frac{1}{\sqrt{12}} \left\{ 2\left[z_5' + z_6'\right] - \left[z_1' + z_2' + z_3' + z_4'\right] \right\} \\
Q_\varepsilon &= \frac{1}{2} \left[z_1' - z_2' + z_3' - z_4'\right]
\end{aligned} \tag{6.80}$$

The meaning of $z_i'$ coordinates ($i = 1 - 6$) is explained in Fig. 6.5. If only $Q_\theta \neq 0$ then $Q_\theta = (2/\sqrt{3})(R_{ax} - R_{eq})$.

## Appendix 2

The expressions of the vibrational wavefunctions corresponding to the vibronic doublet, E and to the A singlet excited state are

$$\chi_{E\theta}(\varphi) = 2\chi_1(\varphi) + \chi_2\left(\varphi - \frac{2\pi}{3}\right) - \chi_3\left(\varphi - \frac{4\pi}{3}\right)$$
$$- 2\chi_4(\varphi - 2\pi) - \chi_5\left(\varphi - \frac{8\pi}{3}\right) + \chi_6\left(\varphi - \frac{10\pi}{3}\right)$$

$$\chi_{E\varepsilon}(\varphi) = +\chi_2\left(\varphi - \frac{2\pi}{3}\right) + \chi_3\left(\varphi - \frac{4\pi}{3}\right) - \chi_5\left(\varphi - \frac{8\pi}{3}\right) - \chi_6\left(\varphi - \frac{10\pi}{3}\right)$$

$$\chi_A(\varphi) = \chi_1(\varphi) - \chi_2\left(\varphi - \frac{2\pi}{3}\right) + \chi_3\left(\varphi - \frac{4\pi}{3}\right)$$
$$- \chi_4(\varphi - 2\pi) + \chi_5\left(\varphi - \frac{8\pi}{3}\right) - \chi_6\left(\varphi - \frac{10\pi}{3}\right)$$

$$(6.81)$$

while the corresponding total wavefunctions are

$$\Psi_A(r,\rho,\varphi) \cong \chi(\rho - \rho_0)\frac{1}{\sqrt{3}}\left\{ |3z^2 - r^2\rangle\chi_1(\varphi) + |3x^2 - r^2\rangle\chi_2\left(\varphi - \frac{2\pi}{3}\right)\right.$$
$$\left. + |3y^2 - r^2\rangle\chi_3\left(\varphi - \frac{4\pi}{3}\right)\right\}$$

$$(6.82)$$

$$\Psi_{E\theta}(r,\rho,\varphi) \cong \chi(\rho - \rho_0)\frac{1}{\sqrt{6}}\left\{ 2|3z^2 - r^2\rangle\chi_1(\varphi) + |3x^2 - r^2\rangle\chi_2\left(\varphi - \frac{2\pi}{3}\right)\right.$$
$$\left. - |3y^2 - r^2\rangle\chi_3\left(\varphi - \frac{4\pi}{3}\right)\right\}$$

The expressions for the vibrational wavefunction in the non-JT case are

$$\chi_{E\theta}(\varphi) = 2\chi_1(\varphi) - \chi_2\left(\varphi - \frac{2\pi}{3}\right) - \chi_3\left(\varphi - \frac{4\pi}{3}\right) + 2\chi_4(\varphi - 2\pi)$$
$$- \chi_5\left(\varphi - \frac{8\pi}{3}\right) - \chi_6\left(\varphi - \frac{10\pi}{3}\right)$$

$$(6.83)$$

$$\chi_{E\varepsilon}(\varphi) = +\chi_2\left(\varphi - \frac{2\pi}{3}\right) - \chi_3\left(\varphi - \frac{4\pi}{3}\right) + \chi_5\left(\varphi - \frac{8\pi}{3}\right) - \chi_6\left(\varphi - \frac{10\pi}{3}\right)$$

$$\chi_A(\varphi) = \chi_1(\varphi) + \chi_2\left(\varphi - \frac{2\pi}{3}\right) + \chi_3\left(\varphi - \frac{4\pi}{3}\right) + \chi_4(\varphi - 2\pi)$$
$$+ \chi_5\left(\varphi - \frac{8\pi}{3}\right) + \chi_6\left(\varphi - \frac{10\pi}{3}\right)$$

## References

1. Bersuker IB (2006) The Jahn-Teller effect. Cambridge University Press, Cambridge
2. Bruce AD (1980) Adv Phys 29:111
3. Millis AJ, Shraiman BI, Mueller R (1996) Phys Rev Lett 77:175

# 6 Dynamic and Static Jahn-Teller Effect in Impurities: Determination of the Tunneling

4. Woodward PM (1997) Acta Crystallogr B 53:44
5. Abragam A, Bleaney B (1970) Electron paramagnetic resonance of transition ions. Claredon Press, Oxford
6. Bill H (1984) In: Perlin YE, Wagner M (eds) The dynamical Jahn-Teller effect in localized systems. Elsevier, Amsterdam
7. Zorita E, Alonso PJ, Alcala R, Spaeth JM, Soethe H (1988) Solid State Commun 66:773
8. Minner E, Lovy D, Bill H (1993) J Chem Phys 99:6378
9. Villacampa B, Cases R, Orera VM, Alcala R (1994) J Phys Chem Solids 55:263
10. Moreno M, Barriuso MT, Aramburu JA, Garcia-Fernandez P, Garcia-Lastra JM (2006) J Phys Condens Matter 18:R315
11. Zorita E, Alonso PJ, Alcala R (1987) Phys Rev B 35:3116
12. Sabbe K, Callens F, Boesman E (1998) Appl Magn Reson 15:539
13. Vercammen H, Schoemaker D, Briat B, Ramaz F, Callens F (1999) Phys Rev B 59:286
14. Vrielinck H, Callens F, Matthys P (2001) Phys Rev B 64:214105
15. Sierro J (1966) J Phys Chem Solids 28:417
16. Coffman RE (1966) Phys Lett 21:381; Coffman RE, Lyle DL, Mattison DR (1968) J Phys Chem 72:1392
17. Boatner LA, Reynolds RW, Abraham MM, Chen Y (1973) Phys Rev Lett 31:7
18. Reynolds RW, Boatner LA, Abraham MM, Chen Y (1974) Phys Rev B 10:3802
19. Riley MJ, Noble CJ, Tregenna-Piggott PLW (2009) J Chem Phys 130:104708
20. Ham FS (1972) In: Geschwind S (ed) Electron paramagnetic resonance. Plenum, New York
21. Bersuker IB (1963) Sov Phys JETP 16:933; Bersuker IB (1963) Sov Phys JETP 17:836
22. Guha S, Chase LL (1974) Phys Rev B 12:1658
23. Schmitz-Du Mont O (1968) Monat für Chemie 99:1285
24. Höchli U, Müller KA, Wysling P (1965) Phys Lett 15:5
25. Garcia-Fernandez P, Sousa C, Aramburu JA, Barriuso MT, Moreno M (2005) Phys Rev B 72:155107
26. Garcia-Fernandez P, Bersuker IB, Aramburu JA, Barriuso MT, Moreno M (2005) Phys Rev B 71:184117
27. Trueba A, Garcia-Lastra JM, De Graaf C, Garcia-Fernandez P, Aramburu JA, Barriuso MT, Moreno M (2006) Chem Phys Lett 430:51
28. Pascual JL, Savoini B, Gonzalez R (2004) Phys Rev B 70:045109
29. Barriuso MT, Garcia-Fernandez P, Aramburu JA, Moreno M (2001) Solid State Commun 120:1
30. Garcia-Fernandez P, Trueba A, Barriuso MT, Aramburu JA, Moreno M (2010) Phys Rev Lett 104:035901
31. Kohn W (1968) In: Many body physics. Gordon and Breach, New York
32. Resta R (2002) J Phys Condens Matter 14:R625
33. Höchli U, Estle TL (1968) Phys Rev Lett 18:128
34. Abragam A, Pryce MHL (1950) Proc R Soc Lond A205:135
35. Ham FS (1968) Phys Rev 166:307
36. Garcia-Fernandez P, Bersuker IB, Boggs JE (2006) J Chem Phys 125:104102
37. Garcia-Lastra JM, Barriuso MT, Aramburu JA, Moreno M (2005) Chem Phys 317:103
38. O'Brien MCM (1964) Proc R Soc London Ser A 281:323
39. Ham FS (1987) Phys Rev Lett 58:725
40. Jasper AW, Kendrick BK, Mead CA, Truhlar DG (2003) In: Yang X, Liu K (eds) Modern trends in chemical reaction dynamics. World Scientific, Singapore
41. Zwanziger JW, Grant ER (1987) J Chem Phys 87:2954
42. Berry MV (1984) Proc R Soc Lond Ser A 392:45
43. Garcia-Fernandez P, Bersuker IB, Boggs JE (2006) Phys Rev Lett 96:163005
44. Meiswinkel R, Köppel H (1990) Chem Phys 144:117
45. Garcia-Fernández P, Senn F, Daul C, Aramburu JA, Barriuso MT, Moreno M (2009) Phys Chem Chem Phys 11:75145
46. Trueba A, Garcia-Fernández P, Senn F, Daul CA, Aramburu JA, Barriuso MT, Moreno M (2010) Phys Rev B 81:075107

47. García-Fernández P, Trueba A, García-Lastra JM, Barriuso MT, Moreno M, Aramburu JA (2009) In: Koeppel H, Yarkony DR, Barentzen H (eds) The Jahn-Teller effect, Springer series of chemical physics. Springer, Heidelberg, pp 415–450
48. Burton W, Cabrera N, Franck FC (1951) Philos Trans R Soc A 2431:299
49. Amelinckx S (1964) The direct observation of dislocations. Academic Press, New York
50. Friedel J (1964) Dislocations. Pergamon Press, Oxford
51. Ma SK (1993) Statistical mechanics. World Scientific, Singapore, ISBN-10: 9971966077
52. Henderson B, Imbush GF (1989) Optical spectroscopy of inorganic solids. Oxford Science Publications, Oxford
53. Bourgoin J, Lannoo M (1983) Point defects in semiconductors. Springer, Berlin
54. Griffith JS (1961) The theory of transition metal ions. Cambridge University Press, Cambridge
55. Jacobsen JM, Tissue BM, Yen WM (1992) J Phys Chem 96:1547
56. Jahren AH, Kruger MB, Jeanloz RJ (1992) J Appl Phys 71:1579
57. Duclos S, Vohra YK (1990) Ruoff AL Phys Rev B 41:5372
58. Gaudry E, Kiratisin A, Sainctavit P, Brouder C, Mauri F, Ramos A (2003) Phys Rev B 67:094108
59. Hjortsberg A, Nygren B, Vallin J, Ham FS (1977) Phys Rev Lett 39:1233
60. Seijo L, Barandiaran Z (1999) In: Leszczynski J (ed) Computational chemistry: reviews of current trends. World Scientific, Singapore
61. Roos BO, Taylor OR, Siegbahn PEM (1980) Chem Phys 48:157
62. Roos BO, Andersson K, Fülscher MP, Malmqvist P-A, Serrano-Andrés L, Pierloot K, Merchán M (1996) Adv Chem Phys 93:219
63. Pou-Amerigo R, Merchan M, Nebot-Gil I, Widmark PO, Roos B (1995) Theor Chim Acta 92:149
64. Hilbrandt N, Martin M (1999) J Phys Chem B 103:4797
65. Asakura K, Iwaswa Y, Kuroda H (1986) J Physique 47:C8–317
66. Sanchez del Rio M, Garcia J, Gonzalez R et al (1989) Physica B 158:527
67. Dick A, Krausz ER, Hadler KS, Noble CJ, Tregenna-Piggott PLW, Riley MJ (2008) J Phys Chem C 112:14555
68. Barriuso MT, Aramburu JA, Moreno M (2002) J Phys Condens Matter 14:6521
69. Schmitz-Du Mont O, Fendel H (1965) Monat für Chemie 96:495
70. Cordischi D, Pepe F, Schiavello M (1973) J Phys Chem 77:1240
71. Raizman A, Barack J, Englman R, Suss JT (1981) Phys Rev B 24:6262
72. Tolparov YN, Bir GL, Sochava LS, Kovalev NN (1974) Sov Phys Solid State 16:573

# Chapter 7
# Experimental Evaluation of the Jahn-Teller Effect Parameters by Means of Ultrasonic Measurements. Application to Impurity Centers in Crystals

**V.V. Gudkov and I.B. Bersuker**

**Abstract** A review is presented of the worked out earlier method that uses ultrasonic experiments to evaluate Jahn-Teller effect (JTE) parameters, mostly linear vibronic coupling constants, in application to impurity centers of dopant crystals. The method employs temperature dependent ultrasound attenuation and phase velocity measurements. Distinguished from previous attempts to detect the JTE parameters by ultrasound, this method does not assume any specific mechanism of relaxation, and hence it can be applied to any JTE problem, irrelevant to its complicated dynamics of distortions. It is shown that by combining the ultrasound results with some additional information about the JT stabilization energy obtained from independent sources, the whole adiabatic potential energy surface of the JT center can be evaluated. Two examples of application of this method relevant to two JTE problems are considered, both for impurity centers in crystals with zinc-blend structure and tetrahedral coordination of the impurity ion: $ZnSe:Fe^{2+}$ with the $E \otimes e$ problem and $ZnSe:Cr^{2+}$ with a T term and the $T \otimes (e + t_2)$ problem.

## 7.1 Introduction

In the overwhelming majority of doped crystals impurity centers in their ground and/or excited states are subject to the Jahn-Teller (JT) effect (JTE) which influences essentially the interpretation of the experimental data and predicts new properties

---

V.V. Gudkov (✉)
Ural Federal University, 19, Mira st., 620002 Ekaterinburg, Russia
e-mail: gudkov@imp.uran.ru

I.B. Bersuker
Institute for Theoretical Chemistry, The University of Texas at Austin, Austin, TX 78712, USA
e-mail: bersuker@cm.utexas.edu

M. Atanasov et al. (eds.), *Vibronic Interactions and the Jahn-Teller Effect: Theory and Applications*, Progress in Theoretical Chemistry and Physics 23, DOI 10.1007/978-94-007-2384-9_7, © Springer Science+Business Media B.V. 2012

(see, e.g., [3]). Investigation of the JTE in crystals doped with transition metals $3d$ impurities was intensified recently in view of their possible applications in optoelectronic and spintronic devices [14, 22]. On the other hand, description of the JTE involves several constants that define the vibronic Hamiltonian, including primary force constants, linear and quadratic vibronic coupling constants, and JT active vibrational frequencies that determine the JT stabilization energy and potential energy barriers. Usually these parameters are obtained by means of model calculations or simulations with optical and magnetic resonance experiments (see, e.g., [3] and more recent calculations for II–VI:$Cr^{2+}$ compounds in [15]). Ultrasonic experiments have not been routinely used so far for investigation of the JTE [20]. The reasons of this are in the lack of standard equipment for the required specific ultrasonic measurements, while the generated in such experiments small energy phonons (as compared with high frequency photons) restrict the treatment to only the lowest energy levels. But there is also the advantage that these particular levels can be studied with ultrasound in more detail.

In crystals containing small concentration of JT impurities the ultrasonic attenuation is seen as a peak of relaxation origin [26]. The system $Al_2O_3$:$Ni^{3+}$ was the first for which information about the ground state of the $3d$ electrons was obtained by means of ultrasonic attenuation measurements [27]. In this system the impurity ion $Ni^{3+}$ $(d^7)$ has an octahedral coordination with an E term ground state subject to the $E \otimes e$ problem. The authors [27] used the temperature dependence of relaxation time obtained from the data of ultrasonic attenuation to estimate the values of the deformation potential $3\beta$, tunneling splitting $3\Gamma$, potential energy barrier $V_0$, and the random strain energy. The simulation procedure in the paper [27] is based on specific assumptions about the mechanisms of relaxation (thermal activation over the assumed potential barrier, tunneling through the barrier, and a two-phonon Raman process) and fitting the model temperature dependence of relaxation time to the experimental one. Some of these mechanisms were also considered by Averkiev et al. [1] in discussion of experiments performed on $p$-GaAs:Mn [19].

A similar to [27] approach was used recently for evaluation of the temperature dependence of relaxation time from attenuation of longitudinal ultrasonic waves propagating along the [110] crystallographic axis in ZnSe:$Cr^{2+}$ [9]. This compound has a zinc-blende structure with tetrahedral surrounding of the chromium ion ($d^4$ configuration, $^5T_2(e^2t^2)$ high-spin ground state [14]), and hence its expected JTE problem is $T \otimes (e + t_2)$ [3]. In the linear approximation the dominant JT distortions in this case can be either of $E$ type (tetragonal) or $T_2$ type (trigonal). These displacements and the angular wave functions that obey the same transformation properties are shown in Figs. 7.1 and 7.2. Keeping in mind that longitudinal waves produce variations of the volume element (i.e., variations of the density), one may assume that the totally symmetric vibrations (breathing mode $A_1$) may be active in this case and it may contribute to the elastic modulus $C_\ell = (C_{11} + C_{12} + 2C_{44})/2$. The totally symmetric mode is described by normal coordinates $Q_a$ that transform as $x^2 + y^2 + z^2$ (Fig. 7.3).

7 Evaluation of the Jahn-Teller Effect Parameters in an Ultrasonic Experiment    145

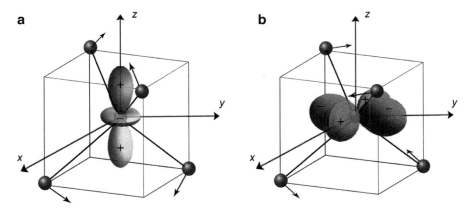

**Fig. 7.1** The $E$-type symmetrized displacements $Q_\vartheta$ (**a**) and $Q_\varepsilon$ (**b**) in $T_d$ symmetry and the angular wave functions $\vartheta \sim 2z^2 - x^2 - y^2$ (**a**) and $\varepsilon \sim x^2 - y^2$ (**b**)

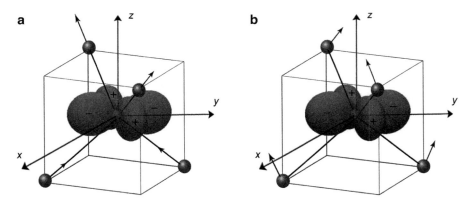

**Fig. 7.2** The $T_2'$ and $T_2''$-type symmetrized displacements $Q_\zeta'$ (**a**) and $Q_\zeta''$ (**b**), respectively, in $T_d$ symmetry and the angular wave functions $\zeta \sim xy$

**Fig. 7.3** Totally symmetric displacements $Q_a$ in tetrahedral $T_d$ symmetry

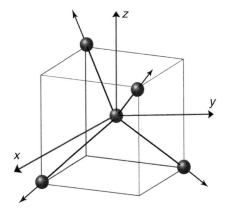

It was shown [10] that the slow shear mode propagating in [110] crystallographic direction in the $ZnSe:Cr^{2+}$ crystal has essential anomalies in temperature dependence: a giant peak of attenuation and reduction of phase velocity. The phase velocity and attenuation of this wave are determined by the real and imaginary parts of the elastic modulus $C_s = (C_{11} - C_{12})/2$. This mode produces tetragonal distortions of the JT complex which are described by $E$-type displacements $Q_\vartheta$ and $Q_\varepsilon$ that obey transformation properties $(2z^2 - x^2 - y^2)$ and $(x^2 - y^2)$, respectively (Fig. 7.2). In contrast to the slow shear mode, the fast shear mode, for which the velocity is determined by the elastic modulus $C_f = C_{44}$, did not display anomalies in temperature dependences of phase velocity and attenuation that can be attributed to the $Cr^{2+}$ impurities. The fast shear mode of [110] propagation direction excites trigonal distortions of the tetrahedral complex. The absence of the anomalies indicates that the trigonal minima of the adiabatic potential energy surface (APES), if any, are not in the ground state. It means that the dominant distortions in this system are tetragonal, and the APES is that of the particular problem $T \otimes e$ [3]. In the other system reviewed here, $ZnSe:Fe^{2+}$, the ion $Fe^{2+}$ ($d^6$) in tetrahedral environment has an $^5E(e^3t^3)$ high-spin ground state subject to the $E \otimes e$ JTE problem with the only tetragonal low-symmetry distortions.

The attenuation and phase velocity of the longitudinal wave are determined by the modulus $C_\ell$, which can be presented in the form of $C_\ell = C_B + C_{44} + C_s/3$, where $C_B = (C_{11} + 2C_{12})$ is the bulk modulus describing the breathing mode. Hence, investigation of the slow shear mode is of particular interest for determining the contribution of all the normal vibronic modes to $C_\ell$ in $ZnSe:Cr^{2+}$ and evaluating the linear vibronic coupling constants of the totally symmetric $F_\alpha$ and tetragonal $F_E$ distortions (in the denotations of [3]).

However, quantitative evaluation of the main JTE parameter, the linear vibronic coupling constant $F_E$, was not achieved in the previous attempts because of the following problems. First, simulation of the relaxation time in the case of negligible Raman processes yields the product $\beta\Gamma$, but not their values independently. Secondly, the concentration of the dopand $n_{Cr} = 10^{20}\,cm^{-3}$ was too large to register the ultrasonic signal of the slow shear mode: its attenuation at 7–20 K (i.e., at the temperatures of particular interest) is too strong. Therefore, measurements have been carried out on another specimen which has lower concentration of chromium ions [13].

In this paper we overview the method [4] and results of evaluation of JTE parameters for impurity centers in crystals by means of ultrasonic measurements. In this method, in particular distinguished from previous attempts, no specific assumptions of relaxation mechanisms are required. This is of special importance in application to JTE problems which have complicated APES with conical intersections, energy troughs, equivalent minima and barriers between them in multidimensional space, where *a priori* reasonable assumptions of relaxation mechanisms may be unreliable. Another important feature of this method is that it makes possible to obtain the modulus of the main JT parameter, the linear vibronic coupling constant, from two independent measurement (attenuation and velocity) thus making the results more reliable. We show also how the combination of the data obtained in this way with some additional information on the system obtained from independent sources allows to estimate the primary force constant and to determine the positions of the orthorhombic and trigonal saddle points in the linear $T \otimes (e + t_2)$ problem for the $Cr^{2+}$ ion in the ZnSe host, and the quadratic vibronic coupling constant for $E \otimes e$ problem of ZnSe:$Fe^{2+}$, thus revealing all the main features of the APES of these two systems.

## 7.2 Experimental Details

The crystals used in the ultrasound experiments were prepared in the Institute of Solid State Physics of the Russian Academy of Sciences using the Bridgman method from a melt under conditions of excess inert gas pressure [18] (ZnSe:$Fe^{2+}$, $n_{Fe} = 2.2 \times 10^{19}\,cm^{-3}$) and in the P.N. Lebedev Physical Institute of the Russian Academy of Sciences, according the technology reported in [16, 17] (ZnSe:$Cr^{2+}$, $n_{Cr} = 3.8 \times 10^{18}\,cm^{-3}$). The concentration of the dopand was evaluated from the absorption spectrum of the crystal. The specimens had the form of parallelepiped with the distance of about 5 mm between the parallel faces.

The experiments were carried out on a setup operating as a variable-frequency bridge [7]. The setup made it possible to measure ultrasonic attenuation $\alpha$ and phase velocity $v$. Longitudinal and shear ultrasonic waves were generated and registered by LiNbO$_3$ piezoelectric transducers with fundamental resonant frequency of about 50 MHz. The ultrasonic waves propagated along the [110] crystallographic axis. The accuracy of the measurements was characterized by 0.2 dB for $\Delta \alpha$ and $2 \times 10^{-5}$ for $\Delta v / v$.

## 7.3 The JTE $E \otimes e$ Problem in ZnSe:$Fe^{2+}$

The symmetry of the high-spin ground state of the $Fe^{2+}$ ion in tetrahedral coordination $^5E(e^3t^3)$ [14] points to the particular JT problem $E \otimes e$ with the two electronic wave functions of $\vartheta \sim 2z^2 - x^2 - y^2$ and $\varepsilon \sim x^2 - y^2$ symmetry and tetragonal displacements $Q_\vartheta$ and $Q_\varepsilon$ shown in Fig. 7.1. The results of the ultrasonic measurements [12] for ZnSe:$Fe^{2+}$ crystal are shown in Fig. 7.4. We can see that the modulus $C_{44}$ has ordinary temperature dependence relevant to cubic crystals (see, e.g., [6]) in contrast to the modulus $C_s$ that manifests significant softening at low temperatures. The attenuation of the slow shear mode $\alpha_s$ for which the phase velocity $v_s$ is determined by this modulus ($v_s = (C_s/\rho)^{1/2}$, where $\rho$ is the density), has a peak in the region of the $C_s$ softening. Attenuation of the fast shear mode, as well as of the longitudinal mode, has no anomalies at low temperatures. It follows that the anomalies in the temperature dependence of ultrasound attenuation and

**Fig. 7.4** Temperature dependences of attenuation of the ultrasonic slow shear wave (**a**) and the elastic moduli $C_{44}$ and $C_s$ (**b**) in ZnSe:Fe$^{2+}$ measured at 52 MHz. $\Delta\alpha_s = \alpha_s(T) - \alpha_{s0}$, $\alpha_{s0} = \alpha_s(T \to 0)$, $\Delta C_i/C_{i0}) = [C_i(T) - C_i(T_0)]/C_i(T_0)$, $T_0 = 90$ K. *Dark symbols* relate to $C_{44}$, *light* ones – to $C_s$. *Solid line* in (**a**) represents simulation of the background attenuation $\alpha_b(T) = \alpha_1 T^2 + \alpha_2 T^4$ with $\alpha_1 = 2 \times 10^{-4}$ dB·cm$^{-1}$K$^{-2}$ and $\alpha_2 = 6.7 \times 10^{-8}$ dB·cm$^{-1}$K$^{-4}$

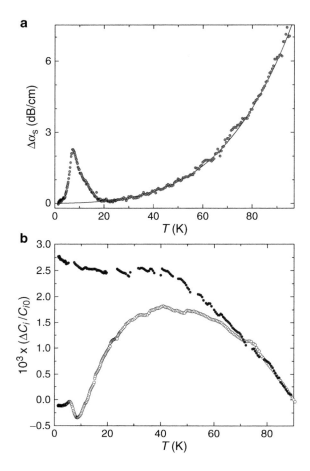

phase velocity are observed only for the mode that generates tetragonal distortions which can reliably be attributed to the JTE as there are no other visible sources of such distortions.

To determine the linear vibronic constant $F_E$ in this JTE problem we use the general relation between the complex wave number and the elastic modulus (also complex). Introducing the contribution of the relaxation process to the wave number $k_\beta$ and total elastic modulus $C_\beta$ as, respectively, $(\Delta k_\beta)_r$ and $(\Delta C_\beta)_r$ we have

$$\frac{(\Delta k_\beta)_r}{k_{\beta 0}} = -\frac{1}{2}\frac{(\Delta C_\beta)_r}{C_{\beta 0}}, \qquad (7.1)$$

where $k_{\beta 0}$ and $C_{\beta 0}$ are the reference values of $k_\beta$ and $C_\beta$.

The contribution of the relaxation process to the dynamic (frequency dependent) elastic modulus can be expressed in terms of relaxation time $\tau$ and relaxed $\Delta C_\beta^R$

7 Evaluation of the Jahn-Teller Effect Parameters in an Ultrasonic Experiment

(isothermal) and unrelaxed $\Delta C_\beta^U$ (adiabatic) elastic moduli [25]:

$$(\Delta C_\beta)_r = -(\Delta C_\beta^U - \Delta C_\beta^R)\frac{1 - i\omega\tau}{1 + (\omega\tau)^2}, \qquad (7.2)$$

where $\omega$ is the cyclic frequency of the wave and the symbol $\Delta$ before $C_\beta^R$ and $C_\beta^U$ indicates their JTE contributions. Space ($\mathbf{r}$) and time ($t$) dependences of the displacements are described by the factor $\exp[i(\omega t - \mathbf{k}_\beta \cdot \mathbf{r})]$. The total elastic modulus may include different relaxation processes; we assume that all the other contributions beyond the JTE have much less pronounced temperature dependence (at least in the investigated temperature interval).

In calculating $(\Delta C_\beta)_r$ we follow the approach suggested by Sturge [26] and consider only the relaxed modulus because the adiabatic one vanishes. Thus,

$$-(\Delta C_\beta^U - \Delta C_\beta^R) = \Delta C_\beta^R. \qquad (7.3)$$

Taking into account that $C_\beta$ can be a linear combination of the moduli defined in Cartesian system, we present the general expression for the tensor component of the elastic moduli in isothermal limit as follows

$$\Delta C_{ijkl}^R = \left(\frac{\partial^2 \mathscr{F}}{\partial \varepsilon_{ij} \partial \varepsilon_{kl}}\right)_{\varepsilon=0}, \qquad (7.4)$$

where $\mathscr{F}$ is the free energy of the JT system, and $\varepsilon_{ij}$ are the components of the strain tensor. For propagation of the slow shear wave we have:

$$\Delta C_s^R = \frac{C_{11}^R - C_{12}^R}{2}. \qquad (7.5)$$

The expression for resonance ultrasonic attenuation by the tunneling-split vibronic levels of the JT system was obtained earlier by one of us [2]. It can be used for relaxation attenuation by transforming it to the limit of infinitesimally small resonance frequencies. In this way we get:

$$\Delta(\alpha_s)_r = \frac{1}{2}\frac{na_0^2 F_E^2}{\kappa T}\frac{k_{s0}}{C_{s0}}\frac{\omega\tau}{1 + (\omega\tau)^2}, \qquad (7.6)$$

where $a_0$ is the $3d$ ion–ligand distance, $n$ is the concentration of the dopand, and ($\kappa$ is the Boltzmann constant. As a result, Eq. 7.1 can be presented in the form:

$$\frac{(\Delta k_s)_r}{k_{s0}} = -\frac{1}{2}\frac{\Delta C_s^R}{C_{s0}}\frac{1 - i\omega\tau}{1 + (\omega\tau)^2}. \qquad (7.7)$$

After separating the real and imaginary parts and substituting $k_s = \omega/v_s - i\alpha_s$ we have

$$\frac{(\Delta \alpha_s)_r}{k_{s0}} = -\frac{1}{2}\frac{\Delta C_s^R}{C_{s0}}\frac{\omega\tau}{1+(\omega\tau)^2}. \tag{7.8}$$

Note that $\alpha_\beta$ characterizes the attenuation of amplitude, not energy. From (7.6) and (7.8) the expression for the relaxed modulus is

$$\Delta C_s^R = \frac{na_0^2 F_E^2}{\kappa T}. \tag{7.9}$$

Hence from attenuation measurements it follows that

$$\frac{(\Delta \alpha_s)_r}{k_{s0}} = \frac{1}{2}\frac{na_0^2 F_E^2}{\kappa T C_{s0}}\frac{\omega\tau}{1+(\omega\tau)^2}, \tag{7.10}$$

while for the phase velocity reduction we get a similar formula:

$$\frac{(\Delta v_s)_r}{v_{s0}} = -\frac{1}{2}\frac{na_0^2 F_E^2}{\kappa T C_{s0}}\frac{1}{1+(\omega\tau)^2}. \tag{7.11}$$

Both expressions have similar dependence on JT vibronic coupling constant, but the measurements of attenuation and phase velocity are independent providing for two independent methods of evaluation of the $|F_E|$ magnitude.

Now we introduce the temperature $T_1$ determined from the condition $\omega\tau(T_1) = 1$. For this temperature the frequency-dependent factors $1/\{1 + [\omega\tau(T_1)]^2\} = \omega\tau(T_1)/\{1 + [\omega\tau(T_1)]^2\} = 1/2$, and the formulas for evaluating $F_E^2$ simplify: they require only the knowledge of either the relaxation contribution to $\alpha_s$, or the velocity reduction $\Delta v_s$ at the temperature $T_1$.

In the case of ZnSe:Fe$^{2+}$ the JT contribution to the modulus $C_s$ emerges as an addition to the temperature dependent background (Fig. 7.4b) which makes it difficult to evaluate $(\Delta v_s)_r$ sufficiently accurate. Therefore, we discuss here the attenuation data only. Figure 7.4a shows that the attenuation of relaxation origin in the form of a peak is superimposed on a temperature dependent background too, but this background attenuation (the solid line in Fig. 7.4a) is well described by the function $\alpha_b(T) = \alpha_1 T^2 + \alpha_2 T^4$ with $\alpha_1 = 2 \times 10^{-4}\,\mathrm{dB \cdot cm^{-1}\,K^{-2}}$ and $\alpha_2 = 6.7 \times 10^{-8}\,\mathrm{dB \cdot cm^{-1}\,K^{-4}}$. As a result, we get $(\Delta\alpha_s)_r = \Delta\alpha_s - \alpha_b$.

Thus, with the parameter values at the temperature where $\omega\tau(T_1) = 1$ and the frequency dependent relaxation term equals 1/2, in Eq. 7.10, we obtain the following expression for the vibronic coupling constant $F_E^2$

$$F_E^2 = \frac{4\kappa T_1 C_{s0}}{na_0^2}\frac{\Delta\alpha_s(T_1)}{k_{s0}}. \tag{7.12}$$

# 7 Evaluation of the Jahn-Teller Effect Parameters in an Ultrasonic Experiment

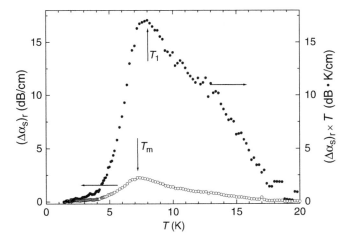

**Fig. 7.5** Temperature dependence of relaxation attenuation $(\Delta\alpha_s(T))_r$ in comparison with the product $(\Delta\alpha_s(T))_r \times T$ which makes it possible to determine the temperature $T_1$ corresponding to the condition $\omega\tau(T) = 1$

To define the temperature $T_1$ we use the method suggested in [11]. It is based on the specificity of the factor $f_1 = \omega\tau/[1+(\omega\tau)^2]$ which enters the expression for attenuation in (7.10): $f_1(T)$ has a maximum at $T = T_1$ which can be seen in the temperature dependence of $\Delta\alpha_s(T) \cdot T$. The procedure of determination of $T_1$ is illustrated in Fig. 7.5. The difference between $T_1$ and the temperature $T_m$ corresponding to peak of attenuation of $(\Delta\alpha_s)_r$ is due to the temperature dependence of the relaxed modulus $\Delta C_s^R \sim 1/T$ which we take into account by multiplying $(\Delta\alpha_s)_r$ by $T$. In the case of ZnSe:Fe$^{2+}$: $T_1 = 8$ K and $T_m = 7.2$ K. With these data and the JT ion–ligand distance $a_0 = 2.46$ Å, as well as $C_{s0} = 1.835 \times 10^{11}$ dyn we obtained the following magnitude for the linear vibronic coupling constant of the JTE in ZnSe:Fe$^{2+}$: $|F_E| = 2.7 \times 10^{-6}$ dyn (0.016 eV/Å). Note that, as mentioned already, in the evaluation of the main JTE constant $F_E$, given above, we do not employ any assumptions about the mechanism of relaxation that may be different in different kinds of JT problems, as they may have essentially different forms of APES with a variety of complicated combinations of equivalent energy minima, energy barriers, conical intersections, etc., induced by the JTE and pseudo JTE [3]. For the majority of these problems a reasonable assumption about the mechanism of relaxation is hardly possible, and in any case the overwhelming mechanism may vary with temperature. Being independent of the specific mechanisms, this method does not require any modeling of the temperature dependence of the relaxation time (as in [27]), and it is thus applicable to any of the JTE situations. Application to another problem, namely, $T \otimes (e+t_2)$, is discussed in Sect. 7.3.

In the linear approximation with respect to the vibronic coupling terms in the Hamiltonian the APES has the form of the "Mexican hat" [3] with free rotation of the tetragonal distortion along the bottom of the circular trough. The latter becomes warped with three equivalent wells alternating regularly with three humps in the

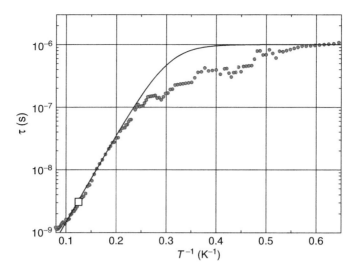

**Fig. 7.6** Temperature dependence of relaxation time in ZnSe:Fe$^{2+}$. *Solid line* represents the dependence $\tau(1/T) = \frac{1}{3}[\tau_t^{-1} + \tau_0^{-1}\exp(-V_0/\kappa T)]^{-1}$, where $\tau_0 = 1.7 \times 10^{-10}$ s, $\tau_t = 3.3 \times 10^{-6}$ s, $V_0 = 22.2$ cm$^{-1}$. *Light square* indicates $\tau(1/T_1)$

quadratic approximation; in this case the static distortions may be observed at low temperatures when the system is "frozen" in the minima of the APES. The latter case may be reflected in the temperature dependence of the relaxation time as a tendency to saturation due to the low frequency of the tunneling through the potential energy barrier.

The procedure of evaluation of the function $\tau(T)$ was initially suggested by Sturge et al. [27] by means of employing Eq. 7.10 and the condition of maximum attenuation $\omega\tau(T_1) = 1$. This leads to the following equations:

$$\frac{(\Delta\alpha_s(T))_r}{k_{s0}} = \frac{1}{2}\frac{na_0^2 F_E^2}{\kappa T C_{s0}} \frac{\omega\tau}{1+(\omega\tau)^2}, \tag{7.13}$$

$$\frac{(\Delta\alpha_s(T_1))_r}{k_{s0}} = \frac{1}{4}\frac{na_0^2 F_E^2}{\kappa T_1 C_{s0}}. \tag{7.14}$$

However, due to temperature dependence of the relaxed modulus $\Delta C_s^R \sim 1/T$, the peak of attenuation does not correspond exactly to $T = T_1$ (Fig. 7.6), and hence the procedure of determination of $T_1$, discussed above, should be employed, in particular, for peaks observed at low temperatures when the relaxed modulus exhibits significant reduction (i.e, at $T < 10$ K).

Solution of the system (7.13–7.14) with respect to $\tau$ yields:

$$\tau(T) = \frac{1}{\omega}\left(\frac{\alpha_r(T_1)T_1}{\alpha_r(T)T} \pm \sqrt{\left(\frac{\alpha_r(T_1)T_1}{\alpha_r(T)T}\right)^2 - 1}\right), \tag{7.15}$$

# 7 Evaluation of the Jahn-Teller Effect Parameters in an Ultrasonic Experiment

where for simplicity we replaced $(\Delta\alpha_s(T))_r$ by $\alpha_r(T)$ and $(\Delta\alpha_s(T_1))_r$ by $\alpha_r(T_1)$, and the sign before the square root is determined by the condition that the relaxation time should be reduced with increasing temperature. Implications of the Eq. 7.15 in the experimental data obtained for $ZnSe:Fe^{2+}$ are shown in Fig. 7.6.

Two mechanisms of relaxation are assumed to contribute to the relaxation rate $\nu = \tau^{-1}$:

$$\tau^{-1} = 3\left(\tau_T^{-1} + \tau_t^{-1}\right), \tag{7.16}$$

where $\tau_T^{-1}$ describes thermal activation over the potential energy barrier $V_0$ (which is determined with respect to the lowest vibronic level in the minima of APES) and $\tau_t^{-1}$ is due to the tunneling through the barrier. Distinguished from [26], this expression contains the factor 3 in the right hand side because all the three equivalent minima of the APES take part in relaxation, and we want to get only one of the three energy barriers between them.

At high temperatures the thermal activation is dominant

$$\tau \approx \frac{1}{3}\tau_T = \frac{1}{3}\tau_0 e^{V_0/\kappa T}, \tag{7.17}$$

where $\frac{1}{3}\tau_0$ is the relaxation time in the high-temperature limit, whereas at low temperatures it is replaced by tunneling through the barrier which is temperature independent. The $V_0$, $\tau_t$ and $\tau_0$ values can be estimated by means of simulation of the curve $\tau(1/T)$. The result of such simulation is shown in Fig. 7.6: $\tau_t = 3.3 \times 10^{-6}$ s, $\tau_0 = 1.7 \times 10^{-10}$ s, and $V_0 = 22.2\,cm^{-1}$. As seen from this figure, at about $T < 2$ K $\tau$ is almost independent of temperature meaning that the relaxation becomes pure tunneling through the barrier $\delta$ between the equivalent minima along the bottom of the trough of the "Mexican hat" of the $E \otimes e$ problem [3],

$$\delta = \frac{4E_{JT}|G_E|}{K_E + 2|G_E|}, \tag{7.18}$$

where $G_E$ is the quadratic vibronic coupling constant, $K_E$ is the primary force constant, $E_{JT}$ is the JT stabilization energy,

$$E_{JT} = \frac{F_E^2}{2(K_E - 2|G_E|)}. \tag{7.19}$$

As the relaxation barrier is read off the zero vibrations, $\delta = \frac{1}{2}\hbar\omega_{rot} + V_0$, where $\omega_{rot}$ is the rotational frequency along the bottom of the trough of the "Mexican hat." Hence in addition to the main linear vibronic coupling constant $F_E$ obtained from the ultrasonic measurements we need the values of at least two out of three constants $E_{JT}$, $K_E$, and $G_E$. In principle, they may be available from calculations or independent optical and/or EPR experiments. Unfortunately, the published data for II–VI:Fe2+ crystals [21, 23] are rather controversial and unreliable; they cannot be used for the construction of the APES of this crystal (as it is done below for ZnS:Cr$^{2+}$ crystal).

## 7.4 The JTE $T \otimes (e+t_2)$ Problem in ZnSe:Cr$^{2+}$

An earlier ultrasonic investigation of the ZnSe:Cr$^{2+}$ crystal with impurity concentration of $n_{Cr} = 10^{20}$ cm$^{-3}$ revealed (1) a peak of attenuation for the longitudinal wave and softening of the corresponding modulus $C_\ell$ [9], (2) no anomalies in temperature dependences of attenuation of the fast shear wave and the modulus $C_{44}$ that can be attributed to the JTE, and (3) a peak of attenuation of the slow shear wave, so large that it was impossible to measure it [10]. The slow shear mode was studied again on a sample of the crystal with $n_{Cr} = 3.8 \times 10^{18}$ cm$^{-3}$ [4]. Figure 7.7 shows the results of the experiment performed with ultrasound at 54 MHz. In addition to the curves related to the doped crystal, we show the data obtained for the undoped crystal [8]. The curves are presented under the assumption that the parameters of the doped and undoped crystals are the same at $T \to 0$, meaning that we assume that the temperature dependent contribution to the relaxation from the JTE is much larger than other possible contributions of the dopand.

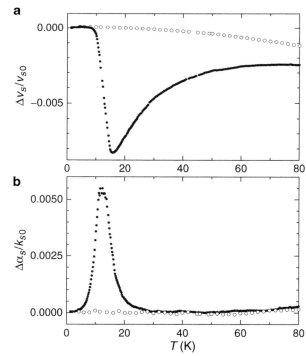

**Fig. 7.7** Temperature dependence of phase velocity (**a**) and attenuation (**b**) of slow shear wave. *Dark symbols* relate to ZnSe:Cr$^{2+}$, *light ones* – to undoped ZnSe. $\Delta v_s(T) = v_s(T) - v_{s0}$, $v_{s0} = v_s(T \to 0)$, $\Delta \alpha_s(T) = \alpha_s(T) - \alpha_{s0}$, $\alpha_{s0} = \alpha_s(T \to 0)$, $f = 54$ MHz

7 Evaluation of the Jahn-Teller Effect Parameters in an Ultrasonic Experiment

The slow shear mode produces tetragonal distortions of the JT complex which are described by $E$-type displacements $Q_\vartheta$ and $Q_\varepsilon$ (Fig. 7.1). The fast shear mode for which the velocity is determined by the elastic modulus $C_f = C_{44}$ did not display anomalies of phase velocity and attenuation that can be attributed to the $Cr^{2+}$ impurities. The fast shear mode propagating along the [110] direction excites trigonal distortions of the tetrahedral complex. The absence of the anomalies in this direction indicates that the system does not have trigonal minima of the APES in the ground state. It means that the dominant distortions in this system are tetragonal, and the APES is that of the particular problem $T \otimes e$ [3]. Since the longitudinal wave produces variations of the volume element (i.e., variations of the density), and this wave has anomalies in temperature dependences of attenuation and phase velocity [9], we may assume that the breathing mode $A_1$ (described by $Q_a$ – totally symmetric normal coordinates shown in Fig. 7.3) may contribute to the attenuation of the longitudinal wave and the elastic modulus $C_\ell = (C_{11} + C_{12} + 2C_{44})/2$. Hence the evaluation of the APES in the linear approximation with respect to the vibronic coupling requires evaluation of the linear vibronic coupling constants $F_\alpha$ and $F_E$.

To explore the influence of the totally symmetric vibrations to the phase velocity and attenuation of the ultrasonic wave we introduce them in terms of real and imaginary parts of corresponding elastic modulus

$$\Delta\alpha_\beta = \frac{1}{2}\frac{\omega}{v_{\beta 0}}\,\mathrm{Im}\frac{\Delta C_\beta}{C_{\beta 0}}, \qquad \frac{\Delta v_\beta}{v_{\beta 0}} = \frac{1}{2}\,\mathrm{Re}\frac{\Delta C_\beta}{C_{\beta 0}}, \qquad (7.20)$$

where the index $\beta = \ell,\, s,$ and $f$ relates, respectively, to the longitudinal, slow, and fast modes propagating along the [110] axis, $v_\beta = \sqrt{C_\beta/\rho}$, $\Delta\alpha_\beta = \alpha_\beta(T) - \alpha_{\beta 0}$, $\Delta v_\beta = v_\beta(T) - v_{\beta 0}$, $\Delta C_\beta = C_\beta(T) - C_{\beta 0}$, $\alpha_{\beta 0} = \alpha_\beta(T_0)$, $v_{\beta 0} = v_\beta(T_0)$, $C_{\beta 0} = C_\beta(T_0)$, and $T_0$ is the reference temperature.

Using the definition of the bulk modulus $C_B$ one can express it in terms of the moduli $C_\ell$, $C_s$, and $C_{44}$ and, taking into account (7.20), derive the expression for the contribution of the breathing mode to the attenuation of the longitudinal wave $\Delta\alpha_\ell$ as follows:

$$\Delta\alpha_B \equiv \frac{\omega}{2C_{\ell 0}v_{\ell 0}}\mathrm{Im}\,(\Delta C_B) = \left(\Delta\alpha_\ell - \frac{v_{f0}^3}{v_{\ell 0}^3}\Delta\alpha_f - \frac{1}{3}\frac{v_{s0}^3}{v_{\ell 0}^3}\Delta\alpha_s\right). \qquad (7.21)$$

In our case $\Delta\alpha_f(T) = 0$. Substitution of the measured $\Delta\alpha_\ell(T)$ and $\Delta\alpha_s(T)$ values yields $\Delta\alpha_B = 0$ within the accuracy of the experiment. Hence, the attenuation of the longitudinal wave written in general form as

$$\Delta\alpha_\ell = \Delta\alpha_B + \Delta\alpha_f + \frac{1}{3}\Delta\alpha_s, \qquad (7.22)$$

is in fact determined exclusively by $\Delta\alpha_s$, i.e., by contribution of the slow shear mode that generates tetragonal distortions. Since the contribution of vibrational modes to the attenuation and phase velocity is proportional to the corresponding squared linear vibronic coupling constant, we conclude that $F_\alpha = 0$, or at least $F_\alpha^2 \ll F_E^2$.

The formulas (7.10) and (7.11) for evaluating the linear vibronic coupling constant $F_E$ are valid in the case of ZnSe:Cr$^{2+}$ as well, but distinguished from the ZnSe:Fe$^{2+}$ system attenuation and phase velocity exhibit much stronger variations in the Cr centers, and hence in addition to attenuation measurements, the evaluation of $F_E$ based on velocity measurements becomes possible. Determination of the attenuation of relaxation origin $(\Delta\alpha_s)_r$ in the ZnSe:Cr$^{2+}$ is relatively not difficult because the background attenuation in this case is approximately constant (see Fig. 7.7b) and hence $(\Delta\alpha_s)_r$ can be obtained by subtracting the constant level (taken, for example, at $T \to 0$) from the total attenuation. Hence, $\Delta\alpha_s(T) \equiv \alpha_s(T) - \alpha(T \to 0) = (\Delta\alpha_s)_r$.

The situation with $(\Delta v_s)_r$ is not so simple as with $(\Delta\alpha_s)_r$. Analyzing the curves for doped and undoped crystals shown in Fig. 7.7a one can see that relaxation contribution to the phase velocity emerges on a temperature dependent background. Assuming that (1) the impurities manifest themselves only in the relaxation contribution to $v_s(T)$ and (2) phase velocities of doped $(v_s)_d$ and undoped $(v_s)_u$ crystals are the same at $T \to 0$, we obtain the relaxation contribution of the JT system as $\Delta(v_s)_r = \Delta(v_s)_d - \Delta(v_s)_u \equiv \delta v_s$, where $\Delta(v_s)_d$ and $\Delta(v_s)_u$ are the measured values and $\delta v_s$ denotes the difference of the velocities in the two crystals, ZnSe:Cr$^{2+}$ and ZnSe at the given temperature.

An expression for the vibronic coupling constant $F_E^2$ with the parameter values at the temperature where the frequency dependent relaxation term in (7.11) at $\omega\tau(T_1') = 1$ equals 1/2, we get

$$F_E^2 = \frac{4\kappa T_1' C_{s0}}{n a_0^2} \frac{\Delta v_s(T_1')}{v_{s0}}. \tag{7.23}$$

The prime in $T_1'$ indicates that it is the temperature corresponding to $\omega\tau = 1$ (like $T_1$ introduced above) but calculated from the velocity measurements, not from attenuation ones.

The magnitude of $T_1'$ can be obtained from (7.11) and the condition $f_2 = 1/[1 + (\omega\tau)^2] = 1/2$ at $T = T_1'$. The curve $f_2(T)$ has two limiting values: $f_2(\omega\tau \to \infty) = 0$ (high frequency or adiabatic limit) and $f_2(\omega\tau \to 0) = 1$ (quasi-static or isothermal limit). Hence to define $T_1'$ one should analyze the curve $[\delta v(T) \cdot T]/|\delta v(T_2)|$ (shown in Fig. 7.8), where $T_2$ represents the quasi-static regime (i.e., $\tau(T_2) << \tau(T_1)$), and find the point corresponding to $[\delta v(T) \cdot T]/|\delta v(T_2)| = -1/2$ on $T$-scale.

The evaluations based on attenuation and velocity measurements at 54 MHz gave similar results: $T_1 = 13.0$ and $T_1' = 13.2$ K.

As a result, the following magnitudes of the linear vibronic constant were obtained from (7.12) and (7.23): (1) $|(F_E)_1| = 5.49 \times 10^{-5}$ dyn – based on the attenuation measurements, and (2) $|(F_E)_2| = 5.57 \times 10^{-5}$ dyn – based on the velocity measurements. In these calculations we used $a_0 = 2.46$ Å, $C_{s0} = 1.835 \times 10^{11}$ dyn/cm, and $v_{s0} = 1.82 \times 10^5$ cm/s. Since the measurements of ultrasonic

# 7 Evaluation of the Jahn-Teller Effect Parameters in an Ultrasonic Experiment

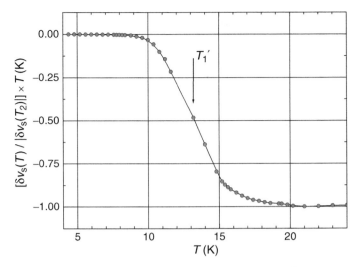

**Fig. 7.8** Temperature dependence of $[\delta v_s(T) \times T]/|\delta v_s(T_2)|$ in ZnSe:Cr$^{2+}$ measured at 54 MHz. $T_2 = 24$ K

attenuation and phase velocity represent absolutely independent procedures, we should emphasize the good agreement between the two values that indirectly supports the method as a whole. In averaging, the value $|F_E| = 5.5 \times 10^{-5}$ dyn can be considered as quite reliable.

The main factors reducing the accuracy of the $|F_E|$ evaluation are the errors in determination of dopand concentration. The errors of $\Delta\alpha(T)$, $\Delta v/v_0$, and $T_1$ measurements are much smaller. We estimate the overall error of evaluation of the vibronic coupling constant $|F_E|$ being approximately 10–20% .The value $|F_E| = (5.5 \pm 1.1) \times 10^{-5}$ dyn $(0.34 \pm 0.07$ eV/Å$)$ is in good agreement with that obtained in EPR stress-alignment study [28]: $|F_E| > 0.4$ eV/Å.

As stated above, following the ultrasound attenuation anomalies in ZnSe:Cr$^{2+}$ obtained earlier [9, 10], the dominant JT active modes are of $E$-type, and at low temperatures the relaxation time becomes approximately constant (at about $10^{-6}$ s). This means that in the $T \otimes (e + t_2)$ problem under consideration the interaction with the tetragonal distortions is dominant reducing the global minima problem to the $T \otimes e$ one [3], i.e., to the coupling of the threefold orbitally degenerate electronic states with the local $E$-type vibrations $Q_\vartheta$ and $Q_\varepsilon$ (Fig. 7.1). To construct the APES in the space of these coordinates we should know the vibronic coupling $F_E$ and primary force constant $K_E$ [3]. The latter cannot be determined form the ultrasonic experiment directly, but if we know the JT stabilization energy $E_{JT}^E$ from independent sources (e.g., calculated or obtained in other type of experiment), then together with the ultrasonic evaluated $|F_E|$ value we can estimate $K_E$ in the linear approximation of the problem [3]:

$$K_E = \frac{F_E^2}{2E_{JT}^E}. \tag{7.24}$$

With the primary force constant known, we can construct the APES $E(Q_\vartheta, Q_\varepsilon)$ related to the ground state minima. It consists of three sheets [3]:

$$E_1 = \frac{1}{2} K_E \left(Q_\vartheta^2 + Q_\varepsilon^2\right) - F_E Q_\vartheta, \tag{7.25}$$

$$E_2 = \frac{1}{2} K_E \left(Q_\vartheta^2 + Q_\varepsilon^2\right) + \frac{1}{2} F_E Q_\vartheta + \frac{\sqrt{3}}{2} F_E Q_\varepsilon, \tag{7.26}$$

$$E_3 = \frac{1}{2} K_E \left(Q_\vartheta^2 + Q_\varepsilon^2\right) + \frac{1}{2} F_E Q_\vartheta - \frac{\sqrt{3}}{2} F_E Q_\varepsilon. \tag{7.27}$$

Again, the sign of the linear vibronic coupling constant should be obtained from independent sources. From EPR experiments [28] it was found to be negative and hence the minima of $E(Q_\vartheta, Q_\varepsilon)$ are located in the points with coordinates (a positive $F_E$ will change the positions of the minima with respect to the axis of the crystal) [3]:

$$\left(-\rho_0^E, 0\right), \left(\frac{1}{2}\rho_0^E, \frac{\sqrt{3}}{2}\rho_0^E\right), \left(\frac{1}{2}\rho_0^E, -\frac{\sqrt{3}}{2}\rho_0^E\right), \tag{7.28}$$

where $\rho_0^E = |F_E|/K_E$ is the value of the symmetrized JT displacement in the minimum; the corresponding atomic displacements may be derived from the transformation tables, e.g., in [3]. The APES shown in Fig. 7.9 was calculated with the use of $F_E = -5.5 \times 10^{-5}$ dyn and $E_{JT}^E = 270\,\text{cm}^{-1}$ (i.e., $K_E = 2.8 \times 10^4$ dyn/cm in accordance with (7.24)). Other magnitudes of $K_E$ and $\rho_0^E$ based on other published values of $E_{JT}^E$ are given in Table 7.1.

Distinguished from the procedure above that does not require the knowledge of the mechanisms of relaxation, the trigonal and orthorhombic extrema points of the APES for the $T \otimes (e + t_2)$ problem under consideration cannot be determined without involving relaxation models. The experimental results [9] show that the mechanisms of relaxation in ZnSe:Cr$^{2+}$ (with $n_{Cr} = 10^{20}\,\text{cm}^{-3}$) should be either tunneling through potential energy barrier or thermal activation over the barrier or both. The activation energy $V_0$ was evaluated there as $38\,\text{cm}^{-1}$. The procedure of such evaluation is described in detail in Sect. 7.3. Figure 7.10 shows the results of processing the data on ultrasonic attenuation in the crystal sample with $n_{Cr} = 3.8 \times 10^{18}\,\text{cm}^{-3}$. In this case we obtained $V_0 = 65\,\text{cm}^{-1}$ ($\approx 8.1\,\text{meV}$) and $\tau_0 = 0.25 \times 10^{-9}$ s, and we believe that these magnitudes are more accurate since they are related to the crystal sample with less impurities which is expected to be more perfect. Moreover, the value of $V_0$ is in good agreement with that $8.5\,\text{meV}$ obtained in EPR experiment from the temperature dependence of the characteristic relaxation time $T_2$ [28].

Since direct transitions between the $e$-type minima of the $T \otimes e$ problem are forbidden by symmetry [3], the observed relaxation times should be related to the transition via the orthorhombic saddle points at which, in addition to the tetragonal displacement, a trigonal coordinate (see Fig. 7.3) is displaced [3]. There are six such equivalent saddle points, and their positions on the APES in the five-dimensional

# 7 Evaluation of the Jahn-Teller Effect Parameters in an Ultrasonic Experiment

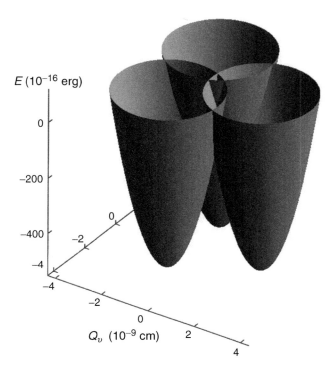

**Fig. 7.9** The adiabatic potential energy surface of the linear JTE $T \otimes e$ problem for the ground state of the $Cr^{2+}$ center in the ZnSe:$Cr^{2+}$ crystal built with the constants $F_E = -5.5 \cdot 10^{-5}$ dyn and $K_E = 2.8 \cdot 10^4$ dyn/cm

**Table 7.1** Force constant $K_E$ (in $10^4$ dyn/cm), JT displacement $\rho_0^E$ (in Å), JT stabilization energies $E_{JT}^O$ and $E_{JT}^T$ (in cm$^{-1}$) derived with the use of $|F_E| = 5.5 \times 10^{-5}$ dyn and the magnitudes of $E_{JT}^E$ (in cm$^{-1}$) from different sources. $E_{JT}^O$ and $E_{JT}^T$ are calculated with the potential energy barrier $\delta = 100$ cm$^{-1}$

| $E_{JT}^E$ | $K_E$ | $\rho_0^E$ | $E_{JT}^O$ | $E_{JT}^T$ |
|---|---|---|---|---|
| 550 [28] | 1.4 | 0.40 | 450 | 417 |
| 370 [24] | 2.1 | 0.27 | 270 | 237 |
| 270 [5]  | 2.8 | 0.19 | 170 | 137 |
| 181 [15] | 4.2 | 0.13 | 81  | 48  |

space of tetragonal plus trigonal coordinates is given by their stabilization energy $E_{JT}^O$ as follows [3]:

$$E_{JT}^O = \frac{1}{4}E_{JT}^E + \frac{3}{4}E_{JT}^T, \tag{7.29}$$

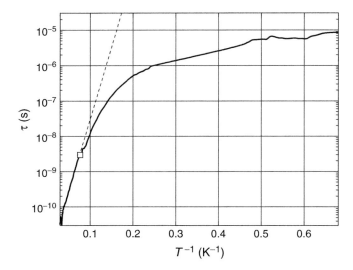

**Fig. 7.10** Temperature dependence of relaxation time in ZnSe:Cr$^{2+}$ obtained in the specimen with $n_{Cr} = 3.8 \times 10^{18}$ cm$^{-3}$. *Dash line* represents the dependence $\tau(1/T) = \tau_0 \exp(V_0/\kappa T)$, where $\tau_0 = 0.25 \times 10^{-9}$ s, $V_0 = 65$ cm$^{-1}$. *Light square* indicates $\tau(1/T_1)$

where $E_{JT}^E$ is given above and $E_{JT}^T$ is stabilization energy of the four trigonal saddle points. With (7.29) we can estimate the potential energy barrier $\delta$ between the minima (7.28) via orthorhombic saddle points:

$$\delta = E_{JT}^E - E_{JT}^O = \frac{3}{4}(E_{JT}^E + E_{JT}^T). \tag{7.30}$$

The potential energy barrier $\delta = \frac{1}{2}\hbar\omega + V_0$ [26]. The frequency of zero vibration $\omega$ is of the order of the frequency of $e$-type vibrations in the $T \otimes e$ problem, i.e., about 70 cm$^{-1}$ [5, 15]. Calculations of $E_{JT}^O$ and $E_{JT}^T$ performed with the use of (7.29) and (7.30) for different values of $E_{JT}^E$ (obtained by different authors) are given in Table 7.1.

## 7.5 Conclusions

An experimental method of evaluation of the parameters of the JTE in impurity centers in crystals by means of ultrasonic measurements was worked out and applied to the ZnSe:Fe$^{2+}$ and ZnSe:Cr$^{2+}$ dopant crystals. The method, in its main part, does not require the knowledge of the specific mechanism of ultrasound relaxation. This is of special importance in application to the JTE problems because of the variety of possible mechanisms of relaxation in the presence of conical intersections, energy troughs, equivalent minima, and barriers between them.

**Acknowledgements** This work was performed within the Program of the Russian Academy of Sciences (project No. 01.2.006 13395), with partial support of the Russian Foundation for Basic Research (grants No. 09-02-01389 and No. 10-02-08455). A part of this research was supported by grant F-100 from the Welch Foundation.

# References

1. Averkiev NS, Gutkin AA, Krasikova OG, Osipov EV, Reshschikov MF (1988) Solid State Commun 68:1025
2. Bersuker IB (1963) Sov Phys JETP 17:1060
3. Bersuker IB (2006) The Jahn-Teller effect. Cambridge University Press, Cambridge
4. Gudkov VV, Bersuker IB, Zhevstovskikh IV, Korostelin YuV, Landman AI (2011) J Phys: Condens Matter 23:115401
5. Bevilacqua G, Martinelli L, Vogel EE, Mualin O (2004) Phys Rev B 70:075206
6. Garber JA, Granato AV (1975) Phys Rev B 11:3990
7. Gudkov VV, Gavenda JD (2000) Magnetoacoustic polarization phenomena in solids. Springer, New York, pp 25–31
8. Gudkov VV, Lonchakov AT, Tkach AV, Zhevstovskikh IV, Sokolov VI, Gruzdev NB (2004) J Electron Mater 33:815
9. Gudkov VV, Lonchakov AT, Sokolov VI, Zhevstovskikh IV (2006) Phys Rev B 73:035213
10. Gudkov VV, Lonchakov AT, Sokolov VI, Zhevstovskikh IV (2007) Low Temp Phys 33:197
11. Gudkov VV, Lonchakov AT, Sokolov VI, Zhevstovskikh IV, Surikov VT (2008) Phys Rev B 77:0155210
12. Gudkov VV, Lonchakov AT, Zhevstovskikh IV, Sokolov VI, Surikov VT, (2009) Low Temp Phys 35:76
13. Gudkov VV, Lonchakov AT, Zhevstovskikh IV, Sokolov VI, Korostelin YV, Landman AI (2010) Phys Stat Sol B 247:1393
14. Kikoin KA, Fleurov VN (1994) Transition metal impurities in semiconductors: electronic structure and physical properties. World Scientific, Singapore
15. Klokishner SI, Tsukerblat BS, Reu OS, Palii AV, Ostrovky SM (2005) J Chem Phys 316:83
16. Korostelin YV, Koslovsky VI, Nasibov AS, Shapkin VP (1996) J Crystal Growth 159:181
17. Korostelin YV, Koslovsky VI (2004) J Alloys Compd 371:25
18. Kulakov MP, Fadeev AV, Kolesnikov NN (1986) Izv AN SSSR Neorg Matteer 22:29
19. Lasmann K, Schad H (1976) Solid State Commun 18:449
20. Lüthi B (2005) Physical acoustics in the solid state. Springer, Berlin
21. Malguth E, Hoffmann A, Phillips MR (2007) Phys Stat Sol B 245:455
22. Matsumoto Y, Murakami M, Shono T, Hasegawa T, Fukumura T, Kawasaki M, Ahmet P, Chikyow T, Koshihara S, Koinuma H (2001) Science 291:854
23. Mualin O, Vogel EE, de Orue MA, Martinelli L, Bevilacqua G (2001) Phys Rev B 65:035211
24. Nigren B, Vallin JT, Slack GA (1972) Solid State Commun 11:35
25. Pomerantz M (1965) Proc IEEE 53:1438
26. Sturge MD (1967) In: Seitz F, Turnbull D, Ehrenteich H (eds) Solid state physics. The Jahn-Teller effect in solids, vol 20. Academic, New York, pp 92–211
27. Sturge MD, Krause JT, Gyorgy EM, LeCraw RC, Merrit FR (1967) Phys Rev 155:218
28. Vallin JT, Watkins GD (1974) Phys Rev B 9:2051

# Chapter 8
# Raman Scattering for Weakened Bonds in the Intermediate States of Impurity Centres

**Imbi Tehver, G. Benedek, V. Boltrushko, V. Hizhnyakov, and T. Vaikjärv**

**Abstract** A theory of the Raman scattering in resonance with an electronic transition causing a strong softening of vibrations is proposed. In this case the potential surface of the excited state has a flat minimum or maximum in the configurational coordinate space. Two cases of the vibronic coupling are considered: (1) the coupling with a single coordinate and (2) the coupling with the phonon continuum. To describe the Raman scattering the Fourier-amplitude method is applied. In the first case the calculations are performed for the pseudo-Jahn-Teller effect in the excited state. In the second case, despite a strong mixing of phonons, the equations for the Raman Fourier amplitudes can be factorized and solved analytically. It is predicted that the second-order Raman scattering will be strongly enhanced. Moreover, the second-order Raman scattering is also enhanced as compared to the first-order scattering. The Raman excitation profiles show a structure caused by the Airy oscillations. The theory is applied to the triplet-triplet optical transition in $Na_2$ molecule confined at the surface of a $^4He$ droplet.

## 8.1 Introduction

Vibronicinteraction is a key factor for determining the properties of the excited states of molecular systems and impurity centres in solids. To study this coupling, the optical probing, such as the measurements of absorption and resonance Raman scattering (RRS), both spontaneous and stimulated may be used [1–8]. The RRS method allows one to obtain more detailed information on vibronic interactions than

---

I. Tehver (✉) • V. Boltrushko • V. Hizhnyakov • T. Vaikjärv
Institute of Physics, University of Tartu, Tartu, Estonia
e-mail: tehver@fi.tartu.ee; hizh@fi.tartu.ee; taavi.vaikjarv@ut.ee

G. Benedek
Donostia International Physics Center, San Sebastian, Spain

Department of Materials Science, University of Milano-Bicocca, Milan, Italy

M. Atanasov et al. (eds.), *Vibronic Interactions and the Jahn-Teller Effect:*
*Theory and Applications*, Progress in Theoretical Chemistry and Physics 23,
DOI 10.1007/978-94-007-2384-9_8, © Springer Science+Business Media B.V. 2012

absorption and luminescence, when measuring the scattering of different vibrations in different geometry. Hereby the coherent Raman scattering like CARS (coherent anti-Stokes Raman scattering) is suitable for the investigation of a small amount of target substance due to the resonance condition of excitation. For a successful application of these methods one needs to know the characteristics of the excitation profiles of the RRS (REPs) and CARS (CEPs) of impurity centres/molecules.

One of the unsolved problems of vibronic interaction, which has arisen in the last years, is connected with the optical spectrum of the triplet-triplet transitions in $Na_2$ molecule residing on the surface of superfluid $^4He$ droplets. According to Stienkemeier et al. [9, 10] this spectrum consists of very peculiar narrow bands corresponding to excitations of different vibration levels of the molecule: the bands have almost rectangular triangle-like shapes with the tip on the red and with steadily diminishing blue side. The bands do not have zero-phonon lines (ZPLs) and the adjacent dips in the phonon wings on the blue side, although these features are clearly present in the spectra of all other optical transitions of the molecule in the nearby spectral region.

The analogous triangle-like narrow bands without ZPLs have also been observed in the low-temperature absorption spectra of a photoproduct of chlorin in glassy polymers [11].To explain the origin of the observed shape of the bands the authors [11] assumed that the main contribution to the optical transition in their system is given by the linear vibronic coupling with one-dimensional (1D) acoustic phonons. The explanation relies on the fact that the low-frequency part of DOS of phonons in 1D case is strongly increased as compared to 3D case. This increase results in a drastic enhancement of the contribution of low-frequency phonons in the optical transition: the corresponding coupling parameter diverges in the low-frequency limit, resulting in the disappearance of the ZPL and in the formation of a rectangular triangle-like (or lambda-like) spectrum.

The strong enhancement of the vibronic interaction with low-frequency phonons may be indeed behind of both mentioned observations of the triangle-like spectra. However, the cause of the enhancement – the reduction of the dimension of the contributing phonons from 3D (or, may be 2D in the case of the surface phonons of $^4He$ droplets) to 1D, is highly questionable. Recently it was suggested [12, 13] that the enhancement may be a result of a strong softening of the phonon dynamics at the excited state. In glassy polymers (as well as in other amorphous dielectrics) there exist areas with weak bonds which are well described by the soft (anharmonic) potential model [14, 15]. The electronic transitions of a centre residing at a soft potential site may easily result in a further weakening of some atomic bonds which could lead to a strong enhancement of the low-energy part of the phonon wing.

Soft phonon dynamics may also be a result of the degeneracy or quasi-degeneracy of the electronic state: the vibronic mixing of the states with close energy may easily lead to a strong change of the elastic constants – a situation known as the Jahn-Teller (JT) effect or pseudo-Jahn-Teller (PJT) effect [16–26]. The well-known example of soft dynamics is observed in the case of the JT effect in a centre with trigonal symmetry: the vibronic interaction in the electronic state of E-representation with vibrations of also E-representation result in the formation of adiabatic potential with three very shallow minima of lower symmetry [18–20] and of the low-energy phonon excitations in this minima. There may also be other reasons for the softening of phonon dynamics, e.g. according to Ref. [27] strong vibrations of $Na_2$ molecule residing on the surface of 4He droplet can lead to the softening of the dynamics of few nearest $^4$He atoms, which would explain the observed in [9, 10] triangle-shaped spectra. However, more definite conclusions about the reasons of the unusual spectra in this and in other systems can be obtained if in addition to absorption spectra also the Raman scattering excitation profiles (REPs), both theoretical and experimental would be available.

The goal of this communication is to consider the manifestations of the soft dynamics in the excited state in the RRS and CARS excitation profiles. First, we present the results of a numerical study of RRS in the case of soft dynamics caused by the pseudo-Jahn-Teller (PJT) effect in the two-fold quasi-degenerate excited state. Here we neglect the phonon continuum and consider the simplest case when the vibronic mixing of two states is carried out by a single mode. We also take into account the change of the frequency of the mode at the transition. The parameters of the vibronic interaction are chosen so that the lower sheet of the potential surface of the mode in the excited state is flat, so that the energies of the lowest excitations for this sheet are small as compared to those in the ground state. Although the case with a linear vibronic interaction is considered, the first-order RRS on the mode under consideration is found to be absent, while the second-order RRS is present and is enhanced (the first-order RRS may be observed on the totally-symmetric modes which do not contribute to the PJT effect). Although this model takes explicitly the pseudo-Jahn-Teller effect into account it gives oversimplified description of the soft vibrational dynamics.

To give a more realistic description of the soft phonon dynamics we consider the resonant excitation of a non-degenerate excited electronic state having strongly reduced one (or few) elastic springs. This model is well-grounded, e.g. in the case of a strong vibronic mixing (PJT effect) of the apart-situated electronic levels of the centre when the excitation is in resonance with the lower sheet of the adiabatic surface having the flat minimum. In this case the excitation is out of resonance with the apart-situated higher sheet of the adiabatic surface which may be neglected. Note that this model works also in case of other than PJT reasons of the soft dynamics mentioned above. To describe the effect of flat minimum (or flat minima) one needs to take into account strong weakening of some atomic bonds at the electronic transition. This means that one needs to include the strong local quadratic vibronic coupling into consideration. This coupling causes the mixing of N phonon modes over the whole spectrum ($N \sim$ Avogadro number), which complicates the problem in an essential way. However, due to the occurrence of a small parameter, the ratio of mean frequencies in the excited and ground states of contributing phonons, the consideration can be made analytically [28]. To find the excitation profiles of RRS and CARS, we use, as a basis, the equations derived in [27, 28] for the Fourier amplitudes of the phonon-assisted transitions. Within this approach, the excitation profiles of the Raman modes are computed directly from the absorption spectrum [29]. The main conclusion of the consideration of the single-mode

PJT model is confirmed: the soft dynamics shows up in the RRS as a strong enhancement of second-order scatterings. No such enhancement is expected in the case of the linear vibronic coupling in non-degenerate excited states in the systems with low dimensionality of phonons.

## 8.2 Fourier Amplitudes of Raman Scattering

In the harmonic and the Condon approximations the Fourier transform of the absorption spectrum is given by the Lax equation (with $\hbar = 1$) $F(t) = \langle \hat{F}(t) \rangle$, where $\langle \ldots \rangle$ is the quantum-statistical averaging,

$$\hat{F}(t) = e^{it\hat{H}_1} e^{-it\hat{H}_2} \qquad (8.1)$$

is the evolution operator of the electronic excitation, $\hat{H}_1$ and $\hat{H}_2 \equiv \hat{H}_1 + V(q)$ are the vibrational Hamiltonians of the ground and the excited electronic state. The vibrational Hamiltonian $\hat{H}_1$ can be presented in the form

$$\hat{H}_1 = \sum_j \omega_{1j} \left( a_{1j}^+ a_{1j} + \frac{1}{2} \right) \qquad (8.2)$$

where $a_{1j}$ and $a_{1j}^+$ denote the annihilation and creation operators of the $j$-th phonon in the electronic ground state. Using these operators, the Fourier amplitudes of the first- and second-order RRS at zero temperature can be defined as follows [27–30]

$$A_j^{(1)}(t) = \left\langle 0 \left| \hat{F}(t) a_{1j}^+ \right| 0 \right\rangle \qquad (8.3')$$

$$A_{jj'}^{(2)}(t) = \frac{1}{\sqrt{1+\delta_{jj'}}} \left\langle 0 \left| \hat{F}(t) a_{1j}^+ a_{1j'}^+ \right| 0 \right\rangle \qquad (8.3'')$$

where $\delta_{jj'}$ is the Kronecker symbol and $|0\rangle$ is the zero-point state of the Hamiltonian $\hat{H}_1$ having the energy $E_0 = \sum_j \omega_{1j}/2$.

The amplitudes of the spontaneous Raman scattering for the first- and second-order processes are given by the Fourier transforms of $A_j^{(1)}(t)$ and $A_{jj'}^{(2)}(t)$:

$$A(\omega) = \frac{1}{2\pi} \int\limits_{-\infty}^{\infty} dt\, e^{i\omega t} A(t).$$

The amplitude of CARS (coherent anti-Stokes Raman scattering) at $T = 0$ can be expressed up to the normalizing factor via the amplitudes of the spontaneous Raman scattering [8, 31]

$$X_{CARS} = A(\omega_0)A(\omega_0 + \Omega) \qquad (8.4)$$

where $\omega_0$ is the excitation frequency and $\Omega$ is the frequency of the Raman-active mode.

8 Raman Scattering for Weakened Bonds in the Intermediate States... 167

The physics of light scattering is all included in the Fourier transform of the operator $\hat{F}(t)$. Its expansion gives terms with all powers of $V(q)$, from zero to infinity, with the zeroth order representing the elastic (Rayleigh) scattering, the first order the one-phonon Raman scattering, and the other terms the multi-phonon Raman scattering. Since $V(q)$ can in turn be expressed as a power expansion with respect to the vibrational coordinate $q$, providing the linear, quadratic, cubic, etc., vibronic interactions, there will be a hierarchy of processes: a one-phonon Raman scattering due to the linear coupling to 1st order; a two-phonon Raman scattering due either to second-order linear coupling or to first order quadratic coupling; etc. The creation operators in Eqs. 8.3 are there to explicitly select one- and two-phonon processes. In the following discussion only linear and quadratic coupling are considered.

## 8.3 Raman Scattering in the Case of PJT Effect in the Excited State

Let us consider first a simple case of soft vibration dynamics associated with PJTE. In this case

$$H_1 = \omega_1 \left( \hat{a}_1^+ \hat{a}_1 + 1/2 \right)$$

$$H_2 = H_1 + \sigma_z \varepsilon + \kappa \sigma_x q + \left( \omega_2^2 - \omega_1^2 \right) q^2/2 \tag{8.5}$$

where $\omega_1$ and $\omega_2$ are the frequencies of the PJTE-active mode in the ground and excited electronic states, respectively, $\varepsilon$ is the half-splitting of the quasi-degenerate state levels, $\kappa$ is the constant of the vibronic coupling, $\sigma_x$ and $\sigma_z$ are the Pauli matrices. The last, quadratic term is added here in order to describe the case when energy quanta in the lower sheet of the adiabatic surface of the excited state are much smaller than in the ground state. This case takes place if $k^2/\varepsilon \approx \omega_2^2 \ll \omega_1^2$.

Using annihilation and creation operators in the excited state, $\hat{a}_2 = ((\omega_1 + \omega_2)\hat{a}_1 + (\omega_2 - \omega_1)\hat{a}_1^+)/2\sqrt{\omega_1 \omega_2}$ and $\hat{a}_2^+$, the Hamiltonian $H_2$ can be presented in the form $H_2 = \omega_2 \left( \hat{a}_2^+ \hat{a}_2 + 1/2 \right) \cdot I + \sigma_z \varepsilon + k\sigma_x \left( \hat{a}_2 + \hat{a}_2^+ \right)$, where $k = \kappa/\sqrt{2\omega_2}$. Usually in PJT theories only the first two terms in Eq. 8.5 are taken into account.

We will use now the vibronic states $|v\rangle$ and energies $E_v$ of the excited state satisfying the stationary Schrödinger equation $(H_1 + V)|v\rangle = E_v|v\rangle$. To find these states we employ the eigenstates $|n\rangle$ of the Hamiltonian $H_2^0 = \omega_2(\hat{a}_2^+ \hat{a}_2 + 1/2)$ as the basis. Then the equation $H_2|v\rangle = E_v|v\rangle$ is reduced to two matrix-equations with the three-diagonal matrices [32]. There are even $(+)$ and odd $(-)$ solutions of the Schrödinger equation which have the form [32]

$$|v_+\rangle = \sum_n C^+_{2n,v} |2n\rangle |0\rangle_e + C^+_{2n+1,v} |2n+1\rangle |1\rangle_e \qquad (8.6)$$

$$|v_-\rangle = \sum_n C^-_{2n+1,v} |2n+1\rangle |0\rangle_e + C^-_{2n,v} |2n\rangle |1\rangle_e \qquad (8.7)$$

where $|0\rangle_e$ and $|1\rangle_e$ are the components of the excited electronic state. We consider only the Fourier transform $F_-(t)$ and the second-order Raman amplitude $A^{(2)}_-(t)$ predominantly describing the contribution of the lower potential branch of the PJTE; the first-order Raman amplitude $A^{(1)}_-(t)$ in the case under consideration is equal zero. We take also into account that the zero-point state of the Hamiltonians $H_1$ can be presented as follows [33]:

$$|0\rangle_1 = \sum_n S_{2n} |2n\rangle \qquad (8.8)$$

where

$$S_{2n} = \sqrt{\frac{(2n-1)!!\sqrt{\omega_1\omega_2}}{2^{n-1}n!(\omega_1+\omega_2)}} \left(\frac{\omega_1-\omega_2}{\omega_1+\omega_2}\right)^n \qquad (8.9)$$

As a result we get the following equations for $F_-(t)$ and $A^{(2)}_-(t)$

$$F_-(t) \cong \sum_v P^2_{0v} e^{iE^-_v t}, \quad A^{(2)}_-(t) \cong \sum_v P_{0v} P_{2v} e^{iE^-_v t}, \qquad (8.10)$$

where

$$P_{0v} = \sum_n C^-_{2n,v} S_{2n}, \quad P_{2v} = \sum_n C^-_{2n,v} S_{2n} \left[1 + 4n\left(\omega^2_1 + \omega^2_2\right)\Big/(\omega_1+\omega_2)^2\right] \qquad (8.11)$$

The calculated absorption spectrum and the second-order REP are presented in Fig. 8.1 (the main parts of the spectra are situated at positive $\omega_0$).

It can be seen that the soft dynamics in the excited state manifests itself, in agreement with the observations [9–11] as the strong enhancement of the low-frequency part of the spectrum; the envelope of the spectrum has the rectangular triangle-like shape with steadily diminishing blue side. The first-order RRS is absent in this model while the second-order RRS is present. As in absorption spectrum, in the REP the intensity of the peaks drops fast to the blue, even faster than in absorption.

8  Raman Scattering for Weakened Bonds in the Intermediate States...

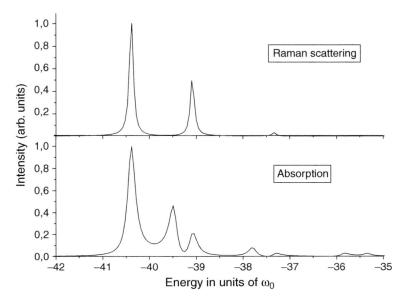

**Fig. 8.1** The excitation profile of the second-order resonance Raman scattering and the corresponding absorption spectrum in the case of the pseudo-Jahn-Teller effect in the excited state; $k^2 = \varepsilon = 40\omega_1$, $\omega_2 = \omega_1/8$ (only low-energy parts of the spectra corresponding to excitation in vicinity of the minimum of the lower sheet of the adiabatic potential are presented)

## 8.4 Soft Phonon Dynamics in the Excited State

### 8.4.1 Quadratic Vibronic Coupling

In the case of a non-degenerate excited state the soft phonon dynamics corresponds to a strong weakening of some elastic springs. This case can be described by the quadratic vibronic interaction $V(q)$ of the form

$$V(q) = a_0 q + \frac{1}{2} b q^2 \tag{8.12}$$

The perturbation $V(q)$ expresses the change in the vibrational potential due to the electronic transition. It contains a linear and a quadratic term, weighed by the respective electron-phonon coupling constants $a_0$ and $b$, and operates on phonon states through the configurational coordinate operator $q$. If in Eq. 8.5 describing the PJT to neglect non-adiabatic terms then for the lower-energy sheet of the potential surface we get Eq. 8.12 with $a_0 = 0$ and $b = 2\omega_2 k^2/\varepsilon + \omega_2^2 - \omega_1^2$. Here we consider the case when all phonons contribute to $q$ taking

$$q = \sum_j e_{1j} \left( \hat{a}_{1j} + \hat{a}_{1j}^+ \right) / \sqrt{2\omega_{1j}} \tag{8.13}$$

where $e_{1j}$ are the components of the corresponding polarization vector in the normal coordinates space.

Note that only one configurational coordinate is included into $V(q)$. The generalization of the theory to the case of several configurational coordinates of different symmetries contributing to $V(q)$ will be made elsewhere.

We consider both Hamiltonians, $H_1$ and $H_2$, in the diagonal representation. In such a representation $H_2$ reads

$$H_2 = \sum_k \omega_{2k} \left( a_{2k}^+ a_{2k} + \frac{1}{2} \right) \tag{8.14}$$

where $\omega_{2k}$ are the phonon frequencies, and $a_{2k}$, $a_{2k}^+$, the annihilation and creation operators of phonons in the excited electronic state. The phonon operators $a_{2k}$, $a_{2k}^+$, and $a_{1j}$, $a_{1j}^+$, are related by the Bogolyubov transformation [30]

$$a_{2k}^+ = \xi_{2k} + \frac{1}{2} \sum_j \frac{c_{jk}}{\sqrt{\omega_{1j}\omega_{2k}}} \left[ (\omega_{1j} + \omega_{2k}) a_{1j}^+ + (\omega_{2k} - \omega_{1j}) a_{1j} \right] \tag{8.15}$$

where $\xi_{2k} = a_0 e_{2k}/\sqrt{2\omega_{2k}^3}$ and $e_{2k}$ are the final-state polarization vectors. The latter are linearly related to the initial-state polarization vectors by $e_{2k} = \sum_j c_{jk} e_{1j}$ where the coefficients

$$c_{jk} = b e_{1j} e_{2k}/\left( \omega_{2k}^2 - \omega_{1j}^2 \right) \tag{8.16}$$

form the orthogonal matrix of the Dushinsky rotation. The back-transformation of the operators $a_{2k}$ and $a_{2k}^+$ into the operators $a_{1j}$ and $a_{1j}^+$ has an analogous form [30].

In the case of soft phonon dynamics in the excited state the mean frequency of vibrations in this state is much smaller than the mean frequency of vibrations in the ground state ($\bar{\omega}_{2k} \ll \bar{\omega}_{1j}$). Therefore, one can apply the approximations

$$c_{jk} \approx -b e_{1j} e_{2k} \omega_{1j}^{-2}, \quad b \approx -f_2^{-1} \tag{8.17}$$

with $f_n = \sum_j e_{1j}^2 \omega_{1j}^{-n}$ denoting the $n$-th negative moment of the phonon spectrum in the initial state.

## 8.4.2 First-Order Raman and CARS Amplitudes

The equation for the first-order Raman Fourier amplitude is obtained from the relationships

$$\hat{F}(t) a_{1j}^+ |0\rangle = \hat{F}(t) \left( a_{1j}^+ + a_{1j} \right) |0\rangle =$$

$$= \hat{F}(t) \left( 2\xi_{1j} + \sum_k \sqrt{\omega_{1j}/\omega_{2k}} c_{jk} \left( a_{2k}^+ + a_{2k} \right) \right) |0\rangle \tag{8.18}$$

8 Raman Scattering for Weakened Bonds in the Intermediate States... 171

with use of transformation (8.15). By commuting $a_{2k}^+$ and $\hat{F}$ with the rule $\hat{F}(t)a_{2k}^+ = e^{it\omega_{2k}}a_{2k}^+\hat{F}(t)$, expressing the operators $a_{2k}$ and $a_{2k}^+$ via $a_{1j}$ and $a_{1j}^+$, and taking into account the relations $\langle 0|a_{1j}^+ = a_{1j}|0\rangle = 0$, one finds the equation [30]

$$A_j^{(1)}(t) = \beta_j F(t) + \sum_{j'} R_{jj'} A_{j'}^{(1)}(t) \tag{8.19}$$

where

$$\beta_j = -a_0 \sqrt{\frac{\omega_{1j}}{2}} \sum_k \frac{c_{jk}e_{2k}}{\omega_{2k}^2} \left(e^{it\omega_{2k}} - 1\right)$$

$$R_{jj'} = \frac{1}{2}\sqrt{\frac{\omega_{1j}}{\omega_{1j'}}} \sum_k \frac{c_{jk}c_{j'k}}{\omega_{2k}} \left(\omega_{2k} - \omega_{1j'}\right)\left(e^{it\omega_{2k}} + 1\right) \tag{8.20}$$

In the case of soft dynamics in the excited electronic state the terms $\propto \omega_{2k}$ in the bracket are small as compared to the terms $\propto \omega_{1j'}$ and can be neglected. Taking also into account that in this case $c_{ij} \approx e_{1j}e_{2k}/f_2\omega_{1j}^2$, one gets [29]

$$A_j^{(1)}(t) \cong \frac{e_{1j}}{\sqrt{f_3\omega_{1j}^3}}A^{(1)}(t) \tag{8.21}$$

where

$$A^{(1)}(t) = \frac{a\sqrt{2}\left(\varphi_2(t) - \varphi_2(0)\right)}{2 + \varphi_1(0) + \varphi_1(t)} F(t) \tag{8.22}$$

is the Fourier amplitude of the total first-order RRS. Here

$$\varphi_n(t) = \omega_1^n \sum_k e_{2k}^2 \omega_{2k}^{-n} e^{it\omega_{2k}} \tag{8.23}$$

is the $n$-th dimensionless spectral function of phonons in the excited state, $a = a_0\omega_1^{-3/2}$, $\omega_1 = f_3/f_2^2$. Equation 8.22 gives the relationship between the Fourier amplitude of the first-order RRS and the Fourier transform of the absorption spectrum in the case under consideration. It also allows one to calculate the Raman scattering amplitude $A^{(1)}(\omega)$ and the corresponding excitation spectrum (Raman excitation profile) $|A^{(1)}(\omega)|^2$. Once this amplitude is obtained, the first-order CARS amplitude

$$X_{CARS}^{(1)}(\omega_0) = A^{(1)}(\omega_0)A^{(1)}(\omega_0 + \Omega)$$

is readily derived.

Note that in the model under consideration, the shapes of the Raman scattering excitation spectra are the same for all vibrations. The reason for this is the strong mode mixing caused by the weakening of the elastic spring.

### 8.4.3 Second-Order Raman and CARS Amplitudes

For the second-order Raman Fourier amplitude we get the following equation [29]:

$$A_{jj'}^{(2)}(t) = d_{jj'}F(t) + \beta_j A_{j'}^{(1)}(t) - \sum_{j''} d_{jj''} \left( A_{j''j'}^{(2)}(t) + \varphi_1(t)\bar{A}_{j''j'}^{(2)}(t) \right) \tag{8.24}$$

where $\bar{A}_{jj'}^{(2)}(t) = \left\langle 0 \left| a_{1j}\hat{F}(t)a_{1j'}^+ \right| 0 \right\rangle$ satisfies the analogous equation

$$\bar{A}_{jj'}^{(2)}(t) = \left( d_{jj'}\varphi_1(t) + \varphi_0(t)\delta_{jj'} \right) F(t)$$

$$+ \beta_j A_j^{(1)}(t) - \sum_{j''} d_{jj''} \left( \bar{A}_{j''j'}^{(2)}(t) + \varphi_1(t)A_{j''j'}^{(2)}(t) \right) \tag{8.25}$$

with $d_{jj'} = \left( e_{1j}e_{1j'} \left/ 2f_3\omega_{1j}^{3/2}\omega_{1j'}^{3/2} \right. \right)\varphi_1(0)$. Taking into account the above-given equations for $A_j^{(1)}(t)$, $\beta_j$ and $d_{jj'}$, one gets the second-order Raman Fourier amplitude (for details see [29])

$$A_{jj'}^{(2)}(t) = \frac{e_{1j}e_{1j'}}{f_3\omega_{1j}^{3/2}\omega_{1j'}^{3/2}}A^{(2)}(t) \tag{8.26}$$

where

$$A^{(2)}(t) \approx \left( 1 + \frac{2a^2\left(\varphi_2(0) - \varphi_2(t)\right)^2}{\left(2 + \varphi_1(0) + \varphi_1(t)\right)^2} - \frac{2\left(2 + \varphi_1(0) + \varphi_0(t)\varphi_1(t)\right)}{\left(2 + \varphi_1(0)\right)^2 - \varphi_1^2(t)} \right) F(t) \tag{8.27}$$

is the Fourier amplitude of the total second-order Raman amplitude. Equation 8.27 gives the relationship between the Fourier amplitude of the second-order RRS and the Fourier transform of the absorption spectrum for the case under consideration. By means of this equation one can calculate the second-order Raman amplitude $A^{(2)}(\omega)$ and the respective second-order CARS amplitude $X^{(2)}{}_{CARS}(\omega_0) = A^{(2)}(\omega_0)A^{(2)}(\omega_0 + \Omega)$.

8 Raman Scattering for Weakened Bonds in the Intermediate States... 173

### 8.4.4 Envelopes

The transform relationships (8.22) and (8.27) can be simplified in the case of optical bands which do not display any specific phonon structure. Then the Fourier transforms are described by short-time asymptotics, which allows one to write $\exp[it\omega_{2k}] \cong 1 + it\omega_{2k}$. The equations for the Fourier amplitudes are then found to be

$$A^{(1)}(t) = \frac{it\omega_1 a}{\sqrt{2}} F(t) \tag{8.28}$$

$$A^{(2)}(t) = -\left(\frac{1}{2}\omega_1^2 a^2 t^2 + \frac{it\omega_1}{2 - it\omega_1}\right) F(t) \tag{8.29}$$

The intensity of the first- and second-order Raman scattering of light of the frequency $\omega_0$ from a vibration of the frequency $\Omega$ equals

$$I^{(1)}(\Omega, \omega_0) = \frac{e_1^2(\Omega)\, a^2 \omega_1^2}{2\Omega^3 f_3} \left| A^{(1)}(\omega_0) \right|^2 \tag{8.30}$$

$$I^{(2)}(\Omega, \Omega', \omega_0) = \frac{e_1^2(\Omega)\, e_1^2(\Omega)}{\Omega^3 \Omega'^3 f_3^2} \left| A^{(2)}(\omega_0) \right|^2 \tag{8.31}$$

where $e_1^2(\Omega) = \sum_j e_{1j}^2 \delta(\Omega - \omega_j)$ is the phonon DOS in the initial electronic state,

$$A^{(1)}(\omega_0) = d\Phi(\omega_0)/d\omega_0 \tag{8.32}$$

$$\Phi(\omega_0) = i\pi\, \kappa(\omega_0) + P \int_{-\infty}^{\infty} \frac{dx\, \kappa(x)}{\omega_0 - x} \tag{8.33}$$

is the normalized complex refractive index, $\kappa(\omega)$, the normalized absorption (extinction) spectrum, and $P$ means the principal value. $A^{(2)}(\omega_0)$ is the Fourier transform of $A^{(2)}(t)$. The full first-order REP equals $|A^{(1)}(\omega_0)|^2$. To find the excitation profiles of the coherent Raman scattering, Eq. 8.4 should be used.

## 8.5 Model Calculations

For calculations we apply the asymptotic form $F(t)$ at small times $t$, which can be expressed analytically (see [27])

$$F(t) = (1 - i\omega_1 t/2)^{-\mu} \exp\left[i\delta t - a^2 \omega_1^2 t^2 (6 - i\omega_1 t)/24\right] \tag{8.34}$$

where $\mu = f_2^3/2f_3^2$, $\delta = -f_1/4f_2 - f_2/4f_3$. As follows from Eq. 8.34, the phonon continuum in the case of a sufficiently small linear interaction parameter $a$ manifests itself in the shape of the spectral band principally through the parameter $\mu$. The latter is chiefly determined by the low-frequency part of the phonon spectrum, which essentially depends on the dimensionality $D$ of the phonon system. When the Debye model is applied (with a unity Debye frequency), then $e_1^2(\omega) = \sum_j e_{1j}^2 \delta(\omega - \omega_{1j}) = (D+2)\omega^{D+1}$. In the $D = 3$ case $\mu = 0.37$, while for $D = 2$ it is $\mu = 0.25$. In the Debye–van Hove model, where $e_1^2(\omega) \propto \omega^{D+1}\sqrt{1 - \omega^2}$, $\mu = 0.35$ for $D = 3$ and $\mu = 0.23$ for $D = 2$. In all cases the values of $\mu$ remarkably differ from $\mu = 0.5$ valid for a single-mode case. The frequency parameter $\omega_1$, which determines the width of the band, is not very sensitive to the details of the phonon spectrum. For example, in the $D = 3$ case it equals 0.9 for the Debye model and 0.85 for the Debye-van Hove model. For $D = 2$ it equals 0.9 for the Debye-van Hove model and 1 for the Debye model. In a single-mode case it also equals 1. The terms $\propto a^2$ in Eq. 8.34 take into account the effect of the acceleration in the final state, whereas the term $\propto ia^2t^3$ describes the Airy oscillations [13,27]. For all given values of $\mu$ the absorption spectrum has a sharp end at the red and steadily diminishing blue side modulated (if $a_0 \neq 0$) by the Airy oscillations.

In the resonance Raman scattering, the soft phonon dynamics manifests itself analogously via a strong enhancement of the low-frequency part of the spectra. The fine structure is caused by Airy oscillations. The second-order RRS as well as CARS is enhanced (strongly for small $a$) and the transform relation is essentially changed. In the second-order REP and CEP these oscillations are more pronounced than in the first-order Raman excitation profiles and in the absorption spectrum.

## 8.6 Application to Na$_2$ Molecule on $^4$He Droplet

Small alkali molecules bound on the surface of helium clusters appear to be unique systems with, possibly, the weakest interatomic interactions in nature. The fine structure observed in the optical spectra of Na$_2$ molecules in the triplet state confined at the surface of small $^4$He droplets provides an example [9, 10]. The excitation spectrum of luminescence in the region between 14,500 and 14,900 cm$^{-1}$ consists of several narrow bands corresponding to different high vibrational levels of the molecule in the excited state. The small, but still remarkable half-width 30 cm$^{-1}$ is a consequence of the interaction of the molecule with the droplet.

In the excited gerade triplet state the molecular bond length is much shorter ($\sim 3.5\,\text{Å}$). The sudden contraction of the molecule in this state yields a sudden decrease of the force constants of the surrounding atoms. Then the motion of these atoms becomes soft after the optical excitation, which leads to a strong increase of the interaction of optical electrons with low energy phonons. As a result the zero-phonon line disappears and the low-energy part of the phonon sideband is strongly enhanced.

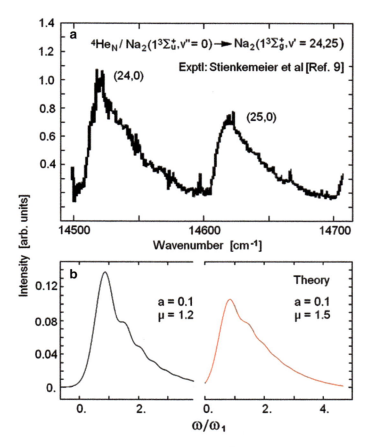

**Fig. 8.2** The optical spectrum of a Na$_2$ molecule on He$^4$ cluster as measured by Stienkemeier et al. [9] (**a**) and calculated with the present theory [27] (**b**). In the excited triplet state the molecular bond length is much shorter ($\sim$3.5 Å). One can suppose that the sudden contraction of the molecule in this state, in comparison with the ground triplet state, yields a sudden decrease of the force constants of the surrounding atoms. The optical excitation makes their dynamics softer afterwards, thus leading to a strong increase of the interaction of the optical electrons with low energy phonons, and a strong enhancement of the low-energy part of the phonon sideband

Finally we note that according to the theory exposed in [27] the case $\mu = 1.5$ $a = 0.1$ (see Fig. 8.2) corresponds to the triplet-triplet transition in a Na$_2$ molecule trapped on the surface of a $^4$He droplet with the excitation to the vibration level # 24 (in [27] it is found that soft dynamics of 4–6 nearest He atoms should correspond to $\mu = 1.5$). Thus, the profiles presented in Fig. 8.3 may be regarded as a prediction of the present theory for the Raman excitation profiles in this particular system.

The other systems where the presented predictions could hold are chlorin-doped glassy polymers [11]. If proposed in [11] model of the linear vibronic coupling with the one-dimensional (1D) acoustic phonons works then the first-order RRS would

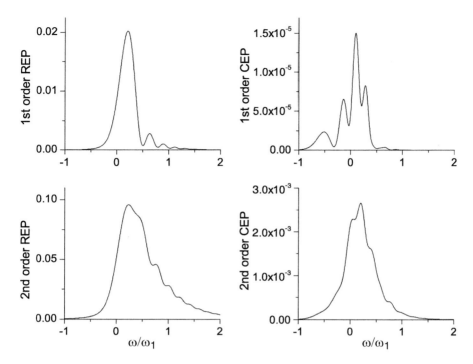

**Fig. 8.3** The manifestation of soft phonon dynamics in the excitation profiles of spontaneous resonant (REP) and coherent anti-Stokes (CEP) Raman scattering ($a = 0.1$, $\mu = 1.5$). In the resonance Raman scattering, the soft phonon dynamics manifests itself via a strong enhancement of the low-frequency part. The excitation profiles of the second-order REP and CEP are strongly enhanced. The fine structure is caused by the Airy oscillations

be stronger than the second-order. However, if the model of soft phonon dynamics in the excited state can be applied then the first-order RRS would be much weaker than the second-order RRS.

**Acknowledgements** The research was supported by the target-financed grant TLOFY0145 and ESF grant GLOFY7741.

## References

1. Long DA (2002) The Raman effect: A unified treatment of the theory of Raman scattering by molecules. Wiley, West Sussex
2. Page JB (1991) In: Cardona M, Güntherodt G (eds) Light scattering in solids VI. Springer, Berlin/Heidelberg/New York, p 17
3. Hizhnyakov V, Tehver I (1967) Phys State Solid 21:755; Hizhnyakov V, Tehver I (1997) J Raman Spectrosc 27:403
4. Champion PM, Albrecht AC (1982) Annu Rev Phys Chem 33:353

5. Siebrand W, Zgierski MZ (1979). In: Excited states, vol 4. Academic, p 1
6. Clark RJH, Stewart B (1979) Struct Bond 36:1
7. Desiderio RA, Hudson BS (1979) Chem Phys Lett 61:445
8. Pfeiffer M, Lau A, Werncke W (1984) J Raman Spectrosc 15:20
9. Stienkemeier F, Ernst WE, Higgins J, Scoles G (1995) J Chem Phys 102:615
10. Stienkemeier F, Higgins J, Ernst WE, Scoles G (1995) Phys Rev Lett 74:3592
11. Kikas J, Suisalu A, Zazubovich V, Vois P (1996) J Chem Phys 104:4434
12. Hizhnyakov V, Benedek G, Tehver I, Boltrushko V (2006) J Non Cryst Solids 352:2558
13. Boltrushko V, Holmar S, Tehver I, Hizhnyakov V (2007) J Mol Struct 838:164
14. Karpov G, Parshin DA (1983) JETP Lett 38:648; (1985) Sov Phys JETP 61:1308
15. Krivoglaz MA (1985) Sov Phys JETP 61:1284
16. Englman R (1972) The Jahn-Teller effect in molecules and crystals. Wiley-Interscience, New York
17. Fischer G (1984) Vibronic coupling. Academic, San Diego
18. Zgierski MZ, Pawlikowski M (1979) J Chem Phys 70:3444
19. Bersuker IB, Polinger VZ (1989) Vibronic interactions in molecules and crystals. Springer, Berlin
20. Bersuker IB (2006) The Jahn–Teller effect. Cambridge University Press, Cambridge
21. Luthi B (2004) Physical acoustics in the solid state. Springer, Berlin
22. Polinger VZ, Bersuker IB (1979) Phys State Solidi (b) 95:403
23. Rosenfeld YuB, Vaisleib AV (1984) Sov Phys JETP 86:1059
24. Gudkov VV, Lonchakov AT, Sokolov VI, Zhevstovskikh IV, Surikov VT (2008) Phys Rev B 77:155210
25. Gudkov VV, Lonchakov AT, Zhevstovskikh IV, Sokolov VI, Surikov VT (2009) Low Temp Phys 35:76
26. Fulton R, Gouterman M (1964) J Chem Phys 35:1059; (1964) 41:2280
27. Hizhnyakov V, Tehver I, Benedek G (2009) Eur Phys J B 70:507
28. Hizhnyakov V, Benedek G (2008) Chem Phys Lett 460:447
29. Hizhnyakov V, Tehver I, Boltrushko V, Benedek G (2010) Eur Phys J B 75:187
30. Hizhnyakov V (1986) J Phys C Solid State Phys 20:6087
31. Hizhnyakov V, Tehver I (1996) J Raman Spectrosc 27:469
32. Loorits V (1979) Projavlenije rezonansa Fermi v opticheskih spektrah psevdo-effekta Jana-Tellera (Akademija Nauk ESSP, Otdelenije fiziko-matematicheskih nauk, Preprint F-11, in Russian)
33. Pae K (2008) Electronic transitions in symmetric systems of strong vibronic coupling. Bachelor thesis, Tartu; http://www.fi.tartu.ee/loputood2009/fyysika/Kaja%20Pae-bakalaureusetoo.pdf

# Chapter 9
# Vibronic Transitions to a State with Jahn-Teller Effect: Contribution of Phonons

**V. Hizhnyakov, K. Pae, and T. Vaikjärv**

**Abstract** A theory of electronic transitions from a non-degenerated to a two-fold degenerated or quasi-degenerated electronic state in an impurity centre of a crystal is proposed. The coupling of optical electron(s) with non-totally symmetrical vibration(s) and with phonons are taken into account. The latter interaction results in the replacement of vibronic lines by phonon-assisted bands. The widths of bands, the rates of the energy relaxation of vibronic states and the shapes of optical spectra are calculated for some values of parameters.

## 9.1  Introduction

The optical probing of the degenerate or quasi-degenerate electronic states of impurity centres in solids allows one to get information about the complicated vibronic dynamics governed by the Jahn-Teller effect (JTE) [1–10]. In the last years a method of considering optical JTE and PJTE of molecular systems - the multi-configurational time-dependent Hartree approximation, has been developed [11–14] (see also [15]), which allows one to study the time evolution of the excited states of these systems) and the manifestations of this evolution in optical spectra [13–20]. The investigations are based on ab-initio coupled cluster calculations of wave packet propagation with typically ten vibrational degrees of freedom for the nuclear motion [15].

Despite its success in the description of molecular systems, this method has limitations in its application to the solid state, as it does not allow one to consider the contribution of phonon continuum. This is not only because of the essential growth of the required amount of calculations and the diminishing of their accuracy with the

---

V. Hizhnyakov (✉) • K. Pae • T. Vaikjärv
Institute of Physics, University of Tartu, Tartu, Estonia
e-mail: hizh@fi.tartu.ee; kaja.pae@gmail.com; taavi.vaikjarv@ut.ee

M. Atanasov et al. (eds.), *Vibronic Interactions and the Jahn-Teller Effect:*
*Theory and Applications*, Progress in Theoretical Chemistry and Physics 23,
DOI 10.1007/978-94-007-2384-9_9, © Springer Science+Business Media B.V. 2012

increasing of the number of contributing coordinates and the calculating time. The principal shortage of this method in its application to solids is the impossibility to describe the relaxation (the energy loss) by the centre at a large time, while the irreversible time evolution is a feature of the systems with a continuous energy spectrum. Another serious shortcoming of the method is the impossibility of taking into account the long-range interactions which essentially determine the vibrations of 3D systems. Here we have in mind the well-known fact that the macroscopic electric field associated with the motion of atoms (ions) essentially determines the phonon dispersion in the bulk. However, this field is not included into consideration in the methods based on the cluster calculations.

The goal of the present study is to present a method which allows one to calculate the optical spectra of impurity centres in the crystals with the JTE and PJTE in the excited state, taking into account the phonons of the bulk crystal. The method works in the case of an arbitrary vibronic interaction with a few local (pseudo-local) modes and a relatively weak interaction with phonons. Unlike the multi-configurational time-dependent Hartree approximation, this method does not require the increasing of the amount of calculations with the increasing of the number of contributing phonons; besides, our method works in the same way in the cases of discrete and continuous phonon spectrum.

The calculations by this method are performed in three steps. In the first step the Hamiltonian of the degenerate or quasi-degenerate electronic state is transformed so that the vibronic coupling, describing the contribution of the non-totally symmetric phonons, is transformed to the totally-symmetric vibrational potential energy $H'$ which constitutes the sum of the products of the coordinate(s) on the basic Jahn-Teller-active mode(s) and of the coordinates of phonons; this term describes the interaction of the Jahn-Teller system with phonons. Then, neglecting $H'$, a preliminary optical spectrum of the system is calculated by using the quantum-mechanical basis of the vibronic states of the Jahn-Teller-active mode(s) (here the results of the earlier publications are used). Finally, the interaction $H'$ of the basic Jahn-Teller mode(s) with phonons is taken into account by applying the cumulant expansion technique. As a result, every vibronic line in the preliminary spectrum is replaced by the phonon-assisted band. Their full widths characterize the phase relaxation due to phonons, while the widths of the replicas at $T = 0$ are determined by the rate of the energy relaxation of the states. Both parameters are calculated for different values of vibronic interaction parameters. Some examples of the numerically computed spectra are also presented.

## 9.2 General

We are considering the optical spectrum of an impurity centre in a crystal corresponding to a vibronic transition from a non-degenerated to the two-fold quasi-degenerated or degenerated electronic state. The latter state belongs to a centre of trigonal symmetry. Here we are interested in the vibronic interaction with

# 9 Vibronic Transitions to a State with Jahn-Teller Effect: Contribution of Phonons

non-totally symmetrical vibrations. In the case of a quasi-degenerate state this interaction leads to PJTE, while in the case of a two-fold degenerate state of $E$-representation it leads to JTE with doubly degenerate vibrations of $e$-symmetry ($E \otimes e$-problem).

In the case under consideration the Fourier transform of the spectrum at zero temperature is determined by the diagonal element(s) of the $2 \times 2$ matrix

$$F(t) = \left\langle 0 \left| e^{it(H+V)} e^{-itH} \right| 0 \right\rangle \tag{9.1}$$

where $|0\rangle$ is the zero-point state of the crystal lattice, $V$ is the $2 \times 2$ matrix of vibronic interaction in the basis $|0\rangle_e$ and $|1\rangle_e$ of the excited electronic states, $H \equiv I \cdot H$ is the phonon Hamiltonian of the initial state which we take in harmonic approximation, $I$ is the $2 \times 2$ unit matrix. We take into account the linear vibronic coupling which in the case of PJTE (corresponds to the quasi-degenerate state) reads [6, 21]

$$V = k\sigma_x Q + \sigma_z \varepsilon \tag{9.2}$$

where Q is the odd configurational coordinate of the dimer describing the quasi-degenerate excited state (PJT), while in the case of JTE, $E \otimes e$-problem it equals [2–6]

$$V = k\left(\sigma_x Q_x + \sigma_y Q_y\right) \tag{9.3}$$

where $Q_x$ and $Q_y$ are the configurational coordinates of the $E$-representation, $\sigma_x$, $\sigma_y$ and $\sigma_z$ are the Pauli matrices, $k$ is the constant of the vibronic coupling, $\varepsilon$ is the half-splitting of the quasi-degenerate state. In polar coordinates $V$ gets the form

$$V = kQ \begin{pmatrix} 0 & e^{-i\varphi} \\ e^{-i\varphi} & 0 \end{pmatrix} \tag{9.3'}$$

where $Q$ is the radial and $\varphi$ is the angular variables of the main e-mode ($Q_x = Q\cos\varphi$, $Q_y = Q\sin\varphi$). The potential curves in the space of the configurational coordinate(s) are shown in Fig. 9.1.

The configurational coordinate(s) $Q$ are the linear combinations of normal coordinates $x_j$. Here we suppose that the main contribution to $Q$ is given by the single normal coordinate $x_0$ (or $x_{0x}$ and $x_{0y}$ in the case of JTE); the contribution of other modes is supposed to be small. In this case

$$Q = \left(1 + \lambda^2\right)^{-1/2} \left(x_0 + \lambda \sum_j e_j x_j\right) \tag{9.4}$$

where $\lambda$ is a small dimensionless parameter, $e_{1j}$ are the normalized polarization vectors, $\sum_j e_j^2 = 1$; the subscripts $x$ and $y$ are omitted. The number of the contributing normal coordinates $j$ (phonons) is considered to be arbitrary.

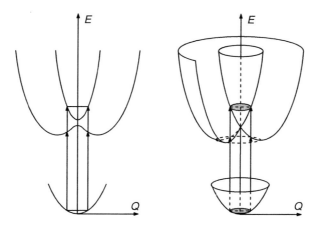

**Fig. 9.1** The scheme of the potentials and electronic transitions in the configurational coordinate space. *Left*: the pseudo-Jahn-Teller effect in the excited state, *right*: the Jahn-Teller effect, $E \otimes e$-problem in the excited state

## 9.2.1 Transformation of Vibrational Hamiltonian

We take now into account that the effects of the main vibronic interaction with the single non-degenerate (PJTE) or doubly degenerate (JTE) normal mode can be found numerically (in the case of PJTE it was done in [22]; in the case of $E \otimes e$-problem it was primarily done in [23, 24]; see also [25]). Our intention is to use these numerical results as the basis for the description of the phonon effects. To this end we transform the phonon Hamiltonian $H$ into a form more convenient for further consideration by introducing the set of configurational coordinates $Q_l (l \geq 2)$ which are mutually orthogonal and simultaneously orthogonal to $x_0$ and $q_1 = \sum_{j \geq 1} e_j x_j$. The last property means that they are also orthogonal to $Q$ and $Q' = (1 + \lambda^2)^{-1/2}(q_1 - \lambda x_0)$. Up to a constant, the potential energy of vibrations in the initial state can now be presented in the form

$$U = \omega_0^2 \, x_0^2/2 + D_{11} q_1^2/2 + \sum_{l \geq 2} D_{1l} q_1 Q_l + \sum_{l,l' \geq 2} D_{ll'} Q_l Q_{l'}/2 \qquad (9.5)$$

Expressing here $x_0$ and $q_1$ via $Q$ and $Q'$ and neglecting the terms up to $\sim \lambda^2$ we get

$$U \approx U_0 + H'$$

where

$$U_0 = \omega_0^2 Q^2/2 + \sum_{l,l' \geq 1} D_{ll'} Q_l Q_{l'}/2$$

# 9 Vibronic Transitions to a State with Jahn-Teller Effect: Contribution of Phonons

is the potential energy of the modes $Q$ and $Q_l$,

$$H' = -\lambda Q \left( \omega_0^2 Q_1 - \sum_{l \geq 2} D_{1l} Q_l \right) \tag{9.6}$$

is the interaction of the modes $Q$ and $Q_l$. We take now into account the relations $Q_l = \sum_{j \geq 1} e_{lj} x_j$ and $\sum_l D_{1l} e_{lj} = \omega_j^2 e_{1j}$. As a result we get (up to a constant)

$$H = H_1 + H' \tag{9.7}$$

where

$$H_1 = H_0 + H_{ph}$$
$$H_0 = \omega_0 a_0^+ a_0$$

is the Hamiltonian, while $a_0^+ = \sqrt{\omega_0/2} Q - \sqrt{1/2\omega_0} \partial/\partial Q$ is the destruction operator of the main mode,

$$H_{ph} = \sum_j \omega_j a_j^+ a_j$$

is the Hamiltonian of phonons,

$$H' = \lambda \sum_j v_j \left( a_0^+ + a_0 \right) \left( a_j + a_j^+ \right) \tag{9.8}$$

is the interaction of the main mode with phonons,

$$v_j = e_j \left( \omega_0^2 - \omega_j^2 \right) / 2\sqrt{\omega_0 \omega_j} \tag{9.9}$$

In the case of JTE in the $E$-state ($E \otimes e$-problem) $H$ is the sum of two terms with the subscripts $x$ and $y$. In what follows we include into consideration only the "resonant" terms $\sim a_0^+ a_j + a_0 a_j^+$ of the interaction Hamiltonian $H'$ and neglect the non-resonant terms $\sim a_0^+ a_j^+ + a_0 a_j$ taking

$$H' \simeq \lambda \sum_j v_j \left( a_0^+ a_j + a_0 a_j^+ \right) \tag{9.10}$$

This is the well-known rotating wave approximation which in our case holds if the interaction with phonons satisfies the condition $\lambda^2 \sum_j v_j^2 (\omega_0 + \omega_j)^{-2} \ll 1$. In the case of $E \otimes e$-problem the equation for mode-phonon coupling can be presented in the form

$$H' \simeq \lambda \sum_{j \geq 1} v_i \left( a_{0+}^+ a_{j-} + a_{0-}^+ a_{j+} + Hc \right) \tag{9.10'}$$

where $a_{j\pm} = (a_{jx} \pm i a_{jy})/\sqrt{2}$ are the operators of the destruction of phonons with a non-zero-momentum.

By using the creation and destruction operators, the vibronic coupling obtains now the form

$$V = \sqrt{D}\sigma_x \left(a_0 + a_0^+\right) + \varepsilon\sigma_z(\text{PJTE}) \tag{9.11}$$

or

$$V = \sqrt{D} \sum_{\alpha=x,y} \sigma_\alpha \left(a_{0\alpha} + a_{0\alpha}^+\right)(\text{JTE}) \tag{9.12}$$

where $D = k^2/2\omega_0$.

In the transformed Hamiltonian the Jahn-Teller active mode(s) $Q$ is a normal mode of the Hamiltonian $H_1$ which is the main part of the vibrational Hamiltonian $H$. If one neglects the remained weak interaction term $H'$ then the problem under consideration reduces to the problem of the optical PJTE or JTE with the single non-degenerate or two-fold degenerate mode contributing to the vibronic coupling. This allows one to use the numerical solutions of the latter problem as the basis and to take into account the interaction with the phonons $H'$ (which is divisible to unit matrix in the electronic basis) approximately by means of the many-body theory methods.

### 9.2.2 Vibronic Basis

We are using the Dyson representation for the evolution operators in the initial and final states:

$$e^{it(H+V)} = \hat{T}\exp\left[i\int_0^t d\tau H'(\tau)_2\right]e^{itH_2}, \ e^{-itH} = e^{-itH_1}\hat{T}\exp\left[-i\int_0^t d\tau H'(\tau)_1\right]$$

where $\hat{T}$ is the time-ordering operator, the time-dependences of the operators $H'(\tau)_1$ and $H'(\tau)_2$ are given by the Hamiltonians $H_1$ and $H_2 = H_1 + V$, respectively. In the rotating wave approximation we have $H'(\tau)_1|0\rangle = 0$. Taking also into account that in our units $H_0|0\rangle = 0$ and that in the rotating wave approximation $H_{ph}|0\rangle = 0$, we find that the Fourier transform of the spectrum is given by the diagonal element(s) of the $2 \times 2$ matrix

$$F(t) = \left\langle 0\left|\hat{T}\exp\left[i\int_0^t d\tau H'(\tau)_2\right]e^{it(H_1+V)}\right|0\right\rangle \tag{9.13}$$

We will use now the vibronic states $|v\rangle$ and the energies $E_v$ of the single (PJTE) or doubly degenerate (JTE) mode satisfying the stationary Schrödinger equation $(H_0+V) = E_v|v\rangle$. To find these states we can use the eigenstates of the Hamiltonian $H_0$ of the active mode(s) as the basis. In the case of PJT these states ($|n\rangle$) have one quantum number: the positive integer $n$ which determines the energy $\omega_0 n$ of the mode. In this basis the equation $(H_0 + V) = E_v|v\rangle$ is reduced to two

# 9 Vibronic Transitions to a State with Jahn-Teller Effect: Contribution of Phonons

matrix-equations with three-diagonal matrices [22]. There are even $(+)$ and odd $(-)$ solutions of these equations which have the form

$$|v_+\rangle = \sum_n C_{2n,v}^+ |2n\rangle |0\rangle_e + C_{2n+1,v}^+ |2n+1\rangle |1\rangle_e \tag{9.14}$$

$$|v_-\rangle = \sum_n C_{2n+1,v}^- |2n+1\rangle |0\rangle_e + C_{2n,v}^- |2n\rangle |1\rangle_e \tag{9.15}$$

In the case of JTE, $E \otimes e$-problem the vibrational Hamiltonian $H_0$ describes the two-dimensional quantum harmonic oscillator. The eigenstates of this Hamiltonian $(|n,m\rangle)$ have two quantum numbers: the positive integer $n$ determining the energy $\omega_0 n$ and the sign-alternating $m$ determining the projection of the rotational momentum on the $z(C_3)$–axis $(|m| \leq n)$ [2]. Analogously, the eigenstates of $H_0 + V$ (vibronic states) have also two quantum numbers: $v$ and $j$, thereat $j = m - 1/2 = \pm 1/2, \pm 3/2, \pm 5/2 \dots$ is the projection of the full (electronic $s = 1/2$ and vibrational) rotational momentum of the vibronic motion on the $z$-axis. At zero temperature only the vibronic states with $j = \pm 1/2$ give a contribution to absorption. Depending on the sign of $j$, these states can be presented in the form

$$|v_+\rangle = \sum_n C_{2n,v}^+ |2n,0\rangle |0\rangle_e + C_{2n+1,v}^+ e^{i\varphi} |2n+1,1\rangle |1\rangle_e \tag{9.16}$$

or

$$|v_-\rangle = \sum_n e^{-i\varphi} C_{2n+1,v}^- |2n+1,-1\rangle |0\rangle_e + C_{2n,v}^- |2n,0\rangle |1\rangle_e \tag{9.17}$$

where the amplitudes $C_{n,v}^\pm$ can be found by diagonalization of three-diagonal matrixes describing the radial motion of the $E \otimes e$-problem [2,23–25]. In both cases the amplitudes $C_{n,v}$ and the energies $E_v$ can easily be calculated [23–25].

Below we consider only the Fourier transform $F_+(t)$ describing the vibronic transitions which involve the even states. The odd states lead to the same result (JTE) or can be found analogously, by replacing the superscripts $+$ and $-$ (PJTE).

By using the vibronic basis $|v_+\rangle$ given above Eq. 9.13 can be presented as follows:

$$F_+(t) = \sum_{vv'} C_{0v}^+ C_{0v'}^+ e^{itE_v^+} \left\langle v_+' \left| \left\langle 0_{ph} \left| \hat{T} \exp\left[ i \int_0^t d\tau H'(\tau)_2 \right] \right| 0_{ph} \right\rangle \right| v_+ \right\rangle \tag{9.18}$$

where $|0_{ph}\rangle$ denotes the zero-point state of phonons. Only the even terms with respect to $H'$ give a contribution in Eq. 9.13.

## 9.2.3 Cumulant Expansion

We are considering the case of a remarkable or a strong vibronic coupling with the main mode and a weak coupling with phonons. In this case the optical band is rather broad, its total width $\sigma_0$ being larger than $\omega_0$. In this case the vibronic structure of the optical band is described by the Fourier transform at $t \gg \sigma_0^{-1}$. Then the main contribution in Eq. 9.18 is given by the diagonal terms $v = v'$. Indeed, if $v = v'$ then in the $2n$-order terms $n$ factors oscillate with different large frequencies $E_v$, while in the case $v \neq v'$ there are at least $n + 1$ factors which oscillate with different large frequencies $E_v$. Therefore the latter terms give smaller contribution to the spectrum than the $v = v'$ terms. Besides, in this case one can restrict oneself only by consideration of the second cumulant in the expansion of diagonal terms. In this approximation we get

$$F_+(t) \cong \sum_v |C_{0v}^+|^2 \exp\left[iE_v^+ t + g_v^+(t)\right] \tag{9.19}$$

where

$$g_v^+(t) = \int_0^t d\tau \int_0^\tau d\tau' \left\langle 0_{ph} \right| \langle v_+ | H'(\tau) \sum_{v'} |v'_-\rangle \langle v'_- | H'(\tau') |v_+\rangle \left| 0_{ph} \right\rangle \tag{9.20}$$

Taking into account that only $\propto a^+$ terms contribute to the left matrix element, while only $\propto a$ terms give a contribution to the right matrix element one finds

$$g_v^+(t) = \lambda^2 \int_0^t d\tau \int_0^\tau d\tau' \, \Phi_v^+(\tau - \tau') D(\tau - \tau') \tag{9.21}$$

where

$$D(t) = \sum_j v_j^2 e^{-i\omega_j t} \tag{9.22}$$

is the phonon correlation function,

$$\Phi_v^+(t) = \sum_{v'} |S_{vv'}^+|^2 e^{i\left(E_v^+ - E_{v'}^-\right)t} \tag{9.23}$$

is the vibronic correlation function,

$$S_{vv'}^+ = \sum_n \left( \sqrt{2n+1} \, C_{2n+1,v}^+ C_{2n,v'}^- + \sqrt{2n+2} C_{2n+2,v}^+ C_{2n+1,v'}^- \right) \tag{9.24}$$

By inserting Eqs. 9.22 and 9.23 into Eq. 9.21, one gets after the integration

$$g_v^+(t) = \lambda^2 \sum_j v_j^2 \sum_{v'} |S_{vv'}^+|^2 \left( \frac{it}{E_v^+ - E_{v'}^- - \omega_j} + \frac{e^{i\left(E_v^+ - E_{v'}^- - \omega_j\right)t} - 1}{\left(E_v^+ - E_{v'}^- - \omega_j\right)^2} \right) \tag{9.25}$$

In the case of a continuous phonon spectrum one should shift the frequencies $\omega_j$ infinitesimally up in the complex plain.

# 9 Vibronic Transitions to a State with Jahn-Teller Effect: Contribution of Phonons

Equation 9.19 together with Eq. 9.25 give a solution of the problem. As it follows from these equations, the effect of phonons is described by the factors $e^{g_V(t)}$. If the contribution of phonons (taking $g_V(t) = 0$) is neglected, then one gets rather a simple equation giving the absorption spectrum as a series of discrete (vibronic) lines. However, if phonons are taken into account, then every vibronic line in the preliminary spectrum is replaced by the phonon-assisted band whose shape is determined by the Fourier transform of $e^{g_V(t)}$.

## 9.3 Numerical

To illustrate the effect of phonons in the optical spectra of the Jahn-Teller systems we use the Debye-Van Hove model

$$e^2(\omega) \equiv \sum_j e_j^2 \delta(\omega - \omega_j) = (15/2)\,\omega^3 \sqrt{\omega_M^2 - \omega^2} \qquad (9.26)$$

where $\omega_M$ is the maximum phonon frequency. The width of the band $v_+$ characterizes the phase relaxation of the vibronic level $v_+$. The quadratic dispersion $\sigma_v^{+2}$ (second central moment) of the band determines the small time asymptotic of $g_v^+(t) = -\sigma_v^{+2} t^2/2$; it equals

$$\sigma_v^{+2} = \lambda^2 \bar{\omega}^2 \sum_n n \left| C_{nv}^+ \right|^2 \qquad (9.27)$$

where $\bar{\omega}^2 = \sum_j v_j^2$. In the Debye-Van Hove model this parameter is as follows: $\bar{\omega}^2 = (15\pi/32)\left(\omega_0^2 - \omega_M^2 + 5\omega_M^4/16\omega_0^2\right)$.

Unlike to the phase relaxation, the energy relaxation is determined by the large time asymptotic of $g_v^+(t)$. Taking into account the relation

$$(1 - \cos xt)/x^2 = \pi\delta(x),\ t \to \infty,$$

we get in the large time limit

$$g_v^+(t) = i\delta_v^+ t - \gamma_v^+ |t|$$

where

$$\delta_v^+ = \lambda^2 \sum_j v_j^2 \sum_{v'} |S_{vv'}^+|^2 / \left(\omega_j + E_{v'}^+ - E_v^-\right) \qquad (9.28)$$

determines the phonon-induced shift of the vibronic sub-band and

$$\gamma_v^+ = \pi\lambda^2 \sum_j v_j^2 \sum_{v'} |S_{vv'}^+|^2 \delta\left(\omega_j + E_{v'}^+ - E_v^-\right) \qquad (9.29)$$

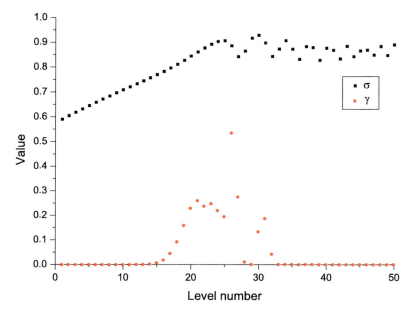

**Fig. 9.2** Energy ($v$) and phase ($\sigma$) relaxation rates of vibronic levels in the case of JTE, $E \otimes e$-problem for $\omega_0 = 1$, $\omega_M = 1.1$, $D = 20$, and $\lambda^2 = 0.01$

describes the broadening of the main line of the sub-band caused by the energy relaxation of the vibronic level $v_+$. Note that Eq. 9.29 coincides with the equation for the decay rate of the vibronic level $v_+$ which can be found by applying the Fermi's golden rule. The results of calculations of $\sigma_v$ and $\gamma_v$ see in Figs. 9.2 and 9.3. Note the non-monotonous dependence of the rates on the level number in the important region $v \approx 30$.

In Figs. 9.4 and 9.5 we present the results of the calculations of the spectra by using the Fourier transform given by the equation

$$F_+(t) \cong \sum_v |G_{0v}^+|^2 \exp\left[iE_v^+ t + i\delta_v^+ t - (\gamma_v^+ + \gamma_0)|t|\right] \qquad (9.30)$$

These spectra include the main lines. The weak phonon-assisted lines are not included supposing that they can bring only some corrections of the periphery of the main lines. Their inclusion requires special calculations which can be made in every concrete case, if needed. One can see that the calculated spectra, despite of rather large value of the vibronic coupling parameter $D$, differ essentially from the ones given by the semi-classical approximation [26]. The details of the spectra depend on the values of the rates of the energy relaxation $\gamma_v^+$ which in their turn in a complicated way depend on the number $v$ of the vibronic level.

9 Vibronic Transitions to a State with Jahn-Teller Effect: Contribution of Phonons 189

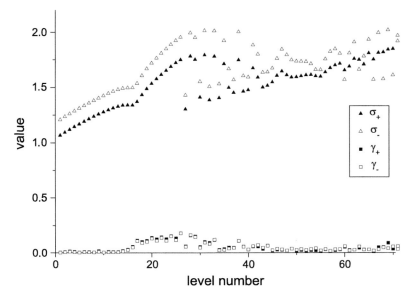

**Fig. 9.3** Energy ($v$) and phase ($\sigma$) relaxation rates of even ($+$) and odd ($-$) vibronic levels of PJTE for $\omega_0 = 1$, $\omega_M = 1.1, \varepsilon = 5$, $D = 20$ and $\lambda^2 = 0.02$

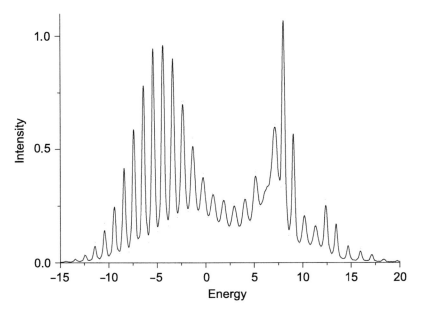

**Fig. 9.4** Optical spectrum in the case of JTE ($E \otimes e$-problem) for $\omega_0 = 1$, $\omega_M = 1.1$ and $D = 20$, $\lambda = 0.1$ (the same parameters as in Fig. 9.2) and $\gamma_0 = 0.1$

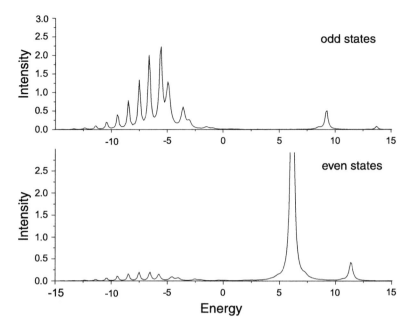

**Fig. 9.5** Optical spectra of odd $(-)$ and even $(+)$ states in the case of PJTE for the same parameters as in Fig. 9.2 and $\gamma_0 = 0.1$

## 9.4 Conclusion

To sum up, the optical spectra of the impurity centres of the crystals, which correspond to the electronic transitions from a non-degenerate to a two-fold-degenerate or quasi-degenerate electronic states, are investigated. A method is proposed which allows one to consider the JTE in the degenerate state and the PJTE in the quasi-degenerate state and simultaneously to take into account the contribution of phonon continuum to JTE and PJTE. In the theory the rotating wave approximation is used, which in our case works if the interaction with the phonon continuum is weak. Only the simplest cases of a single prevailing non-degenerate or doubly degenerate mode were considered. However, one can expect that the method may be extended to the Jahn-Teller systems with a number of prevailing non-totally symmetric modes. Besides, the method may be adopted for the description of the time-evolution of the wave packets of vibrations with the inclusion into consideration the relaxation caused by the creation of bulk phonons.

**Acknowledgements** The research was supported by the ETF grants TLOFY0145 and GLOFY7741.

# References

1. Jahn HA, Teller E (1937) Proc R Soc Lond Ser A 161:220
2. Longuet-Higgins HC, Opik U, Pryce MH, Sack RA (1958) Proc R Soc A 244:1
3. Englman R (1972) The Jahn-Teller effect in molecules and crystals. Wiley-Interscience, New York
4. Bersuker IB (2006) The Jahn–Teller effect. Cambridge University Press, Cambridge
5. Fischer G (1984) Vibronic coupling. Academic, San Diego
6. Bersuker IB, Polinger VZ (1989) Vibronic interactions in molecules and crystals. Springer, Berlin
7. Perrin MH, Gouterman M (1967) J Chem Phys 46:1019
8. van der Waals JH, Berghuis AMD, de Groot MS (1967) Mol Phys 13:301
9. Zgierski MZ, Pawlikowski M (1979) J Chem Phys 70:3444
10. Domcke W, Yarkony DR, Köppel H (eds) (2004) Conical intersections: electronic structure, dynamics and spectroscopy. Word Scientific, Singapore
11. Meyer HD, Manthe U, Cederbaum LS (1990) Chem Phys Lett 165:73
12. Manthe U, Meyer HD, Cederbaum LS (1992) J Chem Phys 97:3199
13. Beck M, Jäckle A, Worth GA, Meyer HD (2000) Phys Rep 324:1
14. Meyer HD, Worth GA (2003) Theor Chem Acc 109:251
15. Köppel H, Yarkony DR, Barentzen H (eds) (2009) Fundamentals and implications for physics and chemistry. Springer, Berlin/Heidelberg
16. Köppel H, Döscher IM, Baldea I, Meyer HD, Szalay P (2002) J Chem Phys 117:2657
17. Gindensperger E, Baldea I, Franz J, Köppel H (2007) Chem Phys 338:207
18. Faraji S, Meyer HD, Köppel H (2008) J Chem Phys 129:074311
19. Nooijen M, Snijders JG (1992) Int J Quantum Chem S 26:55
20. Stanton JF, Gauss J (1994) J Chem Phys 101:8938
21. Fulton R, Gauterman M (1961) J Chem Phys 35:1059; (1964) 41:2280
22. Loorits V (1979) Projavlenij rezonansa Fermi v opticheskih spektrah psevdo-effekta Jana-Tellera, Akademija Nauk ESSP, Otdelenije fiziko-matematicheskih hauk, Preprint F-11
23. Loorits V (1980) Izv Akad Nauk ESSR Fiz-Mat 29:208
24. O'Brien MCM, Evangelou SN (1980) Solid State Commun 36:29
25. Hizhnyakov V, Kristoffel N (1984) Jahn-Teller mercury-like impurities in ionic crystals. In: Perlin YuE, Wagner M (eds) The dynamical Jahn-Teller effect in localized systems. North-Holland, Amsterdam, pp 383–438
26. Toyozawa Y, Inoue W (1966) J Phys Soc Jpn 21:1663

# Chapter 10
# Many-Electron Multiplet Theory Applied to O-Vacancies in (i) Nanocrystalline HfO$_2$ and (ii) Non-crystalline SiO$_2$ and Si Oxynitride Alloys

**Gerry Lucovsky, Leonardo Miotti, and Karen Paz Bastos**

**Abstract** Performance and reliability in semiconductor devices are limited by electronically active defects, primarily O-atom vacancies. Synchrotron X-ray spectroscopy results, interpreted in the context of multiplet theories, have been used to analyze conduction band edge, and O-vacancy defect states in nanocrystalline transition metal oxides such as HfO$_2$, and the non-crystalline oxides including SiO$_2$, and Si$_3$N$_4$ and Si oxynitride alloys. Multiplet theory provides the theoretical foundation for an *equivalent d$^2$ model* for O-vacancy transitions and negative ion states as detected by X-ray absorption spectroscopy in the O K pre-edge regime. Comparisons between theory and experiment have relied on Tanabe-Sugano energy level diagrams for identifying the symmetries and multiplicities of transition energies for an equivalent d$^2$ ground state occupancy. The equivalent d$^2$ model has been applied to nanocrystalline thin films of ZrO$_2$, HfO$_2$, TiO$_2$ and Lu$_2$O$_3$ and provides excellent agreement with X-ray absorption spectroscopy data. The model has also been applied to SiO$_2$ and other Si based dielectrics where very good agreement with multiplet theory has also been demonstrated. The spectra indicate both triplet and singlet final states indicating that the two electrons in the vacancy sites have singlet and triplet ground states that are within a few tenths of an eV of each other. For the transition metal oxides, this is explained by relatively small distortions in the vacancy geometry in which the separation between the respective transition metal atoms is 1.6 times the bond-length in an ideal tetrahedral geometry, or the same factor for two fold coordination in O-atom bonding sites in SiO$_2$ and GeO$_2$. These distortions minimize the exchange energy in triplet spin states, and reduce the radial wave function overlap in singlet spin states.

---

G. Lucovsky (✉) • L. Miotti • K.P. Bastos
Department of Physics, North Carolina State University, Raleigh, NC 27695-8202, USA
e-mail: lucovsky@ncsu.edu

M. Atanasov et al. (eds.), *Vibronic Interactions and the Jahn-Teller Effect: Theory and Applications*, Progress in Theoretical Chemistry and Physics 23, DOI 10.1007/978-94-007-2384-9_10, © Springer Science+Business Media B.V. 2012

## 10.1 Introduction

The primary objective of this paper is to provide a theory based quantitative model for explaining the electronic structure of O-vacancies in nanocrystalline transition metal (TM) oxides, with an extension to non-crystalline $SiO_2$ and Si oxynitride alloys. Soft X-ray absorption and photoemission spectroscopies, XAS and XPS, respectively, have been used to study valence and conduction band edge states, and O-vacancy defect states in (i) nanocrystalline $HfO_2$ and other TM oxides, as well as (ii) non-crystalline $SiO_2$, $Si_3N_4$ and Si oxynitride alloys [1, 2]. Thin films for these studies, typically 2–5 nm thick, were deposited onto passivated (i) oxide, or (ii) nitrided oxide Si and Ge substrates by either remote plasma enhanced chemical vapor deposition (RPECVD) [3], or reactive evaporation [4]. XAS and XPS studies were performed at the Stanford Synchrotron Research Lightsource (SSRL) [1, 2]. Density functional theory (DFT) methods had been previously applied to O-vacancy defects in $HfO_2$ and $ZrO_2$ [5, 6]. The results from these calculations have been used to interpret spectral features in dielectric functions obtained from spectroscopic ellipsometry (SE) measurements [7, 8], and current-voltage characteristics in metal-oxide semiconductor (MOS) test structures [9], but with minimal success. In contrast, this article has continued to refine, and to quantify a more local model, previously proposed for O-vacancies and addressed in the 2008 Jahn Teller symposium proceedings [2]. This approach includes many of the important constraints on two electron terms for d-states that are traditionally cast in terms of Hund's rules for strongly correlated atomic systems [10]. This semi-empirical Hund's rule approach has been quantified, with considerable success, through the introduction of Tanabe-Sugano (T-S) diagrams based on multiplet theory [10]. The new model of this chapter is designated as an *equivalent $d^2$ model*. The rationale for this designation is based on group theory as applied to the term symmetries for occupation of a pair of d-states that are constrained within the confines a neutral O-atom vacancy site, and, most importantly, in a bonding geometry that simultaneously minimizes repulsion, and satisfies Pauli exclusion through a further reduction in the exchange energy.

Section 10.2 develops the *equivalent $d^2$ model*, providing a basis for interpretation of XAS O pre-edge spectra. Sections 10.3 and 10.4 present, respectively, spectroscopic data for (i) nanocrystalline TM oxides, and (ii) non-crystal $SiO_2$, $Si_3N_4$, and Si oxynitrides, $(SiO_2)_{1-x}(Si_3N_4)_x$. Section 10.5 summarizes the application of the equivalent $d^2$ model to X-ray spectroscopy results, including the (i) quantitative detection of spin and symmetry allowed singlet and triplet final state transition energies, including negative ion singlet states play in the interpretation of electrical data for metal-semiconductor-oxide (MOS) gate stacks.

## 10.2 *Equivalent $d^2$ Model* for O-Vacancy Defects

In the ionic limit, removal of a neutral O-atom from a nanocrystalline, or single crystal TM elemental or complex oxide results in two electrons being confined within the vacancy site bounded by TM atoms. In this limit, the TM atoms bordering on the vacancy site as indicated in Fig. 10.1 donate the two electrons. This local approach recognizes qualitative differences between conventional band theory, and the failure of attempts to explain the optical spectra of NiO using traditional one electron theory approaches for periodic solids [11]. Instead the approach taken in this chapter builds on paper of the Falicov group published in 1974 [12]. This paper treated the O 2s and 2p, and the Ni 4p and 4s valence electrons by a conventional one electron band theory. Most importantly, the Ni states at were treated at the atomic level without one-electron theory methods. The Ni d-states were inserted into the energy band gap determined by the O and Ni atomic s and p valence states. Atomic d-state degeneracies were removed by the application of the ligand field of the six O-atom nearest neighbors, and their connectivity within the periodic lattice. The ligand field symmetries of the ground and excited d-state terms activate optical selection rules, so that atomic d-state spherical symmetry constraints for intra d-state, d-d' transitions no longer apply. The Falicov article explained the atomic parentage of experimentally reported optical transitions not only in NiO, but also in other first row TM monoxides including VO, MnO, FeO, and CoO.

The semi-empirical model of this chanter for the O-vacancy is displayed schematically in Fig. 10.1 using tetragonal $HfO_2$ (t-$HfO_2$) as a first example. The two electrons left behind by removal of a neutral O-atom are confined within the vacancy site and reside, with either single or double occupancy on at most two of

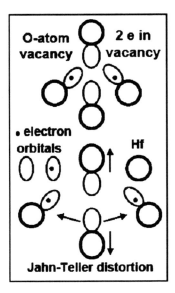

**Fig. 10.1** Schematic representation of O-atom vacancy defect in t-$HfO_2$, including ideal tetrahedral coordination, and a Jahn-Teller distortion that minimizes repulsions between the two singly occupied Hf orbitals

the 5d orbitals of the Hf atoms that border on the site. For an idealized $CaF_2$ lattice for cubic $HfO_2$ prior to a tetragonal distortion, the vacancy site has a tetrahedral symmetry. However in t-$HfO_2$, there is a cooperative Jahn Teller distortion that breaks the cubic $CaF_2$ symmetry with a tetragonal extension in the z-direction, and distortions in the x-y plane that change them from squares to rotated parallelepipeds. These symmetry reducing atomic displacements make the four Hf atoms bordering the vacancy *nonequivalent*. An *equivalent $d^2$ multiplet* model is then developed for the ground and excited state terms of the occupancy of these two electrons. This model, independent of single or double atomic occupancy, includes all of the bonding constraints that are applied to the $d^2$ atomic configuration of a single TM atom, hence Tanabe-Sugano (T-S) diagrams provide a way to characterize the relative energies of allowed and forbidden multiplet transitions of the new model [10].

T-S multiplet term energy level diagrams have been used to identify the symmetries and transition energies associated with occupied and excited TM atom terms in TM elemental oxides and molecular complexes [10]. Multiplet theory predated Falicov's contribution [12], and was first applied to TM oxides in the seminal paper of Hans Bethe in 1929 explaining why NiO was green [13]. Bethe introduced the crystal field concept as well as adapting the quantum state symmetries of Ni to the rock-salt local bonding symmetry. This has evolved to the ligand field description, which applies independent of primary ligand bond ionicity.

The two electrons occupying a neutral O-atom vacancy populate d-state orbitals on two of the four TM atoms bordering the vacancy in t-$HfO_2$ as in Fig. 10.1. This occupancy must minimize repulsion and obey the Pauli exclusion principle. Proceeding in this our approach for defining a two-electron term and using T-S diagrams [10], represents a de facto extension of Hund's rules for the two electrons with *nonequivalent d-orbitals on different atoms* confined within an O-atom vacancy site. The physical separation of these two orbitals in $HfO_2$ is $\sim$0.36 nm in an ideal tetrahedron, and may be large as 0.4–0.45 nm in a distorted tetrahedron that includes relaxations from a $CaF_2$ structure to the distorted tetragonal structure of the HfO2.

Density functional theory (DFT) studies of O-vacancies by the Robertson and Shluger groups minimized repulsion by dispersing the two electrons of a neutral vacancy on all four of the Hf or Zr atom d-state orbitals bordering on the site [5, 6]. Their respective band-structure energy level diagrams indicate this occupancy by placing the two electrons in the same low-spin singlet ground state. This band structure approach is in conflict with strongly correlated d-states as discussed in the Bethe and Falicov studies [12, 13].

The new approach of this research assumes integral atomic d-state occupation with the two electrons residing in either (i) two of four available d-state orbitals for t-$HfO_2$, or (ii) two of the three for rutile $TiO_2$. It is less likely for both electrons to occupy a single d-orbital on one the bordering TM atoms because of increased repulsion.

The ligand field symmetry of the atoms bordering the vacancy plays the determinant role in defining two electron terms. For the nano-grain TM oxides of

**Table 10.1** Term symmetries for singlet and triplet ground to final state transitions octahedral and tetrahedral (cubic) local bonding environments

| Term symmetries | Octahedral coordination | | Tetrahedral and cubic coordination | |
|---|---|---|---|---|
| Ground state terms | Singlet | Triplet | Singlet | Triplet |
| | $^1T_{1g}$ | $^3T_{1g}$ | $^1A_1$ | $^3A_1$ |
| Transition term order | Singlet | Triplet | Singlet | Triplet |
| 1 (lowest eV) | $^1T_{2g}$ | | $^1E_1$ | |
| 2 | $^1E_g$ | | | $^3T_2(3)$ |
| 3 | | $^3T_{2g}(3)$ | $^1A_{1g}$ | |
| 4 | $^1A_{1g}$ | | | $^3T_1(3)$ |
| 5 | | $^3T_{1g}(3)$ | $^1T_{1g}$ | |
| 6 | $^1T_{1g}$ | | $^1T_{2g}$ | |
| 7 | $^1T_{2g}$ | | | $^3T_{1g}(3)$ |
| 8 | | $^3A_{2g}(3)$ | $^1E_g$ | |
| 9 | $^1E_g$ | | $^1T_{2g}$ | |
| 10 (highest eV) | $^1A_{1g}$ | | $^1A_{1g}$ | |

this chapter, the most important aspect of the ligand field symmetry and the ligand field splitting, $\Delta_{LF}$, is the coordination, 6-fold for Ti, 7- or 8-fold for Hf and Zr, and 4-fold for Si. This is included in Table 10.1 in which the ordering of singlet and triplet final state terms has been transported from T-S energy level diagrams at intermediate ligand field strengths of $\sim 25$ to 30 in normalized x-axis units. The spectral terms indicate both tripet and singlet final states indicating that the two electrons in the vacancy sites have singlet and triplet ground states that are within a few tenths of an eV of each other. For the transition metal oxides, this is explained by relatively small distortions in the vacancy geometry in which the separation between the respective transition metal atoms is 1.6 times the bond-length in an ideal tetrahedral geometry, and may be increased to as much as times the bond-length of relaxation which minimizes the exchange energy. Figure 10.2a, b represent the asymmetry of the total two electron wave functions for fermion states. For a singlet spin configuration the radial terms are symmetric with electron density between the two atoms, whilst for the triplet spin configuration the radial terms are asymmetric with no electron density between the atoms. For 6-fold coordinated octahedral bonding within $TiO_2$ nano-grains, and initially with no restrictions for angular momentum or spin for the two Ti atom 3d-states, the two electron is an unrestricted $3d_1^{\,1}3d_2^{\,1}$ configuration [5] The ground state for this configuration includes both high- and low-spin occupancy: $S = 0$ for low-spin singlet states, and $S = 1$ for high-spin triplet states. All values of the total angular momentum, L, are included as well [14–17]. The term symbol that obeys Hund's rules for maximization of L, S, and J for a generalized term, $^{S+1}L_j$, has $L = 4$, $S = 1$, $J = 5$, and is therefore $^3G_5$.

If wave function orthogonality and high-spin occupancy restriction are imposed, then $L = 3$ for the orthogonal d-state character with $l_1 = 2$ and $l_2 = 1$, and $S = 1$ for high-spin occupancy, and the resulting term symbol is $^3F_4$. This is the same term for a $d^2$ state where both electrons are on a single atom in a high-spin

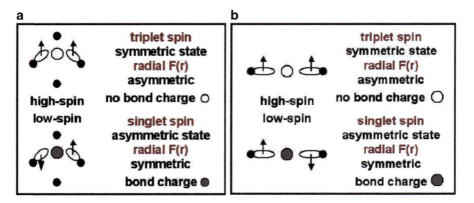

Fig. 10.2 Schematic representations of the fermion wave functions in for singlet and triplet ground states in (**a**) a tetrahedral vacancy site as in $HfO_2$, and (**b**) the 2-fold coordination of O in an $SiO_2$ vacancy

state. In this case, the same T-S multiplet diagrams apply; hence the *equivalent* $d^2$ designation is appropriate. The T-S diagrams interpolate numerically between atomic ionic states, and a maximum ligand field/crystal field for 100% ionicity [10,14,18]. This intermediate regime of T-S diagrams allows the use of T-S diagrams for interpretation of spectroscopic results for O-vacancy defects independent of the specific electron occupancy on the bordering TM atoms [10, 14]. The application of the model is confirmed by the energy range of the O-vacancy transitions which is approximately equal the ligand field splitting that can be obtained independently from XAS spectra: $L_{2,3}$ for Ti and Mn, are other first row TM atoms $M_{2,3}$ for Y and Zr second row atoms transitions, and more generally from, O K edges for the first, second and third row TM oxides addressed in Sect. 10.3.

The $^3F_4$ term is included in the unrestricted and more general $^3G_5$ term that has no restrictions on high- or low-spin, or the angular momentum for two singly occupied d-states. $^3G_5$ includes seventeen different terms: (i) five singlets for the five different angular momentum states, $^1S_0$, $^1P_1$, $^1D_2$, $^1F_3$ and $^1G_4$, and (ii) four triplets as well: $^3P_{0,1,2}$, $^3D_{1,2,3}$, $^3F_{2,3,4}$, and $^3G_{3,4,5}$. with twelve different energies after degeneracy removal for different spin orbit couplings.

The number of terms is reduced significantly when orthogonality and spin restrictions are included in the $^3F_4$ term. This term applies equally for double occupancy in a high-spin $d^2$ configuration, or to orthogonal occupancy for two different d-orbitals at a maximized distance consistent with a relaxed vacancy geometry as in Fig. 10.1. The ordering of the $^3F_4$ free-ion terms in increasing energy are: $^3F_{2,3,4}$, $^1D_2$, $^3P_{0,1,2}$, $^1G_4$ and $^1S_0$ [14]. This ordering is indicated on the y-axis in the T-S diagrams for free ion terms in which $\Delta o$, is equal to the ligand field splitting, $\Delta_{LF}$, and is identically equal to zero [10]. It is important to recognize that even though Hund's rules were developed initially for partially filed valence shells of atoms, they also apply to TM d-orbitals in solids, e.g., $Ti^{3+}$, $Mn^{4+}$, and $Fe^{3+}$, etc, in their respective oxides, or for similar TM ions as impurity atoms or alloy atoms in other TM oxide hosts such as $Ti^{3+}$ substituted for $Ti^{3+}$ in $GdScO_3$ [13]. These same terms, and the ordering of these terms apply exactly when both electrons at a vacancy site are on the same atom, or in close proximity on two different TM atoms bordering a vacancy when have orthogonal angular momenta.

The $d^2$ diagram for cubic symmetry, or equivalently for 4- and 8-fold coordination is the same as the $d^8$ diagram for octahedral symmetry with values of $\Delta o$ and B scaled accordingly [10, 14, 18]. These diagrams apply independent of cooperative Jahn-Teller distortions that change to local bonding symmetry, but not the coordination. This will become apparent in Sect. 10.3.

Proceeding in this way, in the intermediate coupling regime that applies to TM oxides, the ground state term for 6-fold octahedral and distorted octahedral coordination as in $TiO_2$ is $^3T_{2g}$ independent of corner-connection of octahedra as in rutile, or edge-connection as in anatase. For 8-fold coordination as in tetragonal $HfO_2$ with a cooperative Jahn Teller distortion, the corresponding ground state is $^3A_2$. The corresponding ground states for the singlet transitions are $^1T_{2g}$, and $^1A_2$. See Table 10.1 for the singlet and triplet final states.

To facilitate the use of Tanabe-Sugano diagrams for comparing spectroscopic determinations of transition and negative ion state energies, $d^2$ T-S diagrams for 6-fold, and 8-fold TM coordinations have been transformed into a pair schematic tables with an intermediate values of $\Delta o/B$, $25 \pm 2$. These diagrams apply to the three nano-grain TM oxides of Sect. 10.3, octahedral $TiO_2$ and $Lu_2O_3$, and tetragonal $HfO_2$. These are included in Table 10.1 respectively. The energy ordering of the symmetry terms (on the y-axis of the diagram) for spherical symmetry has been noted above, so that for $\Delta o/B = 0$, the ordering is $^3F$, $^1D$, $^3P$, $^1G$ and $^1S$ [10,14,18]. Normalized energies, E/B, for end-member and intermediate ligand field values have been obtained by numerical analysis and variational methods [10, 14]. The diagram for 8-fold coordination also applies for 4-fold tetrahedral coordination, and is applied to O-vacancy defects in $SiO_2$, and N-vacancy defect in $Si_3N_4$, as well. Singlet and triplet states are included in the table consistent with the ground state degeneracy. For octahedral coordination this means that the term energies of the $^1T_{2g}$ and $^3T_{2g}$ are approximately equal, and for the tetrahedral coordination the energies of the $^1A_2$ and $^3A_{2g}$ are also approximately equal.

## 10.3 Nano-Grain Transition Metal Oxides

Figure 10.3a, b display respectively, (i) O K edge spectra for two $TiO_2$ films with rutile and anatase nano-grains dominating, and (ii) the O K pre-edge second derivative spectrum for rutile structured $TiO_2$. These films were deposited on nitrided Ge substrates, and after annealing are in contact with the Ge surface which then serves as a template for grain alignment. Films deposited on nitrided Si, generally always display anatase-structured grains. This difference is evident in O K

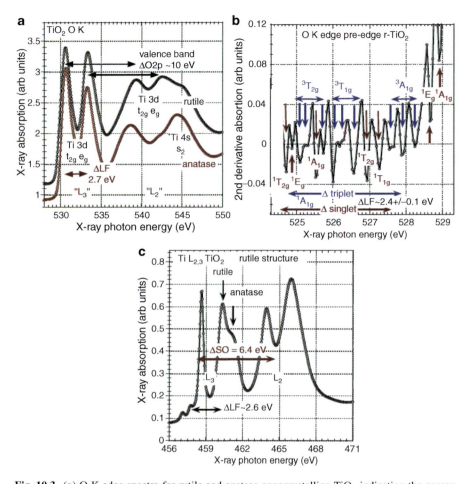

**Fig. 10.3** (a) O K edge spectra for rutile and anatase nanocrystalline $TiO_2$ indicating the energy difference between O 2p valence band molecular orbitals, and ligand field splitting at the Ti sites. (b) Second derivative O K pre-edge spectra for rutile $TiO_2$ indicating transition state and negative ion state term representations, including degeneracy removal of transition state terms. (c) Ti $L_3$ spectrum for rutile $TiO_2$ indicating the spin orbit splitting of the Ti 2p core level and ligand field splitting at the Ti site

and Ti $L_{2,3}$ spectra as well. The primary, 2p to 3d, and satellite, 2p to 4s, transitions are sequential in $TiO_2$ for both rutile and anatase bonding in the $L_{2,3}$ spectra as well as the in O K edge spectra. This explained by a relatively large 3d to 4s energy difference of ∼5 eV, compared with the smaller ∼2.5 ± 0.3 eV for 4d(5d) to 5s(6s) energy differences in $ZrO_2$ and $HfO_2$, where 5s and 6s states, bracket the respective second $t_{2g}$ state features in each of O K spectra [19]. Figure 10.4a, b display the O K and $M_{2,3}$ spectra for $ZrO_2$ where the 5s features bracket the $M_2$ regime in the $M_{2,3}$ spectrum, or $M_2$-derived $t_{2g}$ features in the O K edge spectrum.

10 Many-Electron Multiplet Theory Applied to O-Vacancies in... 201

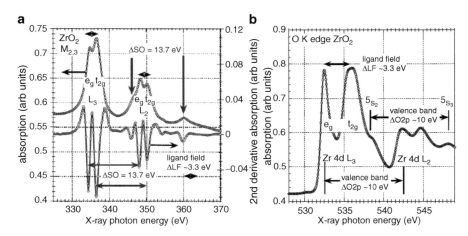

**Fig. 10.4** (a) $M_{2,3}$ XAS and second derivative $M_{2,3}$ spectra for nanocrystalline $ZrO_2$ indicating spin orbit splitting of the Zr 3p core level and ligand field splitting at the Zr site. (b) O K edge spectrum for nanocrystalline monoclinic $ZrO_2$ indicating the energy difference between O 2p valence band molecular orbitals, and ligand field splitting at the Zr site

The local symmetry is different for rutile and anatase nano-grains, and this is reflected in a splitting of the second or $e_g$ feature in the Ti $L_3$ portion of the spectrum as illustrated in Fig. 10.3c. As noted above these two features are found in films prepared on nitrided Ge substrates, and not nitrided Si substrates. For anatase grains in which the c-axis is lengthened the dominant $dz^2$ feature is at a lower energy in the $L_3$ spectrum than for rutile grains in which the c-axis is extended and this feature is at a higher energy. Similar differences in relative energies between rutile and anatase nano-grain films carry over to the O K edge spectra, but these do not change $\Delta_{LF}$, the ligand field scaling parameter for O K pre-edge O vacancy spectral features as displayed in Fig. 10.3b. The experimental value of $\Delta_{LF}$ in nano-grain films $TiO_2$ is the same for each of the unit cell geometries and equal to $\sim 2.6 \pm 0.15$ eV.

Figure 10.3b displays the O K pre-edge spectrum for the 2 nm thick $TiO_2$ thin film with a rutile unit cell structure. Consistent with the T-S derived triplet final state term energies in Table 10.1 for 6-fold octahedral coordination, the spectrum displays nine spin allowed triplet to triplet transitions from the $^3T_{2g}$ ground state term to features with $^3T_{2g}(F)$, $^3T_{1g}(P)$ and $^3A_{2g}(F)$ terms in order of increasing energy. This includes a final state effect degeneracy removal. Figure 10.3b also includes three singlet transitions below the highest triplet transition in order of increasing energy, $^1A_{1g}$, $^1T_{2g}$ and $_1T_{1g}$.

This 6-fold coordination term symmetry includes singlet states above the highest triplet states, which are designated by us as negative ion states with $^1E_{1g}(G)$ and $^1A_{2g}(S)$ terms as indicated in Table 10.1. There are also two degenerate singlet states, $^1T_{2g}$, $^1E_g$, below the lowest triplet final states.

The energy spread of the spin-allowed triplet transitions is $\sim 2.45 \pm 0.15$ eV, and is approximately equal to that of the features in the $L_{2,3}$ spectrum for the

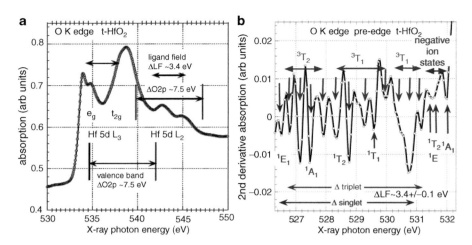

**Fig. 10.5** (a) O K edge spectrum for t-HfO$_2$ indicating the energy difference between O 2p valence band molecular orbitals, and ligand field splitting at the Hf site. (b) Second derivative O K pre-edge spectrum for t-HfO$_2$ indicating (a) transition state and negative ion state term representations, and (b) degeneracy removal of transition state terms

rutile-structured unit cell, and also is the same for anatase-structured unit cells; the spectrum for the thicker film on Ge with anatase structure dominating is not shown. The same energy spread a for spin-allowed singlet states extends from the lowest doubly degenerate state, $^1T_{2g}$, $^1E_g$ to the highest lying states encapsulated with the group of three triplet states.

Figure 10.5 displays spectra of for the 6 nm thick film of nanocrystalline t-HfO$_2$: (a) the O K edge spectrum, and (b) the second derivative spectrum in the O K pre-edge regime. Second derivative analysis of the spectral features in Fig. 10.5a identifies the e$_g$ character of the band edge feature, as well as the ligand field splitting, $\Delta_{LF} = 3.4 \pm 0.15$ eV. The O-atom vacancies yield spectral features between approximately 527 and 531 eV, with the highest negative ion states features ~0.5 eV below the lowest energy 6d-state derived eg features of the conduction band edge. The term symmetry assignments for the features in the pre- edge regime of t-HfO$_2$ between 526.5 and 530.5 eV were made on the basis of comparisons with an energy level scheme in Fig. 10.2b which applies for 8-fold coordination. Each of these dipole allowed triplet transitions is 3-fold degenerate with the degeneracy removed by a cooperative Jahn Teller effect distortion in the vacancy site [1]. The Jahn Teller distortion makes the four Hf-atoms bordering the vacancy nonequivalent, thereby removing the 3-fold degeneracy for the term symbols of the dipole allowed, spin conserving $^3T_{2g}(F)$, $^3T_{1g}(F)$ and $^3T_{1g}(P)$ triplet transitions from the $^3A_{2g}(F)$ ground state. This degeneracy removal occurs at the intermediate strength ligand fields that are reduced from the maximum crystal field splitting corresponding to 100% bond ionicity. The degeneracy removal is included in Fig. 10.2b, with a spacing between these additional states relative the central feature of the order

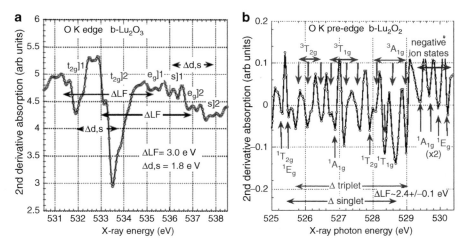

**Fig. 10.6** (a) Second derivative O K edge spectra for bixbyite, b-Lu$_2$O$_3$ indicating sequential pairs of Ti 3d t$_{2g}$]$_1$ and Ti 3d t$_{2g}$]$_2$ and Ti 3d e$_g$]$_1$ and Ti 3d e$_g$]$_2$ features encapsulated between two Ti 4s features: s]$_1$ and s]$_2$.]$_1$ and ]$_2$ notations indicate different local bonding geometries at the two Lu atom sites. (b) Second derivative O K pre-edge spectra for bixbyite, b-Lu$_2$O$_3$ indicating (a) transition state and negative ion state term representations, and (b) degeneracy removal of transition state terms

of ∼0.3 ± 0.1 eV. This is displayed in Fig. 10.5b, where the spacing between the three singlet negative ion states at energies closer to the conduction band edge is also approximately the same. In addition to the triplet final states, there are also singlet final states, two of which are lower in energy that the lowest triplet transition, three of which are encapsulated with the triplet states, and three of which are at higher energies that the highest triplet term and are designated as negative ion states.

The O K edge spectrum in Fig. 10.6a, and second derivative O K pre-edge spectrum in Fig. 10.6b for bixbyite Lu$_2$O$_3$ are consistent with a mixture of ideal octahedral and distorted octahedral 6-fold coordination of Lu, and 4-fold coordination of O in the extended unit cell structure. The bixbyite structure is sometimes also designated as a defect CaF$_2$ structure with 25% of O-atom sites vacant; hence 4-fold coordinated O, and 6-fold coordinated Lu. The analysis of the O K edge spectrum gives an approximate ligand field splitting, $\Delta_{LF}$ of 3.0 ± 0.1 eV between these two inter-penetrating pairs of [t$_{2g}$ and e$_g$]$_{1,2}$ states in a sequential order, t$_{2g}$]$_1$, t$_{2g}$]$_2$, e$_g$]$_1$, e$_g$]$_2$. The derivative O K edge spectrum indicates that the s-state symmetry features, s]$_1$ and s]$_2$ bracket/encapsulate the two e$_g$ d-state features in the following sequence e$_g$]$_1$, s]$_1$, e$_g$]$_2$ and s]$_2$ as also indicated in Fig. 10.6a. This bracketed sequencing by the s-states in Lu$_2$O$_3$ is expected because the 6s-5d energy difference is ∼1.5 eV, and qualitatively similar, and also smaller than that of ZrO$_2$ where the 5s-4d differences is ∼2.7 eV [19]. In ZrO$_2$ as indicated in Fig. 10.4a the 5s-states bracket the M$_2$-derived t$_{2g}$ feature in this oxide with 7-fold coordination. The spacing between these d- and s states, labeled $\Delta$ds, is indicated in Fig. 10.6a, and is equal to 1.5 ± 0.1 eV.

The second derivative spectrum in O K pre-edge regime in Fig. 10.6b is analyzed with the same symmetry assignments as for 6-fold coordinated as $TiO_2$. The second derivative spectrum for $Lu_2O_3$ in the pre-edge regime also includes nine distinct features in the transition region regime consistent with a cooperative Jahn Teller effect removal of degeneracy. These are the spin allowed triplet transitions from the $^3T_{2g}$ ground term to $^3T_{2g}(F)$, $^3T_{1g}(P)$ and $^3A_{2g}(F)$ terms in order of increasing energy. The spacing between these terms is $3.0 \pm 0.1 \, eV$, essentially the same as value of $\Delta_{LF}$ obtained from the O K edge spectrum in Fig. 10.6a. This value is for O-vacancies is determined primarily by the 6-fold coordination of the Lu atoms. It is a mean field property determined from the two different Lu atom symmetries in the periodic bixbyite cubic structure with an extended unit cell.

This 6-fold coordination symmetry term includes two negative ion states with different $^1E_{1g}(G)$ and $^1A_{2g}(S)$ term energies for each of the two nonequivalent Lu atom bonding arrangements. This gives a total of four negative ion features as opposed to two in $TiO_2$ in which all atoms have the same local bonding symmetry. Stated differently, the negative ion state energies are atomic properties reflecting local site symmetry, rather than ligand field average properties like the transition state energies. In $TiO_2$, and $ZrO_2$ and $HfO_2$ the primitive and extended unit cells contain only one local site symmetry, hence the number of negative ion states features in what is predicted by the schematic diagrams in Fig. 10.2a, b, respectively. In $Lu_2O_3$, the bixbyite unit contains two different local Lu atom distortions with different symmetries. Therefore there are four spectral features assigned to two different pairs of negative ion states, one pair of $^1E_{1g}(G)$ and $^1A_{2g}(S)$ terms for each of the two different Lu local site symmetries. Similarly, two pairs of negative ion states are present in corundum-structured $Ti_2O_3$, where the unit cell contains two different local site symmetries. As in bixbyite $Lu_2O_3$, the transitions energies for $Ti_2O_3$ are determined by the average ligand field at the vacancy site that includes both local atomic symmetries. This complementary behavior of the transition energies, and negative ion state energies is a general property of $A_2O_3$ TM oxides, and has been observed by our group in nanocrystalline $Y_2O_3$ and $Gd_2O_3$ as well. All of these oxides, including $Cr_2O_3$, and the normal metal oxide, $Al_2O_3$, have two metals with different local site symmetries, and since negative ion states are metal atom symmetry specific, these oxides will have two pair of negative ion states as well. Finally there are three singlet states encapsulated with the manifold of triplet states, and an addition single state below the lowest triplet state.

## 10.4 Non-crystalline $SiO_2$, $Si_3N_4$ and Si Oxynitrides

The performance, reliability and aggressive scaling of complementary metal-oxide-semiconductor (CMOS) devices are determined by limitations associated with intrinsic bonding defects. $SiO_2$ holds a unique position in the evolution to advanced nano-scale CMOS devices, being the standard against which device properties of alternative dielectrics for passivation and gate applications are compared [20]. Implicit in these comparisons has been an assumption that intrinsic bonding defects in $SiO_2$ are both qualitatively and quantitatively different than those in the Si oxynitride alloys of this section, and the high-k dielectric TM oxides of the preceding Section, and that this was the primary factor for the uniqueness of $SiO_2$. However, this is not the case. The O-vacancy equivalent $d^2$ model introduced in Sect. 10.2 for intrinsic bonding defects in (i) nanocrystalline TM oxides such as $HfO_2$, also extends to (ii) non-crystalline $SiO_2$, $Si_3N_4$ and Si oxynitride alloys, and this is critically important in understanding the unique properties of the Si-based dielectrics. Stated differently, the O-vacancy model applies to all oxides, and fluorides as well, in which a ligand field splitting can be determined spectroscopically. This requires order chemical bonding order beyond the nearest-neighbor ligands and as will discussed later on in this chapter. It will be demonstrated that there is definitive experimental evidence for ligand field splittings in $SiO_2$ that are comparable to those in first row TM oxides. However, the concept of a Jahn Teller distortion is an issue yet to be addressed.

Due to qualitative differences in conduction band edge electronic states, there are also significant differences between the electrical activity of the negative ion states in $SiO_2$ and TM oxides such as $HfO_2$ and some of emerging group IIIB TM oxides such as $Lu_2O_3$ as well. These high-k dielectrics have been addressed in Sect. 10.3. $Gd_2O_3$ also displays a similar pre-edge spectrum with two pair of negative ion states; however, defect state spectra are not included in this article.

It will now be demonstrated that the wave function symmetries of electronic states at the conduction band edge are one-dimensional and "s-like" states in $SiO_2$, but are two or three dimensional and "d-like" in the TM elemental and complex oxides [21]. The symmetry character of the states at the band edges of crystalline Si (c-Si) and non-crystalline $SiO_2$ have been determined from Si $L_{2,3}$ spectra obtained primarily by electron energy loss spectroscopy [21], but also by our group using low energy, $\sim 100 \, eV$ XAS. This aspect of the Si $L_{2,3}$ spectra is directly correlated with differences in the ordering between non-core level occupied and empty atomic states: 3s and 3d for $SiO_2$, but 3d and 4s for $TiO_2$, 4d and 5s in $ZrO_2$, and 5d and 6s in $HfO_2$. The ordering of these states, i.e., their relative energies, is the important factor in determining the symmetry and dimensionality of the ordering and the symmetry of conduction band edge states. Finally, as will be addressed later on in this section when experimental data are presented for, the odd number of p-electrons in N-atoms in $Si_3N_4$ and Si oxynitride alloys, three, introduces a localized $p\pi^*$-state below the conduction band edge in all nitrided oxides. Pre-edge occupied $p\pi$ and virtual $p\pi^*$ states are responsible for Poole-Frenkel transport in $Si_3N4$, and the increased magnitude the negative bias temperature instability (NBTI) Si oxynitride alloys as well.

Contrary to the *conventional wisdom*, there are not a significant number of different defect states within the band-gap of $SiO_2$ [21]. Instead the energies of absorption and luminescent transitions are in this chapter assigned to transition energies of an equivalent $d^2$ high spin state localized on the two Si atoms associated with an O-atom vacancy. When one of these electrons is removed by exposure to UV radiation, g-rays, x-rays, and/or high energy electrons, the remaining singly occupied state is the well-known $E'$ center [21]. When both states are occupied, there is resonance bonding interaction that damps out an electron spin resonance (ESR) signal. O-vacancies in $SiO_2$ are generated in a defect activation process in which a bridging neutral O-atom is removed as in Eq. (10.1) [22];

$$_{3/2}O - Si - O - Si - O_{3/2} =_{3/2} O - Si [0e]^+ +^- [2e] Si - O_{3/2} + O^0. \quad (10.1)$$

The released neutral O-atoms contributes to other ESR active centers in both terminating 1-fold coordination, or bridging 2-fold coordination configurations that each involves two-O-atom sequence [21, 22]. This perspective on defects in $SiO_2$ is revolutionary and not addressed elsewhere, but is supported by the correlation between O K pre-edge spectra and their interpretation in the context of T-S diagrams, and the schematic diagram in Fig. 10.2b.

The local site symmetries of band edge states, which are different that those of $HfO_2$, $TiO_2$, and other TM oxides, play a determinant role in the trapping properties of negative ion states. This explains qualitative differences between $SiO_2$ and $HfO_2$, such as trap-assisted tunneling (TAT). Band edge symmetries have been identified through an interpretation of X-ray absorption spectroscopy (XAS). The band edge states of $SiO_2$, and have been determined using many electron wave functions and charge transfer multiplet theory for interpretation of Si $L_{2,3}$ and O K edge spectra. Si $L_{2,3}$ spectra are dipole-allowed transitions from occupied spin-orbit split Si $2p_{1/2}$ and $2p_{3/2}$ states at $\sim -99\,eV$ ($\Delta_{LF} \sim 0.6\,eV$) to empty Si 3s and 3d states in an order established by the ordering of the atomic states of Si [23]. Figure 10.7 compares Si $L_{2,3}$ spectra of non-crystalline $SiO_2$ and crystalline Si. Band edge states in both materials are "s-like" non-degenerate $^3A_1$ states; however, the extent of the "s-like" regimes significantly different, $\sim 1\,eV$ in c-Si, and approximately $3\,eV$ in non-crystalline $SiO_2$. These "s-like" states define the band edge for $SiO_2$, and this is addressed in Fig. 10.8. However, the conduction band edge of crystalline Si is at $\sim 1.1\,eV$, and the spectral regime between 1.1 and $\sim 2.6\,eV$ is assigned to indirect transitions which involve both optical and acoustical phonon modes. This coupling of electronic and vibrational states generates a non-vanishing optical transition matrix element.

The most important aspect of the spectra in Fig. 10.7 is the extraction of values of $\Delta_{LF}$ that are comparable to those in TM oxides. It was noted above that $\Delta_{LF}$ is primarily a function of coordination and symmetry that includes at least second-neighbors, or medium range order [24], and does not have a strong dependence on bond ionicity. This is origin of the qualitative and quantitative similarities in the defect states of $SiO_2$ and the TM oxides. If $\Delta_{LF}$ is a spectroscopic observable in an oxide, then the equivalent $d^2$ model can be used to describe the charge neutral O-vacancy defect state properties, including (i) occupied ground states, (ii) transition energies, and (iii) empty negative ion states.

Figure 10.8 is the 2nd derivative spectrum of the band edge states in thin film plasma deposited $SiO_2$ annealed at $\sim 950\,°C$. There is a one-to-one correspondence between energy differences (eV) of band edge states as determined by visible and vacuum UV spectroscopies, and O K edge spectra [24]. In order of decreasing

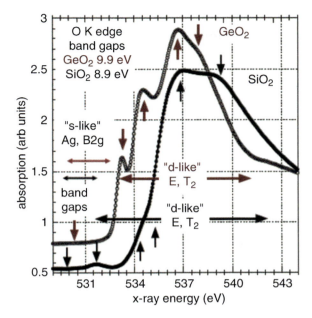

**Fig. 10.7** The Si $L_{2,3}$ spectra of (i) a 10 nm thick non-crystalline nc-SiO$_2$ plasma-deposited film, annealed at 900 °C, and (ii) bulk crystalline c-Si with a (111) orientation. Ligand field splittings, $\Delta$LF for c-Si and nc-SiO$_2$ are indicated

X-ray energy, the features in Fig. 10.8 correspond to: (i) the band-gap, $E_g$, of 8.9 eV, (ii) two bound excitons, $E_2$ and $E_1$, (iii) negative ion states between ∼529.1 and 527.4 eV, and (iv) the symmetry allowed *intra-d state transitions* for that are assigned the O-atom vacancies using the equivalent $d^2$ model. The combination of localized band edge excitonic states, and $^1A_{1g}$ and $^1T_{2g}$ and $^1E_g$ term symmetries of the negative ion states does not support injection or radiative decay into these states. As such, trap-assisted tunneling (TAT) has not been reported for negatively biased n-type Si-SiO$_2$ gate stacks with thickness less than 5 nm. However, the combination of $e_g$ states for band edge ZrO$_2$ and HfO$_2$ states, and the symmetries of negative ion states are responsible for TAT processes in ZrO$_2$ and HfO$_2$ metal-oxide-semiconductor (MOS) test structures [9]. A higher resolution plot, also indicates three singlet state transitions encapsulated with the triplet state transitions.

Figure 10.9a, b indicate respectively, (i) OK band edge, and pre-band edge states of non-crystalline Si$_3$N$_4$, and (ii) expanded x-axis scale plots of the negative ion states of (ii) Si$_3$N$_4$ and a (Si$_3$N$_4$)$_{0.5}$(SiO$_2$)$_{0.5}$ Si oxynitride alloy. There is a significant qualitative difference between Fig. 10.8 for SiO$_2$ and Fig. 10.9a for Si$_3$N$_4$ that is related to the difference in the number of p electrons in the ground states of O- and N-atoms, four for O, and three for N. This correlates with the singly occupied 2p$\pi$ state that gives rise to sharp spectral feature in Si$_3$N$_4$, and Si oxynitride alloys, including compositions that are both SiO$_2$-rich and Si$_3$N$_4$-rich. The final

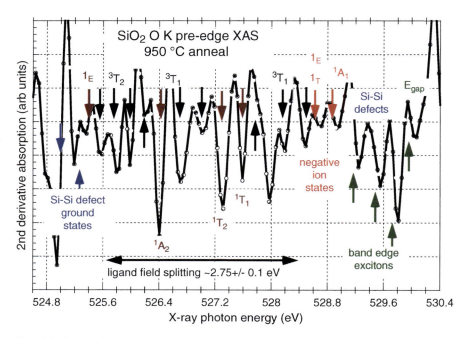

**Fig. 10.8** Second derivative O K XAS spectrum of band edge states in non-crystalline thin film plasma deposited $SiO_2$ annealed at ~950°C

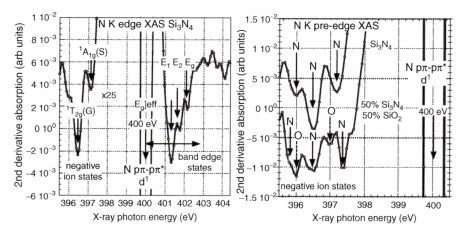

**Fig. 10.9** (a) O K band edge, and pre-band edge states of non-crystalline $Si_3N_4$. (b) Expanded scale plots of the negative ion states of (i) $Si_3N_4$ and (ii) $(Si_3N_4)_{0.5}(SiO_2)_{0.5}$ Si oxynitride alloy

state for this N-atom $2p\pi$ to $2p\pi^*$ transition is at an energy between band edge excitonic states of $Si_3N_4$ and Si oxynitride alloys, and negative ion states of vacancy defects.

The expanded scale plots in Fig. 10.9b support the assignment of the features as negative ion states that are atom specific. Paralleling the assignment made for the negative ion states in $Lu_2O_3$, where there two different Lu local symmetries, the assignment in the $(Si_3N_4)_{0.5}(SiO_2)_{0.5}$ Si oxynitride alloy also reflects different atom-specific negative ion states. In this alloy they are associated with the N- and O-atoms of this alloy.

This difference in ordering of electronic states manifests itself in $Si_3N_4$ thin films by promoting TAT and/or Poole Frenkel transport [25–27]. In addition, this combination of $2p\pi$ and negative ion states also plays a role in the Negative Bias Temperature Instability (NBTI), This defect-related property is much stronger in N-containing dielectrics than $SiO_2$ [28], and is not related to chemical impurity, but rather to differences in the intrinsic properties of the atomic species of the Si oxynitride alloys.

## 10.5 Summary

Based on a universal multiplet approach, the *equivalent $d^2$ model*, has been developed to interpret XAS features for O-vacancy defects in (i) nanocrystalline TM elemental oxides and (ii) non-crystalline Si-containing dielectrics as well. Spin-allowed X-ray triplet transitions that have been detected by a numerical second derivative analysis of features in the O K pre-edge regime beginning below the respective conduction band edges. A *de facto extension* of Hund's rules for strongly correlated electron systems has provided a pathway to the development a high-spin state occupancy *equivalent $d^2$ model*. The atomic configuration of the contributing d-states has an $^3F^4$ term representation with J values of 0, 1, 2, 3 and 4, corresponding to $^1S$, $^3P$, $^1D$, $^3F$ and $^1G$ terms. Applying different T-S diagrams for $d^2$ as required by differences in ligand field symmetry ordering of $e_g$ and $t_{2g}$ crystal field (100% ionicity) limits, the ground state symmetries for 6-fold coordinated Ti in $TiO_2$, and Lu in $Lu_2O_3$ are $^3T_{2g}$, The ground state symmetries for 4-, and 7- and 8-fold coordination, respectively, of Si, and Zr and Hf are $^3A_{2g}$. The T-S diagrams include dipole and spin-allowed triplet-to-triplet transitions, and dipole and spin-allowed singlet-to-singlet transitions. This means that the spacing between singling occupied states within the vacancy are large enough the remove reduce the exchange energy contribution so that singlet and triplet ground states are with at most a few tenths of eV of each other. Negative ion states are empty in neutral O-vacancies. Occupied negative ion states are generated by injection and trapping of electrons into these singlet states above the highest lying triplet to triplet transitions. This occupied states are important in transport and trapping in TM oxides [9], and $Si_3N_4$ [25–27].

This model has been used with considerable success in Sects. 10.3 and 10.4 to interpret O-vacancy features in the O K pre-edge XAS spectra of three representative elemental nanocrystalline thin film TM oxides, $HfO_2$, $TiO_2$, and $Lu_2O_3$, and in non-crystalline $SiO_2$, $Si_3N_4$, and Si oxynitride alloys as well. The most important aspect has the scaling of the spectral width of the allowed transitions with the experimentally determined ligand field splittings obtained from either $L_{2,3}$ and $M_{2,3}$ and/or O K edge spectra. Since the ligand field includes the symmetries of the second-nearest neighbor atoms, this is a new aspect of defect state theory that may have important implications for differences in cooperative Jahn Teller distortions between nanocrystalline and crystalline TM oxides, and non-crystalline $SiO_2$, $Si_3N_4$ and Si oxynitride alloys. As yet unreported results for Ti silicates spanning a composition range from 2% to 100% Ti substitutions for Si indicates evidence for medium range order associated with second neighbor Si-Ti atoms in low composition alloys with 14% Ti, and alloys where the Ti concentration is above the percolation limit of $\sim16.5\%$.

Differences in band edge states combined with the symmetry of the negative ion states (i) account for differences in trap assisted tunneling (TAT) between TM elemental oxides and $SiO_2$ [25–27], and (ii) also explain differences between the low levels of the negative bias temperature instability (NBTI) in $SiO_2$, and the higher levels proportional to the nitrogen content in Si oxynitride alloys [28, 29].

**Acknowledgments** The authors acknowledge partial support from the NSF, SRC, AFOSR, DTRA and ARO. XAS spectra were obtained at the Stanford Synchrotron Lightsource (SSRL). We acknowledge helpful discussions with Professor J. L. Whitten in the Department of Chemistry at NC State University, and Professor D. E. Aspnes in the Department of Physics at NC State University. The $Lu_2O_3$ film was deposited by Professor Heinz Kurz's group at AMO, located at RWTH-Aachen, DE.

# References

1. Lucovsky G, Chung KB, Miotti L et al (2009) Solid State Electron 53:1273
2. Lucovsky G, Seo H, Lee S et al (2007) Jpn J Appl Phys 46:1899; Lucovsky G (2007) J Mol Struct 838:187
3. Niimi H, Lucovsky G (1999) Vac Sci Technol A 17:3185; (1999) Vac Sci Technol B 17:2610
4. Lim S-G, Kriventson S, Johnson TN et al (2002) J Appl Phys 91:4500
5. Xiong K, Robertson J, Clarck SJ (2005) Appl Phys Lett 89:183505
6. Gavartin JL, Muñoz Ramo D, Shluger AL (2006) Appl Phys Lett 89:082908
7. Takeuchi H, Ha D, King TJ (2004) J Vac Sci Technol A 22:1337
8. Price J, Bersuker G, Lysaght PS et al (2009) J Vac Sci Technol B 37:310
9. Autran JL, Munteanu D, Houssa M (2004) In: Houssa M (ed) High-K gate dielectrics. Institute of Physics, Bristol, Chap 3.4
10. Sugano S, Tanabe Y, Kamimura H (1970) Multiplets of transition metal ions. Academic, New York; Tanabe Y, Sugano S (1956) J Phys Soc Jpn 11:864
11. Adler D (1968) Rev Mod Phys 40:714
12. Koiller B, Falicov LM (1974) Phys C Solid State Phys 7:299
13. Bethe H (1929) Ann Physik 3:133
14. Cotton FA (1963) Chemical applications of group theory, 2nd edn. Wiley Interscience, New York, Chap. 8
15. deGroot F, Kotani A (2008) Core level spectroscopy of solids. CRC Press, Boca Raton
16. DeGroot FMF, Fuggle JC, Thoule BT, Swatzky GA (1990) Phys Rev B 41:928
17. DeGroot FMF, Fuggle JC, Thoule BT, Swatzky GA (1990) Phys Rev B 42:5459

# 10 Many-Electron Multiplet Theory Applied to O-Vacancies in...

18. Cox PA (1992) Transition metal oxides. Clarendon, Oxford, Chaps 2, 3 and 5
19. Harrison WA (1999) Elementary electronic structure. World Scientific, Singapore
20. Lucovsky G (2005) In: Massoud HZ et al (eds) Physics and chemistry of $SiO_2$ and the $Si$-$SiO_2$ interface-5. The Electrochemical Society, Pennington, p 3
21. Griscom DL (1974) J Noncryst Solids 24:155
22. Feltz A (1983) Amorphe and glasartige anorganische Festkörper. Akademie, Berlin, pp 241–242
23. Thompson A, Lindau I, Attwood D et al (eds) (2001) X-ray data booklet. Lawrence Berkeley Laboratory, Berkeley, Section 1
24. Lucovsky G (2010) Phys Status Solidi a 207:631
25. Sze SM (2002) Semiconductor devices: physics and technology, 2nd edn. Wiley, New York, Chap 3
26. Arnett PC, Maria DJ (1976) J Appl Phys 47:2093
27. Cheng XR, Liu BY, Chen YC (1987) Appl Surf Sci 30:237
28. Schroeder DK (2007) Microelectron Reliab 47:841
29. Zafar S, Stathis J, Callegari A (2005) In: Massoud HZ et al (eds) Physics and chemistry of $SiO_2$ and the $Si$-$SiO_2$ interface-5. The Electrochemical Society, Pennington, p 127

# Part III
# Fullerenes, Fullerides and Related Systems

# Chapter 11
# $C_{60}$ Molecules on Surfaces: The Role of Jahn–Teller Effects and Surface Interactions

**Janette L. Dunn, Ian D. Hands, and Colin A. Bates[†]**

**Abstract** The molecular orbitals of fullerene molecules on surface substrates can be imaged experimentally using scanning tunnelling microscopy (STM). The observed images are influenced by interactions with the substrate. In addition, for fullerene ions, splitting of the orbitals by the Jahn–Teller (JT) effect also affects the observed images. In this work, we consider the effect of both static and dynamic JT interactions on the images that are expected to be obtained from the fullerene anion $C_{60}^-$, taking into account interactions with the substrate. Our method is to use Hückel molecular orbital (HMO) theory, which is very simple and quick to implement on a desktop computer. Although our approach is not as rigorous as using density functional theory (DFT), the predicted STM images are almost indistinguishable from published DFT images. Furthermore, it readily allows us to explore different situations, such as different adsorption geometries. Our results are also compared with experimental and simulated STM images in the literature.

## 11.1 Introduction

Scanning Tunnelling Microscopy (STM) has become an important tool for probing the structure of molecules with atomic-scale resolution. The fullerene molecule $C_{60}$ is an ideal candidate for STM investigations due to its large size. A number of investigations have been published probing both single molecules [1] and structured monolayers situated directly on various substrates [2–11], and upon a buffer layer to decouple the fullerene molecules from the surface [12, 13]. Experiments recording

---

[†] Deceased

J.L. Dunn (✉) • I.D. Hands • C.A. Bates
School of Physics and Astronomy, University of Nottingham, Nottingham, UK
e-mail: janette.dunn@nottingham.ac.uk; ian.hands@nottingham.ac.uk.

M. Atanasov et al. (eds.), *Vibronic Interactions and the Jahn-Teller Effect:*
*Theory and Applications*, Progress in Theoretical Chemistry and Physics 23,
DOI 10.1007/978-94-007-2384-9_11, © Springer Science+Business Media B.V. 2012

images of fullerene anions are particularly interesting as the centres involved are Jahn–Teller (JT) active. This means that there is the possibility of observing evidence of the JT effect at the molecular level, which is difficult to achieve by other means.

In an STM experiment, a probe containing a very fine tip (often terminated with a single molecule) is scanned over the surface of a sample in the presence of an applied bias. The tip must be sufficiently close to the sample that electrons can tunnel through the gap between the tip and sample. If the bias is positive, electrons will tunnel into empty orbitals and if it is negative they will tunnel from occupied orbitals. There are two common modes of operation; in constant height mode, the distance between the tip and substrate is kept constant and the tunnelling current recorded, and in constant current mode a feedback loop is used so that the tunnelling current is kept constant and the vertical height between the tip and sample is recorded. Both modes result in similar images, and differences between the modes will not be considered in this paper.

According to the standard theory used to interpret STM images due to Tersoff and Hamann [14], the tunnelling current $I$ is taken to be

$$I \propto \sum_v |\psi_v(\mathbf{r})|^2 \delta(E_v - E_F) \tag{11.1}$$

where $\psi_v$ is the wave function of a surface state of energy $E_v$ and $E_F$ is the Fermi energy. Therefore, STM images the local density of states (LDOS) at the Fermi level, which equates to imaging molecular orbitals near the Fermi level. Positive biases image the lowest unoccupied molecular orbital (LUMO) and negative biases image the highest occupied molecular orbital (HOMO). Therefore, whilst STM does not image the positions of atoms directly, it does image the related electron densities.

Most workers use density functional theory (DFT) to simulate the molecular orbitals involved, and there is no doubt that the results obtained using this approach give a good account of the experimental observations. However, it can be computationally-expensive to run DFT simulations (depending on the complexity of the situation considered) and an expert in DFT is required to correctly set up the simulations. This is especially true where electronic degeneracies occur [15], as is the case in the JT-active fullerene anions. Also, it is not a simple matter to add in the effects of additional perturbations, such as interactions between fullerene anions (including co-operative JT effects). In a recent paper [16], we have shown that simulated images of neutral $C_{60}$ molecules which match both experimental and DFT results can be obtained using Hückel molecular orbital (HMO) theory.

HMO theory is an established and well-used theory involving several approximations, the most fundamental of which is that it considers only the available $\pi$-orbitals. Hückel Hamiltonians traditionally involve just two parameters: $\alpha$ (single-atom contribution) and $\beta$ (two-centre contribution between neighbours)— contributions from non-neighbouring atoms are assumed to be negligible and are ignored [17]. After symmetry considerations are used, the Hamiltonian can then be diagonalized. This allows the molecular orbitals and their ordering with respect to energy to be determined very easily and without the need for knowledge of any parameter values (such as $\alpha$ or $\beta$). Despite the approximations made in HMO theory, the simulated STM images obtained using the Hückel orbitals are almost indistinguishable from simulated DFT images [16]. Qualitatively, this is because STM probes the orbitals near the HOMO/LUMO gap only and the details of the inner orbitals are not relevant.

One disadvantage of using HMO theory is that it is not an ab initio method and therefore it is not possible to use it to predict a priori the images that will be observed in a given STM experiment. However, it can be used to say whether a given observation is consistent with a given situation. For example, we could obtain images assuming a JT distortion of a certain symmetry and compare it with experimental results to deduce the appropriateness of the assumed distortion. The same could be done for different types of surface interaction.

In this paper, we first review the application of HMO theory to the interpretation of STM images from neutral $C_{60}$ molecules to show that it is viable to use this method instead of DFT. As the neutral molecule contains a fully-occupied HOMO it is not JT-active, which simplifies the theory somewhat. The calculations are then extended to the mono-anion $C_{60}^-$ in order to determine the likely signature of the JT effect in STM images of this ion. We first consider the idealized case of static JT distortions, where it is assumed that tunnelling between equivalent distorted configurations can be ignored. We then discuss the likely consequences of dynamic JT effects on the observed STM images. Finally, we conclude by making some comparisons between our simulated results and experimental observations involving fullerene anions.

## 11.2 STM of Neutral $C_{60}$ Molecules

The HMOs for $C_{60}$ can be readily obtained, either from first principles or using the results in [18]. In the latter case, the tabulated results assume that single and double carbon–carbon bonds have equal resonance integrals $\beta$, but it is a simple matter to adapt this to the more realistic case where the integrals are different [16]. The $C_{60}$ molecule has icosahedral ($I_h$) symmetry, which supports irreducible representations (irreps.) with up to fivefold degeneracy. The orbitals relevant to STM imaging are the HOMO, LUMO and LUMO+1, which are of symmetry $H_u$, $T_{1u}$ and $T_{1g}$ respectively. Representations of these orbitals are shown in Fig. 11.1, along with a representation of the $C_{60}$ molecule. Note that the LUMO and LUMO+1 appear identical as both have $T_1$ symmetry and the parity of the representation is not relevant in the molecular orbital plots. In drawing these orbitals, we have defined our $z$-axis to be through the centre of a pentagon and the $y$-axis to be through the centre of a double bond, as indicated in the figure.

In modelling $C_{60}$ molecules on a surface, the orientation of the molecule on the surface is clearly very important. As the direction perpendicular to the surface is likely to be affected in a different way to directions in the plane of the surface,

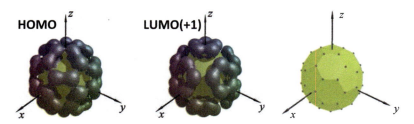

**Fig. 11.1** (*Colour online*) The two *left-hand* images show plots of the HOMO and LUMO/ LUMO+1 respectively of isolated $C_{60}$ molecules. The *right-hand* image represents the $C_{60}$ molecule, showing our choice of axes for adsorption with a pentagon prone to the surface

we find it easiest to determine the effect of a surface interaction by defining our $z$-axis to be perpendicular to the surface. For the case of a molecule adsorbed with a pentagon prone to the surface, this choice coincides with the axes used to plot the orbitals in Fig. 11.1. For adsorption in other orientations, however, we need to redefine our molecular basis. Nevertheless, this is easier than defining the surface to be in different planes, where the direction perpendicular to the surface would be a combination of $x$, $y$ and $z$.

For the case of adsorption with a pentagon prone to the surface, where the $z$-axis coincides with that in Fig. 11.1, an STM image of the HOMO could be expected to pick out five lobes of electron density spreading out radially from the $z$-axis, and images of the LUMO and LUMO+1 (at different biases) would pick out a five-sided ring of electron density. However, experimental images show rather more complicated features than these. This confirms that interactions between the $C_{60}$ molecule and a surface do need to be considered.

Depending on where the molecule sits on the surface, the resulting symmetry can be one of ten subgroups of $I_h$ [16]. None of the site symmetries supports degeneracies greater than two, so all will split the molecular orbitals. The characters of the symmetry operations can be used to decompose the irreps. of the icosahedral group in terms of irreps. of the lower symmetry and hence determine the splitting of the molecular orbitals in the lower symmetry. It has been found that results can be obtained which match experimental observations by assuming the site symmetry is the highest supported, namely $C_{6v}$ [16]. Plots of the molecular orbitals split into $A$, $E$ and $E'$ components of $C_{6v}$ symmetry are shown in Fig. 11.2 for the case of a molecule adsorbed with a hexagon facing the surface. Note that the $z$-axis is through the centre of a hexagon, which is different to Fig. 11.1 where the $z$-axis was through the centre of a hexagon. It should also be noted that if we sum the images in each row then, as expected, we will obtain the same images as in Fig. 11.1.

The resolution of experimental STM images is limited by various factors in the STM apparatus. Figure 11.3 shows our STM images of the HOMO, LUMO and LUMO+1 for a hexagon-prone $C_{60}$ molecule in an idealized, high-current simulation. It also shows one possible set of results for the $A$-component of the split orbitals under lower-current conditions. Whilst the low-current results are more realistic, it is very difficult to predict precisely the conditions required to

# 11 C$_{60}$ Molecules on Surfaces: The Role of Jahn–Teller Effects and Surface...

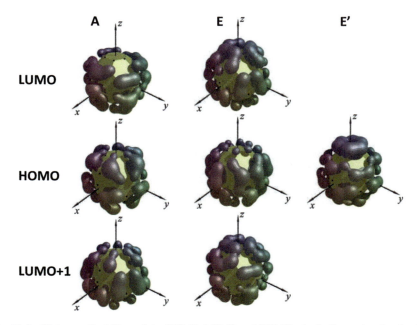

**Fig. 11.2** (*Colour online*) Plots of the HOMO, LUMO and LUMO+1 of a C$_{60}$ molecule adsorbed with a hexagon facing a surface in the $x$–$y$ plane, when the orbitals have been split into components $A$, $E$ and $E'$ due to reduction to $C_{6v}$ symmetry (Reprinted from Fig. 8 of [16])

reproduce images from a given apparatus. Furthermore, it is possible to deduce much of the information that would appear in a low-current image from the idealized higher-current case by considering only the central portion of the image and blurring sharply-changing features. Therefore, in the remainder of this paper, we will concentrate on results for the idealized, high-current case.

We can now compare our HMO results with existing experimental images and DFT simulations. In Fig. 11.4, the top two rows show experimental results and DFT simulations of the HOMO of an isolated C$_{60}$ molecule adsorbed on a Si(111) surface, which are reproduced from [4]. The bottom row shows our corresponding HMO simulations for a $C_{6v}$ surface interaction. The column on the left shows results for when the C$_{60}$ molecule is oriented such that a hexagon is in contact with the surface, the middle column is for a double-bond aligned with the surface and the rightmost column is for a pentagon aligned with the surface. In all cases, the theoretical images have been arbitrarily rotated to match the experimental observations. It can be seen that the matches between the HMO and DFT results are extremely close in all three cases. The very small differences are impossible to resolve in the experimental data.

Comparisons have been made between our HMO images and published experimental STM images on a variety of substrates [such as Si(111) and Au(111)] and for molecules in various orientations, and good matches found in almost all cases. Further details are given in [16].

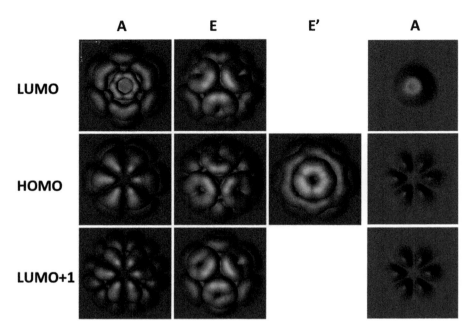

**Fig. 11.3** The *leftmost* three columns show idealized STM simulations of the split molecular orbitals of the hexagon-prone C$_{60}$ molecule depicted in Fig. 11.2. The *right-hand* column shows equivalent low-current simulations of the A component

## 11.3 STM of C$_{60}^{-}$ Anions

We have shown that simulations using the HMO approach can accurately reproduce observed STM images on neutral C$_{60}$ molecules. We will now consider charged states, where the JT effect must be taken into account in addition to surface interactions. We consider the C$_{60}^{-}$ ion only, as the C$_{60}$ molecule will much more readily gain electrons than lose them, and the mono-anion is the simplest JT system to treat. However, the methods described here can be applied to other charge states with suitable modifications. We will also ignore intermolecular interactions, although they could be added in at a later stage.

The JT effect, because of coupling between electronic and vibrational motion, induces the C$_{60}^{-}$ ion to distort into a configuration which has lower energy than the undistorted $I_h$ case. However, there will be multiple configurations with the same energy, characterized by minima in the lowest adiabatic potential energy surface (APES). If the JT coupling is very strong, the system may effectively remain localized in one of the distorted configurations. This is known as the static JT effect. At more realistic values of coupling, the system will tunnel between the equivalent configurations in what is known as the dynamic JT effect. We will first consider the static case, before discussing the likely implications of dynamic effects.

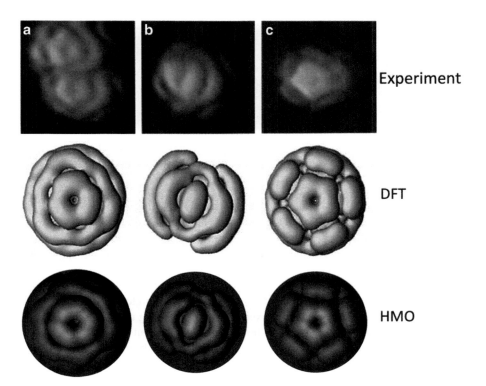

**Fig. 11.4** The *top* two rows show a comparison of experimental STM images and DFT simulations of the HOMO of isolated C$_{60}$ molecules on a Si(111) surface (Reproduced with permission from [4]). The *bottom* row shows equivalent simulations for a neutral C$_{60}$ molecule using HMO theory. The three columns refer to the C$_{60}$ molecule adsorbed with a hexagon, double bond and pentagon prone to the surface respectively

## 11.3.1 Static Jahn–Teller Effects

The HOMO of the C$_{60}^-$ ion is subject to a $T_{1u} \otimes h_g$ JT effect, where the $T_{1u}$ electronic state is coupled to $h_g$-type vibrations. There are eight different $h_g$ modes, but for these purposes it is sufficient to consider a single effective mode [19]. In linear JT coupling, the lowest APES contains a multi-dimensional trough of equivalent-energy points [20]. Due to the non-simple reducibility of the product $H \otimes H$, there are two distinct types of quadratic coupling, each characterized by a quadratic coupling constant. These (and higher orders of coupling) warp the trough to give either 6 equivalent minima of $D_{5d}$ symmetry, 10 minima of $D_{3d}$ symmetry, 15 minima or $D_{2h}$ symmetry or (for a very small range of quadratic coupling constants) 30 minima of $C_{2h}$ symmetry. Although some estimates have been made of the values of the JT coupling constants for the C$_{60}^-$ ion, values for the quadratic constants in particular are not known. Therefore it is necessary to consider all of these possible distortion symmetries.

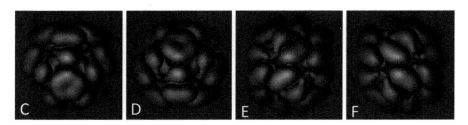

**Fig. 11.5** Simulated STM images of the ground state of the four states labelled C to F of a $C_{60}^-$ ion when distorted to $D_{5d}$ symmetry by the JT effect and adsorbed with a double bond prone to the surface

Whilst the JT effect will displace the atoms in the $C_{60}$ ion from the positions corresponding to the pure icosahedral symmetry of the neutral molecule, these displacements will be very small compared to the overall size of the molecule. Hence it will be very unlikely that shifts in the atomic positions could be observable in an STM image. We will therefore neglect these displacements. However, we expect a static JT effect to split the $T_{1u}$ orbital into a singlet ground state and either a singly or doubly degenerate excited state, depending upon the distortional symmetry. As the components will be imaged at different biases in an STM experiment, the effect of this splitting is expected to be observable, in the same way that splittings due to surface interactions are observed for the neutral molecule.

Each of the allowed distortions corresponds to a distortion along a particular axis. These axes will all be oriented differently with respect to the viewing plane of an STM experiment, so images of each distortion will be different. To illustrate this, consider JT distortions to $D_{5d}$ symmetry. There are six equivalent minima in the APES, which we will label from A to F [21]. Each corresponds to a distortion along a different fivefold ($C_5$) symmetry axis through the centres of the six pairs of mutually-opposite pentagons that form a truncated icosahedron. This symmetry results in a singlet $A$-type ground state and a doublet $E$-type excited state. Thus a static JT effect could be expected to result in six different images for the ground state and another six images for the excited state. Some images can be obtained by rotations of other images, but others will appear rather different. For example, Fig. 11.5 shows simulated images of the ground state of wells C to F for a molecule adsorbed with a double-bond prone to the surface. It can be seen that the images for wells C and D are related by a rotation about 180° in the viewing plane, as are the images for wells E and F. However, the images for C and D are rather different to those for E and F.

At this point, we have not taken the surface interaction into account. However, we expect it to be important, as it was for the neutral molecule. A surface interaction can be expected to have different effects on each of the wells and hence remove their equivalences.

# 11 C$_{60}$ Molecules on Surfaces: The Role of Jahn–Teller Effects and Surface...

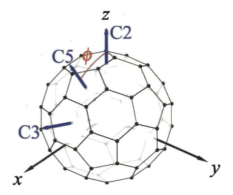

**Fig. 11.6** (*Colour online*) Definition of the angle $\phi$ used to characterize the orientation of the C$_{60}$ molecule on a surface

As we want to probe the likely effect of JT interactions on the STM images, we will consider the JT effect to be dominant and treat the surface interaction as a perturbation. We do this by following [22] and introducing a model Hamiltonian to describe the surface splitting of the form

$$\mathcal{H}_S = -\frac{\Delta_S}{6} \begin{bmatrix} 1 - 3\cos 2\phi & 0 & 3\sin 2\phi \\ 0 & -2 & 0 \\ 3\sin 2\phi & 0 & 1 + 3\cos 2\phi \end{bmatrix} \quad (11.2)$$

where $\Delta_S$ is a constant governing the strength of the interaction. In addition, $\phi$ is an angle characterizing the orientation of the molecule on the surface which, with respect to the choice of axes in Fig. 11.1, is defined to be the angle from the $z$-axis towards the $x$-axis in the $x$–$y$ plane. This is shown in Fig. 11.6. We are particularly interested in orientations in which a hexagon is prone to the surface, which is the angle to the $C_3$-axis in Fig. 11.6, a pentagon prone to the surface for which $\phi$ is the angle to the $C_5$-axis, and for a double bond prone to the surface, for which $\phi = 0$. This model Hamiltonian will split the molecular orbitals in a manner which is consistent with the splittings obtained using the group-theoretical arguments discussed for the neutral molecule in Sect. 11.2.

The change in energy $\Delta E$ due to the effect of $\mathcal{H}_S$ as a perturbation acting on each of the states associated with the wells can easily be calculated and hence we can determine which wells correspond to the lowest-energy configuration in the presence of the surface interaction. This shows, for example, that the effect of the surface interaction on a molecule oriented with a double-bond prone to the surface ($\phi = 0$) is to lower the energy of wells C and D if $\Delta_S$ is positive, or to lower wells E and F if $\Delta_S$ is negative. We could therefore expect the system to favour these wells. This means that we would expect to observe either the first two images in Fig. 11.5 or the last two. A comparison with experimentally-observed images will be given in Sect. 11.4.

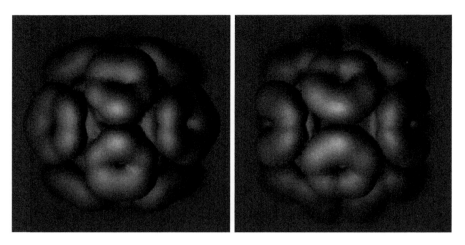

**Fig. 11.7** Superposition of images of equivalent JT distortions to $D_{5d}$ symmetry. The *left-hand* image would be obtained if $\Delta_S > 0$ and the *right-hand* image if $\Delta_S < 0$

## 11.3.2 Dynamic Jahn–Teller Effects

The timescale of an STM experiment is slow compared to dynamic effects; the time to image a single $C_{60}$ molecule is believed to be the order of milliseconds, whereas the time taken to pseudorotate between equivalent distortions is the order of pico or femtoseconds [22]. Thus, it is unlikely that images like those described in Sect. 11.3.1 will be observed in practice, unless there are additional mechanisms to stabilize one distorted configuration over another.

If the quadratic JT coupling is strong, which means that the barriers between equivalent configurations are high, then it may be a reasonable approximation to assume that the system spends most of its time in the distorted wells and only very short times tunnelling between the wells. In this case, the images observed would be a simple sum of the equivalent configurations discussed in Sect. 11.3.1. Figure 11.7 shows the resultant images for distortion to $D_{5d}$ symmetry for positive and negative values of the surface interaction parameter $\Delta_S$. The image on the left shows the superposition of the images corresponding to a distortion to wells C and D as shown in Fig. 11.5, and the image on the right shows the equivalent result for wells E and F.

It is immediately obvious that the combined images in Fig. 11.7 are much more similar to the images obtained in the absence of a JT effect than the images corresponding to individual wells, with fivefold pentagonal rings again appearing. However, the rings are no longer fully symmetric, with some parts appearing more strongly than others. The differences are sufficiently great that we could expect to be able to distinguish between the JT and non-JT cases in high-resolution STM experiments. However, the detail could become lost in lower resolution experiments.

If the barriers between wells are small then the observed images may also contain contributions from configurations corresponding to the tunnelling paths between wells. This was illustrated for the $E \otimes e$ JT problem in [23]. Whilst the conversion from one well to another is far too fast to be observable in even the fastest STM experiment, it is interesting to examine what would be observed if the inter-conversion could be observed. The supplementary material to this paper contains animations for the conversions between both the wells C and D and the wells E and F shown in Fig. 11.5.

## 11.4 Comparison with Experiment

As mentioned previously, various experiments have been published showing images of $C_{60}$ molecules on different surfaces. Most are on Ag, Au or Si surfaces, but a few are on surfaces with a buffer layer to isolate the $C_{60}$ molecules from the substrate. This can be expected to reduce the surface interaction, so will be the most relevant for comparison of our images, which have been obtained assuming that the surface interaction is small compared to the JT effect.

Figure 11.8 shows images of $C_{60}$ molecules on a self-assembled monolayer of alkylthiol on an Au(111) support, reproduced from [24]. By examining our simulated images for various adsorption geometries and JT distortion symmetries, we are able to find possible explanations for the observed results if the images are due to $C_{60}^-$ anions. Two examples of the simulations we obtain assuming a low-current situation are also shown in Fig. 11.8. These matches are from a JT effect favouring distortion to $D_{3d}$ symmetry. The uppermost domain in the experimental

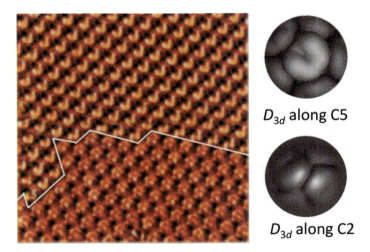

**Fig. 11.8** (*Colour online*) STM image of a layer of $C_{60}$ molecules on an alkylthiol self-assembled monolayer on an Au(111) support (Reprinted from part of Fig. 2 of [24] with permission from Macmillan Publishers Ltd, Copyright (2001). Also shown are possible matches to the experimental images from our simulations)

results bears a strong resemblance to the results we obtain assuming each $C_{60}$ molecule is adsorbed with a $C_5$ axis pointing to the surface. The distinctive 'v'-shape appears because there is a node at one of the atoms of the uppermost pentagon. This is a more convincing match than that provided by the authors themselves. The images in the lower domain can also be described by a $D_{3d}$ JT distortion, this time for adsorption with a double-bond facing the surface.

It should be noted that we have considered isolated $C_{60}$ molecules. We have included surface interactions but not $C_{60}$–$C_{60}$ interactions. Hence, at this stage, we cannot explain the regular ordering observed in the two domains. It may be that the JT interaction favours a specific orientation such that the same image is observed from each molecule in a domain. However, it is more likely that the regular ordering is due to additional intermolecular interactions, such as co-operative JT effects. Furthermore, there is a possibility that the preference of certain wells over others that we have attributed to surface interactions are actually due to intermolecular interactions. This could occur if the lifting of equivalences between wells due to intermolecular interactions follows the same pattern as that due to surface interactions.

Reference [25] contains images of $C_{60}$ molecules on the same surface as [24] which are clearly different to those of [24]. We reproduce those results in Fig. 11.9, along with two possible low-current simulations. The images in the uppermost domain could be matched to a $D_{3d}$-distorted molecule (as was the case in [24]) with a double-bond prone to the surface, although the match is to a different state to that needed to match the results of [24] and the current is an order of magnitude larger than that used for our other lower-current simulations in Figs. 11.8 and 11.9. The images in the lower domain can also have a JT explanation, this time involving an excited state of a $D_{2h}$-distorted ion. In fact the match is better than that provided by the non-JT explanation in [25] (see inset to experimental image) as the distribution is somewhat narrower. However, it is questionable whether the bias used to obtain the image could image an excited state.

We have also analyzed experimental images of $C_{60}$ molecules adsorbed directly on Ag, Au and Si surfaces. Here, the surface interaction is expected to be more important. Also, the degree of charge transfer is unknown, and may be more significant than for adsorption onto a buffer layer. It is possible to obtain explanations involving JT distortions, but non-JT explanations can also be found. Further details are given in [22]. Also, we have matched images using a static JT model, even though we would expect dynamic effects to cause some averaging so that such images would not be directly observable. However, as matches can be made, it may be that other interactions which we have not considered explicitly may stabilize certain images so that they are equivalent to those that would be observed with a static JT interaction alone.

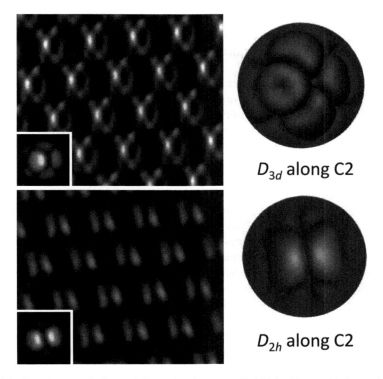

**Fig. 11.9** STM images of a layer of $C_{60}$ molecules on an alkylthiol self-assembled monolayer on an Au(111) support (Reprinted from Fig. 1 of [25] with permission from the American Chemical Society, Copyright (2003)). Also shown are possible matches to the experimental images from our simulations

## 11.5 Conclusions and Future Work

We have simulated STM images using HMO theory, rather than the DFT approach more usually used. Ours is a simple approach which is very easy to implement without specialist knowledge, and is simple to adapt to different situations. Good agreement with DFT images has been obtained for neutral $C_{60}$ molecules, which gives confidence that the approach is valid. The method used contains free parameters whose values are not intrinsically known, such as the JT coupling constants and the size of the surface splitting. This means that we can't predict in advance the likely outcome of a given experimental investigation. However, by matching experimental data and simulated images we are able to provide possible interpretations of the results obtained. In this sense, our approach should be considered complementary to that of DFT.

Having applied our computational method to neutral $C_{60}$, we have used the same approach to consider $C_{60}^-$ anions, where the JT effect must be included in the formalism. Surface interactions have also been included, but have been assumed to be weak compared to JT interactions. By comparing our simulated images with experimental images of $C_{60}$ ions isolated from a substrate by a buffer layer, we have found some indication that there may be JT distortions apparent in the experimental data.

In this paper, there are a number of factors we have not considered which could form the basis of further work. We have not explicitly considered charge transfer between the substrate and $C_{60}$ molecule. The charge on each $C_{60}$ might vary with time, and the effective charge may not even be an integer. However, even a partial electron transfer will yield some JT effect and hence splitting of the MOs. Partial occupancy of the MOs between 0 and 1 would affect the tunneling current but the appearance of STM images, if determined by the JT effect, would be the same as those we give. Also, we have only considered singly-charged anions $C_{60}^{-}$. Interesting images have been obtained experimentally from ions in higher charge states, in particular from $C_{60}^{3-}$ and $C_{60}^{4-}$ ions in the $A_3C_{60}$ and $A_4C_{60}$ fullerides [6]. The former ions are of particular interest as these materials can be superconducting at relatively high temperatures, whereas the latter are insulators. However, to simulate STM images from these ions, it is necessary to account for intramolecular JT effects incorporating interactions due to electron–electron Coulombic repulsion, as well as potentially considering intermolecular (co-operative) JT interactions between $C_{60}^{n-}$ ions. Both of these are far from trivial tasks. Finally, there are other factors, such as interactions with the tip and other intramolecular interactions which could be important, and may even provide an alternative explanation to the experimentally-observed images.

**Acknowledgments** We wish to thank the UK Engineering and Physical Sciences Research Council for supporting this work (Grant EP/E030106/1).

# References

1. Lu X, Grobis M, Khoo KH, Louie SG, Crommie MF (2004) Phys Rev B 70:115418
2. Altman EI, Colton RJ (1993) Phys Rev B 48:18244
3. Hashizume T, Motai K, Wang XD, Shinohara H, Saito Y, Maruyama Y, Ohno K, Kawazoe Y, Nishina Y, Pickering HW, Kuk Y (1993) Sakurai T Phys Rev Lett 71:2959
4. Pascual JI, Gómez-Herrero J, Rogero C, Baró AM, Sánchez-Portal D, Artacho E, Ordejón P, Soler JM (2000) Chem Phys Lett 321:78
5. Grobis M, Lu X, Crommie MF (2002) Phys Rev B 66:161408(R)
6. Wachowiak A, Yamachika R, Khoo KH, Wang Y, Grobis M, Lee DH, Louie SG, Crommie MF (2005) Science 310:468
7. Casarin M, Forrer D, Orzali T, Petukhov M, Sambi M, Tondello E, Vittadini A (2007) J Phys Chem C 111:9365
8. Schull G, Berndt R (2007) Phys Rev Lett 99:226105
9. Wang Y, Yamachika R, Wachowiak A, Grobis M, Khoo KH, Lee DH, Louie SG, Crommie MF (2007) Phys Rev Lett 99:086402
10. Wang Y, Yamachika R, Wachowiak A, Grobis M, Crommie MF (2008) Nat Mater 7:194
11. Li HI, Pussi K, Hanna KJ, Wang LL, Johnson DD, Cheng HP, Shin H, Curtarolo S, Moritz W, Smerdon JA, McGrath R, Diehl RD (2009) Phys Rev Lett 103:056101

12. Frederiksen T, Franke KJ, Arnau A, Schulze G, Pascual JI, Lorente N (2008) Phys Rev B 78:233401
13. Silien C, Pradhan NA, Ho W, Thiry PA (2004) Phys Rev B 69:115434
14. Tersoff J, Hamann DR (1985) Phys Rev B 31:805
15. Bersuker IB (1997) J Comput Chem 18:260
16. Hands ID, Dunn JL, Bates CA (2010) Phys Rev B 81:205440
17. Atkins P, Friedman R (2005) Molecular quantum mechanics. Oxford University Press, Oxford
18. Deng Y, Yang CN (1992) Phys Lett A 170:116
19. O'Brien MCM (1983) J Phys C Solid State Phys 16:85
20. Dunn JL, Eccles MR, Liu YM, Bates CA (2002) Phys Rev B 65:115107
21. Dunn JL, Bates CA (1995) Phys Rev B 52:5996
22. Hands ID, Dunn JL, Bates CA (2010) Phys Rev B 82:155425
23. Hands ID, Dunn JL, Rawlinson CSA, Bates CA (2009) In: Köppel H, Yarkony DR, Barentzen H (eds) The Jahn–Teller effect. Springer series in chemical physics, vol 97. Springer, Heidelberg, pp 517–551
24. Hou JG, Yang JL, Wang HQ, Li QX, Zeng CG, Yuan LF, Wang B, Chen DM, Zhu QS (2001) Nature 409:304
25. Yuan LF, Yang J, Wang H, Zeng C, Li Q, Wang B, Hou JG, Zhu Q, Chen DM (2003) J Am Chem Soc 125:169

# Chapter 12
# The Quadratic $p^3 \otimes h$ Jahn–Teller System as a Model for the $C_{60}^{3-}$ Anion

**Andrew J. Lakin, Ian D. Hands, Colin A. Bates[†], and Janette L. Dunn**

**Abstract** The fullerene trianion, $C_{60}^{3-}$, and the compounds associated with it are known to have properties that differ significantly from the other fullerene ions. For example, compounds of the form $A_3C_{60}$ (where A is an alkali metal) which contain this ion, are known to be superconductors up to around 40 K, whereas the alkali metal fullerenes containing $C_{60}^{2-}$ and $C_{60}^{4-}$ are found to be insulators, properties often attributed to the Jahn–Teller effect. In spite of this, little work has been undertaken analysing the Jahn–Teller effect in the trianion. In this work, the symmetry reduction caused by this effect is investigated by introducing quadratic terms into the Hamiltonian to model the Jahn–Teller interaction. It is found that, unlike the previously investigated ions of $C_{60}$, an electronic degeneracy remains if the molecular distortion were to be described by either the $D_{3d}$ or $D_{5d}$ point groups. Thus, a further reduction in symmetry is expected, and it is found that the distorted molecule is actually described by either the $C_{2h}$ or $D_{2h}$ group. A distortion of $C_{2h}$ symmetry in a fullerene molecule has previously undergone little analysis, and so it is this that is then investigated by considering a set of distortional axes relating to the minimum energy wells formed under a quadratic interaction.

## 12.1 Introduction

Since the discovery of the $C_{60}$ Buckminsterfullerene molecule by Kroto et al. [1], much work has been undertaken on the various properties of the systems associated with it and its ions. Due to the high symmetry ascribed to the molecule via its truncated icosahedral structure, one of the areas of research that $C_{60}$ lends itself

---

[†] Deceased

A.J. Lakin (✉) • I.D. Hands • J.L. Dunn
School of Physics and Astronomy, University of Nottingham, Nottingham, UK
e-mail: ppxal@nottingham.ac.uk; ian.hands@nottingham.ac.uk; janette.dunn@nottingham.ac.uk

M. Atanasov et al. (eds.), *Vibronic Interactions and the Jahn-Teller Effect:*
*Theory and Applications*, Progress in Theoretical Chemistry and Physics 23,
DOI 10.1007/978-94-007-2384-9_12, © Springer Science+Business Media B.V. 2012

particularly well to is the investigation of symmetry lowering interactions such as the Jahn–Teller (JT) effect. This high symmetry results in highly degenerate vibrational levels and electronic states which is the source of the JT interaction in this molecule [2].

When in its neutral state, the highest occupied molecular orbital (HOMO) of $C_{60}$ is fully occupied, meaning there is no electronic degeneracy and no JT effect is permitted. However, when the molecule is ionised a degeneracy is formed via either the removal of one or more electrons from the fivefold degenerate HOMO to form a cation, or the addition of one or more electrons to the threefold degenerate lowest unoccupied molecular orbital (LUMO) to form an anion. This degeneracy allows the JT effect to be exhibited, and as such a reduction in molecular symmetry is expected to occur for the fullerene ions.

Much work has been undertaken investigating the form of this symmetry reduction for a variety of the charge states, of which particular emphasis has been placed on the negatively charged anions [3–7]. Thus far, the $C_{60}^-$ and $C_{60}^{2-}$ anions, (and hence, via electron-hole equivalence, the $C_{60}^{5-}$ and $C_{60}^{4-}$ anions) have been extensively investigated, with the symmetry reduction of these being fully characterized [3–5, 7]. However, the $C_{60}^{3-}$ trianion is yet to undergo the same level of investigation, with all previous work focusing on linear coupling effects only [2, 6, 8]. Under these conditions the minimum adiabatic potential energy surface (APES) takes the form of a trough incorporating a continuum of isoenergetic points of different symmetry. In this situation, the molecule would undergo free pseudorotation, the term given to the rotation of a molecular distortion. It is only with the addition of quadratic level terms that this APES warps to form minimum energy wells of a lower symmetry which will describe the distorted molecule. In the case of weak coupling, this will result in a hindered pseudorotation, where the molecular distortion would tunnel between lowest energy configurations. For strong coupling, the molecular distortion may be considered locked into one of the lowest energy wells, resulting in a distortion associated with specific molecular vibrations.

Although little work has been done on the trianion, it is of particular interest as it possesses properties that differ from the other fullerene anions. Perhaps the most intriguing of these is that when this ion is present with an alkali metal in the form $A_3C_{60}$, the resulting alkali metal fulleride is found to be a superconductor up to a relatively high critical temperature of around $T_C \approx 40K$ [9]. This is in contrast with the $C_{60}^{2-}$ and $C_{60}^{4-}$ anions which are both found to be insulators [10, 11]. These properties are often attributed to the JT effect, although the mechanism behind which they occur is not yet known. Hence, an understanding of the way in which the trianion distorts may provide a good basis for future investigations in this area.

In this paper, a JT interaction Hamiltonian will be constructed which considers quadratic level electron–nuclei coupling. Restrictions will then be made on this Hamiltonian to imply a molecular distortion of a certain symmetry, where it is assumed that the distortion of lowest energy will represent the true distortion of the molecule. The possible implications of these results will then be discussed in relation to the pattern of nearest and next nearest neighbour wells, which give rise to a set of distortional axes which describe the molecular distortion.

## 12.2 The Hamiltonian

When the $C_{60}^{3-}$ anion is formed, the three additional electrons populate the threefold degenerate $T_{1u}$ (or $p$ in angular momentum notation) LUMO [6]. These electrons couple to $h_g$-type molecular vibrations in what is referred to as the $p^3 \otimes h$ JT effect [8]. The full interaction must incorporate all eight of the vibrational molecular modes individually, giving the full form as $p^3 \otimes 8h$. However, for many problems it is possible to model these modes as a single effective mode [12] and it is this formulation that is utilised here.

It is found that the only JT active electronic states arising from the $p^3$ ($T_{1u} \otimes T_{1u} \otimes T_{1u}$) electronic configuration is the coupling between the low spin $T_{1u}$ and $H_u$ states [2, 8]. Due to selection rules for half-filled orbitals (as is the case for the trianion), it is not necessary to consider coupling within these states as all the resultant matrix elements are found to be zero [13].

The Hamiltonian for the trianion system takes the form:

$$\mathcal{H} = \mathcal{H}_{JT} + \mathcal{H}_{term} \tag{12.1}$$

where $\mathcal{H}_{term}$ represents the term splitting Hamiltonian describing the Coulombic interaction between the $T_{1u}$ and $H_u$ states, and $\mathcal{H}_{JT}$ is the JT interaction Hamiltonian which may be divided further as:

$$\mathcal{H}_{JT} = \mathcal{H}_{vib} + \mathcal{H}_{lin} + \mathcal{H}_{quad}^{(1)} + \mathcal{H}_{quad}^{(2)}. \tag{12.2}$$

Here, $\mathcal{H}_{vib}$ represents the vibrational Hamiltonian and takes the same form as given in Ref. [3], $\mathcal{H}_{lin}$ represents the linear JT interaction Hamiltonian, and $\mathcal{H}_{quad}^{(1)}$ and $\mathcal{H}_{quad}^{(2)}$ represent the two forms of the quadratic interaction Hamiltonian brought about by the non-simple reducibility of the $h \otimes h$ Kronecker product, as discussed in Refs. [3, 14, 15].

Each component of the Hamiltonian is represented in matrix form via an $8 \times 8$ matrix corresponding to the eightfold, $T_{1u} \oplus H_u$ electronic basis. The term splitting Hamiltonian ($\mathcal{H}_{term}$) is constructed by setting the energy difference between the $T_{1u}$ and $H_u$ states equal to $\Delta$, and as such the matrix representation of $\mathcal{H}_{term}$ has the first three diagonal elements, corresponding to the $T_{1u}$ orbital, equal to $\Delta$, and all other elements exactly zero.

The interaction Hamiltonian may be constructed using the same method as given in Refs. [3, 5, 6, 15, 16] modified for the $p^3 \otimes h$ system. A set of five coordinates, $\{Q_\theta, Q_\varepsilon, Q_4, Q_5, Q_6\}$, are utilised to describe the fivefold degenerate $h_g$ effective mode, where $Q_4, Q_5$ and $Q_6$ transform as the $d_{yz}, d_{zx}$ and $d_{xy}$ hydrogen-like spherical harmonics respectively, and $Q_\theta$ and $Q_\varepsilon$ are related to the linear combinations:

$$Q_\theta = \sqrt{\frac{3}{8}} d_{3z^2-r^2} + \sqrt{\frac{5}{8}} d_{x^2-y^2}$$

$$Q_\varepsilon = \sqrt{\frac{3}{8}} d_{x^2-y^2} - \sqrt{\frac{5}{8}} d_{3z^2-r^2}. \tag{12.3}$$

As no coupling exists within each state, the interaction Hamiltonian is of the form:

$$\mathscr{H}_{\text{lin}} = V_1 \begin{bmatrix} 0 & M \\ M^T & 0 \end{bmatrix} \tag{12.4}$$

where $V_1$ is the linear coupling coefficient, the zeros represent the zero block matrix of the appropriate dimensions, and $M$ is a $3 \times 5$ matrix representing the interaction between the $T_{1u}$ and $H_u$ state. The form of this matrix may be obtained from the Clebsch–Gordan coefficients given in Ref. [14] using a method analogous to that given in Refs. [3] and [6] to relate these coefficients to the normal mode coordinates, with the result that:

$$M = \sqrt{\frac{3}{10}} \begin{bmatrix} -\frac{\phi^2}{\sqrt{2}}Q_4 & \frac{\sqrt{3}}{\sqrt{2}\phi}Q_4 & \xi_1 & -Q_6 & Q_5 \\ \frac{1}{\sqrt{2}\phi^2}Q_5 & -\sqrt{\frac{3}{2}}\phi Q_5 & Q_6 & \xi_2 & -Q_4 \\ \sqrt{\frac{5}{2}}Q_6 & \sqrt{\frac{3}{2}}Q_6 & -Q_5 & Q_4 & \xi_3 \end{bmatrix} \tag{12.5}$$

where:

$$\xi_1 = \frac{\phi^2}{\sqrt{2}}Q_\theta - \frac{\sqrt{3}}{\sqrt{2}\phi}Q_\varepsilon,$$

$$\xi_2 = \sqrt{\frac{3}{2}}\phi Q_\varepsilon - \frac{1}{\sqrt{2}\phi^2}Q_\theta,$$

$$\xi_3 = \sqrt{\frac{3}{2}}Q_\varepsilon + \sqrt{\frac{5}{2}}Q_\theta, \tag{12.6}$$

and where $\phi$ is the golden mean, $\frac{1+\sqrt{5}}{2}$.

With the linear Hamiltonian defined, the two forms of the quadratic interaction matrices can be found in terms of this matrix by making the same substitutions as given in Ref. [7], where: $\mathscr{H}_{\text{quad}}^{(1)} = \mathscr{H}_{\text{lin}}(Q \mapsto A, V_1 \mapsto V_2)$ and $\mathscr{H}_{\text{quad}}^{(2)} = \mathscr{H}_{\text{lin}}(Q \mapsto B, V_1 \mapsto V_3)$, where:

$$A_\theta = \frac{1}{2\sqrt{6}}\left(3Q_\theta^2 - 3Q_\varepsilon^2 - Q_4^2 - Q_5^2 + 2Q_6^2\right),$$

$$A_\varepsilon = \frac{-1}{2\sqrt{2}}\left(2\sqrt{3}Q_\theta Q_\varepsilon - Q_4^2 + Q_5^2\right),$$

$$A_4 = \frac{-1}{\sqrt{6}}\left(Q_\theta Q_4 - \sqrt{3}Q_\varepsilon Q_4 + 2\sqrt{2}Q_5 Q_6\right),$$

$$A_5 = \frac{-1}{\sqrt{6}}\left(Q_\theta Q_5 + \sqrt{3}Q_\varepsilon Q_5 + 2\sqrt{2}Q_4 Q_6\right),$$

$$A_6 = \frac{2}{\sqrt{6}}\left(Q_\theta Q_6 - \sqrt{2}Q_4 Q_5\right), \tag{12.7}$$

and

$$B_\theta = \frac{1}{2\sqrt{2}} \left( 2Q_\theta Q_\varepsilon + \sqrt{3}Q_4^2 - \sqrt{3}Q_5^2 \right),$$

$$B_\varepsilon = \frac{1}{2\sqrt{2}} \left( Q_\theta^2 - Q_\varepsilon^2 + Q_4^2 + Q_5^2 - 2Q_6^2 \right),$$

$$B_4 = \frac{1}{\sqrt{2}} \left( Q_\varepsilon + \sqrt{3}Q_\theta \right) Q_4,$$

$$B_5 = \frac{1}{\sqrt{2}} \left( Q_\varepsilon - \sqrt{3}Q_\theta \right) Q_5,$$

$$B_6 = -\sqrt{2}Q_\varepsilon Q_6. \tag{12.8}$$

With the full form of the Hamiltonian constructed, the unitary transform method devised by Bates et al. [17] may be utilised to shift the phonon coordinates to the points $Q_j - \alpha_j\hbar$ on the assumption that in the extreme of strong coupling, the distorted molecule is locked into minima at these displaced coordinates. By looking at the case of infinite coupling the position of the minima are defined, which in turn defines the symmetry of the distorted molecule. With this assumption relaxed, the positions of the minima will still have the same ratio, although tunnelling between the wells will be permitted. It is possible to investigate this situation using symmetry-adapted states, as used in Refs. [3, 15, 18], however, as this work is only interested in the reduction of symmetry, finding the positions of the minima is sufficient.

As the system is being considered in the ground state, the Hamiltonian can be simplified in the same way as Ref. [3] by making the substitutions $Q_j \mapsto -a_j$, $V_2 \mapsto V_2', V_3 \mapsto V_3'$, omitting $V_1$, and making the vibrational Hamiltonian equal to $\frac{1}{2}\sum a_i^2$. The terms used in these transformations are defined as:

$$a_j = \frac{\mu\hbar\omega^2}{V_1}\alpha_j, \quad E' = \frac{\mu\omega^2}{V_1^2}E, \quad V_2' = \frac{V_2}{\mu\omega^2}, \quad V_3' = \frac{V_3}{\mu\omega^2} \tag{12.9}$$

where $\mu$ is the reduced mass of the nuclei, $\omega$ is the vibrational frequency, and $E$ is the energy. Making these substitutions is equivalent to applying the unitary transform to the interaction Hamiltonian, providing a form of the Hamiltonian that is a function of the transformed '$a$' coordinates.

## 12.3 The Energy of the System

With the unitary transform applied, easier manipulation of the Hamiltonian is allowed. In order to obtain the minimum energy for this system it is necessary to diagonalise this transformed Hamiltonian and minimise the resultant expressions in terms of the displaced '$a$' coordinates to obtain which of the expressions produces the lowest value.

However, before this is done, it is possible to simplify the Hamiltonian still further if a form of distorted symmetry is assumed. It has been shown in Refs. [2] and [19] that the JT effect must reduce the initial $I_h$ symmetry to either the $D_{5d}$, $D_{3d}$, $D_{2h}$ or $C_{2h}$ point groups. It is found that for each of these point groups there is a minimum energy point where the $a_4$ and $a_5$ coordinates are both exactly zero [16]. Furthermore, the relationships given in this reference add further restrictions for the $D_{5d}$, $D_{3d}$ and $D_{2h}$ cases such that:

$$D_{5d} : a_6 \text{ is independent, } a_\varepsilon = -\sqrt{\frac{2}{3}} a_6 \text{ and } a_\theta = 0,$$

$$D_{3d} : a_6 \text{ is independent, } a_\theta = -\sqrt{2} a_6 \text{ and } a_\varepsilon = 0,$$

$$D_{2h} : a_\theta \text{ and } a_\varepsilon \text{ are independent and } a_6 = 0,$$

$$C_{2h} : a_\theta, a_\varepsilon \text{ and } a_6 \text{ are all independent.} \qquad (12.10)$$

These relationships can be substituted into the Hamiltonian to obtain a form of the Hamiltonian specific for each of the given symmetries, and hence each form of the molecular distortion can be examined independently. For the $D_{5d}$ and $D_{3d}$ cases the minimisation is one dimensional as the Hamiltonian is only dependent on one coordinate ($a_6$). For the $D_{2h}$ point group it is two dimensional as both $a_\theta$ and $a_\varepsilon$ are independent, and for $C_{2h}$ a three dimensional minimisation is required.

### 12.3.1 $D_{5d}$ and $D_{3d}$ Molecular Distortions

By making the necessary substitutions into the Hamiltonian for either $D_{5d}$ or $D_{3d}$ symmetry, the resulting matrix is simple to diagonalise analytically using a suitable computer package. It is found that four of the eight eigenvalue expressions obtained via this diagonalisation are not JT active, and simply minimise the vibrational coordinates to zero. Of the remaining four expressions, there is a twofold degeneracy of two expressions, which for the $D_{5d}$ case are of the form:

$$E' = \frac{1}{6} \left( 5a_6^2 + 3\Delta \pm 3\sqrt{a_6^2 \left( 8V_3'^2 a_6^2 + 8V_3' \sqrt{3} a_6 + 6 \right) + \Delta^2} \right), \qquad (12.11)$$

and for the $D_{3d}$ case take the form:

$$E' = \frac{1}{10} \left( 15a_6^2 + 5\Delta \pm \sqrt{5} \sqrt{a_6^2 \left( 72V_2'^2 a_6^2 + 72V_2' \sqrt{3} a_6 + 54 \right) + 5\Delta^2} \right). \qquad (12.12)$$

This degeneracy in these expressions is not permitted by the JT effect, whereby any electronic degeneracy will be removed via a molecular distortion, and as such it can be deduced that the symmetry must be reduced further still. The degeneracy of a distortion of these symmetries has not been noted in any of the previously investigated anions, with the $D_{5d}$ and $D_{3d}$ point groups representing the symmetry of the other ions under the vast majority of conditions, and as such this is a significant result.

## 12.3.2 $D_{2h}$ and $C_{2h}$ Symmetry

As the molecular distortion may not be represented by either the $D_{5d}$ or $D_{3d}$ point groups, it remains to investigate the possibility of a distortion of $D_{2h}$ or $C_{2h}$ symmetry. However, the additional degrees of freedom incorporated into these Hamiltonians by the extra independent variables in Eq. 12.10, make the analytical expressions for the two sets of unminimised eigenvalues considerably more difficult to manipulate. In fact an analytical expression for the minimised coordinates, and hence the minimum energy, has not been obtainable except for the simplest case where $D_{2h}$ symmetry is assumed and the term splitting is set to zero.

Nevertheless, considerable information about the system can still be gained by examining the expressions from a numerical perspective. It is found that by choosing specific values for the quadratic coupling constants $V_2'$ and $V_3'$, and the term splitting $\Delta$, the eigenvalue expressions can be minimised with respect to the transformed coordinates for both forms of symmetry.

If the case where $\Delta = 0$ is considered initially, the symmetry of the distorted molecule can be found under various strengths of quadratic coupling. The results of this are shown in Fig. 12.1, where the dark grey (orange online) regions represents a molecular distortion of $D_{2h}$ symmetry, the light grey (yellow online) region represents a molecular distortion of $C_{2h}$ symmetry, and the region outside of this is outside the bounds of the system where the energy function is unbounded. Physically, this is interpreted as the region where the coupling would cause the nuclear vibrations to be sufficiently strong as to break down the molecule.

The black line in Fig. 12.1 is given by the expression $V_2' = \frac{3}{\sqrt{5}} V_3'$, a ratio that has been encountered before in the investigation of other $C_{60}$ ions [3, 5, 16, 20]. Previously it has represented a crossover where the two forms of the quadratic interaction effectively cancel each other out, and the spherical APES is restored to its undistorted shape. Taking quadratic coupling constants of this ratio for this ion does indeed produce an undistorted APES, and thus the same representation is also applicable to this system.

This plot has been produced numerically by substituting the appropriate coordinates for $C_{2h}$ symmetry into the Hamiltonian as given in Eq. 12.10, whereby $a_\theta, a_\varepsilon$ and $a_6$ are all independent. The eigenvalues of this Hamiltonian have then be found computationally. This results in eight eigenvalue expressions, six of which represent cases which are Jahn–Teller active. Of these six, only three may represent the minimum energy of the system as the expressions consist of three pairs of the form $\chi \pm \sqrt{\zeta}$, where $\chi$ and $\zeta$ are both complicated functions of $V_2', V_3', a_\theta, a_\varepsilon$ and $a_6$.

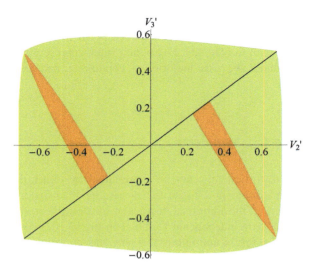

**Fig. 12.1** (*Colour online*) Symmetry reduction of the trianion for varying strengths of quadratic coupling when $\Delta = 0$. The region in dark grey (*orange online*) represents a molecular distortion of $D_{2h}$ symmetry, the region in light grey (*yellow online*) represents a $C_{2h}$ distortion, and the area surrounding the plot is outside the bounds of the system. The *black line* passing through the graph represents the ratio where $V_2' = \frac{3}{\sqrt{5}} V_3'$

These expressions were then analysed numerically by taking specific values for $V_2'$ and $V_3'$, and minimising the '$a$' coordinates computationally at each point for the three possible energy expressions. The coordinates corresponding to the expression of lowest energy were then taken, and the symmetry determined by examining the $a_6$ coordinate. When this coordinate was found to minimise to zero, the molecular distortion is described by $D_{2h}$ symmetry, as shown by the $D_{2h}$ well in Eq. 12.10. For all other values the symmetry is described by the $C_{2h}$ point group.

To gain more confidence in the results, a number of extremal points were analysed by constructing the full five dimensional Hessian matrix, and using this to confirm that the points are indeed extrema in the five dimensional space. This was undertaken for approximately 100 points, all of which were indeed found to be minima.

The values of the dimensionless quadratic coupling constants were taken between $-0.7$ and $0.7$ at intervals of $0.001$. It is of note that by using such a small interval between the analysed coupling values, Fig. 12.1 is very highly resolved and gives the appearance of analytical regions. It is worthwhile however to acknowledge that the regions are not obtained analytically, but produced numerically.

Figure 12.1 shows that for most combinations of quadratic coupling, the molecule will distort to $C_{2h}$ symmetry, with two regions of $D_{2h}$ symmetry also possible. At first, this would seem to be a limited example in that the term splitting is neglected when obtaining these results. However, it is interesting to note that when the term splitting is introduced, and it remains within the approximate bounds of

±2 (in the transformed energy units), it appears to have negligible effect on the symmetry of the distorted molecule, even though the magnitude of $\Delta$ *does* affect the transformed coordinates.

As the magnitude of $\Delta$ increases above this, a region forms at low quadratic coupling (centred on the origin of Fig. 12.1) where the transformed coordinates minimize to zero, indicating no JT effect is exhibited. At higher levels of quadratic coupling, a JT effect is still permitted, and the regions of symmetry follow the same pattern as shown in Fig. 12.1. As the term splitting becomes larger, this region expands from the origin, until the magnitude is approximately three, at which point no JT effect is observed. This result is intuitive as with a sufficiently large gap between the states, as is represented by higher values of the term splitting, the quadratic coupling between these states would have to be stronger to overcome the increased energy difference.

## 12.4 Division of the $C_{2h}$ Region

When the APES distorts under the quadratic interaction, wells are formed at the positions associated with the transformed coordinates. It is possible to map these minima into three dimensions by considering them as a distortion of a sphere [21]. As the wells are isoenergetic, this distortion can be modelled by considering a set of axes aligned with the minima, along which the sphere is distorted by an equal amount.

For the $D_{5d}$, $D_{3d}$ and $D_{2h}$ distortions it is possible to associate these distortional axes with specific points on the molecule. For the $D_{5d}$ and $D_{3d}$ cases the distortional axes are found in the centre of each pentagon ($C_5$ rotational axis), and hexagon ($C_3$ rotational axis), respectively, with the $D_{2h}$ distortional axes found through the middle of each double bond ($C_2$ rotational axis). However, for a $C_{2h}$ distortion a particular axis may fall on any point in a plane with (or perpendicular to) any double bond of the molecule (indicated in Fig. 12.2). Thus, it is anticipated that different forms of a $C_{2h}$ distortion should exist.

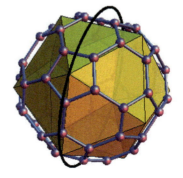

**Fig. 12.2** Possible positions of a single distortional axis for a $C_{2h}$ molecular distortion relative to the $C_{60}$ molecule. The icosahedron is shown as a guide to the symmetry of the molecule

To investigate these different forms it is necessary to look at the pattern of nearest and next nearest neighbour wells associated with a particular minima. The distance between the wells is simply found by their separation in the original five dimensional space, i.e:

$$\left| \mathbf{a}^{(w1)} - \mathbf{a}^{(w2)} \right| = \sqrt{\sum_i \left( a_i^{(w1)} - a_i^{(w2)} \right)^2} \qquad (12.13)$$

where $\mathbf{a}^{(w1)}$ and $\mathbf{a}^{(w2)}$, are two different wells, and $a_i$ is the $i$th coordinate of $\mathbf{a}$ where $i$ sums over $\theta, \varepsilon, 4, 5, 6$. This expression can then be used to calculate the nearest and next nearest neighbours for each well.

By following a path of nearest neighbours, (i.e. a path from one well to a nearest neighbour, and from this well to another nearest neighbour and so on), it is possible to categorise four different types of $C_{2h}$ distortion. It is found these paths will cycle back to the original starting well for virtually all positions of the distortional axes (and hence minima) relating to a $C_{2h}$ distortion. For the simplest case of a small deviation from a $C_2$ rotational axis, where each well has only one nearest neighbour, the path of nearest neighbours will cycle between these two wells in what will be referred to as a twofold precession. Likewise, a threefold precession is found where three distortional axes congregate around a $C_3$ rotational axis, and a fivefold precession is found with five distortional axes around a $C_5$ rotational axis. The fourth and final form of $C_{2h}$ distortion is found when the distortional axes are equidistant from each other. In this instance a path of nearest neighbours can be found from one well to any of the others without cycling, however this only occurs at very specific points where there is a transition between the different types of precession.

The interpretation of these precessions is considered in Fig. 12.3 where the distortional axes are represented as black points on the molecule. It can be seen that by rotating the distortional axes away from the $C_2$ axis in a plane with the double bond, different congregations occur.

Using the same numerical approach as was used to create Fig. 12.1 a similar plot is shown in Fig. 12.4 with the addition that the $C_{2h}$ region has been divided up into regions corresponding to each type of precession. Again, the term splitting has been assumed to be zero, although like the previous case, the pattern appears to be unaffected by the introduction of the term splitting within its bounds.

It can be seen from Fig. 12.4 that the combinations of quadratic coupling leading to a $C_{2h}$ distortion either side of the $D_{2h}$ region result in twofold precessions, with the $C_{2h}$ form either extending to a threefold or fivefold precession depending on the direction moved away from the $D_{2h}$ region. It is proposed therefore that moving away from the $D_{2h}$ region towards the fivefold precession is equivalent to a rotation of the minima away from the $C_2$ axis along the line of the double bond in three dimensions, and moving towards the threefold precession region is equivalent to a rotation perpendicular to the double bond. Thus, the further away from the $D_{2h}$ region the quadratic coupling constants are taken, the greater the congregation around the appropriate symmetry axis (the $C_5$ axes for a fivefold precession and a $C_3$ axes for a threefold precession). For the twofold precession, the most congregation occurs closest to the $D_{2h}$ region as would be expected.

12 The Quadratic $p^3 \otimes h$ Jahn–Teller System as a Model for the $C_{60}^{3-}$ Anion 241

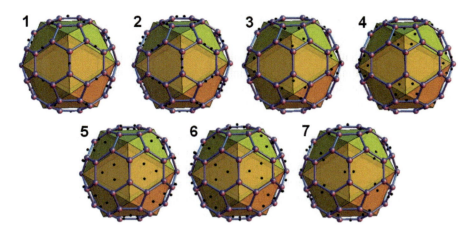

**Fig. 12.3** (*Colour online*) The different forms of a $C_{2h}$ distortion. The images are ordered to show a continued rotation away from the line of the double bond of the distortional axes. In order, they represent: **1.** A $D_{2h}$ distortion with the distortional axes along the $C_2$ rotational axes, **2.** a twofold precession, **3.** and **4.** fivefold precessions, **5.** and **6.** threefold precessions, and **7.** a twofold precession

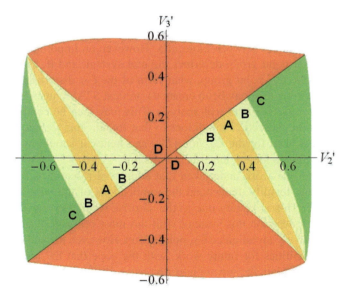

**Fig. 12.4** (*Colour online*) The region of $C_{2h}$ symmetry may be divided up as shown, with *region A* representing the distortion of $D_{2h}$ symmetry, *region B* representing a twofold precession of $C_{2h}$ symmetry, *C* giving a threefold precession and *D* a fivefold precession. The *black line* shown represents the ratio where the APES remains undistorted

## 12.5 Conclusions and Future Work

It has been shown that, unlike any of the other previously investigated ions of $C_{60}$, a molecular distortion of $D_{5d}$ or $D_{3d}$ symmetry would not remove the electronic degeneracy in the trianion system. As such the distorted molecule must be described by even lower symmetry, with the majority of quadratic coupling values giving a $C_{2h}$ molecular distortion, and certain combinations giving a $D_{2h}$ distortion.

Analysis into a $C_{2h}$ type distortion has not been undertaken previously, and as such it is this that has undergone the most investigation. Using the idea of distortional axes relating to the positions of the minima of the APES, these $C_{2h}$ distortions have been characterized in terms of the pattern of nearest neighbours, with the result that different forms of a $C_{2h}$ distortion is permitted. The implications of these different forms may provide an ideal starting point for further investigation into this anion.

As mentioned, this anion has superconducting properties when present in the form $A_3C_{60}$, and although nothing from this investigation may be directly related to this phenomena, the symmetry reduction is a second unique property of the trianion. The superconductivity of these compounds is often attributed to the JT effect, and as such, future work could look in to whether the superconductivity and symmetry reduction of the molecule are intrinsically linked.

There are a number of ways in which this may be undertaken, possibly by looking at a cooperative JT effect between molecules. However, an important step may be to determine which point group describes the real system, and if, as would seem most likely, it is the $C_{2h}$ point group, what form of the $C_{2h}$ distortion is found. A method of determining this may be to create theoretical STM images showing how the molecule may appear under real analysis. This may then be compared to real images to match the theoretical conditions to that found. This technique has been successfully undertaken for the $C_{60}^-$ ion [21], and hence it should be possible to extend this technique to look at the effect in this ion.

It is clear from the work undertaken that additional research is required into the trianion system to fully understand the implications of the way in which the molecule distorts. This is largely due to the surprise result that the $D_{5d}$ and $D_{3d}$ groups may *never* describe this system, with the most probable distortion being described by the poorly investigated $C_{2h}$ point group. However, the work has opened a number of possibilities for future work, with the hope that the unique features found may help to explain the superconducting nature of the compounds associated with this ion.

## References

1. Kroto HW, Heath JR, O'Brien SC, Curl RF, Smalley RE (1985) Nature 318:162
2. Chancey CC, O'Brien MCM (1997) The Jahn-Teller effect in $C_{60}$ and other icosahedral complexes. Princeton University Press, Princeton

3. Dunn JL, Bates CA (1995) Phys Rev B 52:5996
4. Hands ID, Dunn JL, Bates CA (2010) Phys Rev B 81:205440
5. Sookhun S, Dunn JL, Bates CA (2003) Phys Rev B 68:235403
6. Dunn JL, Li HM (2005) Phys Rev B 71:115411
7. Sindi LM, Hands ID, Dunn JL, Bates CA (2007) J Mol Struct 838:78
8. O'Brien MCM (1996) Phys Rev B 53:3775
9. Capone M, Fabrizio M, Castellani C, Tosatti E (2009) Rev Mod Phys 81:943
10. Granath M, Östlund S (2003) Phys Rev B 68:205107
11. Iwasa Y, Takenobu T (2003) J Phys Condens Mat 15:R495
12. O'Brien MCM (1983) J Phys C Solid State 16:85
13. Ceulemans A (1994) Top Curr Chem 171:27
14. Fowler PW, Ceulemans A (1985) Mol Phys 54:767
15. Hands ID, Diery WA, Dunn JL, Bates CA (2007) J Mol Struct 838:66
16. Hands ID, Dunn JL, Diery WA, Bates CA (2006) Phys Rev B 73:115435
17. Bates CA, Dunn JL, Sigmund E (1987) J Phys C: Solid State 20:1965
18. Dunn JL (1989) J Phys Condens Mat 1:7861
19. Ceulemans A, Vanquickenborne LG (1989) Struct Bond 71:125
20. Manini N, De los Rios P (2000) Phys Rev B 62:29
21. Hands ID, Dunn JL, Bates CA (2010) Phys Rev B 82:155425

# Chapter 13
# Estimation of the Vibronic Coupling Constants of Fullerene Monoanion: Comparison Between Experimental and Simulated Results

**Naoya Iwahara, Tohru Sato, Kazuyoshi Tanaka, and Liviu F. Chibotaru**

**Abstract** The vibronic coupling constants of $C_{60}^-$ are derived experimentally and theoretically. The experimental constants are obtained by simulating the photoelectron spectrum measured by Wang et al. (J Chem Phys 123:051106, 2005). The vibronic states are calculated with the Lanczos method and second order perturbation theory. We find that the coupling constants are underestimated with the perturbation method. The couplings are calculated based on the density-functional method and state-averaged complete active space self-consistent-field methods. The vibronic coupling constants obtained from the gradient of the total electronic energy of $C_{60}^-$ using the B3LYP functional are close to the experimental values. On the other hand, the coupling constants are overestimated with the derivative of the lowest unoccupied Kohn–Sham level of the neutral $C_{60}$ which is sometimes used in the literatures.

---

N. Iwahara • K. Tanaka
Department of Molecular Engineering, Graduate School of Engineering, Kyoto University, Kyoto 615-8510, Japan
e-mail: iwaharanaoya@t03.mbox.media.kyoto-u.ac.jp; ktanaka@moleng.kyoto-u.ac.jp

T. Sato (✉)
Department of Molecular Engineering, Graduate School of Engineering, Kyoto University, Kyoto 615-8510, Japan

Fukui Institute for Fundamental Chemistry, Kyoto University, Takano-Nishihiraki-cho 34-4, Sakyo-ku, Kyoto 606-8103, Japan
e-mail: tsato@scl.kyoto-u.ac.jp

L.F. Chibotaru (✉)
Division of Quantum and Physical Chemistry, University of Leuven, Celestijnenlaan 200F, B-3001 Leuven, Belgium
e-mail: Liviu.Chibotaru@chem.kuleuven.be

M. Atanasov et al. (eds.), *Vibronic Interactions and the Jahn-Teller Effect: Theory and Applications*, Progress in Theoretical Chemistry and Physics 23, DOI 10.1007/978-94-007-2384-9_13, © Springer Science+Business Media B.V. 2012

## 13.1 Introduction

The vibronic coupling is an important concept in molecules and solids, and has been studied extensively. Many authors have derived the coupling constants in many systems experimentally and theoretically [1,2]. One of the most interesting systems is fullerene monoanion ($C_{60}^-$) since the molecule exhibits the Jahn–Teller effect because of the high symmetry ($I_h$) [3]. Although the vibronic coupling constants of $C_{60}^-$ have been estimated experimentally and theoretically [4–11], the disagreement between the theoretical and experimental constants has been unsolved [5,7–9,11]. Moreover, the theoretical coupling constants depend on the methods.

Recently, we reported the vibronic coupling constants of $C_{60}^-$ [12] that are derived from the photoelectron spectrum measured by Wang et al. at low temperature with high resolution [13]. Furthermore, we found that the experimental coupling constants agree well with those calculated with the B3LYP functional. Nevertheless the difference of the methods were not discussed in detail in Ref. [12].

The purpose of this work is to compare several methods to evaluate the vibronic coupling constants in detail.

## 13.2 Vibronic Hamiltonian

We define the vibronic coupling constants and the vibronic Hamiltonian of fullerene monoanion.

First of all, we must choose a reference geometry $\mathbf{R}_0$ [1, 14]. Throughout this article, we consider the process of photoelectron spectroscopy of $C_{60}^-$, that is, one electron which occupies a singly-occupied molecular orbital (SOMO) is emitted from the irradiated $C_{60}^-$. Thus we obtained the equilibrium geometry $\mathbf{R}_0$ which possesses $I_h$ symmetry and normal modes from the neutral $C_{60}$. The vibronic coupling constant is a matrix element of the derivative of the electronic Hamiltonian $H_e$ with respect to a mass-weighted normal coordinate [14]. The ground electronic state of $C_{60}^-$ at $\mathbf{R}_0$ is $T_{1u}$, and the $T_{1u}$ electronic state linearly couples with two $a_g$ and eight $h_g$ vibrational modes:

$$[T_{1u}^2] = a_g \oplus h_g. \tag{13.1}$$

The linear vibronic coupling constants of the $a_g$ and $h_g$ modes are defined by the following equations:

$$V_{a_g(i)} = \langle T_{1u}\gamma_e| \left( \frac{\partial H_e(\mathbf{r},\mathbf{R})}{\partial Q_{a_g(i)}} \right)_0 |T_{1u}\gamma_e\rangle, \tag{13.2}$$

$$V_{h_g(v)} = \sqrt{\frac{2}{5} \sum_\gamma \left[ \langle T_{1u}\gamma_e| \left( \frac{\partial H_e(\mathbf{r},\mathbf{R})}{\partial Q_{h_g(v)\gamma}} \right)_0 |T_{1u}\gamma_e\rangle \right]^2}, \tag{13.3}$$

where $\mathbf{r}$ is the electronic coordinate, $\mathbf{R}$ is the nuclear coordinate, $Q_{a_g(i)}$ is the mass-weighted normal coordinate of the $a_g(i)$ $(i = 1, 2)$ mode, $Q_{h_g(v)\gamma}$ is the mass-weighted normal coordinate of the $h_g(v)\gamma$ $(v = 1, 2, \cdots, 8)$ mode, $\gamma_e$ and $\gamma$ are multiplicity of the $T_{1u}$ and $H_g$ representation respectively, and $(,)_0$ means that $\mathbf{R}$ is substituted with $\mathbf{R}_0$. The factor $\sqrt{2/5}$ in Eq. 13.3 is introduced to make $V_{h_g(v)}$ corresponds to the widely used coupling constants [3, 4, 15, 16] defined by O'Brien [17]. In this article, we use the dimensionless vibronic coupling constants $g_{\Gamma(v)\gamma}$ $(\Gamma = a_g, h_g)$ and stabilization energies $E_s$ and $E_{\text{JT}}$. The dimensionless coupling constants are defined by

$$g_{\Gamma(v)\gamma} = \frac{V_{\Gamma(v)\gamma}}{\sqrt{\hbar \omega_{\Gamma(v)}^3}}. \tag{13.4}$$

The stabilization energy due to the $h_g(v)$ and $a_g(i)$ modes are

$$E_{\text{JT},v} = \frac{V_{h_g(v)}^2}{2\omega_{h_g(v)}^2}, \tag{13.5}$$

$$E_{s,i} = \frac{V_{a_g(i)}^2}{2\omega_{a_g(i)}^2}. \tag{13.6}$$

The stabilization energy after eight $h_g$ modes $E_{\text{JT}}$ and that after two $a_g$ modes $E_s$ are defined as

$$E_{\text{JT}} = \sum_{v=1}^{8} E_{\text{JT},v}, \tag{13.7}$$

$$E_s = \sum_{i=1}^{2} E_{s,i}. \tag{13.8}$$

The stabilization energy is the depth of the adiabatic potential energy surface from the electronic energy at the reference geometry $\mathbf{R}_0$.

Neglecting the higher order couplings and off-diagonal couplings between the ground and excited electronic states, the vibronic Hamiltonian (the linear $T_{1u} \otimes (2a_g \oplus 8h_g)$ Jahn–Teller Hamiltonian) is written as follows:

$$H = \sum_{i=1}^{2} \left( \frac{1}{2} P_{a_g(i)}^2 + \frac{\omega_{a_g(i)}^2}{2} Q_{a_g(i)}^2 + V_{a_g(i)} Q_{a_g(i)} \right) \hat{I}$$

$$+ \sum_{v=1}^{8} \sum_{\gamma} \left( \frac{1}{2} P_{h_g(v)\gamma}^2 + \frac{\omega_{h_g(v)}^2}{2} Q_{h_g(v)\gamma}^2 \right) \hat{I} + H_{\text{JT}}. \tag{13.9}$$

Here, $P_{a_g(i)}$ is the conjugate momentum of $Q_{a_g(i)}$, $P_{h_g(v)\gamma}$ is the conjugate momentum of $Q_{h_g(v)\gamma}$, $\hat{I}$ is the $3 \times 3$ unit matrix, and $H_{JT}$ is the vibronic coupling term.

The Hamiltonian is the same as that defined by O'Brien [17]. Note that the normal coordinate includes the reduced mass.

$H_{JT}$ is written using two basis sets. In the calculation of the eigenstates (vibronic states) of the vibronic Hamiltonian (13.9), we use the complex basis: the $T_{1u}$ electronic states $|m_e\rangle$ and $h_g$ normal coordinates $Q_{h_g(v)m}$ transform as spherical harmonic functions $Y_{1m_e}$ ($m_e = -1,0,1$) and $Y_{2m}$ ($m = -2,-1,0,1,2$) respectively [16, 18]. The vibronic term is

$$H_{JT} = \sum_{v=1}^{8} V_{h_g(v)} \begin{pmatrix} \frac{1}{2} Q_{h_g(v)0} & \frac{\sqrt{3}}{2} Q_{h_g(v)1} & \sqrt{\frac{3}{2}} Q_{h_g(v)2} \\ -\frac{\sqrt{3}}{2} Q_{h_g(v)-1} & -Q_{h_g(v)0} & -\frac{\sqrt{3}}{2} Q_{h_g(v)1} \\ \sqrt{\frac{3}{2}} Q_{h_g(v)-2} & \frac{\sqrt{3}}{2} Q_{h_g(v)-1} & \frac{1}{2} Q_{h_g(v)0} \end{pmatrix}. \quad (13.10)$$

The Clebsch–Gordan coefficients are taken from Ref. [19]. The form of the vibronic term (13.10) is convenient to calculate the vibronic states with the use of $SO(3)$ symmetry of the Hamiltonian (It is easier to make the symmetry-adapted vibronic basis with the complex basis [Eq. 13.16]) [18, 20]. On the other hand, a real basis is useful to consider the structural distortion of the fullerene. We show the results of the density-functional calculations using the real basis. The relation between the complex electronic basis and the real basis is as follows:

$$|x\rangle = \frac{1}{\sqrt{2}} (|-1\rangle - |1\rangle),$$

$$|y\rangle = \frac{i}{\sqrt{2}} (|-1\rangle + |1\rangle),$$

$$|z\rangle = |0\rangle. \quad (13.11)$$

The real mass-weighted normal coordinates are represented by the complex coordinates as follows:

$$Q_{h_g(v)\theta} = Q_{h_g(v)0},$$

$$Q_{h_g(v)\varepsilon} = \frac{1}{\sqrt{2}} \left( Q_{h_g(v)-2} + Q_{h_g(v)2} \right),$$

$$Q_{h_g(v)\xi} = \frac{i}{\sqrt{2}} \left( Q_{h_g(v)-1} + Q_{h_g(v)1} \right),$$

$$Q_{h_g(v)\eta} = \frac{1}{\sqrt{2}} \left( Q_{h_g(v)-1} - Q_{h_g(v)1} \right),$$

$$Q_{h_g(v)\zeta} = \frac{i}{\sqrt{2}} \left( Q_{h_g(v)-2} - Q_{h_g(v)2} \right). \quad (13.12)$$

13 Estimation of the Vibronic Coupling Constants of Fullerene Monoanion... 249

Using Eqs. 13.11 and 13.12, $H_{\mathrm{JT}}$ (13.10) is rewritten as [3, 17]

$$H_{\mathrm{JT}} = \sum_{v=1}^{8} V_{h_g(v)} \begin{pmatrix} \frac{1}{2}Q_{h_g(v)\theta} - \frac{\sqrt{3}}{2}Q_{h_g(v)\varepsilon} & -\frac{\sqrt{3}}{2}Q_{h_g(v)\zeta} & -\frac{\sqrt{3}}{2}Q_{h_g(v)\eta} \\ -\frac{\sqrt{3}}{2}Q_{h_g(v)\zeta} & \frac{1}{2}Q_{h_g(v)\theta} + \frac{\sqrt{3}}{2}Q_{h_g(v)\varepsilon} & -\frac{\sqrt{3}}{2}Q_{h_g(v)\xi} \\ -\frac{\sqrt{3}}{2}Q_{h_g(v)\eta} & -\frac{\sqrt{3}}{2}Q_{h_g(v)\xi} & -Q_{h_g(v)\theta} \end{pmatrix}.$$

$$(13.13)$$

## 13.3 Method: Estimation of the Vibronic Coupling Constants

The methods of derivation of the vibronic coupling constants are explained. We explain the method of the simulation of the photoelectron spectrum and the methods of the theoretical calculations.

In the simulation of the photoelectron spectrum, we use the Lanczos diagonalization and perturbation theory. We show the formula of the transition intensity based on the second order perturbation theory to calculate the vibronic states. The formula of the vibronic coupling constants which is suitable for ab initio calculations is introduced.

### 13.3.1 Vibronic States and Photoelectron Spectrum

To simulate the vibronic progression in the photoelectron spectrum, we have to calculate the eigenstates (vibronic states) of the vibronic Hamiltonian (13.9). It is known that in the linear $T_{1u} \otimes (8h_g)$ Jahn–Teller Hamiltonian, the magnitude of the vibronic angular momentum $\mathbf{J}$, and the $z$ component of the angular momentum $J_z$ commute because of the symmetry of the Hamiltonian [1, 17, 20]. Therefore, each vibronic state is indicated by a set of the quantum numbers $\{v_1, v_2, k, J, M\}$, where $v_1$ and $v_2$ are the energy of $a_g$ modes, $J$ is the magnitude of the angular momentum, $M$ is the $z$ component of the angular momentum, and $k$ distinguishes eigenstates which have the same $J$ and $M$. The vibronic state which belongs to the eigenenergy (vibronic level) $E_{kJ} + \sum_{i=1}^{2} \hbar\omega_{a_g(i)} - E_s$ is written as

$$|\Psi_{v_1 v_2 kJM}\rangle = \sum_{m_e=-1}^{1} \sum_{\mathbf{n}_1,\cdots\mathbf{n}_8} \sum_{v_1', v_2'} |m_e\rangle|\mathbf{n}_1,\cdots\mathbf{n}_8\rangle|v_1', v_2'\rangle$$

$$\times C_{m_e,\mathbf{n}_1,\cdots\mathbf{n}_8;kJM}^{\mathrm{JT}} S_{v_1' v_1}(g_{a_g(1)}) S_{v_2' v_2}(g_{a_g(2)}), \qquad (13.14)$$

where $\mathbf{n}_v$ is a set of the vibrational quantum numbers of the $h_g(v)$ mode $n_{vm}$ ($m = -2, -1, 0, 1, 2$), and $v_i$ ($i = 1, 2$) is the quantum number of the displaced oscillator. $S_{v'v}(g)$ is calculated analytically.

$$S_{v'v}(g) = \sqrt{\frac{v'!v!}{2^{v'-v}}} \exp\left(-\frac{g^2}{4}\right) \sum_{k=k_{min}}^{m} \left(-\frac{1}{2}\right)^k \frac{g^{2k+v'-v}}{k!(v-k)!(v'-v+k)!}, \quad (13.15)$$

where $k_{min}$ is $m$ when $v' \geq v$, and $k_{min} = v - v'$ for $v' < v$. On the other hand, $C^{JT}_{m_e,\mathbf{n}_1,\cdots\mathbf{n}_8:kJM}$ are not known. We obtain the coefficients by using the Lanczos diagonalization [1] and the second order perturbation theory.

In the diagonalization of the $T_{1u} \otimes (8h_g)$ Jahn–Teller Hamiltonian numerically, we use a set of the products of the electronic states and vibrational states as the basis of the Hamiltonian matrix:

$$\left\{ |m_e\rangle|\mathbf{n}_1,\cdots\mathbf{n}_8\rangle \,\middle|\, \sum_{v=1}^{8}\sum_{m=-2}^{2} n_{vm} \leq 6, \; m_e + \sum_{v=1}^{8}\sum_{m=-2}^{2} mn_{vm} = 0 \right\}. \quad (13.16)$$

The former condition in Eq. 13.16 is introduced to truncate the vibronic basis set. The quantity of the latter condition is the $z$ component of the vibronic angular momentum. We calculate the vibronic states with $M = 0$ to reduce the dimension of the Hamiltonian matrix [20]. In this case, the number of the basis states is 2,944,155. The vibronic states whose $J$'s are from 0 to 7 are taken into account. The frequencies of the fullerite measured by Raman scattering experiment [21] are used in our calculation.

In the perturbation calculation, we consider the ground vibronic state ($J = 1$) and the lowest sets of the excited vibronic states ($J = 1, 2, 3$). The unperturbed state of the ground vibronic state is the product of an electronic state and the ground vibrational state. The unperturbed states of the lowest excited vibronic states are the symmetry adapted vibronic states which are represented as linear combinations of the products of the electronic and vibrational states which satisfy the condition

$$\sum_{m=-2}^{2} n_{vm} = \delta_{1v}. \quad (13.17)$$

Here $\omega_{h_g(1)}$ is assumed to be the lowest frequency in the eight $h_g$ modes.

Using Eq. 13.14, the transition intensity of the photoelectron spectrum $I$ is written as follows:

$$I(\Omega) \propto \sum_{kJ}\sum_{v_1,v_2} P^{JT}_{kJ} P^{a_g}_{v_1} P^{a_g}_{v_2} \sum_{m_e=-1}^{1}\sum_{\mathbf{n}_1,\cdots\mathbf{n}_8}\sum_{v'_1,v'_2}$$

$$\times \left| C^{JT}_{m_e,\mathbf{n}_1,\cdots\mathbf{n}_8:kJ0} S_{v'_1 v_1}(g_{a_g(1)}) S_{v'_2 v_2}(g_{a_g(2)}) \right|^2$$

$$\times \delta\left( \frac{E_0 + E_s - E_{kJ}}{\hbar} + \sum_{v=1}^{8}\sum_{m=-2}^{2} \omega_{h_g(v)} n_{vm} + \sum_{i=1}^{2} \omega_{a_g(i)}(v'_i - v_i) - \Omega \right),$$

$$(13.18)$$

where $\Omega$ is the electron binding energy, and $P_{v_i}^{a_g}$ and $P_{kJ}^{JT}$ are the Boltzmann factors,

$$P_{v_i}^{a_g} = \frac{1}{Z_{a_g(i)}} \exp\left(-\hbar \omega_{a_g(i)} v_i \beta\right), \tag{13.19}$$

$$P_{kJ}^{JT} = \frac{2J+1}{Z_{JT}} \exp\left(-E_{kJ}\beta\right). \tag{13.20}$$

$2J+1$ in Eq. 13.20 is the degree of the degeneracy of the vibronic state, and $Z_{a_g}$ and $Z_{JT}$ are the normalization factors. Derivation of the formula of the transition intensity (13.18) are shown in the Appendix.

Within the second order perturbation method, the transition intensity $I$ is given as follows:

$$I(\Omega) \propto \int d\Omega' \sum_{k,J} \sum_{v_1,v_2} P_{v_1}^{a_g(1)} P_{v_2}^{a_g(2)} \left| S_{v_1'v_1}(g_{a_g(1)}) S_{v_2'v_2}(g_{a_g(2)}) \right|^2$$

$$\times P_{kJ}^{JT} I_{kJ} \left(\frac{E_{kJ}}{\hbar} + \Omega'\right) \delta\left(\Omega' + \frac{E_0 + E_s}{\hbar} + \sum_{i=1}^{2} \hbar \omega_{a_g(i)} (v_i' - v_i) - \Omega\right),$$

$$\tag{13.21}$$

where $\sum_{k,J}$ is taken over the states considered. $I_{11}$ is

$$I_{11}(\Omega) = N_{11}\left[\delta(\Omega) + \sum_{v_1=1}^{8} \frac{5}{4} g_{h_g(v_1)}^2 \delta\left(\omega_{h_g(v_1)} - \Omega\right) + \sum_{v_1=1}^{8} \frac{55}{128} g_{h_g(v_1)}^4 \right.$$

$$\times \delta\left(2\omega_{h_g(v_1)} - \Omega\right) + \sum_{1 \le v_1 < v_2 \le 8} \frac{5}{16} g_{h_g(v_1)}^2 g_{h_g(v_2)}^2$$

$$\left. \times \frac{5\omega_{h_g(v_1)}^2 + \omega_{h_g(v_1)}\omega_{h_g(v_2)} + 5\omega_{h_g(v_2)}^2}{(\omega_{h_g(v_1)} + \omega_{h_g(v_2)})^2} \delta\left(\omega_{h_g(v_1)} + \omega_{h_g(v_2)} - \Omega\right)\right].$$

$$\tag{13.22}$$

The other $I_{kJ}$ $((k,J) \ne (1,1))$ in Eq. 13.21 is written as

$$I_{kJ}(\Omega) = N_{kJ}\left[A_{kJ}^{(0)}\delta(\Omega) + \delta\left(\omega_{h_g(1)} - \Omega\right) + \sum_{1 < v_1 \le 8} \frac{g_{h_g(1)}^2 g_{h_g(v_1)}^2}{64}\right.$$

$$\left. \times \left(\frac{A_{kJ}^{(1)}\omega_{h_g(1)} + B_{kJ}^{(1)}\omega_{h_g(v_1)}}{\omega_{h_g(1)} - \omega_{h_g(v_1)}}\right)^2 \delta\left(\omega_{h_g(v_1)} - \Omega\right)\right.$$

$$+A_{kJ}^{(2)}g_{h_g(1)}^2\delta\left(2\omega_{h_g(1)}-\Omega\right)+\sum_{1<v_1\leq8}\frac{5}{4}g_{h_g(v_1)}^2\delta\left(\omega_{h_g(1)}+\omega_{h_g(v_1)}-\Omega\right)$$

$$+A_{kJ}^{(3)}g_{h_g(1)}^4\delta\left(3\omega_{h_g(1)}-\Omega\right)$$

$$+\sum_{1<v_1<8}\frac{g_{h_g(1)}^2g_{h_g(v_1)}^2}{32}\frac{B_{kJ}^{(3)}\omega_{h_g(1)}^2+C_{kJ}^{(3)}\omega_{h_g(1)}\omega_{h_g(v_1)}+D_{kJ}^{(3)}\omega_{h_g(v_1)}^2}{(\omega_{h_g(1)}+\omega_{h_g(v_1)})^2}$$

$$\times\delta\left(2\omega_{h_g(1)}+\omega_{h_g(v_1)}-\Omega\right)+\sum_{1<v_1<8}\frac{55}{128}g_{h_g(v_1)}^4\delta\left(\omega_{h_g(1)}+2\omega_{h_g(v_1)}-\Omega\right)$$

$$+\sum_{1<v_1<v_2<8}\frac{5}{16}g_{h_g(v_1)}^2g_{h_g(v_2)}^2\frac{5\omega_{h_g(v_1)}^2+\omega_{h_g(v_1)}\omega_{h_g(v_2)}+5\omega_{h_g(v_2)}^2}{(\omega_{h_g(v_1)}+\omega_{h_g(v_2)})^2}$$

$$\times\delta\left(\omega_{h_g(1)}+\omega_{h_g(v_1)}+\omega_{h_g(v_2)}-\Omega\right)\Bigg]. \tag{13.23}$$

$E_{kJ}$'s are

$$E_{11}=-\frac{5}{4}\sum_{v=1}^{8}\hbar\omega_{h_g(v)}g_{h_g(v)}^2, \tag{13.24}$$

$$E_{13}=\hbar\omega_{h_g(1)}-2\hbar\omega_{h_g(1)}g_{h_g(1)}^2-\frac{5}{4}\sum_{1<v\leq8}\hbar\omega_{h_g(v)}g_{h_g(v)}^2, \tag{13.25}$$

$$E_{12}=\hbar\omega_{h_g(1)}-\frac{7}{8}\hbar\omega_{h_g(1)}g_{h_g(1)}^2-\frac{5}{4}\sum_{1<v\leq8}\hbar\omega_{h_g(v)}g_{h_g(v)}^2, \tag{13.26}$$

$$E_{21}=\hbar\omega_{h_g(1)}-\frac{1}{8}\hbar\omega_{h_g(1)}g_{h_g(1)}^2-\frac{5}{4}\sum_{1<v\leq8}\hbar\omega_{h_g(v)}g_{h_g(v)}^2. \tag{13.27}$$

$N_{11}$ and $N_{kJ}$ which appear in Eqs. 13.22 and 13.23 are the normalization constants of the wave functions:

$$N_{11}=\Bigg[1+\frac{5}{4}\sum_{v_1=1}^{8}g_{h_g(v_1)}^2+\frac{55}{128}\sum_{v_1=1}^{8}g_{h_g(v_1)}^4$$

$$+\frac{5}{16}\sum_{1\leq v_1<v_2\leq8}g_{h_g(v_1)}^2g_{h_g(v_2)}^2\frac{5\omega_{h_g(v_1)}^2+\omega_{h_g(v_1)}\omega_{h_g(v_2)}+5\omega_{h_g(v_2)}^2}{(\omega_{h_g(v_1)}+\omega_{h_g(v_2)})^2}\Bigg]^{-1},$$

$$\tag{13.28}$$

# 13 Estimation of the Vibronic Coupling Constants of Fullerene Monoanion...

**Table 13.1** The coefficients which appear in Eqs. 13.23 and 13.29 for each set $(k,J)$

| $(k,J)$ | $A_{kJ}^{(0)}$ | $A_{kJ}^{(1)}$ | $B_{kJ}^{(1)}$ | $A_{kJ}^{(2)}$ | $A_{kJ}^{(3)}$ | $B_{kJ}^{(3)}$ | $C_{kJ}^{(3)}$ | $D_{kJ}^{(3)}$ |
|---|---|---|---|---|---|---|---|---|
| $(1,3)$ | 0 | 6 | 0 | 2 | 47/64 | 53 | 16 | 80 |
| $(1,2)$ | 0 | 3 | 0 | 7/8 | 49/128 | 62 | 7 | 35 |
| $(2,1)$ | 5/4 | $-1$ | 10 | 11/8 | 21/32 | 73 | 11 | 55 |

$$
N_{kJ} = \left\{ 1 + A_{kJ}^{(0)} + A_{kJ}^{(2)} g_{h_g(1)}^2 + A_{kJ}^{(3)} g_{h_g(1)}^4 \right.
$$

$$
+ \sum_{1 < v_1 \leq 8} \left[ \frac{5}{4} g_{h_g(v_1)}^2 + \frac{55}{128} g_{h_g(v_1)}^4 + \frac{g_{h_g(1)}^2 g_{h_g(v_1)}^2}{64} \left( \frac{A_{kJ}^{(1)} \omega_{h_g(1)} + B_{kJ}^{(1)} \omega_{h_g(v_1)}}{\omega_{h_g(1)} - \omega_{h_g(v_1)}} \right)^2 \right.
$$

$$
\left. + \frac{g_{h_g(1)}^2 g_{h_g(v_1)}^2}{32} \frac{B_{kJ}^{(3)} \omega_{h_g(1)}^2 + C_{kJ}^{(3)} \omega_{h_g(1)} \omega_{h_g(v_1)} + D_{kJ}^{(3)} \omega_{h_g(v_1)}^2}{(\omega_{h_g(1)} + \omega_{h_g(v_1)})^2} \right]
$$

$$
\left. + \sum_{1 < v_1 < v_2 \leq 8} \frac{5}{16} g_{h_g(v_1)}^2 g_{h_g(v_2)}^2 \frac{5\omega_{h_g(v_1)}^2 + \omega_{h_g(v_1)} \omega_{h_g(v_2)} + 5\omega_{h_g(v_2)}^2}{(\omega_{h_g(v_1)} + \omega_{h_g(v_2)})^2} \right\}^{-1}.
$$

$$(13.29)$$

$A_{kJ}^{(0)}$, $A_{kJ}^{(1)}$, $B_{kJ}^{(1)}$, $A_{kJ}^{(2)}$, $A_{kJ}^{(3)}$, $B_{kJ}^{(3)}$, $C_{kJ}^{(3)}$, and $D_{kJ}^{(3)}$ are shown in Table 13.1.

The shape of the photoelectron spectrum $F(\Omega)$ is calculated using a Gaussian function with the standard deviation $\sigma$:

$$
F(\Omega) = \int d\Omega' I(\Omega') \exp\left( \frac{(\Omega - \Omega')^2}{2\sigma^2} \right).
$$

$$(13.30)$$

## 13.3.2 Ab Initio Calculation

In the theoretical calculation of the vibronic coupling constants, we use the real electronic basis and vibrational modes. The coupling constants defined by Eqs. 13.2 and 13.3 are rewritten using the Hellmann–Feynman theorem [22]:

$$
V_{a_g(i)} = \left( \frac{\partial E(\mathbf{R})}{\partial Q_{a_g(i)}} \right)_0 = \sum_{A=1}^{60} \left( \frac{\partial E(\mathbf{R})}{\partial \mathbf{R}_A} \right)_0 \cdot \frac{\mathbf{u}_A^{a_g(i)}}{\sqrt{M}},
$$

$$(13.31)$$

$$
V_{h_g(v)} = \sqrt{\sum_\gamma \left[ \left( \frac{\partial E(\mathbf{R})}{\partial Q_{h_g(v)\gamma}} \right)_0 \right]^2} = \sqrt{\sum_\gamma \left[ \sum_{A=1}^{60} \left( \frac{\partial E(\mathbf{R})}{\partial \mathbf{R}_A} \right)_0 \cdot \frac{\mathbf{u}_A^{h_g(v)\gamma}}{\sqrt{M}} \right]^2}, \quad (13.32)
$$

where $E(\mathbf{R})$ is the total electronic energy of $C_{60}^-$ at the geometry $\mathbf{R}$, $\mathbf{R}_A$ is the Cartesian coordinate of the carbon atom $A$ $(A = 1, 2, \cdots 60)$, $M$ is the mass of carbon atom, and $\mathbf{u}_A^\alpha$ is the mass-weighted vibrational vector of the $\alpha$ mode. The equation is invariant under mixing of the degenerate electronic states and vibrational modes. The reference geometry $\mathbf{R}_0$ and the mass-weighted vibrational vectors $\mathbf{u}_A^\alpha$ are obtained from the calculation of the neutral $C_{60}$ with the density-functional theory (DFT) method and Hartree–Fock method. The frequencies of $C_{60}$ are taken from the experimental results [21]. The gradient of the total electronic energy of $C_{60}^-$ $(\partial E(\mathbf{R})/\partial \mathbf{R}_A)_0$ is obtained with the coupled-perturbed Kohn–Sham method and coupled-perturbed state-averaged complete active space self-consistent-field (CASSCF) method. In the case of the DFT calculation, we use the electron density whose spatial and spin symmetries are broken.

Some authors use the derivative of LUMO level of the neutral $C_{60}$ $(\varepsilon_{t_{1u}\gamma_e})$ instead of the total energy of $C_{60}^-$ [8–11]:

$$\left( \frac{\partial E(\mathbf{R})}{\partial Q_{\Gamma(v)\gamma}} \right)_0 \simeq \left( \frac{\partial \varepsilon_{t_{1u}\gamma_e}(\mathbf{R})}{\partial Q_{\Gamma(v)\gamma}} \right)_0 . \tag{13.33}$$

The total electronic energy in Eqs. 13.31 and 13.32 is replaced by the LUMO level.

$$V_{a_g(i)}^{\text{LUMO}} = \left( \frac{\partial \varepsilon_{t_{1u}\gamma_e}(\mathbf{R})}{\partial Q_{a_g(i)}} \right)_0 , \tag{13.34}$$

$$V_{h_g(v)}^{\text{LUMO}} = \sqrt{ \frac{1}{3} \sum_{\gamma_e} \sum_{\gamma} \left[ \left( \frac{\partial \varepsilon_{t_{1u}\gamma_e}(\mathbf{R})}{\partial Q_{h_g(v)\gamma}} \right)_0 \right]^2 } . \tag{13.35}$$

The Eq. 13.35 is invariant under arbitrary unitary transformation of the degenerate LUMO's and vibrational modes.

The gradient of the $t_{1u}$ LUMO level is computed by distorting the neutral $C_{60}$. The distorted structure of $C_{60}$ is written as

$$\mathbf{R}_{\Gamma(v)\gamma}^{\pm} = \mathbf{R}_0 \pm \delta Q \frac{\mathbf{u}^{\Gamma(v)\gamma}}{\sqrt{M}} , \tag{13.36}$$

The derivative of the LUMO level is computed numerically:

$$\left( \frac{\partial \varepsilon_{t_{1u}\gamma}(\mathbf{R})}{\partial Q_{\Gamma(v)\gamma}} \right)_0 = \frac{\varepsilon_{t_{1u}\gamma}\left( \mathbf{R}_{\Gamma(v)\gamma}^+ \right) - \varepsilon_{t_{1u}\gamma}\left( \mathbf{R}_{\Gamma(v)\gamma}^- \right)}{2\delta Q} . \tag{13.37}$$

In our calculation, we choose $\delta Q$ as 0.05 (amu$^{1/2}$ $a_0$) [23].

We checked that the value of $\delta Q$ gives results with sufficient significant-digits and is small enough to give linear gradients.

In this work, we use GAUSSIAN 03 package to calculate the geometry, vibrational modes, and electronic states [24].

## 13.4 Results and Discussion

The vibronic coupling constants are obtained from the photoelectron spectrum measured by Wang et al. [13] and ab initio calculations. First, we determine the coupling constants of $C_{60}^-$ from the experimental spectrum with the method described in Sect. 13.3.1. The results of the two methods are compared. We employ the theoretical calculation within the DFT method and state-averaged CASSCF method. Furthermore, we use two methods to evaluate the coupling constants.

### 13.4.1 Simulation of the Photoelectron Spectrum

The photoelectron spectrum measured by Wang et al. [13] is simulated using the Lanczos method and perturbation theory. The experimental and simulated spectra are shown in Fig. 13.1. The simulation is performed at 70 K which corresponds to the experimental condition. The standard deviation of $\sigma = 80\,\mathrm{cm}^{-1}$ is obtained from the width of the experimental zero phonon line. The dimensionless vibronic coupling constants and stabilization energies derived from the spectrum are shown in Tables 13.2 and 13.3, respectively. The statistical weights of the Jahn–Teller part $P_{kJ}^{\mathrm{JT}}$ at 70 K is ca. 98% in both methods.

We obtain several sets of the coupling constants (i), (ii), and (iii) from the spectrum. This ambiguity originates from the $a_g(2)$, $h_g(7)$, and $h_g(8)$ modes whose frequencies are close to each other. In fact, without significant change of the shape of the peak around $1{,}500\,\mathrm{cm}^{-1}$, $g_{a_g(2)}$ ($E_{s,2}$) can be increased from 0.1 to 0.3 (0.9–8.2 meV) if $g_{h_g(7)}$ and $g_{h_g(8)}$ ($E_{\mathrm{JT},7}$ and $E_{\mathrm{JT},8}$) are decreased at the same time. The

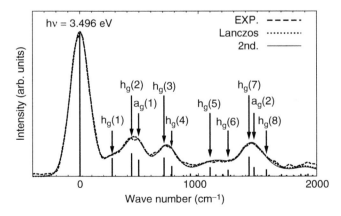

**Fig. 13.1** The experimental photoelectron spectrum measured by Wang et al. (*broken line*), the simulated spectrum using Lanczos method (*dotted line*), and the simulated spectrum using second order perturbation theory (*solid line*). The simulation is performed at 70 K with $\sigma = 80\,\mathrm{cm}^{-1}$. The set of vibronic coupling constants (i) in Table 13.2 are used for the Lanczos calculation

**Table 13.2** Absolute values of dimensionless vibronic coupling constants derived from the photoelectron spectrum measured by Wang et al. [13]. The sets of coupling constants (i), (ii), and (iii) are derived with the Lanczos diagonalization method (Lanczos). The set (iv) is obtained with the second order perturbation theory (2nd)

| | Frequency $(cm^{-1})$ | Lanczos | | | 2nd |
|---|---|---|---|---|---|
| | | (i) | (ii) | (iii) | (iv) |
| $a_g(1)$ | 496 | 0.505 | 0.505 | 0.500 | 0.485 |
| $a_g(2)$ | 1,470 | 0.100 | 0.200 | 0.300 | 0.355 |
| $h_g(1)$ | 273 | 0.500 | 0.500 | 0.490 | 0.320 |
| $h_g(2)$ | 437 | 0.525 | 0.520 | 0.515 | 0.360 |
| $h_g(3)$ | 710 | 0.465 | 0.460 | 0.455 | 0.325 |
| $h_g(4)$ | 774 | 0.310 | 0.310 | 0.300 | 0.240 |
| $h_g(5)$ | 1,099 | 0.285 | 0.280 | 0.280 | 0.225 |
| $h_g(6)$ | 1,250 | 0.220 | 0.230 | 0.235 | 0.190 |
| $h_g(7)$ | 1,428 | 0.490 | 0.470 | 0.435 | 0.330 |
| $h_g(8)$ | 1,575 | 0.295 | 0.285 | 0.260 | 0.195 |

**Table 13.3** Stabilization energy of each active mode (meV) obtained from the photoelectron spectrum measured by Wang et al. Sets (i), (ii), and (iii) are obtained from the photoelectron spectrum using the Lanczos method [12], (iv) is obtained by using the second order perturbation theory

| | Frequency $(cm^{-1})$ | Lanczos | | | 2nd |
|---|---|---|---|---|---|
| | | (i) | (ii) | (iii) | (iv) |
| $a_g(1)$ | 496 | 7.8 | 7.8 | 7.7 | 7.2 |
| $a_g(2)$ | 1,470 | 0.9 | 3.6 | 8.2 | 11.5 |
| $h_g(1)$ | 273 | 4.2 | 4.2 | 4.1 | 1.7 |
| $h_g(2)$ | 437 | 7.5 | 7.3 | 7.2 | 3.5 |
| $h_g(3)$ | 710 | 9.5 | 9.3 | 9.1 | 4.6 |
| $h_g(4)$ | 774 | 4.6 | 4.6 | 4.3 | 2.8 |
| $h_g(5)$ | 1,099 | 5.5 | 5.3 | 5.3 | 3.4 |
| $h_g(6)$ | 1,250 | 3.9 | 4.1 | 4.3 | 2.8 |
| $h_g(7)$ | 1,428 | 21.3 | 19.6 | 16.8 | 9.6 |
| $h_g(8)$ | 1,575 | 8.5 | 7.9 | 6.6 | 3.7 |
| $E_s$ | | 8.7 | 11.4 | 15.9 | 18.7 |
| $E_{JT}$ | | 65.0 | 62.3 | 57.7 | 32.1 |
| $E_s + E_{JT}$ | | 73.7 | 73.7 | 73.6 | 50.8 |

same problem was reported by some authors [5, 7]. However, the range of the uncertainties in the coupling constants of these modes obtained by Gunnarsson et al. were larger: Gunnarsson et al. varied $E_{s,2}$ from 0 to 45 meV, while in the present work the range of $E_{s,2}$ is from 0.9 to 8.2 meV. Therefore, our results are improved compared with the previous result.

In comparison with the couplings of the Lanczos calculations, the second order perturbation theory gives the weaker couplings of $h_g$ modes: the stabilization by the perturbation theory is obtained smaller by approximately 50% than that of the

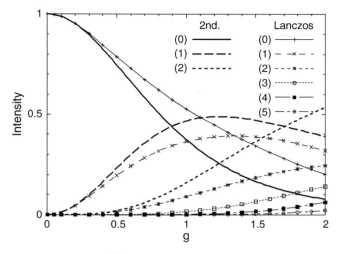

**Fig. 13.2** The transition intensity $I^{(n)}$ from the ground vibronic state. 2nd and Lanczos mean the results with the second order perturbation theory and the Lanczos method

Lanczos calculation despite of the agreement of their spectra. This fact shows that the vibronic couplings in $C_{60}^-$ are too strong to apply the perturbation theory.

To see the difference between the vibronic wave functions obtained by the Lanczos calculation and by the second order perturbation theory, we calculate the transition intensity of the single-mode $T_{1u} \otimes h_g$ Jahn–Teller model [16, 17]. In this model, the transition intensity is written as $I(\Omega) = \sum_n I^{(n)} \delta \left( n\omega_{h_g} - \Omega \right)$, where $n = \sum_{m=-2}^{2} n_m$. The Lanczos calculation is employed with the basis set that the sum of the vibrational quantum numbers is less than 50 ($\sum_\gamma n_\gamma \leq 50$). Figure 13.2 shows the transition intensity $I^{(n)}$ from the ground vibronic state. With the second order perturbation theory, $I^{(0)}$ is overestimated, while $I^{(1)}$ is underestimated. In other words, the dimensionless vibronic coupling $g_{h_g}$ is underestimated. In the case of the single-mode system, the result of the perturbation method is reliable when $g_{h_g}$ is less than 0.3.

The linear vibronic coupling constants of $C_{60}^-$ have been estimated from experiments [5, 7]. The stabilization energies are tabulated in Table 13.4. Gunnarsson et al. obtained the couplings from the photoelectron spectrum of $C_{60}^-$ in gas phase [5]. The coupling of the $h_g(2)$ mode is the strongest in the case of Gunnarsson et al., while we obtained the coupling of the $h_g(7)$ mode is the strongest. In Ref. [12], we compared both the experimental conditions and the methods of the simulation of the spectra, and we concluded that the present constants are better than Gunnarsson et al.'s constants. Hands et al. derived the strengths from the shifts in the frequencies of Raman spectra of solid state fullerene complexes using the approximate formula based on Allen's theory [7]. The distribution of couplings obtained by Hands et al. is similar to that of Gunnarsson et al. [5], which is different from the distribution of the present coupling constants.

**Table 13.4** Previously reported stabilization energies of $C_{60}^-$ (meV)

| | Frequency (cm$^{-1}$) | PES Gun [5] | Raman Han [7] | LDA Man [8] | GGA Fre [10] | B3LYP Sai[a] [9] | Laf [11] |
|---|---|---|---|---|---|---|---|
| $a_g(1)$ | 496 | 0.6 | – | 0.2 | 1.5 | – | 1.8 |
| $a_g(2)$ | 1,470 | 16.4 | – | 2.7 | 11.0 | – | 16.4 |
| $h_g(1)$ | 273 | 11.4 | 7.2±2.4 | 2.7 | 2.8 | 3.6 | 3.5 |
| $h_g(2)$ | 437 | 24.0 | 24.0±2.6 | 6.3 | 7.0 | 6.6 | 6.5 |
| $h_g(3)$ | 710 | 7.8 | 0.7±4.1 | 5.5 | 6.1 | 6.6 | 7.1 |
| $h_g(4)$ | 774 | 10.8 | 12.1±1.8 | 2.4 | 2.4 | 3.0 | 3.1 |
| $h_g(5)$ | 1,099 | 7.2 | 1.4±0.8 | 2.6 | 2.6 | 3.0 | 3.0 |
| $h_g(6)$ | 1,250 | 3.0 | −1.1±1.1 | 1.5 | 1.9 | 1.8 | 1.3 |
| $h_g(7)$ | 1,428 | 10.2 | 5.6±1.0 | 9.0 | 9.0 | 13.2 | 13.8 |
| $h_g(8)$ | 1,575 | 13.8 | −2.0±1.7 | 8.2 | 8.8 | 10.2 | 10.6 |
| $E_s$ | | 17.0 | | 2.9 | 12.5 | – | 18.2 |
| $E_{JT}$ | | 88.2 | | 38.2 | 40.6 | 48.0 | 48.9 |
| $E_s + E_{JT}$ | | 105.2 | | 41.1 | 53.1 | – | 67.1 |

[a]The stabilization energies of the $a_g$ modes are not reported.

Hands et al. have estimated the couplings from the visible and near-infrared spectrum of $C_{60}^-$ assuming the static distortions of $C_{60}^-$ [25]. Their model includes both the linear and quadratic couplings, and they have found that the linear couplings are overestimated without the quadratic couplings. Though, in this article, we neglect the quadratic couplings, we performed the exact diagonalization of the Hamiltonian without assuming any static distortions.

## 13.4.2 Ab Initio Calculations of the Coupling Constants

We calculate the vibronic coupling constants of $C_{60}^-$ by using the DFT method and CASSCF method. The functionals B3LYP [26], PW91PW91, and PW91LYP [27] are used. The fraction of the Hartree–Fock exchange-correlation functional in the B3LYP functional is varied from the original fraction (20%) to 30% by 5%. In the CASSCF calculation we use two active spaces. The first active space includes one electron and $t_{1u}$ LUMO, which we call (1,3). In second active space (11,8), eleven electrons in $h_u$ HOMO and $t_{1u}$ LUMO are included. The basis set is the triple zeta 6-311G(d).

The stabilization energies $E_{s,i}$, $E_s$, $E_{JT,v}$, and $E_{JT}$ are shown in Table 13.5. The set (viii) is obtained from the gradient of the LUMO of $C_{60}$ (Eqs. 13.34 and 13.35). The others are calculated by following the definition of the vibronic couplings (Eqs. 13.31 and 13.32).

The total stabilization energy $E_s + E_{JT}$ is overestimated with the CASSCF calculations. The values are twice as large as the experimental results. The PW91 functionals give smaller values than the experimental values. This result agrees with the previous results within the LDA [8] and GGA [10].

**Table 13.5** Stabilization energies calculated with the DFT methods and CASSCF method (meV). The sets of coupling constants (v) through (x) are evaluated using the DFT calculation, the sets (xi) and (xii) are obtained using the CASSCF calculation. The set (viii) is obtained from the gradient of the LUMO level (Eqs. 13.34 and 13.35). Rest of the sets are obtained from the gradient of the total electronic energy of $C_{60}^-$. (viii)' is calculated based on the incorrect formula of the electron–phonon coupling (Eqs. 13.40 and 13.41)

| | | B3LYP[a] | | | | | PW91[b] | | CASSCF[c] | |
|---|---|---|---|---|---|---|---|---|---|---|
| | Frequency $(cm^{-1})$ | 20 % (v) | 25 % (vi) | 30 % (vii) | 20 % (viii) | 20 % (viii)' | PW91 (ix) | LYP (x) | (1,3) (xi) | (11,8) (xii) |
| $a_g(1)$ | 496 | 2.5 | 2.3 | 2.2 | 1.8 | 1.8 | 2.7 | 2.6 | 1.6 | 1.5 |
| $a_g(2)$ | 1,470 | 15.7 | 18.0 | 19.3 | 16.2 | 16.2 | 11.3 | 11.2 | 47.8 | 52.7 |
| $h_g(1)$ | 273 | 3.2 | 3.2 | 3.3 | 7.8 | 3.1 | 2.3 | 2.4 | 6.4 | 7.3 |
| $h_g(2)$ | 437 | 6.7 | 6.9 | 7.0 | 15.4 | 6.1 | 6.1 | 5.7 | 5.5 | 6.6 |
| $h_g(3)$ | 710 | 7.7 | 9.5 | 10.0 | 17.9 | 7.1 | 6.0 | 5.7 | 8.4 | 8.5 |
| $h_g(4)$ | 774 | 3.2 | 2.8 | 2.8 | 7.6 | 3.1 | 2.6 | 2.8 | 1.6 | 2.2 |
| $h_g(5)$ | 1,099 | 3.0 | 3.7 | 4.0 | 7.5 | 3.0 | 2.6 | 2.6 | 6.5 | 7.3 |
| $h_g(6)$ | 1,250 | 1.2 | 2.2 | 2.5 | 3.2 | 1.3 | 1.5 | 1.3 | 1.2 | 1.8 |
| $h_g(7)$ | 1,428 | 14.0 | 15.2 | 16.6 | 34.2 | 13.7 | 9.3 | 9.6 | 20.9 | 19.3 |
| $h_g(8)$ | 1,575 | 11.2 | 11.0 | 11.6 | 26.8 | 10.7 | 8.1 | 7.8 | 13.7 | 13.5 |
| $E_s$ | | 18.2 | 20.3 | 21.5 | 18.0 | 18.0 | 14.0 | 13.8 | 49.4 | 54.2 |
| $E_{JT}$ | | 50.2 | 54.4 | 57.8 | 120.4 | 48.1 | 38.5 | 37.9 | 64.2 | 66.5 |
| $E_s + E_{JT}$ | | 68.4 | 74.7 | 79.3 | 138.4 | 66.1 | 52.5 | 51.7 | 113.6 | 120.7 |

[a]20%, 25%, and 30% are fractions of the Hartree–Fock exchange-correlation energy in B3LYP functional.
[b]PW91 and LYP mean correlation energy functionals.
[c](1,3) and (11,8) mean the size of the active space.

In the present results, the total stabilization energies (v) and (vi) with the B3LYP functional are close to the present experimental values (i), (ii), and (iii) in Table 13.5. On the other hand, the stabilization energies obtained with the gradient of LUMO of the neutral $C_{60}$ (viii) are larger than those from the total electronic energy of $C_{60}^-$ (v). The discrepancy is not surprising because the gradient of the LUMO level of the neutral $C_{60}$ $V_{\Gamma(\nu)\gamma}^{LUMO}$ does not correspond to the vibronic coupling constant (see the Appendix in Ref. [12]). Furthermore, the gradient of the LUMO of $C_{60}$ does not include the electronic relaxation which is crucial in the calculation of the vibronic coupling constant [28].

Saito and Laflamme Janssen et al. estimated the stabilization energies from the gradient of the LUMO of the neutral $C_{60}$ [9, 11] (Sai and Laf in Table 13.4). However, the total stabilization energies $E_s + E_{JT}$ by Saito and Laflamme Janssen et al. appear to be close to our present result (v) derived from the total electronic energy. The agreement between their results and present values comes from their method of the calculation of the electron–phonon coupling. The electron–phonon couplings $\lambda_\Gamma/N(0)$ are defined as [4, 15]:

$$\frac{\lambda_{a_g(i)}}{N(0)} = \frac{1}{9\omega_{a_g(i)}^2} \sum_{\gamma_e, \gamma_e'} \sum_{\gamma} \left[ \langle \gamma_e | \left( \frac{\partial H_e(\mathbf{R})}{\partial Q_{a_g(i)}} \right)_0 | \gamma_e' \rangle \right]^2 = \frac{2}{3} \frac{V_{a_g(i)}^2}{2\omega_{a_g(i)}^2}, \quad (13.38)$$

$$\frac{\lambda_{h_g(v)}}{N(0)} = \frac{1}{9\omega_{h_g(v)}^2} \sum_{\gamma_e,\gamma_e'} \sum_\gamma \left[ \langle \gamma_e | \left( \frac{\partial H_e(\mathbf{R})}{\partial Q_{h_g(v)\gamma}} \right)_0 |\gamma_e'\rangle \right]^2 = \frac{5}{3} \frac{V_{h_g(v)}^2}{2\omega_{h_g(v)}^2}. \quad (13.39)$$

Here, $N(0)$ is the density of states per spin and per $C_{60}$ at the Fermi level. To evaluate the electron–phonon couplings, they replace the sum of the matrix elements $\sum_{\gamma_e,\gamma_e'} \langle \gamma_e | (\partial H_e(\mathbf{R})/\partial Q_{\Gamma(v)\gamma})_0 | \gamma_e' \rangle$ with $\sum_{\gamma_e} [(\partial \varepsilon_{t_{1u}\gamma_e}^{\mathrm{LUMO}}/\partial Q_{\Gamma(v)\gamma})_0]^2$ [9–11]:

$$\frac{\lambda_{a_g(i)}^{\mathrm{LUMO}}}{N(0)} = \frac{1}{9\omega_{a_g(i)}^2} \sum_{\gamma_e} \sum_\gamma \left[ \left( \frac{\partial \varepsilon_{t_{1u}\gamma_e}}{\partial Q_{a_g(i)}} \right)_0 \right]^2 = \frac{2}{3} \frac{\left( V_{a_g(i)}^{\mathrm{LUMO}} \right)^2}{2\omega_{a_g(i)}^2}, \quad (13.40)$$

$$\frac{\lambda_{h_g(v)}^{\mathrm{LUMO}}}{N(0)} = \frac{1}{9\omega_{h_g(v)}^2} \sum_{\gamma_e} \sum_\gamma \left[ \left( \frac{\partial \varepsilon_{t_{1u}\gamma_e}^{\mathrm{LUMO}}}{\partial Q_{h_g(v)\gamma}} \right)_0 \right]^2 = \frac{2}{3} \frac{\left( V_{h_g(v)}^{\mathrm{LUMO}} \right)^2}{2\omega_{h_g(v)}^2}. \quad (13.41)$$

In the case of the $a_g$ mode, the form of the electron–phonon coupling is the same as Eq. 13.38. On the other hand, Eq. 13.41 is $2/5$ times smaller than the correct formula (13.39) because the off-diagonal elements of the vibronic couplings are not taken into account. Therefore, the stabilization energies by Saito, Laflamme Janssen et al., and Frederiksen et al. should be $2/5$ times smaller than the gradient of LUMO level. In fact, by multiplying $E_{\mathrm{JT},v}$ (viii) with $2/5$ in Table 13.5, the resulting $E_{\mathrm{JT},v}$ ((viii)' in Table 13.5) are similar to the values of Saito and Laflamme Janssen et al. (Sai and Laf in Table 13.4).

## 13.5 Summary

We simulated the photoelectron spectrum by using the Lanczos method, and derived the coupling constants of $C_{60}^-$. The formula of the transition intensity in the photoelectron spectroscopy based on the second order perturbation theory is given. However, we found that the vibronic coupling constants are underestimated with the formula. The experimental values were compared with the constants using several theoretical methods. Among these theoretical methods, the gradient of the total electronic energy with the B3LYP functional gives the closest values to the experimental result. On the other hand, the vibronic coupling constants are overestimated with the gradient of LUMO level of the neutral $C_{60}$.

## Appendix: Derivation of the Intensity of the Photoelectron Spectrum

To derive the intensity of the photoelectron spectrum, we use following assumptions:

1. Fullerene monoanions are in a thermal equilibrium state.
2. Photoelectron does not interact with $C_{60}$ (Sudden approximation) [29].
3. Orientation of $C_{60}$ in the laboratory system is completely random.
4. Strength of the electronic part of the transition is constant.

As the wave length of the light of 355 nm [13] is large enough compared with the size of the fullerene, the coupling between the electronic state of $C_{60}^-$ and light is written as follows:

$$V = \sum_{m'=-1}^{1} -F\mathbf{e}_{-m'}\mu_{m'}\exp\left(-i\Omega_{in}t\right),\qquad(13.42)$$

where $F$ is the strength of the electric field, $\mathbf{e}$ is the polarization of the electric field, $\mathbf{e}_{m'}$ is a component of $\mathbf{e}$, $\mu$ is the electronic dipole moment, $\mu_{m'}$ is a component of $\mu$, and $\Omega_{in}$ is the wave number of the incoming light. We use the complex basis for $\mathbf{e}$ and $\mu$. The irreducible representation $j'$ of $\mathbf{e}$ and $\mu$ is $j' = 1$ in the case of $C_{60}^-$, hence $m' = -1, 0, 1$. From the first assumption, the initial state is expressed by the ensemble density operator:

$$\rho = \sum_{v_1,v_2}\sum_{kJM} P_{v_1}^{a_g} P_{v_2}^{a_g} P_{kJ}^{\mathrm{JT}} |\Psi_{v_1 v_2 kJM}\rangle\langle\Psi_{v_1 v_2 kJM}|.\qquad(13.43)$$

As the photoelectron does not couple with $C_{60}$, a final state $|\Phi_{\alpha j_f m_f, \mathbf{n}_1, \cdots \mathbf{n}_8, v_1', v_2'}\rangle$ is the product of an electronic state $|\alpha j_f m_f; E_0\rangle$ and a vibrational state $|\mathbf{n}_1 \cdots \mathbf{n}_8\rangle|v_1', v_2'\rangle$. Here $E_0$ is the energy gap between the electronic levels of $C_{60}$ and $C_{60}^-$ at $\mathbf{R}_0$, $j_f, m_f$ are the magnitude of the angular momentum and $z$ component respectively, and $\alpha$ distinguishes the quantum states which have the same $j_f$ and $m_f$.

Including up to first order of $V$, the transition intensity per unit time $w$ which is proportional to the intensity of the spectrum is given by [30]

$$w = \frac{2\pi}{\hbar^2}\sum_{kJM}\sum_{v_1,v_2}\sum_{\alpha,j_f}\sum_{m_f=-j_f}^{j_f}\sum_{v_1',v_2'}\sum_{\mathbf{n}_1,\cdots\mathbf{n}_8} P_{v_1}^{a_g(1)} P_{v_2}^{a_g(2)} P_{kJ}^{\mathrm{JT}}$$

$$\times\left|\sum_{m'=-1}^{1}\langle\Phi_{\alpha j_f m_f,\mathbf{n}_1,\cdots\mathbf{n}_8,v_1',v_2'}|F\mathbf{e}_{-m'}\mu_{m'}'|\Psi_{v_1 v_2 kJM}\rangle\right|^2$$

$$\times\delta\left(\frac{E_0+\varepsilon_\alpha+E_s-E_{kJ}}{\hbar}+\sum_{v=1}^{8}\sum_{m=-2}^{2}\omega_{h_g(v)}n_{vm}+\sum_{i=1}^{2}\omega_{a_g(i)}(v_i'-v_i)-\Omega_{in}\right),$$

$$(13.44)$$

where $\varepsilon_\alpha$ is the energy of the photoelectron. As the orientation of the fullerene in laboratory coordinate system is random, the polarization of the light should be averaged.

$$\frac{1}{4\pi} \int_0^\pi \sin\theta d\theta \int_0^{2\pi} d\phi \left| \sum_{m'=-1}^1 \langle \alpha j_f m_f; E_0 | \mathbf{e}_{-m'} \mu_{m'} | j_e m_e \rangle \right|^2 \tag{13.45}$$

$$= \frac{1}{3} \sum_{m'=-1}^1 |\langle \alpha j_f m_f; E_0 | \mu_{m'} | j_e m_e \rangle|^2 \tag{13.46}$$

$$= \frac{|\langle \alpha j_f \| \mu_j \| j_e \rangle|^2}{3(2j_f+1)} \sum_{m'=-1}^1 |\langle j_f m_f | j' m' j_e m_e \rangle|^2 \tag{13.47}$$

Here we focus on the electronic part of the matrix element of $\sum_{m'=-1}^1 \mathbf{e}_{-m'} \mu_{m'}$ and use the Wigner–Eckart theorem [19]. Using a symmetry relation of the Clebsch–Gordan coefficient (Eq. 3.5.16 in Ref. [19])

$$\frac{1}{\sqrt{2j_f+1}} \langle j_f m_f | j' m' j_e m_e \rangle = \frac{1}{\sqrt{2j_e+1}} \langle j_e m_e | j_f m_f j' - m' \rangle \tag{13.48}$$

and the unitarity of the Clebsch–Gordan coefficient, $w$ is written as

$$w = \sum_{\alpha, j_f} \frac{2\pi F^2 |\langle \alpha j_f \| \mu_j \| j_e \rangle|^2}{3\hbar^2 (2j_e+1)} \sum_{kJM} \sum_{v_1, v_2} \sum_{m_e=-1}^1 \sum_{\mathbf{n}_1, \cdots \mathbf{n}_8} \sum_{v_1', v_2'} P_{v_1}^{a_g(1)} P_{v_2}^{a_g(2)} P_{kJ}^{\mathrm{JT}}$$

$$\times \left| C_{m_e, \mathbf{n}_1, \cdots \mathbf{n}_8; kJM}^{\mathrm{JT}} S_{v_1' v_1}(g_{a_g(1)}) S_{v_2' v_2}(g_{a_g(2)}) \right|^2$$

$$\times \delta \left( \frac{E_0 + \varepsilon_\alpha + E_s - E_{kJ}}{\hbar} + \sum_{v=1}^8 \sum_{m=-2}^2 \omega_{h_g(v)} n_{vm} + \sum_{i=1}^2 \omega_{a_g(i)} (v_i' - v_i) - \Omega_{in} \right).$$

$$\tag{13.49}$$

Using the last assumption that the transition intensity does not depend on the final electronic state, and replacing the electron binding energy $\Omega_{in} - \varepsilon_\alpha$ with $\Omega$, we obtain Eq. 13.18.

**Acknowledgments** We would like to thank X.B. Wang and L.S. Wang for sending us unpublished data. N.I. would like to thank the research fund for study abroad from the Research Project of Nano Frontier, graduate school of engineering, Kyoto University. T.S. and N.I. are grateful to the Division of Quantum and Physical Chemistry at the University of Leuven for hospitality. Theoretical calculations were partly performed using the Supercomputer Laboratory of Kyoto University and Research Center for Computational Science, Okazaki, Japan. This work was

13 Estimation of the Vibronic Coupling Constants of Fullerene Monoanion... 263

supported by a Grant-in-Aid for Scientific Research, priority area Molecular theory for real systems (20038028) from the Japan Society for the Promotion of Science (JSPS). This work was also supported in part by the Global COE Program International Center for Integrated Research and Advanced Education in Materials Science (No. B-09) of the Ministry of Education, Culture, Sports, Science and Technology (MEXT) of Japan, administered by the JSPS. Financial support from the JSPS–FWO (Fonds voor Wetenschappelijk Onderzoek–Vlaanderen) bilateral program is gratefully acknowledged.

# References

1. Bersuker IB, Polinger VZ (1989) Vibronic interactions in molecules and crystals. Springer, Berlin
2. Bersuker IB (2005) The Jahn–Teller effect. Cambridge University Press, Cambridge
3. Chancey CC, O'Brien MCM (1997) The Jahn–Teller effect in $C_{60}$ and other icosahedral complexes. Princeton University Press, Princeton
4. Varma CM, Zaanen J, Raghavachari K (1991) Science 254:989
5. Gunnarsson O, Handschuh H, Bechthold PS, Kessler B, Ganteför G, Eberhardt W (1995) Phys Rev Lett 74:1875
6. Breda N, Broglia RA, Colò G, Roman HE, Alasia F, Onida G, Ponomarev V, Vigezzi E (1998) Chem Phys Lett 286:350
7. Hands ID, Dunn JL, Bates CA (2001) Phys Rev B 63:245414
8. Manini N, Carso AD, Fabrizio M, Tosatti E (2001) Philos Mag B 81:793
9. Saito M (2002) Phys Rev B 65:220508(R)
10. Frederiksen T, Franke KJ, Arnau A, Schulze G, Pascual JI, Lorente N (2008) Phys Rev B 78:233401
11. Laflamme Janssen J, Côté M, Louie SG, Cohen ML (2010) Phys Rev B 81(7):073106
12. Iwahara N, Sato T, Tanaka T, Chibotaru LF (2010) Phys Rev B 82:245409
13. Wang XB, Woo HK, Wang LS (2005) J Chem Phys 123:051106
14. Sato T, Tokunaga K, Iwahara N, Shizu K, Tanaka K (2009) In: Köppel H, Yarkony DR, Barentzen H (eds) The Jahn–Teller effect: fundamentals and implications for physics and chemistry. Springer, Berlin, p 99
15. Lannoo M, Baraff GA, Schlüter M, Tomanek D (1991) Phys Rev B 44:12106
16. Auerbach A, Manini N, Tosatti E (1994) Phys Rev B 49:12998
17. O'Brien MCM (1969) Phys Rev 187:407
18. O'Brien MCM (1971) J Phys C Solid State Phys 4:2524
19. Edmonds AR (1974) Angular momentum in quantum mechanics. Princeton University Press, Princeton
20. Manini N, Tosatti E (1995) In: Kadish KM, Ruoff RS (eds) Recent advances in the chemistry and physics of fullerenes and related materials, vol 2. The Electrochemical Society, Pennington, p 1017
21. Bethune DS, Meijer G, Tang WC, Rosen HJ, Golden WG, Seki H, Brown CA, de Vries MS (1991) Chem Phys Lett 179:181
22. Feynman RP (1939) Phys Rev 56:340
23. In our preliminary calculation, we obtained essentially the same constants even if we used larger $\delta Q$'s. Thus our choice of $\delta Q$ is good enough to calculate the gradient of the LUMO level
24. Frisch MJ et al (2004) Gaussian 03, Revision E.01. Gaussian Inc., Wallingford, CT
25. Hands ID, Dunn JL, Bates CA, Hope MJ, Meech SR, Andrews DL (2008) Phys Rev B 77:115445

26. Becke AD (1993) J Chem Phys 98:5648
27. Perdew JP, Burke K, Wang Y (1996) Phys Rev B 54(23):16544
28. Sato T, Tokunaga K, Tanaka K (2008) J Phys Chem A 112:758
29. Hedin L, Lundqvist S (1969) In: Ehrenreich H, Turnbull D, Seitz F (eds) Solid state physics, vol 23. Academic, New York, p 1
30. Messiah A (1962) Quantum mechanics, vol 2. North-Holland, Amsterdam

# Chapter 14
# Investigations of the Boron Buckyball $B_{80}$: Bonding Analysis and Chemical Reactivity

**Jules Tshishimbi Muya, G. Gopakumar, Erwin Lijnen, Minh Tho Nguyen, and Arnout Ceulemans**

**Abstract** The boron fullerene $B_{80}$ is a spherical network of 80 boron atoms, which has a shape similar to the celebrated $C_{60}$. The 80 Bs span two orbits: while the first contains 60 atoms localised on the vertices of a truncated icosahedron like $C_{60}$, the second includes 20 extra B atoms capping the hexagons of the frame. Quantum chemical calculations showed that $B_{80}$ is unusually stable and has interesting physical and chemical properties. Its geometry is slightly distorted from $I_h$ to $T_h$ symmetry. However, the boron buckyball is only observed in *silico*, so far the synthesis of this molecule is only a remote possibility. Using DFT at the B3LYP/SVP level, we have analyzed the chemical bonding in $B_{80}$, the possibility of methyne substitution and the stability of endohedral boron buckyball complexes. A symmetry analysis revealed a perfect match between the occupied molecular orbitals in $B_{80}$ and $C_{60}$. The cap atoms transfer their electrons to the truncated icosahedral frame, and they contribute essentially to the formation of $\sigma$ bonds. The frontier MOs have $\pi$ character and are localised on the $B_{60}$ truncated icosahedral frame. The boron cap atoms can be replaced by other chemical groups, such as methyne (CH), which are also able to introduce three electrons in the cage. Symmetrical substitutions of the boron cap atoms by methyne groups in $T$ and $T_h$ symmetries revealed two stable endo methyne boron buckyballs, endo-$B_{80-x}(CH)_x$, with $x = 4, 8$. The stability of these compounds seems to be due to the formation of six boron 4-centre bonding motifs in between the substituted

---

J. Tshishimbi Muya (✉) • E. Lijnen • M.T. Nguyen • A. Ceulemans
Department of Chemistry and INPAC, Institute for Nanoscale Physics and Chemistry,
Katholieke Universiteit Leuven, B-3001 Leuven, Belgium
e-mail: Jules.Tshishimbi@chem.kuleuven.be; Erwin.Lijnen@chem.kuleuven.be;
Minh.Nguyen@chem.kuleuven.be; Arnout.Ceulemans@chem.kuleuven.be

G. Gopakumar
Max-Planck Institut für Kohlenforschung, Kaiser-Wilhelm-Platz 1,
45470 Mülheim an der Ruhr, Germany
e-mail: gopakumar@kofo.mpg.de

M. Atanasov et al. (eds.), *Vibronic Interactions and the Jahn-Teller Effect:*
*Theory and Applications*, Progress in Theoretical Chemistry and Physics 23,
DOI 10.1007/978-94-007-2384-9_14, © Springer Science+Business Media B.V. 2012

hexagons. These localized bonding motifs are at the basis of the observed symmetry lowering, via a pseudo-Jahn-Teller effect. The methyne hydrogen atoms in the two endohedral fullerenes can be replaced by other atoms, which can lead to cubane or tetrahedral endohedral boron fullerenes. Theoretical study on encapsulated small bases molecules, tetrahedral and cubane like clusters of Group V atoms, showed that the boron buckyball is a hard acid and prefers hard bases like $NH_3$ or $N_2H_4$, to form stables off-centred complexes with $B_{80}$. Tetrahedral and cubane like clusters of this family are usually metastable in the encapsulated state, due to steric strain. The most favorable clusters are mixed tetrahedral and cubane clusters formed by nitrogen and phosphorus atoms such as $P_2N_2@B_{80}$, $P_3N@B_{80}$ and $P_4N_4@B_{80}$. The boron cap atoms act as electrophilic centres, which react with nucleophilic sites rich in electrons.

## 14.1 Introduction

Boron shares with carbon an important property that it can form a covalently bonded network (catenation) with in principle unlimited size [1]. Recently, Jemmis et al. [2] argued that the chemistries of boron, boranes, macropolyhedral boranes, elemental boron and boron-rich solids form an unknown continent to be explored and understood. The boron nanotubes are as fascinating as the carbon counterparts by their properties, which can be modified with small changes in diameter and chirality. In the search for new fullerenes, the boron buckyball is probably one of the most interesting molecules coming out of current quantum chemical calculations [3] (Fig. 14.1). It contains 80 boron atoms forming two orbits. While the first is characterized by 60 boron atoms seated on the vertices of a truncated icosahedron, the second orbit consists of 20 borons localized on the centre of hexagonal faces forming a dodecahedron frame. The shape of $B_{80}$ resembles to that of the celebrated $C_{60}$ [4]. It is also related to the energetically most stable boron structure, the α-sheet [5].

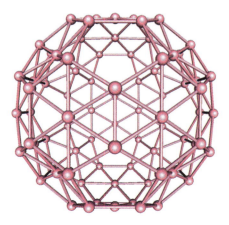

**Fig. 14.1** Optimized geometry of $B_{80}$

The initial report on the stability of the boron buckyball has given rise to an intense computational afterglow. It was shown by Gopakumar et al. that its equilibrium geometry is slightly distorted from $I_h$ to $T_h$ symmetry [6] (at B3LYP/6-31G(d)). This distortion yields two isomers characterised by eight atoms going inward and twelve going outwards of the hexagonal plane or vice-versa. However, the difference in energy between both structures is very small, and appears to be sensitive to the quantum chemical method used [7]. Recently, an alternative hollow cage having the texture of a 'volleyball', was proposed [8]. It differs from the standard 'buckyball' by a redistribution of the capping atoms from hexagons to pentagons. A different set of structures was also proposed, based on the expectation that stuffing may increase the stability of boron fullerenes. Along these lines, some so-called core-shell $B_{80}$ forms were investigated containing a nucleus of icosahedral boron $B_{12}$ [9, 10]. Clearly, these structures point toward a different clustering process, and are intermediate steps to the ß solid state.

In the present paper, we offer an overview of several chemical and physical properties of the boron buckyball $B_{80}$. As far as chemical bonding is concerned, we analyse in detail the contention of an isoelectronic equivalence between $C_{60}$ and $B_{80}$, and compare their valence shells. In this perspective we also address the $I_h$ to $T_h$ symmetry breaking from the point of view of a pseudo Jahn-Teller effect (PJTE), and identify the active mode. We subsequently show that methyne groups, which introduce three electrons into the cage and are able to form hexacoordinate bonds in boron aromatic systems, can substitute the boron cap atoms to form methyne boron buckyballs. We further reveal that the cap boron atoms are hard acids and constitute preferential sites of nucleophilic attack.

The boron buckyball has an internal radius of 4.14 Å, nearly 0.7 Å higher than that of $C_{60}$, which makes it more suitable to encapsulate small molecules in the cage. The encapsulation of tetrahedral and cubic molecules of the group V family is investigated. For some binary clusters containing nitrogen and phosphorus, endohedral encapsulation is found to be exothermic.

According to recent results, $B_{80}$ should present promising prospects for electronic transmission, hydrogen storage capacity, electronic conductivity of solid state $B_{80}$ doped by Mg and alkali metals, and tuneable magnetic properties when it encapsulates transition metals such as Co or Ni [11–15]. We hope that all these promising prospects will stimulate the quest for its actual synthesis.

## 14.2 Methods

Density functional theory (DFT) calculations using the hybrid B3LYP functional in conjunction with different basis sets including STO-3G, SVP, def2-TZVPP and 6–31G(d) were used for geometry optimizations. Harmonic vibrational frequencies were first computed using the small basis set B3LYP/STO-3G, and then with the larger basis sets SVP or 6–31G(d) in order to characterize the nature of the stationary points, and to determine the zero-point vibrational energy corrections by analytical evaluation of the second derivatives of the energy with respect to nuclear displacements. All calculations were carried out using the Gaussian 03 and TURBOMOLE-V-5-10 packages. The gOpenmol and gmolden programs were used to visualize the total electron densities, molecular orbitals and equilibrium geometries. The distribution of charges on different atoms in the endohedral boron fullerenes has been calculated at B3LYP/6-31G(d) level by the NBO program. We have analysed the contribution of atomic orbitals in the construction of molecular orbitals of different symmetries, and project the PJT active mode. For methyne boron buckyballs, the isomers were considered in the $T_h$ and $T$ symmetry, which correspond to the symmetry and subgroup of the lowest energy structure of $B_{80}$. Starting geometries in the case of encapsulation of bases in $B_{80}$ were usually based on symmetry considerations with heteroatoms either pointing towards the endo-boron caps, or towards the exo-caps. Symmetry constraints were relaxed during the optimization processes. The frequency analysis of high symmetry stationary points was performed at a large basis set, whereas the frequencies of the lower symmetry endohedral boron fullerenes were calculated using smaller basis sets.

## 14.3 Results and Discussion

### 14.3.1 Analysis of Chemical Bonding and PJTE in $B_{80}$

#### 14.3.1.1 Chemical Bonding

Both $B_{80}$ and $C_{60}$ have the same number of valence electrons and the same soccer ball shape. This argument is however not sufficient to consider that $B_{80}$ is isoelectronic to $C_{60}$. It is further required that the group symmetries of the occupied valence orbitals should match. A detailed symmetry analysis along these lines has recently been published [16]. Here we summarize the results and show a plot, which demonstrates the equivalence of the valence orbitals. NBO calculations show that the cap B atoms act as electron donors and the B atoms on the frame are electron acceptors. The number of electrons transferred from $B_{20}$ to $B_{60}$ is nearly three electrons. The valence shell of B consists of 2s and 2p orbitals. The 2p orbitals on a spherical cluster form a radial component and two tangential components denoted by $p_\sigma$ and $p_\pi$. The irreducible symmetry representations of the MOs spanned by these components using induction theory are given in following expressions:

$$\begin{aligned}
\Gamma_\sigma^{60} &= A_g + T_{1g} + 2T_{1u} + T_{2g} + 2T_{2u} + 2G_g + 2G_u + 3H_g + 2H_u \\
\Gamma_\pi^{60} &= A_g + A_u + 3T_{1g} + 3T_{1u} + 3T_{2g} + 3T_{2u} + 4G_g + 4G_u + 5H_g + 5H_u \\
\Gamma_\sigma^{20} &= A_g + T_{1u} + T_{2u} + G_g + G_u + H_g \\
\Gamma_\pi^{20} &= T_{1g} + T_{1u} + T_{2g} + T_{2u} + G_g + G_u + 2H_g + 2H_u
\end{aligned} \tag{14.1}$$

The 90 bonds in $B_{80}$ are spread over two orbits, $O_{60} + O_{30}$. The $O_{60}$ orbit contains the 5–6 bonds adjacent to a pentagon and a hexagon, whereas the $O_{30}$ orbit is the set of the 6–6 bonds adjacent to two hexagons. The 60 $\sigma$-bonds along the 5–6 edges span exactly the same $\Gamma_\sigma^{60}$ symmetry as the 2s and $2p_\sigma$ orbitals on the $O_{60}$ boron atoms. The 30 $\sigma$-bonds along the 6–6 edges transform as $\Gamma_\sigma^{30}$:

$$\Gamma_\sigma^{30} = A_g + T_{1u} + T_{2u} + G_g + G_u + 2H_g + H_u \tag{14.2}$$

The site symmetry in the centre of a 6–6 edge is $C_{2v}$. The induced representation of anti-bonds localized along the 6–6 edges has symmetries as:

$$\Gamma_{\sigma^*}^{30} = T_{1g} + T_{1u} + T_{2g} + T_{2u} + G_g + G_u + H_g + H_u \tag{14.3}$$

Rehybridization between $O_{30}$ and $O_{20}$ sites gives rise to the relation:

$$\Gamma_\sigma^{20} + \Gamma_\pi^{20} = \Gamma_\sigma^{30} + \Gamma_{\sigma^*}^{30} \tag{14.4}$$

In $C_{60}$, the hybridized $sp^2$ orbitals provide a bonding network along 90 edges, which accounts for 180 electrons. The remaining radial $p_\sigma$ orbitals form a conjugated $\pi$-bonding network, which is mainly localized along the 6–6 edges, and accommodates another 60 electrons. In total the 240 electrons of $C_{60}$ occupy 120 orbitals with the following symmetries:

$$\Gamma_{bond}(C_{60}) = \Gamma_\sigma^{60} + 2\Gamma_\sigma^{30} \tag{14.5}$$

From Eqs. 14.1 and 14.2, the symmetries of the occupied orbitals of the valence shell of $C_{60}$ are thus identified as follows:

$$\Gamma_{bond}(C_{60}) = 3A_g + T_{1g} + 4T_{1u} + T_{2g} + 4T_{2u} + 4G_g + 4G_u + 7H_g + 4H_u \tag{14.6}$$

If we now compare the list in Eq. 14.6 with the symmetries of valence orbitals of $B_{80}$, there is a perfect match. The detailed classification of these bonding orbitals can be based on the spherical parentage, which is provided in Table 14.1, and applied to both sub-bands of $\sigma$ and $\pi$ character.

The orbitals are assigned spherical labels in order of their appearance in the MO sequence of the valence shell. Only approximate assignments as either $\sigma$ or $\pi$ can be made in view of strong hybridization between the two sub-bands. The results are shown in Fig. 14.2, which displays a comparison between the boron and carbon buckyballs. Several results are apparent from Fig. 14.2. Overall, the $\sigma$ sub-band, comprising 90 valence orbitals, is lying below the $\pi$ sub-band, which consists of the remaining 30 orbitals. Furthermore, there is no doubt that the valence shells in the B and C balls are correlated, but hybridization between the two sub-bands is clearly more pronounced for B than for C.

A further confirmation of the isomorphism between both electronic structures can be obtained by analyzing the contributions of the cap atoms over the irreducible representations of the icosahedral group, as given in Table 14.2.

**Table 14.1** Spherical shells and σ- and π-band of $B_{80}$

| L | $I_h$ | σ-band | π-band |
|---|---|---|---|
| 0 | $A_g$ | $3a_g$ | $5a_g$ |
| 1 | $T_{1u}$ | $4t_{1u}$ | $7t_{1u}$ |
| 2 | $H_g$ | $5h_g$ | $10h_g$ |
| 3 | $T_{2u} + G_u$ | $4t_{2u} + 4g_u$ | $7t_{2u} + 7g_u$ |
| 4 | $G_g + H_g$ | $4g_g + 6h_g$ | $7g_g + 11h_g$ |
| 5 | $T_{1u} + T_{2u} + H_u$ | $5t_{1u} + 5t_{2u} + 3h_u$ | $6h_u$ |
| 6 | $A_g + T_{1g} + G_g + H_g$ | $4a_g + 2t_{1g} + 5g_g + 7h_g$ | |
| 7 | $T_{1u} + T_{2u} + G_u + H_u$ | $6t_{1u} + 6t_{2u} + 5g_u + 4h_u$ | |
| 8 | $T_{2g} + G_g + 2H_g$ | $2t_{2g} + 6g_g + 8h_g + 9h_g$ | |
| 9 | $T_{1u} + T_{2u} + 2G_u + H_u$ | $6g_u + 5h_u$ | |
| | $B_{80}$ | $C_{60}$ | |

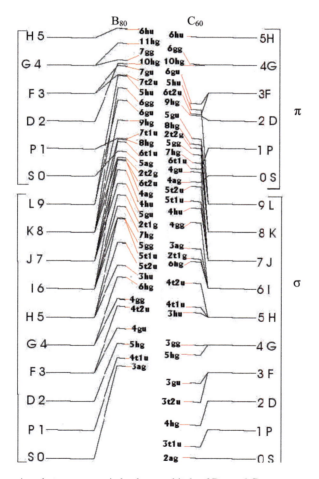

**Fig. 14.2** Comparison between occupied valence orbitals of $B_{80}$ and $C_{60}$

**Table 14.2** Valence contributions of 2s and 2p orbitals on the cap atoms (Taken from [16])

| MOs | 2s | 2p | 2s+2p | $\Gamma_\sigma^{30}$ |
|---|---|---|---|---|
| $a_u$ | 0 | 0 | 0 | 0 |
| $a_g$ | 0.51 | 0.31 | 0.82 | 1 |
| $t_{1g}$ | 0 | 0.94 | 0.94 | 0 |
| $t_{1u}$ | 1.21 | 2.40 | 3.62 | 3 |
| $t_{2g}$ | 0 | 0.40 | 0.40 | 0 |
| $t_{2u}$ | 1.84 | 1.05 | 2.89 | 3 |
| $g_g$ | 2.06 | 2.12 | 4.18 | 4 |
| $g_u$ | 1.53 | 1.03 | 2.56 | 4 |
| $h_g$ | 2.87 | 6.35 | 9.21 | 10 |
| $h_u$ | 0 | 4.68 | 4.68 | 5 |
| Total | 10.01 | 19.28 | 29.29 | 30 |

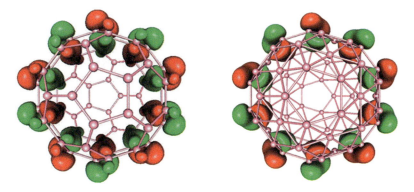

**Fig. 14.3** One of the $6h_u$ HOMO components for $C_{60}$ (*left*) and $B_{80}$ (*right*)

It is seen that the distribution of these orbitals closely matches the content of the $\Gamma_\sigma^{30}$ group representation. This proves that the capping atoms together inject 60 electrons in the 6–6 bonds, to compensate for the electron deficiency of the boron frame.

A detailed analysis of the chemical bonds in $B_{80}$ is reported in Ref. [16]. The point groups of the MOs in $B_{80}$ match with those in $C_{60}$. The HOMO in both structures ($6h_u$) has in fact the same symmetry (Fig. 14.2).

In Fig. 14.3, we compare one component of the HOMO's of both buckyballs to show their similarity. Also note that the cap atoms are almost not contributing, as they are mainly involved in the $\sigma$ sub-bands.

$B_{80}$ and $C_{60}$ share the same shape and the same number of valence electrons. The fact that the molecular orbitals in both systems are similar reveals that the two molecules are isoelectronic. Hence, the properties observed in $C_{60}$ are expected being seen in $B_{80}$. However, the symmetry of the most stable $B_{80}$ differs slightly to that of $C_{60}$. The icosahedral $I_h$-$B_{80}$ is metastable and undergoes a distortion that leads to $T_h$ symmetry via a second order Jahn-Teller effect.

**Fig. 14.4** Pseudo-JT distortion vector

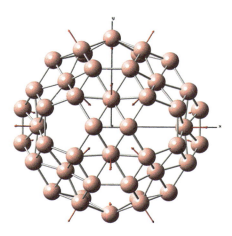

### 14.3.1.2 Pseudo Jahn-Teller Effect in $B_{80}$

A detailed vibrational frequency analysis of the $I_h$ structure shows the presence of seven imaginary frequencies, transforming as $g_g + t_{2u}$. Geometry relaxation leads to a stable energy minimum of $T_h$ symmetry. The distortion vector is defined as the difference between the $I_h$ and $T_h$ structures:

$$\vec{\Delta} = \left[ \begin{pmatrix} x_i - x'_i \\ y_i - y'_i \\ z_i - z'_i \end{pmatrix} \right]_{i=1,80} \tag{14.7}$$

Projection of this vector onto the normal modes of the icosahedral $B_{80}$ shows that one of the $g_g$ modes contributes about 90% to the distortion vector. Clearly this mode corresponds to the PJT coordinate (8), which is displayed in Fig. 14.4:

$$\vec{\Delta} = \sum_k c_k \cdot \vec{g}_g^k \Rightarrow c_k = \frac{\vec{\Delta} \cdot \vec{g}_g^k}{\vec{g}_g^k \cdot \vec{g}_g^k} \tag{14.8}$$

where $c_k$ is the coefficient of the $\vec{g}_g^k$ vector in this linear equation.

It is evident that the $t_{2u}$ mode cannot be involved, since its presence should destroy the inversion symmetry, and thus cannot be compatible to a symmetry lowering to $T_h$.

As stated above, the energy difference between both $I_h$ and $T_h$ structures is very small. An orbital analysis shows that the distortion does not lead to appreciable vibronic mixing between HOMO ($h_u$) and LUMO ($t_{2u}$). This is also not expected since the distortion vector mainly affects the cap atoms: 8 cap atoms move inside the cage, whereas the 12 other caps are moved outside, as shown in Fig. 14.4. These 12 exo-boron atoms tend to form six pairs of four-centre bonds, which ultimately stabilize the molecule in $T_h$ symmetry, as can be illustrated by the total density

**Fig. 14.5** Formation of four-centre bond

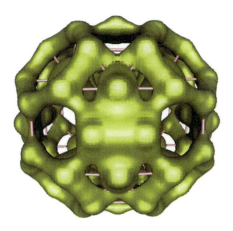

maps (Fig. 14.5). This result suggests that the vibronic coupling extends deep into the valence shell, involving also the σ-frame.

There is clearly a need for a detailed vibronic analysis over a broad excitation range to obtain a better understanding of the symmetry breaking process.

## 14.3.2 Reactivity of $B_{80}$

The pseudo Jahn-Teller effect exerts mainly forces on the cap atoms. Besides of a vibronic analysis this effect can also be probed by inserting other chemical groups. One way is to replace *endo* or *exo* caps by substituents that have the same symmetry lowering effect as the radial cap movements. This is an application of the so-called 'substitutional' Jahn-Teller effect [17]. Methyne turns out to be a suitable substituent for the cap atoms.

The other way is to insert endohedral chemical groups, which may form bonds with the cap atoms, or exert a low-symmetry pressure on the cage.

### 14.3.2.1 Methyne Capping

We have replaced the cap boron atoms symmetrically by methyne groups, which can also share three electrons with the cage [18]. Methyne groups are also used in substitution of boron atoms in carborane cages. Carbon has been known to have the capacity to form a hexacoordinate compound as in aromatic boron to compensate the electron deficiency of boron [19]. Hence the three $B_{80-x}(CH)_x$ substitutions with $x = 4, 8, 12$ in both $T$ or $T_h$ symmetries have been examined. A substitution of all 20 cap atoms introduces too much strain. The methyne groups are oriented in the direction of the radius vector from the centre of the cluster, with the hydrogens either inside or outside of the cage. We denote them as *endo* and *exo*-orientations,

**Table 14.3** Calculated frontier MO gaps and cohesive energies of $B_{80-x}(CH)_x$ with x = 4, 8, 12 in $T_h$ symmetry at the B3LYP/SVP level. N represents the number of imaginary frequencies and C.E the cohesive energy

| Isomers | sym | HOMO-LUMO (eV) | C.E/e | N |
|---|---|---|---|---|
| Endo-76 | T | 2.3 | −4.0 | 0 |
| Exo-76 | T | 1.9 | −3.9 | 8 |
| Endo-72 | $T_h$ | 2.6 | −3.9 | 0 |
| Exo-72 | $T_h$ | 1.8 | −3.7 | 11 |
| Endo-68 | $T_h$ | 2.1 | −3.8 | 12 |
| Exo-68 | $T_h$ | 2.1 | −3.6 | 20 |
| $B_{80}$ | $T_h$ | 1.9 | −4.2 | 0 |

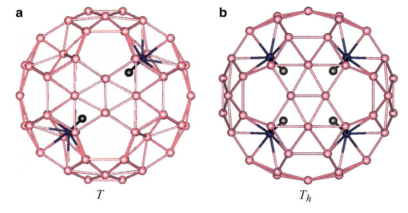

**Fig. 14.6** Optimized geometry of (**a**) *endo*-$B_{76}(CH)_4$, and (**b**) *endo*-$B_{72}(CH)_8$ computed at B3LYP/SVP

respectively. The three substitutions thus give rise to six structures. Out of these only two were found true minima, namely *endo*-72 and *endo*-76 in $T_h$ and $T$ symmetries respectively. (Table 14.3) Their optimized geometries are depicted in Fig. 14.6a, b.

These two structures are characterized by the isolation of methyne groups (each carbon is surrounded by pentagons and capped hexagons, cf. Fig. 14.7), and by the formation of four–centre bonds (see Fig. 14.5). The latter is also observed in the most stable $B_{80}$ in $T_h$ but not in $I_h$ symmetry, and can lend a further support for the Pseudo Jahn-Teller effect mechanism in $B_{80}$.

The B-B and B-C bond lengths in both compounds are in the same range with other compounds such as boron buckyball $B_{80}$, carboranes and different viable planar hexacoordinated carbon molecules suggested by Ito et al. [20].

The electronic structures of the *endo*-76 and *endo*-72 show similarities of valence MOs of methyne boron buckyballs and its boron buckyball counterpart $B_{80}$. These structures are isoelectronic to $C_{60}$ (Fig. 14.8). The presence of a $B_4$ bond in these compounds reinforces their stability.

14 Investigations of the Boron Buckyball B$_{80}$... 275

**Fig. 14.7** Isolated methyne group in B$_{72}$(CH)$_8$ and B$_{76}$(CH)$_4$. The carbon atom at the middle of the scheme in blue color is surrounded by pentagons and hexagons and linked to a H atom

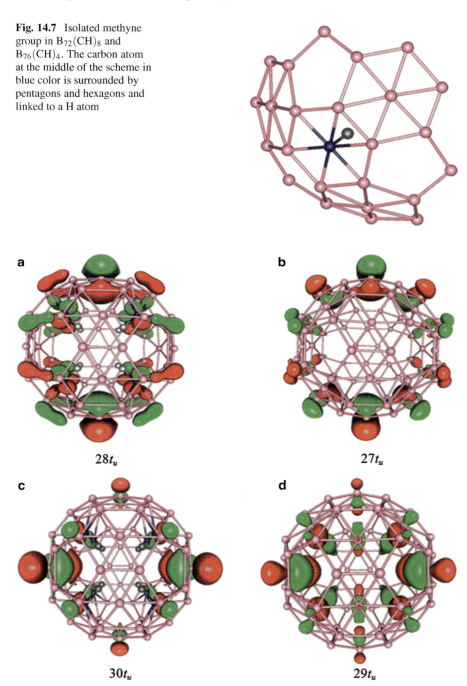

**Fig. 14.8** Components of frontier orbitals of endo-72 and B$_{80}$: (**a**) HOMO of *endo*-72, (**b**) HOMO of B$_{80}$, (**c**) LUMO of *endo*-72, and (**d**) LUMO +2 of B$_{80}$ (isocontour values ±0.03)

**Table 14.4** Comparison of bond distances of $T_h$ $B_{80-x}(CH)_x$ isomers $B_{80}$ and carborane molecules (Table taken from [18])

| Molecules | $R_{B-B}(\text{Å})$ | $R_{B-C}(\text{Å})$ | $R_{C-H}(\text{Å})$ | BCH angle |
|---|---|---|---|---|
| Endo-72 | 1.63–1.84 | 1.82–1.84 | 1.10 | 115.1–115.8 |
| $B_{80}$ | 1.65–1.76 | - | - | - |
| $1,12\text{-}C_2I_2B_{10}H_{10}$ | 1.77–1.78 | 1.72–1.74 | 1.10 | 115.2–118.5 |
| $C_8H_{16}B_{10}$ | 1.76–1.80 | 1.68–1.74 | | 106.3–108.3 |

The hydrogen sites can be substituted by other elements in order to include cubane or tetrahedral endohedral clusters. There is a strong pyramidalization of C off the hexagonal face, which leads to an elongation of B-C bond distances (Table 14.4).

### 14.3.2.2 Encapsulation of Base Molecules in $B_{80}$

We present here a short summary of our calculations on endohedral compounds. A detail account is given in Ref. [21].

As the diameter of $B_{80}$ is larger than that of $C_{60}$, small molecules can be encapsulated. We have inserted inside the hollow cage of $B_{80}$ some ligands of group V such as $NH_3$, $PH_3$, $AsH_3$, $H_4N_2$, $H_3NBH_3$ and $(CH_3)_3P$. Most of the bases move to endo cap atoms to form strong B-N or P-N bonds, except for the $AsH_3$, which prefers to stay at the middle of the cage. The complex $NH_3@B_{80}$ is the most stable with a large HOMO-LUMO gap of 1.82 eV, and a complexation energy of $-27.1$ kcal/mol. The least favorable value is observed for $H_3NBH_3@B_{80}$, which amounts to $-0.15$ kcal/mol. Since $H_3NBH_3$ is less basic and ammonia is a harder base than $PH_3$ and $AsH_3$, the cap boron atoms are presumably hard acids and tend to interact strongly with hard bases.

Tetrahedral clusters of group V such as $N_4$, $P_4$, $As_4$, and $Sb_4$ are encaged in $B_{80}$. Most of these complexes are thermodynamically endothermic and tend to form bidentate links with two borons of $B_{80}$. Further, we have encapsulated tetrahedral mixed clusters of nitrogen and phosphorus atoms namely $N_2P_2$ and $NP_3$ molecules, which lead to exothermic and thermodynamically neutral complexes, respectively.

The orientation of the heteroatom has an influence on the electron donating properties of the encaged cluster. If the heteroatom is oriented in the direction of a single boron cap, its ability of giving electrons to the boron cage is reduced as compared to that having a bidentate orientation.

The $Sb_4$ cluster is too large to seat comfortably in the centre of the cage. To take advantage of the eight *endo* B atoms, we have inserted some cubane like structures such as $P_8$, $P_4N_4$, $B_4N_4$ in $B_{80}$ and $P_4C_4$ in *endo*-76.

All cubic donor-acceptor complexes considered are not favorable, except for the $P_4N_4@B_{80}$ complex, which is characterized by formation energy of $-32.9$ kcal/mol. (Table 14.5). The $P_8$ is apparently too large and exerts much strain on the cage and thereby may introduce frustration in the PJTE forces exerting

**Table 14.5** Mean bond distances between heteroatom and nearest cap boron atom (X-B, Å), HOMO-LUMO gaps (eV), and lowest frequencies (cm$^{-1}$), and formation energies (FE, kcal/mol) at the B3LYP/SVP level

| Compounds | X − B | Gap | Low freq | FE |
|---|---|---|---|---|
| P$_4$C$_4$@B$_{76}$C$_4$ (T) | 1.47 (C–C) | 2.57 | 163 | 23.84 |
|  | 1.81 (B–P) |  |  |  |
| P$_4$N$_4$@B$_{80}$ (T) | 1.57 (B–N) | 1.96 | 174 | −32.87 |
|  | 1.805 (B–P) |  |  |  |
| B$_4$N$_4$@endo76 (T) | 1.50 (B–N) | 0.23 | 133 | 187.17 |
|  | 1.615 (B–C) |  |  |  |
| P$_8$@B$_{80}$ (C$_1$) | 1.77 | 2.04 | 72[a] | 401.83 |
| B$_{80}$$^2$ (T$_h$) |  | 1.96 |  |  |

[a]Frequency calculation at the B3LYP/STO-3G level

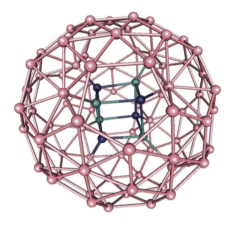

**Fig. 14.9** Optimized geometry of P$_4$N$_4$@B$_{80}$ (B3LYP/6-31G(d))

in the B$_{80}$ cage, as it tends to move eight caps outwards, while the PJTE drives eight caps inwards. However, the P$_8$ cube transfers the largest amount of electrons to the cage, being nearly five electrons, through 8 B-P bonds, which however cannot cancel the resulting strain energies coming from its large size.

The B$_{80}$ cage stabilizes the unusual cubic P$_4$N$_4$ molecule. The latter is usually observed in a cyclic form (Fig. 14.9).

The phosphorus centres are highly reactive for electrophilic attack. The stability of the P$_4$N$_4$@B$_{80}$ complex is due to its size, and also its ability to share electrons with the electron deficient cap boron atoms. The B$_4$N$_4$@B$_{80}$ complex has a small HOMO-LUMO gap, which suggests its high reactivity, whereas the P$_4$C$_4$@B$_{76}$C$_4$ turns out to be endothermic with a positive formation energy of ∼ 24 kcal/mol.

## 14.4 Conclusion

$B_{80}$ is isoelectronic to $C_{60}$. The cap atoms transfer electrons from hybrid $sp^2$ orbitals to the boron atoms on the truncated icosahedral frames. The vibrational $g_g$ mode is a major component of the distortion vector under PJTE from $I_h$ to $T_h$ symmetry. The symmetrical substitution of cap boron by methyne groups leads to two viable endohedral boron buckyballs, *endo*-76 and *endo*-72. The formation of boron 4-centre bond appears to be a hallmark of stability. The observed pseudo-Jahn-Teller distortion and the substitutional symmetry breaking by methyne insertion thus point to the same electronic relaxation mechanism. Cap boron atoms are hard acids and prefer to react with hard bases. Mixed N and P atoms in tetrahedral and cubane clusters form thermodynamically stable endohedral complexes. The volume of the $B_{80}$ hollow is rather sufficient to store small molecules, and to perform some reactions inside. From a catalytic point of view, the cavity inside $B_{80}$ proves to be a special micro-environment, which can activate unusual bonding phenomena.

**Acknowledgments** The authors thank the Fund for Scientific Research Flanders (FWO) for the continued support.

## References

1. Grimes RN (2004) J Chem Educ 81:657
2. Jemmis ED, Prasad DLVK (2008) Curr Sci 95:1277
3. Szwacki NG, Sadrzadeh A, Yakobson BI (2007) Phys Rev Lett 98:166804
4. Kroto HW (1987) Nature 329:529
5. Tang H, Ismail-Beigi S (2007) Phys Rev Lett 99:115501
6. Gopakumar G, Nguyen MT, Ceulemans A (2008) Chem Phys Lett 450:175
7. Szwacki NG, Tymczak CJ (2010) Chem Phys Lett 494:80
8. Wang X (2010) Phys Rev B 82:153409
9. Li H,Shao N, Shang B, Yuan L-F, Yang J, Zeng XC (2010) Chem Commun 46:3878
10. Zhao J, Wang L, Li F, Chen Z (2010) J Phys Chem A 114:9969
11. Liu AY, Zope RR, Pederson MR (2008) Phys Rev B 78:155422
12. Boustani I, Karna SP (2010) J Phys Chem C 114:4149
13. Botti S, Castro A, Lanthiotakis NL, Andrade X, Marques MAL (2009) Phys Chem Chem Phys 11:4523
14. Li M, Li Y, Zhou Z, Shen P, Chen Z (2009) Nano Lett 9:1944
15. Wu G, Wang J, Zhang X, Zhu L (2009) J Phys Chem C 113:7052
16. Ceulemans A, Muya JT, Gopakumar G, Nguyen MT (2008) Chem Phys Lett 461:226
17. Ceulemans A, Fowler PW (1992) J Chem Soc Faraday Trans 88:2797
18. Muya JT, Nguyen MT, Ceulemans A (2009) Chem Phys Lett 483:101
19. Exner K, Schleyer PvR (2008) Science 290:1937
20. Ito K, Chen Z, Corminboeuf C, Wannere CS, Zhang XH, Li QS, Schleyer PvR (2007) J Am Chem Soc 129:1510
21. Muya JT, Lijnen E, Nguyen MT, Ceulemans A (2011) J Phys Chem A 115:2268

# Part IV
# Conical Intersections and Interplay Between Jahn-Teller Coupling and Spin-Orbit Coupling: Theory and Manifestations in Magnetism and Spectroscopy

# Chapter 15
# Adiabatic to Diabatic Transformation and Nuclear Dynamics on Diabatic Hamiltonian Constructed by Using *Ab Initio* Potential Energy Surfaces and Non-adiabatic Coupling Terms for Excited States of Sodium Trimer

**Amit Kumar Paul, Somrita Ray, and Satrajit Adhikari**

**Abstract** The non-adiabatic coupling terms (NACTs) among the electronic states $2^2E'$ and $1^2A_1'$ of $Na_3$ system demonstrate the numerical validity of so called "Curl Condition" and thus such states closely form a sub-Hilbert space. For this subspace, we employ the NAC terms to solve the "Adiabatic–Diabatic Transformation (ADT)" equations to obtain the functional form of the transformation angles and pave the way to construct the continuous and single valued diabatic potential energy surface matrix. Nuclear dynamics has been carried out on those diabatic surfaces to reproduce the experimental spectrum for system **B** of $Na_3$ cluster and thereby, to explore the numerical validity of the theoretical development on beyond Born–Oppenheimer approach for adiabatic to diabatic transformation.

## 15.1 Introduction

The adiabatic Potential Energy Surfaces (PESs) and the Non-Adiabatic Coupling (NAC) terms are the outcomes of the Born–Oppenheimer (BO) treatment, which introduces the distinction between the fast moving electrons and the slow moving nuclei [1, 2]. Such NAC terms along with adiabatic PESs are the origin for the driving forces to govern the motion of the atoms in molecular system. If a molecular process is assumed to occur "exclusively" on the ground adiabatic surface, the relevant Schrödinger equation (SE) within the *BO approximation* is essentially dictated only by Coulombic interactions and is expected to provide enough accurate solutions for the observables – such as reactive /non-reactive cross sections or spectroscopic quantities. On the contrary, even if a molecular process "occurs" on the ground state, the excited electronic states can affect the ground state dynamics very strongly due to the presence of so called "non-adiabatic coupling" terms and

---

A.K. Paul • S. Ray • S. Adhikari (✉)
Indian Association for the Cultivation of Science, Jadavpur, Kolkata - 700032, India
e-mail: pcakp@iacs.res.in; pcsr@iacs.res.in; pcsa@iacs.res.in

M. Atanasov et al. (eds.), *Vibronic Interactions and the Jahn-Teller Effect:*
*Theory and Applications*, Progress in Theoretical Chemistry and Physics 23,
DOI 10.1007/978-94-007-2384-9_15, © Springer Science+Business Media B.V. 2012

the *BO approximate* equation fails to calculate correct transition probabilities no matter how much sophisticated numerical algorithm is being used for its solution. The necessity to include the effects of one or more than one electronic states on the other state could be important while studying the molecular processes, namely, in case of charge transfer [3, 4], the higher electronic states extensively interfere the ground state or vice versa.

Longuet-Higgins (LH) [5, 6] and colleagues [7] discovered one of the most fundamental features in molecular physics related to the BO adiabatic eigenfunctions. While surrounding a point of degeneracy in the Configuration Space (CS), these functions may acquire a phase and flip their sign. Herzberg and Longuet-Higgins (HLH) [8] demonstrated this feature of multi-valuedness explicitly for the Jahn–Teller (JT) Conical Intersection (CI) model and corrected the deficiency by multiplying a complex phase factor, known as Longuet-Higgins phase, leading to a single valued wavefunction. Such 'modification' of the electronic adiabatic eigenfunctions is not an outcome of any first principle based theory but imposed in an *ad hoc* manner. M. Baer derived [9] a set of differential equations to obtain the ADT matrix elements for a tri-atomic system and suggested to solve those equations into their integral form along a two-dimensional contour. Moreover, a necessary and sufficient condition was derived that guarantees the existences and uniqueness of the solution of those integral equation along a contour line – such condition was latter termed as "Curl Condition" [10].

Mead and Truhlar [11] (MT) introduced a vector potential in the nuclear Hamiltonian to generalize the single surface ordinary BO equation, which is reminiscent of the complex phase factor treatment of HLH. Moreover, the LH phase had received a tremendous boost with the exposure of the novel adiabatic phase termed as topological (Berry) phase – a natural feature of a system containing fast moving part(s) (e.g. electron) driven by a slow moving field(s) (e.g. nuclei).

The development of any first principle based theory by considering BO treatment includes the fact that slow-moving nuclei is distinguishable from fast-moving electrons in molecular systems and intends to impose the *BO approximation* by neglecting the effect of upper electronic state(s) on the lower with the implication – the non-adiabatic coupling (NAC) elements are negligibly small. Such approximation has been assumed to be independent of the eigenspectrum of the system and thereby, the ordinary BO equations are being frequently used for calculations even for systems with large NAC terms. If the components of the total wavefunction on the upper electronic states are negligibly small at enough low energies, the products between the large NAC terms and the amplitudes of the excited state(s) wavefunctions could be finite in magnitude leading to the breakdown of *BO approximation*. Moreover, the Hellmann–Feynman theorem [12] and its extension by Epstein [13] predicts the presence of singularity in the NAC terms and these singularities arise due to the fact that electronic states are degenerate at certain points or along a line (seam) in the configuration space [14]. Baer and Englman [15] pointed out the LH phase as identical with ADT angle upto an additive constant for two electronic state Hilbert space and presented their version of approximate beyond BO single surface equation in terms of NAC elements, whereas Varandas and Xu [16] derived the approximate extended BO equation for two state system in terms of ADT angle. Baer et al. [17] calculated the transition probabilities of a two arrangement channel Jahn–Teller model by employing their formulated beyond Born–Oppenheimer single surface equation and have shown the huge effect of NAC terms on reactive cross section. Moreover, Baer et al. introduced [10, 18] the quantization of NAC matrix and formulated the approximate extended BO equation for three state sub-Hilbert space, whereas Sarkar et al. performed [19–22] a generalized approach for beyond BO treatment for any three/four state sub-Hilbert space, demonstrated the validity of "Curl Condition" in terms of ADT angles and presented the necessary condition among the ADT angles to derive the approximate extended BO equation.

Thus, it is a matter of contemporary research how elegantly one can handle the NAC terms instead of neglecting them forcibly. Since the definition of NAC terms appear in the adiabatic representation of SE and those terms are usually very sharp functions of nuclear coordinates with singularity in the configuration space, one need to perform an unitary transformation to obtain the diabatic representation of those SEs, where couplings among the electronic states are slowly varying functions of nuclear coordinates and therefore, the dynamical calculations on the diabatic PESs are numerically accurate and stable. On the other hand, the transformation from adiabatic to diabatic representation of SEs for a given sub-Hilbert space is guaranteed only when the NAC terms being vector fields satisfy the so called Curl Conditions [10]. Therefore, the nature of the Curls of the NAC terms is a crucial aspect to explore in order to carry out the first principle based theoretical development on BO treatment.

In this article, we perform calculations in two different ways to explore the numerical validity of beyond Born–Oppenheimer theories, which exploits theoretical means to construct diabatic PESs starting from adiabatic PESs and NAC terms. At first, we consider a three state model diabatic Hamiltonian matrix representing the excited states ($2^2E'-1^2A_1'$) of $Na_3$ cluster [23] at the pseudo-Jahn–Teller (PJT) model situation and calculate [24] the non-adiabatic coupling elements, show the validity of Curl Conditions, evaluate the ADT angles and transform back to the starting diabatic potential matrix. Secondly, we carry out *ab initio* calculations to evaluate the adiabatic PESs and NAC terms, solve Adiabatic–Diabatic transformation equations to obtain [25] the ADT angles, demonstrate the validity of Curl Condition and construct the diabatic PESs as functions of nuclear coordinates. Finally, we have persuaded the nuclear dynamics on those diabatic surfaces to reproduce the experimental photoabsorption spectrum [26,27] for system **B** of $Na_3$ cluster.

## 15.2 Theoretical Development on the BO Treatment of Sub-Hilbert Space

The details of the theoretical development was discussed elsewhere [19–22]. Therefore, only a brief discussion is presented here.

Since we assume these three states as either decoupled or approximately decoupled from rest of the states of a molecular system, the BO expansion of the wavefunction for this subspace of the Hilbert space along with the total electron – nuclei Hamiltonian in the adiabatic representation is presented as:

$$\Psi(\mathbf{n}, \mathbf{e}) = \sum_{i=1}^{3} \psi_i(\mathbf{n}) \xi_i(\mathbf{e}, \mathbf{n}),$$

$$\hat{H} = \hat{T}_n + \hat{H}_e(\mathbf{e}, \mathbf{n}),$$

$$\hat{T}_n = -\frac{\hbar^2}{2m} \sigma_n \nabla_n^2,$$

$$\hat{H}_e(\mathbf{e}, \mathbf{n}) \xi_i(\mathbf{e}, \mathbf{n}) = u_i(\mathbf{n}) \xi_i(\mathbf{e}, \mathbf{n}), \tag{15.1}$$

where the eigenfunction $[\xi_i(\mathbf{e}, \mathbf{n})]$ of the electronic Hamiltonian, $\hat{H}_e(\mathbf{e}, \mathbf{n})$, is defined by the sets of nuclear $(\mathbf{n})$ and electronic $(\mathbf{e})$ coordinates with nuclear coordinate dependent eigenvalue, $u_i(\mathbf{n})$. $\hat{T}_n$ is the nuclear kinetic energy (KE) operator and the expansion coefficient, $\psi_i(\mathbf{n})$, shall appear as nuclear wavefunction.

The kinetically coupled nuclear Schrödinger equations (adiabatic representation) for a three state sub-Hilbert space can be written as:

$$-\frac{\hbar^2}{2m} \begin{pmatrix} \nabla & \tau_{12} & \tau_{13} \\ -\tau_{12} & \nabla & \tau_{23} \\ -\tau_{13} & -\tau_{23} & \nabla \end{pmatrix}^2 \begin{pmatrix} \psi_1 \\ \psi_2 \\ \psi_3 \end{pmatrix} + \begin{pmatrix} u_1 - E & 0 & 0 \\ 0 & u_2 - E & 0 \\ 0 & 0 & u_3 - E \end{pmatrix} \begin{pmatrix} \psi_1 \\ \psi_2 \\ \psi_3 \end{pmatrix} = 0, \tag{15.2}$$

where the NAC matrix $[\tau (\equiv \tau_{ij}^{(1)} = \langle \xi_i(\mathbf{e}, \mathbf{n}) | \nabla \xi_j(\mathbf{e}, \mathbf{n}) \rangle)]$ is defined as,

$$\tau = \begin{pmatrix} 0 & \tau_{12} & \tau_{13} \\ -\tau_{12} & 0 & \tau_{23} \\ -\tau_{13} & -\tau_{23} & 0 \end{pmatrix}. \tag{15.3}$$

Since the three states constitute the sub-Hilbert space (i.e., a complete space for the present case), it is possible to transform ($\Psi = \mathbf{A}\Psi^d$) the adiabatic nuclear SE (Eq. 15.2) to the diabatic one as below,

$$\begin{pmatrix} -\frac{\hbar^2}{2m}\nabla^2 - E & 0 & 0 \\ 0 & -\frac{\hbar^2}{2m}\nabla^2 - E & 0 \\ 0 & 0 & -\frac{\hbar^2}{2m}\nabla^2 - E \end{pmatrix} \begin{pmatrix} \psi_1^d \\ \psi_2^d \\ \psi_3^d \end{pmatrix} + \begin{pmatrix} W_{11} & W_{12} & W_{13} \\ W_{21} & W_{22} & W_{23} \\ W_{31} & W_{32} & W_{33} \end{pmatrix} \begin{pmatrix} \psi_1^d \\ \psi_2^d \\ \psi_3^d \end{pmatrix} = 0, \tag{15.4}$$

15 Adiabatic to Diabatic Transformation and Nuclear Dynamics...

where $\mathbf{W} = \mathbf{A}^\dagger \mathbf{U} \mathbf{A}$ with $U_{ij} = u_i \delta_{ij}$, under the condition:

$$\nabla \mathbf{A} + \tau \mathbf{A} = 0. \tag{15.5}$$

This equation was first formulated elsewhere [9], and is known as Adiabatic–Diabatic Transformation (ADT) condition. In order to obtain a meaningful solution of Eq. 15.5, we have to ensure that the chosen form of $\mathbf{A}$ matrix has the following features: (a) It is orthogonal at any point in configuration space; (b) Its elements are cyclic functions with respect to a parameter, i.e., starting with an unit diagonal matrix, the chosen form of $\mathbf{A}$ matrix has to generate a diagonal matrix with even number of $(-1)$s after completing the cycle.

In case of three-dimensional Hilbert space, there are nine elements in the ADT matrix $(\mathbf{A})$. Since the model form of $\mathbf{A}$ has to be an orthogonal matrix and the ortho-normality conditions demand the fulfillment of six relations, three independent variables namely Euler like angles of rotation $[\theta_{12}(\mathbf{n}), \theta_{23}(\mathbf{n})$ and $\theta_{13}(\mathbf{n})]$, commonly called ADT angles, are the natural requirement to construct the three state $\mathbf{A}$ matrix by taking the product of three $3 \times 3$ rotation matrices, $\mathbf{A}_{12}(\theta_{12})$, $\mathbf{A}_{23}(\theta_{23})$, and $\mathbf{A}_{13}(\theta_{13})$. Let us define these three $3 \times 3$ rotation matrices $[\mathbf{A}_{12}(\theta_{12})$, $\mathbf{A}_{23}(\theta_{23})$, and $\mathbf{A}_{13}(\theta_{13})]$ and one of the ways of their product $(\mathbf{A})$ as:

$$\mathbf{A}(\theta_{12}, \theta_{23}, \theta_{13}) = \mathbf{A}_{12}(\theta_{12}) \cdot \mathbf{A}_{23}(\theta_{23}) \cdot \mathbf{A}_{13}(\theta_{13})$$

$$= \begin{pmatrix} \cos\theta_{12} & \sin\theta_{12} & 0 \\ -\sin\theta_{12} & \cos\theta_{12} & 0 \\ 0 & 0 & 1 \end{pmatrix} \begin{pmatrix} 1 & 0 & 0 \\ 0 & \cos\theta_{23} & \sin\theta_{23} \\ 0 & -\sin\theta_{23} & \cos\theta_{23} \end{pmatrix} \begin{pmatrix} \cos\theta_{13} & 0 & \sin\theta_{13} \\ 0 & 1 & 0 \\ -\sin\theta_{13} & 0 & \cos\theta_{13} \end{pmatrix}$$

$$= \begin{pmatrix} \cos\theta_{12}\cos\theta_{13} & \sin\theta_{12}\cos\theta_{23} & \cos\theta_{12}\sin\theta_{13} \\ -\sin\theta_{12}\sin\theta_{13}\sin\theta_{23} & & +\sin\theta_{12}\cos\theta_{13}\sin\theta_{23} \\[2mm] -\sin\theta_{12}\cos\theta_{13} & \cos\theta_{12}\cos\theta_{23} & -\sin\theta_{12}\sin\theta_{13} \\ -\cos\theta_{12}\sin\theta_{13}\sin\theta_{23} & & +\cos\theta_{12}\cos\theta_{13}\sin\theta_{23} \\[2mm] -\sin\theta_{13}\cos\theta_{23} & -\sin\theta_{23} & \cos\theta_{13}\cos\theta_{23} \end{pmatrix}.$$

$$\tag{15.6}$$

When we substitute the above model form of $\mathbf{A}$ matrix Eq. 15.6 and the anti-symmetric form of $\tau$ matrix Eq. 15.3 in Eq. 15.5, the simple manipulation as performed by Top and Baer [25] leads to the following equations for ADT angles:

$$\nabla\theta_{12} = -\tau_{12} + \tan\theta_{23}(\tau_{13}\cos\theta_{12} - \tau_{23}\sin\theta_{12}), \tag{15.7a}$$

$$\nabla\theta_{23} = -(\tau_{13}\sin\theta_{12} + \tau_{23}\cos\theta_{12}), \tag{15.7b}$$

$$\nabla\theta_{13} = -\frac{1}{\cos\theta_{23}}(\tau_{13}\cos\theta_{12} - \tau_{23}\sin\theta_{12}), \tag{15.7c}$$

which in turn brings the explicit form of $\tau$ matrix elements in terms of ADT angles:

$$\tau_{12} = -\nabla\theta_{12} - \sin\theta_{23}\nabla\theta_{13}, \tag{15.8a}$$

$$\tau_{23} = \sin\theta_{12}\cos\theta_{23}\nabla\theta_{13} - \cos\theta_{12}\nabla\theta_{23}, \tag{15.8b}$$

$$\tau_{13} = -\cos\theta_{12}\cos\theta_{23}\nabla\theta_{13} - \sin\theta_{12}\nabla\theta_{23}. \tag{15.8c}$$

Once the adiabatic PESs $u_1$, $u_2$ and $u_3$ and the non-adiabatic coupling elements $\tau_{12}$, $\tau_{23}$ and $\tau_{13}$ are evaluated by using *ab initio* calculation for a particular nuclear configuration, the solution of Eq. 15.7 provides the ADT angles for the same nuclear configuration and those ADT angles can be used to construct the diabatic potential energy matrix (See **Appendix**). On the other hand, if we have the total electron–nuclear Hamiltonian of a molecular system in the diabatic representation, one can calculate the ADT matrix by diagonalizing the **W** matrix Eq. 15.4 and thereby, obtain the NAC elements through Eq. 15.5.

A Curl Condition [10] for each NAC element, $\tau_{ij}$, has been derived and proved to exist for an isolated group of states (sub-Hilbert space) by considering the analyticity of the ADT matrix **A** for a pair of nuclear degrees of freedom,

$$\frac{\partial}{\partial p}\tau_{ij}^q - \frac{\partial}{\partial q}\tau_{ij}^p = (\tau^q\tau^p)_{ij} - (\tau^p\tau^q)_{ij},$$

$$\tau_{ij}^p = \langle \xi_i | \nabla_p \xi_j \rangle, \quad \tau_{ij}^q = \langle \xi_i | \nabla_q \xi_j \rangle, \tag{15.9}$$

where the Curl due to vector product of NAC elements and the analyticity of ADT matrix are given by $C_{ij} = (\tau^q\tau^p)_{ij} - (\tau^p\tau^q)_{ij}$ and $Z_{ij} = \frac{\partial}{\partial p}\tau_{ij}^q - \frac{\partial}{\partial q}\tau_{ij}^p$, respectively in terms of *Cartesian coordinates* $p$ and $q$ with $\nabla_p = \frac{\partial}{\partial p}$ and $\nabla_q = \frac{\partial}{\partial q}$.

Thus, the explicit form of Curl equation in terms of ADT angles for each NAC element is obtained by using Eqs. 15.8 and 15.9 as below:

$$\text{Curl } \tau_{12}^{pq} = C_{12} = Z_{12} = -\cos\theta_{23}(\nabla_q\theta_{23}\nabla_p\theta_{13} - \nabla_p\theta_{23}\nabla_q\theta_{13}), \tag{15.10a}$$

$$\text{Curl } \tau_{23}^{pq} = C_{23} = Z_{23} = \cos\theta_{12}\cos\theta_{23}(\nabla_q\theta_{12}\nabla_p\theta_{13} - \nabla_p\theta_{12}\nabla_q\theta_{13})$$

$$- \sin\theta_{12}\sin\theta_{23}(\nabla_q\theta_{23}\nabla_p\theta_{13} - \nabla_p\theta_{23}\nabla_q\theta_{13})$$

$$+ \sin\theta_{12}(\nabla_q\theta_{12}\nabla_p\theta_{23} - \nabla_p\theta_{12}\nabla_q\theta_{23}), \tag{15.10b}$$

$$\text{Curl } \tau_{13}^{pq} = C_{13} = Z_{13} = \sin\theta_{12}\cos\theta_{23}(\nabla_q\theta_{12}\nabla_p\theta_{13} - \nabla_p\theta_{12}\nabla_q\theta_{13})$$

$$+ \cos\theta_{12}\sin\theta_{23}(\nabla_q\theta_{23}\nabla_p\theta_{13} - \nabla_p\theta_{23}\nabla_q\theta_{13})$$

$$- \cos\theta_{12}(\nabla_q\theta_{12}\nabla_p\theta_{23} - \nabla_p\theta_{12}\nabla_q\theta_{23}). \tag{15.10c}$$

## 15.3 Adiabatic to Diabatic Transformation (ADT) for a Three State Model Hamiltonian of Na₃ System

Cocchini et al. [23] have shown that some of the excited states of Na₃ exhibit three state ($\phi^\pm$ and $\phi^0$) interactions. It appears that the additional states ($1^2A'_1$ and $2^2A'_1$ under $D_{3h}$ symmetry) are rotationally invariant, and show three state interaction, namely, between $1^2A'_1$ and $2^2E'$ states as well as between $2^2A'_1$ and $3^2E'$ states. When the system attain obtuse and acute-angled equilibrium geometries, the vibrationally resolved experimental spectrum for those geometries are termed as Na₃(**B**) and Na₃(**C**) system, respectively. For an equilateral triangular geometry, the normal modes are defined as bending $Q_x$, antisymmetric stretching $Q_y$ and symmetric stretching $Q_s$ (see Fig. 15.1). When the coupling to the symmetric coordinate $Q_s$ is ignored and the electronic Hamiltonian is being expanded in terms of the active modes (complex linear combination of $Q_x$ and $Q_y$, where $Q^+ = Q_x + iQ_y = \rho e^{i\phi}$ and $Q^- = Q_x - iQ_y = \rho e^{-i\phi}$) considering small distortion from the equilateral triangular geometry ($D_{3h}$), the system presumably includes *all the nuclear configurations* leading to the following general form of the diabatic Hamiltonian matrix (3 × 3) for the electron–nuclear BO system [23, 28]:

$$\hat{H}(\rho,\phi) = \hat{T}_N \cdot 1 + \hat{H}_e(\rho,\phi), \qquad \hat{T}_N = -\frac{\hbar^2}{2\mu}\left(\frac{1}{\rho}\frac{\partial}{\partial \rho}\rho\frac{\partial}{\partial \rho} + \frac{1}{\rho^2}\frac{\partial^2}{\partial \phi^2}\right),$$

$$\hat{H}_e(\rho,\phi) = \begin{pmatrix} \frac{\rho^2}{2} & K\rho e^{i\phi} + \frac{1}{2}g\rho^2 e^{-2i\phi} & P\rho e^{-i\phi} + \frac{1}{2}f\rho^2 e^{2i\phi} \\ K\rho e^{-i\phi} + \frac{1}{2}g\rho^2 e^{2i\phi} & \frac{\rho^2}{2} & P\rho e^{i\phi} + \frac{1}{2}f\rho^2 e^{-2i\phi} \\ P\rho e^{i\phi} + \frac{1}{2}f\rho^2 e^{-2i\phi} & P\rho e^{-i\phi} + \frac{1}{2}f\rho^2 e^{2i\phi} & \varepsilon_0 + \frac{d\rho^2}{2} \end{pmatrix},$$

(15.11)

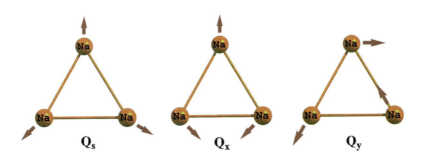

**Fig. 15.1** The three normal modes of Na₃ system, $Q_s$, $Q_y$ and $Q_x$ where $Q_y$ and $Q_x$ are degenerate

which under the unitary transformation matrix,

$$
\hat{U}_e = \begin{pmatrix} \dfrac{1}{2} & -\dfrac{i}{\sqrt{2}} & \dfrac{1}{2} \\[2mm] \dfrac{1}{2} & \dfrac{i}{\sqrt{2}} & \dfrac{1}{2} \\[2mm] -\dfrac{1}{\sqrt{2}} & 0 & \dfrac{1}{\sqrt{2}} \end{pmatrix},
$$

takes the following form:

$$
\hat{H}_e(\rho,\phi) = \begin{pmatrix} \dfrac{\rho^2}{2} + \Delta + \dfrac{1}{2}U_1 - \sqrt{2}W_1 & -\dfrac{U_2}{\sqrt{2}} - W_2 & -\Delta + \dfrac{1}{2}U_1 \\[3mm] -\dfrac{U_2}{\sqrt{2}} - W_2 & \dfrac{\rho^2}{2} - U_1 & -\dfrac{U_2}{\sqrt{2}} + W_2 \\[3mm] -\Delta + \dfrac{1}{2}U_1 & -\dfrac{U_2}{\sqrt{2}} + W_2 & \dfrac{\rho^2}{2} + \Delta + \dfrac{1}{2}U_1 + \sqrt{2}W_1 \end{pmatrix},
$$

$$(15.12)$$

with

$$
U_1 = K\rho\cos\phi + \frac{1}{2}g\rho^2\cos(2\phi),
$$

$$
U_2 = K\rho\sin\phi - \frac{1}{2}g\rho^2\sin(2\phi),
$$

where $K$ and $g$ are taken as the characteristic coupling parameters. $\frac{1}{2}\rho^2$ and $\varepsilon_0 + \frac{1}{2}d\rho^2$ are the energies for the E- and $A_1$-type states, respectively, whereas $U_i$ and $W_i$ ($i = 1,2$) are the interaction potentials among those states. The potentials, $W_i$'s, are of the same form as the $U_i$'s but defined in terms of different set of coupling parameters $P$ and $f$, which replace $K$ and $g$, respectively. If $K$ and $g$ are zeros, the Hamiltonian represents the pseudo Jahn–Teller interaction and the eigenvalues of the matrix takes the following form [23, 28]:

$$
u_1(\rho,\phi) = \frac{1}{2}\left(\varepsilon_0 + \frac{d\rho^2}{2} + \frac{\rho^2}{2}\right) - \sqrt{\Delta^2 + 2w}, \tag{15.13a}
$$

$$
u_2(\rho,\phi) = \frac{\rho^2}{2}, \tag{15.13b}
$$

$$
u_3(\rho,\phi) = \frac{1}{2}\left(\varepsilon_0 + \frac{d\rho^2}{2} + \frac{\rho^2}{2}\right) + \sqrt{\Delta^2 + 2w}, \tag{15.13c}
$$

It is important to note that the coordinate, $\rho$, is a dimensionless variable and values of $K$ and $g$ are taken as 4.90 and 0.035 (dimensionless), respectively as

15 Adiabatic to Diabatic Transformation and Nuclear Dynamics... 289

introduced by Cocchini et al. [23]. The relations, $K = \sqrt{2}P$ and $g = \sqrt{2}f$, have been introduced [23] to define the values of $P$ and $f$, respectively.

With $\Delta = \frac{1}{2}\left(\varepsilon_0 + \frac{d\rho^2}{2} - \frac{\rho^2}{2}\right)$, $w = W_1^2 + W_2^2 = P^2\rho^2 + \frac{1}{4}f^2\rho^4 + Pf\rho^3\cos(3\phi)$ and the columns of the following matrix are the three electronic basis functions:

$$
\mathbf{A} =
\begin{pmatrix}
-\dfrac{(W_1 + W_2)}{\sqrt{2w}} & \dfrac{(W_1 - W_2)}{\sqrt{2w + \{\Delta - \sqrt{\Delta^2 + 2w}\}^2}} & \dfrac{(W_1 - W_2)}{\sqrt{2w + \{\Delta + \sqrt{\Delta^2 + 2w}\}^2}} \\[2em]
\dfrac{(W_1 - W_2)}{\sqrt{2w}} & \dfrac{(W_1 + W_2)}{\sqrt{2w + \{\Delta - \sqrt{\Delta^2 + 2w}\}^2}} & \dfrac{(W_1 + W_2)}{\sqrt{2w + \{\Delta + \sqrt{\Delta^2 + 2w}\}^2}} \\[2em]
0 & \dfrac{\Delta - \sqrt{\Delta^2 + 2w}}{\sqrt{2w + \{\Delta - \sqrt{\Delta^2 + 2w}\}^2}} & \dfrac{\Delta + \sqrt{\Delta^2 + 2w}}{\sqrt{2w + \{\Delta + \sqrt{\Delta^2 + 2w}\}^2}}
\end{pmatrix}.
$$

$$(15.14)$$

Köppel et al. [28, 29] had persuaded two highly relevant investigations on the $Na_3(\mathbf{B})$ system concerning the transition between pseudo-JT and JT model with non-zero energy gap and the details of rovibrational spectrum. Alijah and Baer [30] investigated the same model Hamiltonian on $Na_3$ cluster (Cocchini et al.) by evaluating the electronic basis functions through numerical diagonalization and those basis functions were being used to obtain the non-adiabatic coupling terms and the corresponding ADT angles and thereby, they explore the nature of those quantities. We use the above $3 \times 3$ ADT matrix Eq. 15.14 and evaluate the analytical forms of $\rho$ and $\phi$ components of the NAC matrix elements:

$$
\tau_{12}^{\rho} = -\frac{Pf\rho^2\sin(3\phi)}{[2w(2w + \{\Delta - \sqrt{\Delta^2 + 2w}\}^2)]^{1/2}}, \tag{15.15a}
$$

$$
\tau_{13}^{\rho} = -\frac{Pf\rho^2\sin(3\phi)}{[2w(2w + \{\Delta + \sqrt{\Delta^2 + 2w}\}^2)]^{1/2}}, \tag{15.15b}
$$

$$
\tau_{23}^{\rho} = \frac{\Delta \cdot (2P^2\rho + f^2\rho^3 + 3Pf\rho^2\cos(3\phi))}{2\sqrt{2w}(\Delta^2 + 2w)}, \tag{15.15c}
$$

and

$$
\tau_{12}^{\phi} = \frac{[2P^2\rho^2 - f^2\rho^4 - Pf\rho^3\cos(3\phi)]}{[2w(2w + \{\Delta - \sqrt{\Delta^2 + 2w}\}^2)]^{1/2}}, \tag{15.16a}
$$

$$
\tau_{13}^{\phi} = \frac{[2P^2\rho^2 - f^2\rho^4 - Pf\rho^3\cos(3\phi)]}{[2w(2w + \{\Delta + \sqrt{\Delta^2 + 2w}\}^2)]^{1/2}}, \tag{15.16b}
$$

$$
\tau_{23}^{\phi} = -\frac{\Delta \cdot 3Pf\rho^3\sin(3\phi)}{2\sqrt{2w}(\Delta^2 + 2w)}. \tag{15.16c}
$$

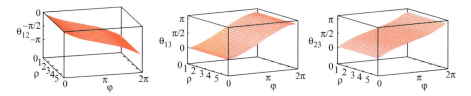

**Fig. 15.2** The functional form of the ADT angles, $\theta_{12}$, $\theta_{13}$ and $\theta_{23}$

With the above analytical form of NAC terms, their Curls [10] appear as:

$$\mathbf{Curl}\ \tau_{12}^{\rho\phi} = \frac{\Delta}{4w(\Delta^2 + 2w)[2w + \{\Delta + \sqrt{\Delta^2 + 2w}\}^2]^{1/2}}$$

$$\times \left[ -3P^2 f^2 \rho^5 \sin^2(3\phi) + \{2P^2\rho^2 - f^2\rho^4 - Pf\rho^3 \cos(3\phi)\} \right.$$

$$\left. \{2P^2\rho + f^2\rho^3 + 3Pf\rho^2 \cos(3\phi)\} \right], \qquad (15.17\text{a})$$

$$\mathbf{Curl}\ \tau_{13}^{\rho\phi} = \frac{\Delta}{4w(\Delta^2 + 2w)[2w + \{\Delta - \sqrt{\Delta^2 + 2w}\}^2]^{1/2}}$$

$$\times \left[ 3P^2 f^2 \rho^5 \sin^2(3\phi) - \{2P^2\rho^2 - f^2\rho^4 - Pf\rho^3 \cos(3\phi)\} \right.$$

$$\left. \{2P^2\rho + f^2\rho^3 + 3Pf\rho^2 \cos(3\phi)\} \right], \qquad (15.17\text{b})$$

$$\mathbf{Curl}\ \tau_{23}^{\rho\phi} = 0. \qquad (15.17\text{c})$$

It is important to note that in case of pseudo-JT model situation for each and every $\rho$ and $\phi$ points in the configuration space: (a) at $\Delta = 0$, *all the Curl elements become identically zero* (**Curl** $\tau = \mathbf{0}$) *even with non-zero $\tau$-components*; (b) at $\Delta \neq 0$, *even with diabatic Hamiltonian, at least two Curl elements show up theoretically non-zero contribution.*

Those NAC terms have then been incorporated in the first order differential equations of ADT angles (15.7) and solved by stiff equation solver method in order to obtain the functional form of such angles. It may be noted that there are six coupled differential equations needed to solve simultaneously to find out the $\rho$ and $\phi$ dependence of the three ADT angles. Figure 15.2 presents those ADT angles, namely, $\theta_{12}$, $\theta_{13}$ and $\theta_{23}$.

Once the ADT angles have been computed, one can obtain the functional form of the diabatic potential energy matrix elements by inserting those angles into

15 Adiabatic to Diabatic Transformation and Nuclear Dynamics... 291

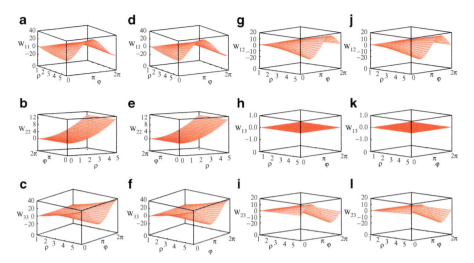

**Fig. 15.3** Comparison of the diabatic PES elements in polar coordinate. Panel (**a–c**), (**g**), (**h**) and (**i**) are evaluated using Eq. 15.12, whereas panel (**d–f**), (**j**), (**k**) and (**l**) are the functional form of Eq. 15.22 (see Appendix). The entire calculation was carried out by taking $\Delta = 0$ under PJT condition

Eq. 15.22 (see **Appendix**). The diabatic potential energy matrix elements which have been generated via ADT angles following Eq. 15.22 matches exactly with the original matrix elements (15.12) while comparing. Figure 15.3 is demonstrating such comparison in polar coordinate.

## 15.4 Ab Initio Calculations of PESs and NAC Terms

The calculations employed a (3s,3p) Gaussian basis set. The basis set has been augmented with a single set of $d$ functions ($\zeta_d = 0.1$) to better describe the electron correlation effects and two diffuse $s$ functions ($\zeta_{4s} = 0.008736$, $\zeta_{5s} = 0.00345$) [23] to improve the excited state description. State Averaged Complete Active Space SCF (SA-CASSCF) calculation has been performed to obtain the PESs of symmetry $2\ ^2E'$ and $1\ ^2A'_1$. Though the calculation has been employed with 33 electrons spread over 27 orbitals, 15 out of that 27 orbitals are chosen as closed shell and the remaining three(3) electrons are distributed over 12 active orbitals. The nonadiabatic coupling matrix elements have been evaluated by using the numerical methods called DDR. All those above performances were executed by MolPro quantum chemistry package [31].

In this article we present the calculated results only at the stretching coordinate ($Q_s$) equal to 3.4Å with other two vibrational coordinate ($Q_x$ and $Q_y$) ranging from

$(-0.9\text{Å})$ to $(0.9\text{Å})$. Those three normal coordinates are related to the Cartesian coordinate by the following orthogonal transformation:

$$
\begin{pmatrix} x_1 \\ y_1 \\ z_1 \\ x_2 \\ y_2 \\ z_2 \\ x_3 \\ y_3 \\ z_3 \end{pmatrix} = \frac{1}{2\sqrt{3}} \begin{pmatrix} -2 & -2 & 0 \\ 0 & 0 & 2 \\ 0 & 0 & 0 \\ 1 & 1 & \sqrt{3} \\ -\sqrt{3} & \sqrt{3} & -1 \\ 0 & 0 & 0 \\ 1 & 1 & -\sqrt{3} \\ \sqrt{3} & \sqrt{3} & -1 \\ 0 & 0 & 0 \end{pmatrix} \begin{pmatrix} Q_s \\ Q_x \\ Q_y \end{pmatrix}.
\tag{15.18}
$$

The above transformation was used to evaluate the Cartesian coordinates of the three Na atoms for each value of the normal modes coordinates followed by the evaluation of x/y $(= Q_x/Q_y)$ component of the NAC matrix elements.

Figure 15.4 presents the *ab initio* PESs for $2\,{}^2E'$ and $1^2A'_1$ states. In Fig. 15.4a, the lower surface of $2^2E'$ is presented for a larger range of $Q_x$ and $Q_y$, namely, from $-0.8\text{Å}$ to $0.8\text{Å}$ with $Q_s = 3.4\text{Å}$, where a kind of "Mexican hat" feature is visible. On the other hand, in Fig. 15.4b, c, we show both the lower and upper surface of $2^2E'$ symmetry for a smaller range of $Q_x$ and $Q_y$ $(-0.15\text{Å}$ to $0.15\text{Å})$ and locate four conical intersections between the surfaces of same symmetry. It is worth to mention that among those four CIs, the CI at $\{Q_x,Q_y\} = 0$ is of $D_{3h}$ symmetry and the other three are either of $C_{2v}$ (at $\{Q_x,Q_y\} \neq 0$) or $C_s$ symmetry (at $Q_x = 0$ and $Q_y \neq 0$). The surface of $1^2A'_1$ symmetry simply appears like the two dimensional Harmonic Oscillator (Fig. 15.4d).

Though the PESs of $2\,{}^2E'$ symmetry (Figs. 15.4b, c) indicate the possibility of the existence of four conical intersections, the (1,2) elements of non-adiabatic elements could bring even more clear understanding on those four points. Figure 15.5 presents the x and y components of *ab initio* NAC terms where panel (a) and (d) are for $\tau_x^{12}$ and $\tau_y^{12}$, respectively. The calculated magnitudes of those NAC terms demonstrate the same possibility with the four conical intersections between two surfaces of $2\,{}^2E'$ symmetry. The functional forms of the other NAC elements are displayed in rest of the panels [(b), (c), (e) and (f)] as discussed in the figure caption.

Those *ab initio* NAC elements are then incorporated into the first order differential equations of ADT angles (15.7) and solved by stiff equation solver technique to obtain the function form of the ADT angle, viz, $\theta_{12}$, $\theta_{13}$ and $\theta_{23}$. When we find the ADT angles, it is easy to construct the diabatic potential energy matrix elements through Eq. 15.22. Figure 15.6 presents those diabatic PESs in polar coordinates.

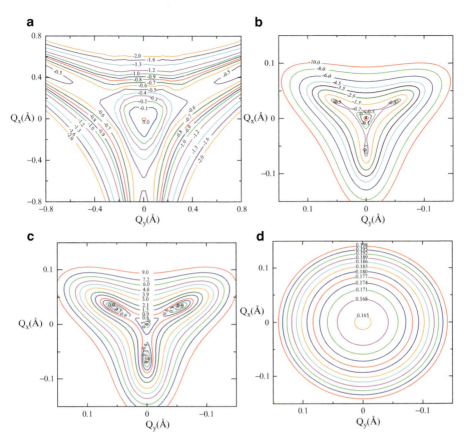

**Fig. 15.4** The *ab initio* $2\,^2E'$ and $1^2A'_1$ PESs at $Q_s = 3.4$ Å. Panel (**a**) presents the lower state of $2^2E'$ symmetry in region $Q_x$ and $Q_y$ equal to $-0.8$Å to $0.8$Å. Panels (**b**), (**c**) and (**d**) refer to lower state of $2\,^2E'$, higher state of $2\,^2E'$ symmetry and $1^2A'_1$ state, respectively within the range $-0.15$Å to $0.15$Å. All embedded energy values are written in eps unit with factors $10^{-1}$ for panel (**a**), $10^{-3}$ for panel (**b**), and $10^{-3}$ for panel (**c**)

## 15.5 Nuclear Dynamics

The dynamics has been performed on those diabatic surfaces constructed by using the *ab initio* NAC terms and adiabatic PESs as described in **Appendix** (15.22).

We consider a Gaussian wavepacket (GWP) as the time-independent solution of the Hamiltonian ($\hat{\mathbf{H}}(\rho,\phi) = \hat{\mathbf{T}}_\mathbf{N} \cdot \mathbf{1} + \frac{1}{2}\rho^2$) and perform the Adiabatic–Diabatic transformation with the ADT matrix to obtain diabatic wavefunctions with different initial position of the GWP in the column matrix. As for example, the diabatic wavefunction starting with the initial wavefunction located on the ground state is given by: $\psi^{dia} = A\psi^{adia}$ with $\psi^{adia} = (\text{GWP}, 0, 0)$.

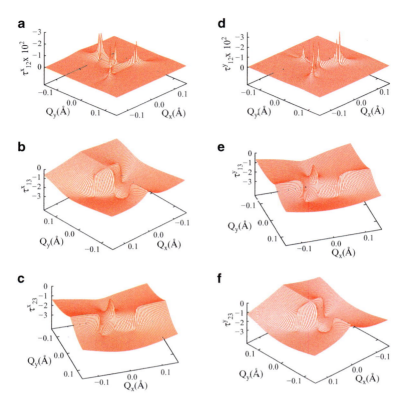

**Fig. 15.5** The x and y ($Q_x$ and $Q_y$) components of NAC elements. Panel (**a**) and (**d**) refers to $\tau_x^{12}$ and $\tau_y^{12}$, panel (**b**) and (**e**) refers to $\tau_x^{13}$ and $\tau_y^{13}$ and panel (**c**) and (**f**) refers to $\tau_x^{23}$ and $\tau_y^{23}$ elements, respectively

The nuclear wavefunction obtained by FFT-Lancozs method is used to calculate the autocorrelation function ($C(t)$) and the Fourier transform of $C(t)$ gives photoabsorption spectra,

$$I(\omega) \propto \omega \int_{-\infty}^{\infty} C(t) \exp(i\omega t) dt, \qquad (15.19)$$

where

$$C(t) = \langle \Psi(t) | \Psi(0) \rangle, \qquad (15.20)$$

$$= \left\langle \Psi^\star \left( \frac{t}{2} \right) \middle| \Psi \left( \frac{t}{2} \right) \right\rangle. \qquad (15.21)$$

Equation 15.21 is more accurate, computationally faster and convenient to implement than the previous equation (15.20). On the other hand, Eq. 15.21 is valid only if the initial wavefunction is real and the Hamiltonian is symmetric.

# 15 Adiabatic to Diabatic Transformation and Nuclear Dynamics...

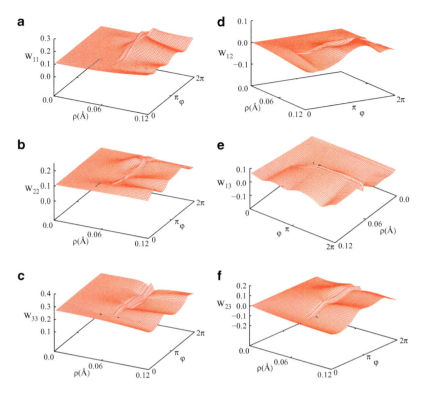

**Fig. 15.6** The diagonal and off-diagonal elements of diabatic potential energy matrix. Panel (**a**), (**b**) and (**c**) are for the diagonal (1,1), (2,2) and (3,3) elements, whereas panels (**d**), (**e**) and (**f**) refer to (1,2), (1,3) and (2,3) elements, respectively

Finally, Fig. 15.7 presents the direct comparison of the calculated photo absorption spectra with two experimental outcome. One of the experimental spectra was evaluated by G. Delacrétaz and L. Wöste considering [26] the resonant two photon ionization (TPI) technique, while the other was obtained by W. E. Ernst and O. Golonzka [27] through *cw* laser excitation followed by laser ionization and mass selective detection in a molecular beam experiment.

## 15.6 Summary

In $2^2E'$ states of $X_3$ JT system, CIs of symmetry other than the conventional $D_{3h}$ have been addressed in the literature both for model and *ab initio* calculations. J.W. Zwanziger and E.R. Grant [32] performed a model calculation on linear plus quadratic $E \otimes e$ JT system and predicted three more CIs (apart from that at the origin) at three equivalent points with a fixed radius. While working with $Li_3$ system,

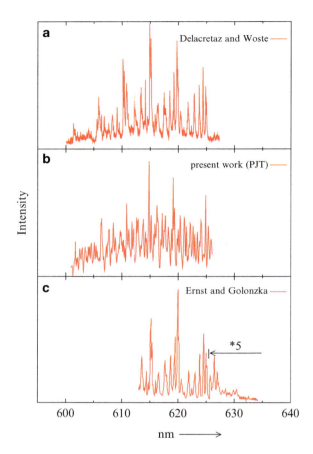

**Fig. 15.7** The theoretical photoabsorption spectra under PJT condition of the present work (panel [**b**]) in comparison with two experimental spectra (Taken from Ref. [26] (panel [**a**]) and Ref. [27] (panel [**c**]))

R. G. Sadygov and D. R. Yarkony have located [33] those CIs either in $D_{3h}$ or $C_{2v}$ symmetry. Bersuker et al. reported [34] the four-point conical intersection between the $2^2E'$ states of $Na_3$ system considering a model Hamiltonian. We have also presented and confirmed [35] the existence of same number of conical intersections with $D_{3h}$, $C_{2v}$ and $C_s$ symmetry through beyond Born–Oppenheimer theories, namely, sign changes of the diagonal element of adiabatic to diabatic transformation matrix and Longuet-Higgings' theorem.

In this article, at first, we have demonstrated the workability of Adiabatic–Diabatic transformation equation (15.5) considering the model Hamiltonian by Cocchini et al. under pseudo Jahn–Teller situation and have shown that each element of the diabatic potential energy matrix in Eq. 15.22 exactly reproduces the corresponding element of the model Hamiltonian (15.12) (see Fig. 15.3). Then we have evaluated the *ab initio* Non-Adiabatic coupling terms among the states, $2\ ^2E'$ and $1^2A'_1$ along with the PESs of the same states. Those NAC terms are incorporated into Eq. 15.7 and solved to obtain the ADT angles which are finally used to calculate the diabatic potential energy matrix elements through Eq. 15.22 (see

# 15 Adiabatic to Diabatic Transformation and Nuclear Dynamics...

Fig. 15.6). The nuclear dynamics on those diabatic surfaces provided encouraging results (Fig. 15.7) for system **B** of $Na_3$ with peak-by-peak matching to the existing experimental spectra. Such *ab initio* calculation, ADT and nuclear dynamics explore the numerical validity of the theoretical development on beyond Born–Oppenheimer approach.

## Appendix

In case of three-state sub-Hilbert space, while transforming the adiabatic Hamiltonian matrix to the diabatic representation, the potential matrix elements appear in terms of ADT angles as follows:

$$
\begin{aligned}
W_{11} = {} & \frac{1}{4}\left[u_3 + \frac{3}{2}(u_1 + u_2)\right] + \frac{1}{8}(u_1 - u_2)\cos 2\theta_{12} + \frac{1}{4}\left[\frac{1}{2}(u_1 + u_2) - u_3\right]\cos 2\theta_{13} \\
& - \frac{1}{4}\left[\frac{1}{2}(u_1 + u_2) - u_3\right]\cos 2\theta_{23} + \frac{3}{8}(u_1 - u_2)\cos 2\theta_{12}\cos 2\theta_{13} \\
& + \frac{1}{4}\left[\frac{1}{2}(u_1 + u_2) - u_3\right]\cos 2\theta_{13}\cos 2\theta_{23} + \frac{1}{8}(u_1 - u_2)\cos 2\theta_{12}\cos 2\theta_{23} \\
& - \frac{1}{8}(u_1 - u_2)\cos 2\theta_{12}\cos 2\theta_{13}\cos 2\theta_{23} - \frac{1}{2}(u_1 - u_2)\sin 2\theta_{12}\sin 2\theta_{13}\sin \theta_{23},
\end{aligned}
\tag{15.22a}
$$

$$
\begin{aligned}
W_{22} = {} & \frac{1}{2}\left[u_3 + \frac{1}{2}(u_1 + u_2)\right] - \frac{1}{4}(u_1 - u_2)\cos 2\theta_{12} + \frac{1}{2}\left[\frac{1}{2}(u_1 + u_2) - u_3\right]\cos 2\theta_{23} \\
& - \frac{1}{4}(u_1 - u_2)\cos 2\theta_{12}\cos 2\theta_{23},
\end{aligned}
\tag{15.22b}
$$

$$
\begin{aligned}
W_{33} = {} & \frac{1}{4}\left[u_3 + \frac{3}{2}(u_1 + u_2)\right] + \frac{1}{8}(u_1 - u_2)\cos 2\theta_{12} - \frac{1}{4}\left[\frac{1}{2}(u_1 + u_2) - u_3\right]\cos 2\theta_{23} \\
& - \frac{3}{8}(u_1 - u_2)\cos 2\theta_{12}\cos 2\theta_{13} - \frac{1}{4}\left[\frac{1}{2}(u_1 + u_2) - u_3\right]\cos 2\theta_{13}\cos 2\theta_{23} \\
& + \frac{1}{8}(u_1 - u_2)\cos 2\theta_{12}\cos 2\theta_{23} + \frac{1}{8}(u_1 - u_2)\cos 2\theta_{12}\cos 2\theta_{13}\cos 2\theta_{23} \\
& + \frac{1}{2}(u_1 - u_2)\sin 2\theta_{12}\sin 2\theta_{13}\sin \theta_{23} - \frac{1}{4}\left[\frac{1}{2}(u_1 + u_2) - u_3\right]\cos 2\theta_{13},
\end{aligned}
\tag{15.22c}
$$

$$W_{12} = W_{21} = \frac{1}{2}\left[u_3 - \frac{1}{2}(u_1 + u_2)\right]\sin\theta_{13}\sin 2\theta_{23}$$

$$+ \frac{1}{2}(u_1 - u_2)\sin 2\theta_{12}\cos\theta_{13}\cos\theta_{23} + \frac{1}{4}(u_1 - u_2)\cos 2\theta_{12}\sin\theta_{13}\sin 2\theta_{23},$$

$$(15.22d)$$

$$W_{13} = W_{31} = -\frac{1}{4}\left[u_3 - \frac{1}{2}(u_1 + u_2)\right]\sin 2\theta_{13} + \frac{3}{8}(u_1 - u_2)\cos 2\theta_{12}\sin 2\theta_{13}$$

$$- \frac{1}{4}[u_3 - (u_1 + u_2)]\sin 2\theta_{13}\cos 2\theta_{23} - \frac{1}{8}(u_1 - u_2)\cos 2\theta_{12}\sin 2\theta_{13}\cos 2\theta_{23}$$

$$+ \frac{1}{2}(u_1 - u_2)\sin 2\theta_{12}\cos 2\theta_{13}\sin\theta_{23},$$

$$(15.22e)$$

$$W_{23} = W_{32} = -\frac{1}{2}\left[u_3 - \frac{1}{2}(u_1 + u_2)\right]\cos\theta_{13}\sin 2\theta_{23}$$

$$+ \frac{1}{2}(u_1 - u_2)\sin 2\theta_{12}\sin\theta_{13}\cos\theta_{23} - \frac{1}{4}(u_1 - u_2)\cos 2\theta_{12}\cos\theta_{13}\sin 2\theta_{23}.$$

$$(15.22f)$$

# References

1. Born M, Oppenheimer JR (1927) Ann Phys (Leipzig) 84:457
2. Born M. Huang K (1954) Dynamical theory of crystal lattices. Oxford University Press, New York
3. Baer M, Niedner-Schatteburg G, Toennies JP (1989) J Chem Phys 91:4169; Baer M, Liao C-L, Xu R, Flesch GD, Nourbaksh S, Ng CY (1990) J Chem Phys 93:4845; Last I, Gilibert M, Baer M (1997) J Chem Phys 107:1451
4. Aguilon F, Sizun M, Sidis V, Billing GD, Markovic N (1996) J Chem Phys 104:4530; Sizun M, Aguilon F, Sidis V, Zenevich V, Billing GD, Markovic N (1996) Chem Phys 209:327; Grimbert D, Sidis V, Cobut V (1998) J Chem Phys 108:6331
5. Longuet-Higgins HC (1961) Adv Spectrosc (NY) 2:429
6. Longuet-Higgins HC (1975) Proc R Soc Lond Ser A 344:147
7. Child MS (2002) Adv Chem Phys 124:1; Manolopoulos DE, Child MS (1999) Phys Rev Lett 82:2223
8. Herzberg G, Longuet-Higgins HC (1963) Discuss Faraday Soc 35:77
9. Baer M (1975) Chem Phys Lett 35:112
10. Baer M (2002) Phys Rep 358:75
11. Mead CA, Truhlar DG (1979) J Chem Phys 70:2284
12. Hellmann H (1937) Einfuhrang in die Quantenchemie. Franz Deutiche, Leipzig, Germany; Feynman R (1939) Phys Rev 56:340
13. Epstein ST (1954) Am J Phys 22:613
14. Baer M, Billing GD (2002) The role of degenerate states in chemistry. Advances in chemical physics, vol 124. Wiley, New York

15. Baer M, Englman R (1992) Mol Phys 75:293
16. Varandas AJC, Xu ZR (2000) J Chem Phys 112:2121
17. Baer R, Charutz D, Kosloff R, Baer M (1996) J Chem Phys 105:9141
18. Baer M (2006) Beyond Born-Oppenheimer: conical intersections and electronic nonadiabatic coupling terms. Wiley, Hoboken
19. Sarkar B, Adhikari S (2006) J Chem Phys 124:074101
20. Sarkar B, Adhikari S (2007) Indian J Phys 81(9):925
21. Sarkar B, Adhikari S (2009) Int J Quan Chem 109:650
22. Sarkar B, Adhikari S (2008) J Phys Chem A 112:9868
23. Cocchini F, Upton TH, Andreoni W (1988) J Chem Phys 88:6068
24. Paul AK, Sardar S, Sarkar B, Adhikari S (2009) J Chem Phys 131:124312; Paul AK, Sarkar B, Adhikari S (2010) In: Chaudhuri RK, Mekkaden MV, Raveendran AV, Narayanan AS (eds) Recent advances in spectroscopy. Astrophysics and space science proceedings. Springer, Berlin, p 63
25. Top ZH, Baer M (1977) J Chem Phys 66:1363
26. Delacrétaz G, Wöste L (1985) Surf Sci 156:770
27. Ernst WE, Golonzka O (2004) Phys Scr T112:27
28. Meiswinkel R, Köppel H (1990) Chem Phys 144:117
29. Mayer M, Cederbaum LS, Köppel H (1996) J Chem Phys 104:8932
30. Alijah A, Baer M (2000) J Phys Chem A 104:389
31. Werner H-J, Knowles PJ et al (2009) MOLPRO, Version 2009.1, a package of ab initio programs
32. Zwanziger JW, Grantt ER (1987) J Chem Phys 87:2954
33. Sadygov RG, Yarkony DR (1999) J Chem Phys 110:3639
34. Koizumi H, Bersuker IB (1999) Phys Rev Letts 83:3009
35. Paul AK, Ray S, Mukhopadhyay D, Adhikari S (2011) Chem Phys Letts. doi:10.1016/j.cplett.2011.03.087

# Chapter 16
# Jahn–Teller Effect and Spin-Orbit Coupling in Heavy Alkali Trimers

**Andreas W. Hauser, Gerald Auböck, and Wolfgang E. Ernst**

**Abstract** Triatomic alkali-metal clusters in their high-spin manifolds of electronically excited states provide the chance to investigate the spectroscopic consequences of the combination of Jahn–Teller effect and spin-orbit coupling with powerful methods of quantum chemistry such as open-shell coupled cluster approaches and multireference Rayleigh-Schroedinger perturbation theory. With respect to available experimental data the $2^4E' \leftarrow 1^4A_2'$ transitions are selected to document the quenching of the paradigmatic $E \otimes e$ Jahn–Teller distortion with increasing spin-orbit coupling. The simulated spectra for potassium, rubidium and cesium trimers are provided together with all relevant parameters such as harmonic frequencies, Jahn–Teller parameters and spin-orbit splittings obtained from the ab initio approach. Beside that, the molecular geometries and formation energies of these van der Waals molecules are also listed in this chapter.

## 16.1 Introduction

Equilateral triangle triatomics are the simplest systems with a threefold axis of symmetry, allowing twofold degenerate E terms. Following the well-known review article of I.Bersuker. [1], $Na_3$ is the most studied species of this type, both theoretically and experimentally, due to its relatively simple electronic structure and existence in the vapour phase. Bersuker once suggested that this molecule may

---

A.W. Hauser (✉) • W.E. Ernst
Institute of Experimental Physics, Graz University of Technology, Petersgasse 16,
A-8010 Graz, Austria
e-mail: andreas.w.hauser@gmail.com; wolfgang.ernst@tugraz.at

G. Auböck
Ecole Polytechnique Fédérale de Lausanne, Institut des sciences et ingénierie chimiques,
CH-1015 Lausanne, Switzerland
e-mail: gerald.aubock@epfl.ch

---

M. Atanasov et al. (eds.), *Vibronic Interactions and the Jahn-Teller Effect:*
*Theory and Applications*, Progress in Theoretical Chemistry and Physics 23,
DOI 10.1007/978-94-007-2384-9_16, © Springer Science+Business Media B.V. 2012

be regarded as a *"probe system of the Jahn–Teller and Pseudo-Jahn–Teller effects which can be used as a base for the treatment of more complicated systems."*

Today, almost 10 years later, we may expand this idea with respect to highly topical challenges, preferably located in the area of relativistic quantum chemistry. Seen from a theoretical point of view, the alkali-metal trimers, now investigated in their high-spin states, still turn out to be perfect benchmark systems: First, an adequate treatment of their weak van der Waals-type bonding requires an extensive inclusion of correlation energy. Second, the high symmetry of the molecules leads to nonadiabatic coupling effects such as Jahn–Teller (JT) distortions in the electronically excited states. Third, the inclusion of heavy atoms necessitates the inclusion of spin-orbit (SO) coupling, which is, as will be shown in this chapter, the crucial factor for the correct interpretation of currently available experimental data.

Based on previous experimental and theoretical investigations of our group [2–5] we restrict the effort in this chapter to one specific electronic transition of $K_3$, $Rb_3$ and $Cs_3$, namely the excitation from the $1A_2'$ state ($D_{3h}$ point group nomenclature), which is the lowest electronic state in the quartet manifold, to the $2^4E'$ state, which is afflicted by an $E \otimes e$ JT distortion. This choice comes in handy for theoreticians as well as experimentalists, since the transitions lie within an experimentally traceable range of 10,000 to 12,000 $cm^{-1}$ and the upper JT states involved are reasonably far apart from other excited states of the same symmetry, which allows to treat them quasi-adiabatically, i.e., in the common picture of a non-adiabatic two-state coupling scheme. Note that the lower states of the chosen transition are not the ground states of the alkali-metal trimers, which may be labeled as $1^2B_2$ states in the $C_{2v}$ point group. They are affected by a static JT distortion leading to an isosceles, obtuse triangle geometry. The first few excited states of the low-spin manifold of $K_3$ are treated in Ref. [6]. A JT analysis of the ground states of $K_3$ and $Rb_3$ can be found in Ref. [7]. The fact that the alkali-metal trimers prefer to be covalently bound in their doublet ground states makes it difficult to obtain experimental spectra of quartet transitions. Here the helium nanodroplet isolation spectroscopy (HENDI) [8] has proven to be a powerful tool for the production of alkali trimers in high-spin states [3].

Beside the aspect of reviewing former results obtained for the $2^4E' \leftarrow 1^4A_2'$ transitions in $K_3$ and $Rb_3$, it is the aim of this work to provide the reader with the first spectra simulations of the same transition in $Cs_3$ at vibrational resolution. This will be done in a similar way to our previous work on $K_3$ and $Rb_3$: The essential parameters are extracted from ab initio results, and the spectrum is then generated within a framework of JT effect theory and SO coupling. Finally, for the sake of an improved overall understanding of the coupling mechanisms, the spectra of all three species, $K_3$, $Rb_3$ and $Cs_3$, are compared.

This chapter is structured as follows: The next section is dedicated to the ab initio computation of the involved potential energy surfaces for $Cs_3$. In Sect. 16.3 a brief overview of the relativistic JT effect theory is given, and parameters for the calculation of the $2^4E' \leftarrow 1^4A_2'$ transition in $Cs_3$ are extracted. The obtained simulated spectra are then presented in Sect. 16.4 and compared to former results for $K_3$ and $Rb_3$. Concluding remarks are given in Sect. 16.5.

## 16.2 The Computation of Potential Energy Surfaces

According to previous calculations for $K_3$ and $Rb_3$ [5] we choose the nine valence-electron basis set ECP46MDF of the Stuttgart/Köln group [9] in uncontracted form for cesium. It has the scheme (12s,11p,5d,3f,2g). Following the recommendation in Ref. [9], we added an effective core-polarization potential (CPP) as demonstrated by Müller et al. [10], with a static dipole polarizability $\alpha_D$ of 0.6935 and a cutoff-parameter $\delta$ of 1.4702.

In a first step, the geometry of $Cs_3$ is optimized in its lowest quartet state. Here we employ a single-reference partially spin restricted open-shell variant of the Coupled Cluster (CC) method with single and double excitations plus perturbative triples [RHF-RCCSD(T)] [11,12]. No frozen shells are used. The equilateral triangle structure has a bond length of 5.793 Å. A comparison with $K_3$ and $Rb_3$ is given in Table 16.1. Bond lengths, molecular formation energies $E_{form}$ and the dissociation energies $E_{diss}$ for the reaction path $X_3 \rightarrow X_2 + X$ are listed. To probe the influence of the CPP we also repeat the optimization without inclusion of the additional core polarization term, obtaining a slightly larger bond length of 5.896 Å ($\approx 2$ percent), but almost the same relative energies. The difference between $E_{form}$ and $E_{diss}$ corresponds to the dissociation energy of the triplet dimer. Our values of 237 (+CPP) and 238 (no CPP) are in reasonable agreement with the experimental value of $279.349\,cm^{-1}$ presented in Ref. [13]. The corresponding equilibrium bond lengths of the dimer are 6.396 (+CPP) and 6.446 Å (no CPP), the experimental value is 6.235 Å. Slightly better theoretical results may be found in Ref. [14], where a tailor-made large-core pseudopotential was combined with a full CI calculation for the two valence electrons. A preliminary study extending this approach to the cesium trimer may be found in Ref. [15].

### 16.2.1 Symmetry-Adapted Internal Coordinates

To provide convenient cuts through the three-dimensional energy surface of the trimers the symmetry-adapted internal coordinates $Q_s$ (breathing mode), $Q_x$

**Table 16.1** Comparison of $K_3$, $Rb_3$ and $Cs_3$ lowest quartet states

| Species | Basis set | Bond length (Å) | E (hartree) | $E_{form}^a$ $(cm^{-1})$ | $E_{diss}^b$ $(cm^{-1})$ |
|---|---|---|---|---|---|
| $K_3$ | ECP10MDF + s,p,d$^c$ | 5.049 | −84.87617 | 1,260 | 1,001 |
| $Rb_3$ | ECP28MDF + s,p,d$^c$ | 5.500 | −72.127286 | 939 | 734 |
| $Cs_3$ | ECP46MDF + CPP | 5.793 | −60.671710 | 1,144 | 907 |
| $Cs_3$ | ECP46MDF | 5.896 | −60.251732 | 1,146 | 908 |

[a]Difference between minimum energy and the sum of the atomic energies calculated at the same level of theory
[b]Dissociation energy for the reaction $X_3 \rightarrow X_2 + X$, calculated at the same level of theory
[c]Augmentation of diffuse functions to improve the description of excited states

(symmetric mode) and $Q_y$ (fully asymmetric mode) are chosen, which describe the planar vibrational modes of the systems in $D_{3h}$ symmetry. They are related to the six in-plane, space-fixed Cartesian coordinates by the equations [6, 16, 17]

$$Q_x = \frac{1}{\sqrt{3}}\left(-x_1 + \left[\frac{1}{2}x_2 + \frac{\sqrt{3}}{2}y_2\right] + \left[\frac{1}{2}x_3 - \frac{\sqrt{3}}{2}y_3\right]\right) \tag{16.1}$$

$$Q_y = \frac{1}{\sqrt{3}}\left(y_1 + \left[\frac{\sqrt{3}}{2}x_2 - \frac{1}{2}y_2\right] + \left[-\frac{\sqrt{3}}{2}x_3 - \frac{1}{2}y_3\right]\right) \tag{16.2}$$

$$Q_s = \frac{1}{\sqrt{3}}\left(-x_1 + \left[\frac{1}{2}x_2 - \frac{\sqrt{3}}{2}y_2\right] + \left[\frac{1}{2}x_3 + \frac{\sqrt{3}}{2}y_3\right]\right) \tag{16.3}$$

In the following sections the involved potential surfaces are plotted as a function of $Q_x$, with $Q_s$ fixed to the energetic minimum value given in Table 16.1. Note that for an equilateral structure the value of $Q_s$ coincides with the bond length of the triangle. $Q_y$ is set to zero, restricting the molecular point group to $C_{2v}$ at least.

## 16.2.2 The $1^4A_2'$ States

For the sake of a direct comparison all potential surface scans are done at the same level of theory. For the static part of the correlation energy the CASSCF method in the version of Werner and Knowles is applied [18–21], followed by a modified version of CASPT2 (complete active space with second-order perturbation theory) to account for dynamic correlation [22]. This method is referred to as "RS2C" in the MOLPRO package [23]. The active space consists of 18 hand-picked orbitals to ensure smooth potential surfaces. According to its internal orbital ordering in the $C_{2v}$ point group ($A_1/B_1/B_2/A_2$) the $1^4A_2'$ state occupation scheme is 5/2/4/1 doubly occupied plus 2/0/1/0 singly occupied orbitals, corresponding to the electron configuration $(6a_1)^1 (7a_1)^1 (5b_2)^1$. The doubly occupied orbitals are fully optimized in the CASSCF approach, but do not form part of the active space. In the follow-up CASPT2 calculation the three lowest orbitals 2/0/1/0, corresponding to the atomic 5s shells, are kept frozen. Energies shifts of 0.2 hartree are applied to avoid intruder state problems [24]. Energies reported here are internally corrected for this shift by MOLPRO.

The obtained potential curve for the $1^4A_2'$ state $Cs_3$ is plotted in Fig. 16.1, together with the results of previous scans of $K_3$ and $Rb_3$ taken from Ref. [5]. The ab initio data points are interpolated with cubic splines. In all three cases an almost harmonic shape of the curve is evident, having its minimum at $Q_x = 0$,

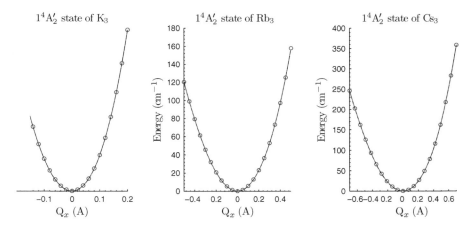

**Fig. 16.1** Potential energy curves at the RS2C level for the lowest quartet states of $K_3$, $Rb_3$ and $Cs_3$, scanned along the $Q_x$ coordinate at $\{Q_s, Q_y\} = \{5.05, 0\}$, $\{5.50, 0\}$ and $\{5.79, 0\}$, respectively. According to the $D_{3h}$ point group nomenclature, which applies as the energetic minimum ($Q_x = 0$), the states are of $A'_2$ symmetry (The *left* and *middle* pictures are reprinted with permission from Ref. [5]. Copyright 2010, American Institute of Physics)

the equilateral triangle structure. Harmonic fits yield the frequencies 33, 21, and 18 cm$^{-1}$ for $\omega^A_{Q_x/Q_y}$. For the energy surface as a function of $Q_s$ (not shown here), we obtain the harmonic frequencies 38, 30 and 18 cm$^{-1}$.

### 16.2.3 The $2^4E'$ States

For the vertical excitation from the $1^4A'_2$ to the $2^4E'$ states we obtain an energy of 10,157 cm$^{-1}$. The reader may locate the positions of these states in Fig. 16.2, which gives an overview of the quartet manifold for $Cs_3$. At $Q_x = 0$, the electronic states are labeled with respect to the $D_{3h}$ point group. The colors and the state labels near the energy axis are chosen to represent the $C_{2v}$ symmetry labels. In Ref [25], where helium nanodroplets were doped with cesium atoms, three peaks at 11,144, 12,265 and 12,585 cm$^{-1}$ in the photoionization spectrum were assigned to quartet transitions of $Cs_3$. According to the results of our $Q_x$-scan shown in Fig. 16.2 we suggest that these transitions may be related to the $3^4E'$ state at 12,460 cm$^{-1}$, which is affected by a Pseudo-Jahn–Teller coupling with a higher $5^4A_1$ state (not shown in the figure). Furthermore, in Ref. [26], transitions at 15,016 and 15,248 cm$^{-1}$ were also assumed to belong to the quartet manifold of $Cs_3$. A tentative inclusion of higher excited states into our ab initio approach yields an E state at 15,155 cm$^{-1}$ ($6^4E'$) with a strong linear JT distortion.

In the picture of single electron excitation the $2^4E'$ state may be described as follows [5, 27]: Using $C_{2v}$ labels, the electronically excited $4^4B_2$ state can be interpreted as the electron promotion $9a_1 \leftarrow 7a_1$, whereas the $3^4A_1$ state can be

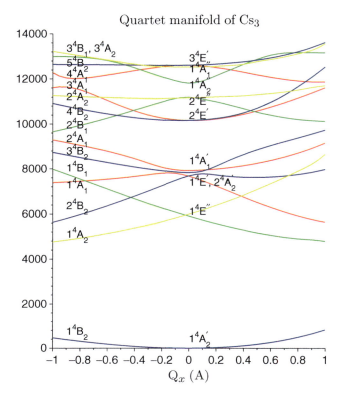

**Fig. 16.2** The lowest quartet states of Cs$_3$, scanned along the $Q_x$ coordinate at $\{Q_s, Q_y\} = \{5.79, 0\}$

seen as 9a$_1$ ← 5b$_2$. At $Q_x = 0$ the orbitals 7a$_1$ and 5b$_2$ are energetically degenerate and form the $2^4E'$ state. This degeneracy is removed with increasing values of $|Q_x|$, causing the paradigmatic E⊗e JT splitting as it is shown in Fig. 16.3 in more detail. Again we compare the result for Cs$_3$ with former scans of K$_3$ and Rb$_3$ [5].

The $4^4B_2$ states are plotted in blue, the $3^4A_1$ states in red. Each pair of states undergoes a conical intersection at $Q_x = 0$. Comparing the different curvatures of the three species one finds that the generally small linear JT coupling is largest for Rb$_3$. The degeneracies are not symmetrically lifted, indicating the influence of higher coupling terms in all three cases.

Since electronic E states can have a nonvanishing projection of their angular momentum onto the C$_3$ axis of the molecule, the pairs of states involved in the E⊗e JT distortion may be also severely affected by SO coupling. To account for this influence, we apply the state-interacting method as implemented in MOLPRO. $H_e + H_{SO}$, the sum of electronic and SO-Hamiltonian, is diagonalized in a basis of eigenfunctions of $H_e$. The matrix elements of $H_{SO}$ are calculated by the ECP-LS technique from the corresponding SO pseudopotentials [9] at the CASSCF level,

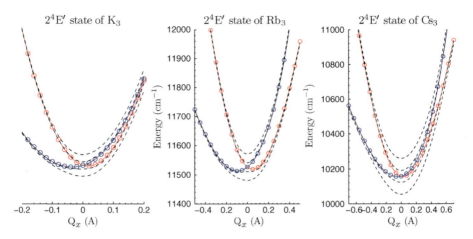

**Fig. 16.3** Potential energy curves at the RS2C level for the electronically excited $2^4E'$ states of $K_3$, $Rb_3$ and $Cs_3$, scanned along the $Q_x$ coordinate, again at $\{Q_s, Q_y\} = \{5.05, 0\}$, $\{5.50, 0\}$ and $\{5.79, 0\}$, respectively. The double degeneracy at $Q_x = 0$ is removed when the symmetry is lowered to $C_{2v}$, leading to pairs of the states $4^4B_2$ (*blue*) and $3^4A_1$ (*red*). After inclusion of SO coupling four potential surfaces are obtained (*black, dashed*), and the conical intersection becomes an avoided crossing (The *left* and *middle* pictures are reprinted with permission from Ref. [5]. Copyright 2010, American Institute of Physics)

but the diagonal entries are replaced by the precomputed, qualitatively better RS2C energies before the diagonalization. The corrected potential surfaces are shown as black dashed lines in Fig. 16.3. In the quartet manifold each pair of states splits up into four curves showing avoided crossings instead of conical intersections. Even in the case of $K_3$, where the SO splitting is smallest, it is still large enough to fully quench the characteristic double minimum curvature of a linear $E \otimes e$ distortion.

## 16.3 The Application of the Relativistic E⊗e Jahn–Teller Effect Theory

The analytical model applied to the $2^4E'$ state is based on the detailed review article of Barckholtz and Miller [28], which describes the inclusion of SO coupling for doublet states [29–31]. JT Hamiltonians for higher spin multiplicities (3, 4 and 5) can be found in Ref. [32]. This section gives only a brief overview, since our approach has been described in great detail elsewhere [5, 7]. We restrict our analysis to the JT active modes $Q_x$ and $Q_y$, since the influence of the total symmetric vibrations on the JT parameters is known to be small for alkali-metal trimers. Their influence on E⊗e JT coupling in general is discussed in Ref. [33].

### 16.3.1 The Relativistic E⊗e Jahn–Teller Effect

The electronic degeneracy of the E state is lifted due to the interaction with the doubly degenerate vibrational e mode. We may write the full Hamiltonian of the system as

$$\hat{H} = \hat{T}_N + \hat{H}_e + \hat{H}_{so} \tag{16.4}$$

where $\hat{T}_N$, $\hat{H}_e$ and $\hat{H}_{SO}$ denote the kinetic energy operator, the electronic Hamiltonian and the SO coupling term, respectively. For the calculation of vibronic spectra we introduce a diabatic vibronic basis set written as

$$|\Lambda \Sigma\rangle|vl\rangle \tag{16.5}$$

in product space notation. The first part contains two quantum numbers $\Lambda$ and $\Sigma$ for the description of the electronic part: $\Lambda$ corresponds to the direction of the electronic angular momentum projection on the $C_3$ axis of the trimer ($\Lambda = \pm 1$), $\Sigma$ to the spin projection ($\Sigma = \pm 3/2, \pm 1/2$). The second part describes the nuclear vibration in a basis of harmonic oscillator eigenfunctions defined by $v = 0, 1, 2, \ldots$ and $l = -v, v+2, \ldots, v-2, v$, the vibration and the vibrational angular momentum quantum numbers, respectively. The vibronic levels are then obtained by diagonalization of $H$ in the chosen basis.

### 16.3.2 Parameter Extraction

For the representation of the Hamiltonian in the chosen basis we refer to Refs. [4, 5, 28] (linear/quadratic JT coupling) and Ref. [7] (anharmonicity), where the matrix elements are listed explicitly. However, for the explanation of our parameter fitting strategy, it is helpful to look at least at the electronic Hamiltonian. It has the form

$$H_e = \begin{pmatrix} V_{2a}r^2 & V_{1e}e^{-i\phi}r + V_{2e}e^{2i\phi}r^2 \\ V_{1e}e^{i\phi}r + V_{2e}e^{-2i\phi}r^2 & V_{2a}r^2 \end{pmatrix} \otimes \mathbf{1}_4 \tag{16.6}$$

where $\mathbf{1}_4$ denotes the $4 \times 4$ unity matrix, since $H_e$ is diagonal in the spin subspace. This analytical expression is obtained from a series expansion of $H_e$ up the second order in complex polar coordinates

$$Q_\pm = (Q_x \pm iQ_y)/\sqrt{2} = re^{\pm i\phi}. \tag{16.7}$$

## 16 JT Effect and SO Coupling in Heavy Alkali Trimers

**Table 16.2** JT Parameters for the $2^4E'$ state of $Cs_3$, based on different expansions of the vibronic Hamiltonian in terms of $r$. SO-splitting $\Delta$ is 34.3 cm$^{-1}$. Uncertainty is given in squared brackets

| Expansion | $V_{2a}$ | $V_{1e}$ | $V_{2e}$ | $V_{3e}$ | $V_{3a}$ |
|---|---|---|---|---|---|
| lin | 1,797 [66] | 329 [41] | – | – | – |
| +quad | 1,797 [38] | 324 [25] | −266 [43] | – | – |
| +cubic | 1,797 [33] | 182 [62] | −261 [37] | 425 [168] | – |
| +anharm | 1,797 [13] | 182 [24] | −261 [15] | 425 [66] | 223 [21] |

The notation of the series expansion terms is taken from Ref. [34]: $V_{2a}$ is the elastic force constant, $V_{1e}$ is the linear and $V_{2e}$ the quadratic JT-coupling parameter. To effectively include SO coupling, we assume a fully diagonal SO Hamiltonian

$$\{H_{SO}\}_{ij} = \left\langle \{\Lambda\Sigma\}_i \middle| H_{SO} \middle| \{\Lambda\Sigma\}_j \right\rangle = \mathscr{A}\zeta\Lambda_i\Sigma_i\delta_{ij}, \tag{16.8}$$

describing the coupling of $\Sigma$ to $\Lambda$. $\mathscr{A}$ is the SO-coupling constant, and will be, together with the angular momentum projection $\zeta$, absorbed into the energy splitting parameter $\Delta = \mathscr{A}\zeta/2$, which we will used to characterize the SO contribution henceforth. The diagonalization of $H_e + H_{SO}$, the potential part of the full vibronic Hamiltonian, yields four eigenvalues:

$$\varepsilon_{\frac{1}{2}\pm} = V_{2a}r^2 \pm \left[\Delta^2 + V_{1e}^2 r^2 + 2V_{1e}V_{2e}\cos(3\phi)r^3 + V_{2e}^2 r^4\right]^{\frac{1}{2}}$$

$$\varepsilon_{\frac{3}{2}\pm} = V_{2a}r^2 \pm \left[(3\Delta)^2 + V_{1e}^2 r^2 + 2V_{1e}V_{2e}\cos(3\phi)r^3 + V_{2e}^2 r^4\right]^{\frac{1}{2}}. \tag{16.9}$$

Each of them is doubly degenerate due to the molecular-fixed coordinate system and describes one branch of the SO corrected, adiabatic potential energy surfaces plotted in Fig. 16.3 as black dashed lines.

The parameters for the $2^4E'$ state of $Cs_3$ are obtained from nonlinear fits of the SO-corrected ab initio curves. They are listed in Table 16.2. It shows that higher order JT terms are essential for an adequate description of the curvature in the fit range of $\pm 0.7$ Å. By stepwise inspection of the analytical functions we find that the inclusion of quadratic coupling is sufficient to describe the two upper surfaces. With inclusion of cubic JT coupling also the two lower surfaces are well reproduced in the range between $-0.4$ and $0.4$ Å. The final inclusion of anharmonicity yields an adequate description of all surfaces in the full plotting range of Fig. 16.3. We further mention, that the inclusion of SO coupling slightly affects the values obtained for $V_{1e}$ and $V_{3e}$. Without SO coupling the linear JT coupling term is reduced by 16%, whereas the cubic term increases by 18%. The remaining parameters are almost equal.

For the comparison with other triatomic systems the conversion into dimensionless parameters is useful. A new energy variable is defined in multiples of $\hbar\omega$ and

**Table 16.3** Dimensionless linear JT stabilization energies ($D$), linear ($k$) and quadratic ($g$) vibronic coupling constants, SO-splittings ($\Delta'$) and harmonic frequencies (in cm$^{-1}$) for the JT-problem in the $2^4E'$ states of K$_3$, Rb$_3$ and Cs$_3$

| Species | $D$ | $k$ | $g$ | $\Delta'$ | $\omega_{Q_x/Q_y}$ (cm$^{-1}$) | $\omega_{Q_s}$ (cm$^{-1}$) |
|---|---|---|---|---|---|---|
| K$_3^a$ | 0.0202 | 0.2010 | $-0.3110$ | 0.0656 | 67 | 52 |
| Rb$_3^a$ | 0.2500 | 0.7071 | $-0.1774$ | 0.3578 | 42 | 31 |
| Cs$_3$ | 0.1526 | 0.5525 | $-0.1452$ | 1.1360 | 30 | 24 |

[a] Reprinted with permission from Ref. [5]. Copyright 2010, American Institute of Physics

a new length in multiples of $\sqrt{\hbar/m\omega}$, using the vibrational frequency $\omega$ and the reduced mass $m$, which is equal to the atomic mass in the case of an equilateral, homonuclear trimer. With these transformations the Eq. 16.9 reduce to

$$\varepsilon_{\frac{1}{2}\pm} = \frac{1}{2}r^2 \pm \left[\Delta'^2 + k^2r^2 + kg\cos(3\phi)r^3 + \frac{g^2}{4}r^4\right]^{\frac{1}{2}}$$

$$\varepsilon_{\frac{3}{2}\pm} = \frac{1}{2}r^2 \pm \left[(3\Delta')^2 + k^2r^2 + kg\cos(3\phi)r^3 + \frac{g^2}{4}r^4\right]^{\frac{1}{2}} \tag{16.10}$$

where we have switched to the common notation [1, 35] of dimensionless linear (k) and quadratic (g) coupling parameters. In Table 16.3 our results for Cs$_3$ are compared with those of K$_3$ and Rb$_3$. The linear JT-stabilization energy is defined as $D = k^2/2$.

# 16.4 Spectra Simulations for the $2^4E' \leftarrow 1^4A_2'$ Transition

For a detailed description of the spectra generation we refer to articles [4, 5]. This section contains only a brief review of the main steps. The $n$th eigenvector, obtained from the diagonalization of the Hamiltonian given in Eq. 16.4 in the chosen basis, may be written as

$$|\phi_n\rangle = \sum_\Lambda \sum_\Sigma \sum_v \sum_l^\infty c_{\Lambda\Sigma vl}^n |\Lambda\Sigma\rangle |vl\rangle. \tag{16.11}$$

The vibronic ground state, which has no electronic angular momentum (no projection for the $1^4A_2'$ state) and is in the vibrational ground state ($v = l = 0$), may be denoted as

$$|\phi_0^A\rangle = |0\Sigma'\rangle|00\rangle \tag{16.12}$$

With the approximation of a constant A → E dipole transition, the following relation can be obtained for the intensity of the $n$th state:

$$I_n \propto \sum_\Lambda \sum_\Sigma \sum_v^\infty \sum_l (c^n_{\Lambda \Sigma vl})^2 |\langle \Sigma'|\Sigma \rangle|^2 |\langle 00|vl \rangle|^2 = \sum_\Lambda (c^n_{\Lambda \Sigma'00})^2. \qquad (16.13)$$

The last simplification shown in Eq. 16.13 holds only if the harmonicity in the lower and upper electronic state is assumed to be the same. Only then one would obtain the necessary $\delta_{v0}$ relation. However, the frequencies for $Cs_3$ are 18 and $30\,\mathrm{cm}^{-1}$ for the A and E state, respectively. To account for this difference, we replace $|\langle 0|v \rangle|^2$ by the Franck–Condon factors 0.9375 for $v = 0$, 0.0586 for $v = 2$ and set the rest to zero.

In the case of $K_3$ and $Rb_3$ we could identify $I_n$ of Eq. 16.13 as the laser-induced fluorescence (LIF) signal measured in Ref. [4]. To help with the peak assignment, the experimentalists of our group complemented their measurements by the technique of magnetic circular dichroism (MCD) spectroscopy. Here the difference between left-circularly and right-circularly polarized radiation propagating along the direction of an externally applied magnetic field is measured. The advantage of this technique is that spin degeneracies in the lower and the upper vibronic levels are removed. Although the Zeeman splittings are far too small to be experimentally resolved with the HENDI method, the obtained MCD spectra contain additional information which allows to distinguish the spectroscopic contributions of the classical $E \otimes e$ JT distortion from the effects caused by SO coupling. The simulation of MCD spectra is laid out clearly in Refs.[4, 5], so the formalism will be skipped here.

### 16.4.1 Results

The parameters of Table 16.3 are used to calculate the matrix elements of the Hamiltonian in the chosen basis (see Eq. 16.5). The nuclear part of the wavefunction is truncated at $v = 8$, including all angular momentum states $l = -8, -6, \ldots, 8$. Only the linear JT parameter $k$ and SO coupling $\Delta'$ are taken into consideration, since the higher orders do not cause significant changes of the spectra. Following our procedure in Ref. [5] we further increase $\Delta'$ by a factor of 1.3 to correct for the underestimation of the SO coupling obtained with the ECP-LS method: For the Cs atom only 70% of the experimental splitting between the $^2P$ terms could be reproduced with our computational approach.

The results of our LIF and MCD spectra simulations for the $2\,^4E' \leftarrow 1\,^4A'_2$ transition in $Cs_3$ are presented in Fig. 16.4, together with previous measurements and simulations of the corresponding transitions in $K_3$ and $Rb_3$. The numerical eigenvalues and intensities are given as black stick spectra. To simulate the

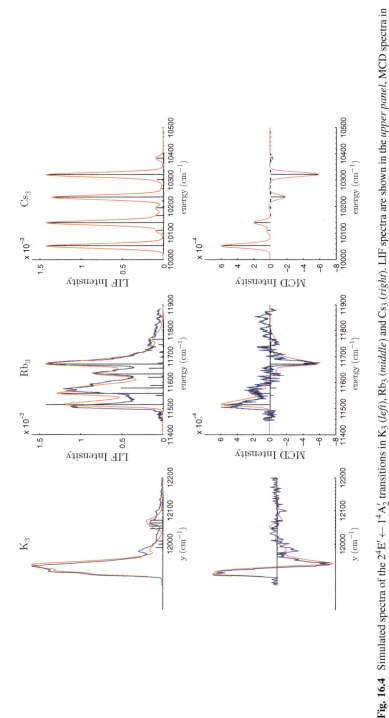

**Fig. 16.4** Simulated spectra of the $2^4E' \leftarrow 1^4A_2'$ transitions in $K_3$ (*left*), $Rb_3$ (*middle*) and $Cs_3$ (*right*). LIF spectra are shown in the *upper panel*, MCD spectra in the *lower panel*. Experimental data are plotted in *blue*, the calculated stick spectra are printed in *black*. For a better comparison the convolutions of the theoretical values with respect to the known broadening effects of the HENDI experimental technique (see text) are given as *red traces*. For $Cs_3$ no experimental spectra are available. The harmonic frequencies, JT parameters and SO splittings used for the simulation are summarized in Table 16.3. Note that the SO splittings $\Delta'$ are multiplied by a correction factor of 1.3 for all three trimer species (The spectra for $K_3$ and $Rb_3$ (*Left* and *middle row*) are reprinted with permission from Ref. [5]. Copyright 2010, American Institute of Physics)

broadening effects of the experimental method, empirically motivated asymmetric lineshape functions [5] were used to generate convolutions (red trace) in the case of $K_3$ and $Rb_3$. Experimental data, only available for $K_3$ and $Rb_3$, are plotted in blue.

As guide to the eye, we also provide a convolution for $Cs_3$, which is based on gaussian profiles with a FWHM of $20\,cm^{-1}$. Figure 16.4 clearly illustrates the increasing effect of SO coupling on the LIF and MCD spectra. In the case of $K_3$ the three larger experimental peaks are identified as states with the vibronic quantum numbers $v = 0, 1$ and 2, respectively. Each peak consists of four finer components due to the SO splitting. They are not resolved in the experimental LIF spectrum, but the corresponding MCD spectrum reveals that each of these groups has two components belonging to the lower SO-corrected energy surfaces and two belonging to the upper ones. In other words, the vibronic coupling is only slightly distorted by SO coupling effects. The experimental peak structure for $Rb_3$ is significantly different: Four strong peaks occur, and can be identified as the four SO components of the lowest vibronic quantum number $v = 0$. However, they still show a slight interaction with the SO components of the next higher vibronic quantum numbers. This interpretation is underlined by the MCD spectrum, which mainly consists of two positive and two negative peaks, again corresponding to the four significantly split surfaces of the spin projections $\Sigma = -3/2, -1/2, 1/2$ and $3/2$. For $Cs_3$, the SO coupling is fully dominating. The four main peaks of almost equal intensity can be clearly identified as SO components belonging to the vibronic quantum number $v = 0$, and it makes sense now to talk of vibronic substates for each spin projection. Each of these large peaks is followed by a smaller contribution of its corresponding $v = 1$ substate. For the chosen gaussian lineshape function, the second vibronic substate of the 1/2 spin projection is covered by the main peak of the 3/2 spin projection.

To summarize, the situation of a classical $E \otimes e$ JT effect which is slightly distorted by SO-coupling has inverted its character to a situation of very strong SO-coupling coupling with minor, almost negligible JT distortions. The peak structure of $Cs_3$ is fully insensitive to variations of JT parameters, as the JT coupling is strongly quenched due to the large SO splitting of the surfaces. A variation of the linear coupling term only slightly modifies the blue shoulders of the main peaks. The inclusion of quadratic coupling does not change the spectrum at all.

## 16.5 Conclusions and Remarks

In this chapter we demonstrated the influence of spin-orbit coupling on the $E \otimes e$ JT coupling in the states of $K_3$, $Rb_3$ and $Cs_3$. We applied the JT effect theory and extracted parameters from the ab initio surfaces of the $2^4E'$ states, fitting them with an analytical expression for the electronic energy as a function of the symmetric bending coordinate $Q_x$. As can be seen from the inspection of the calculated

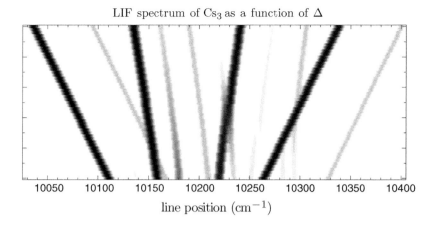

**Fig. 16.5** LIF spectra of the $2^4E' \leftarrow 1^4A'_2$ transition in Cs$_3$ (*right*), plotted as a function of the SO splitting $\Delta$. Intensity information is coded into the grayscale. The four spin components are dominating the peak structure throughout the observed range. Only for $\Delta$ values around 35 cm$^{-1}$ some intensity borrowing can be observed due to a coupling between the $v = 0$ vibronic substate of the third spin component (*strong line* starting at 10,220 cm$^{-1}$) and the $v = 3$ vibronic substate of the second spin component (*weak line* starting at 10,235 cm$^{-1}$). The linear JT coupling parameter $k$ is fixed to a value of 0.5525. The stick spectra are slightly broadened to improve the readability of the illustration (gaussian profile convolution, FHWM is set to 5 cm$^{-1}$)

SO-corrected potential surfaces, the inclusion of spin-orbit coupling reduces the strength of the nonadiabatic coupling. This effect starts to become qualitatively important for Rb$_3$, where the peak structure is mainly determined by the spin-orbit splitting, and is of essential importance in the case of Cs$_3$, where only a slight spectroscopic influence of the JT parameters remains. However, the spin-orbit splitting constant obtained from the ab initio calculations turns out to be the weak point of the approach, and had to be readjusted with respect to the experimental data available for K$_3$ and Rb$_3$. Although this shows that for an improved description of the SO coupling in heavier alkali-metal trimers a more sophisticated ab initio approach might be necessary, the JT parameter results and harmonic frequencies presented here remain valid, since they do not depend (or show only a weak dependence, as it is the case for the linear and the cubic JT coupling) on the tentatively adjusted spin-orbit splitting value. To account for this, we also provide an illustration which contains the simulated spectra of Cs$_3$ as a function of the SO splitting parameter. Figures 16.5 and 16.6, with the LIF and MCD intensities plotted as a function of line position and SO splitting parameter, together with Table 16.3, should help future interpretations of experimental spectra obtained for the $2^4E' \leftarrow 1^4A'_2$ transition in Cs$_3$.

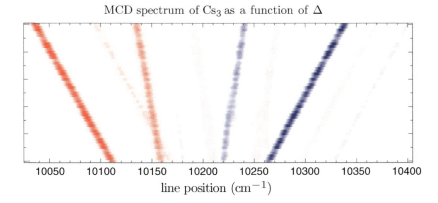

**Fig. 16.6** MCD spectra of the $2^4E' \leftarrow 1^4A'_2$ transition in Cs$_3$, plotted as a function of the SO splitting $\Delta$. Positive MCD signal is scaled in red, negative signal in *blue*. Since the four main peaks are well separated over the observed range, no essential overlapping effects occur, and the structure is almost equal to the LIF structure of Fig. 16.5. The linear JT coupling parameter $k$ is fixed to a value of 0.5525. The stick spectra are slightly broadened to improve the readability of the illustration (gaussian profile convolution, FHWM is set to 5 cm$^{-1}$)

**Acknowledgments** AWH gratefully acknowledges support from the Graz Advanced School of Science, a cooperation project between TU Graz and the University of Graz, and the Austrian Science Fund (FWF, Grant No. P19759-N20).

# References

1. Bersuker IB (2001) Chem Rev 101:1067
2. Nagl J, Auböck G, Hauser AW, Allard O, Callegari C, Ernst WE (2008) Phys Rev Lett 100(6):063001. doi:10.1103/PhysRevLett.100.063001
3. Nagl J, Auböck G, Hauser AW, Allard O, Callegari C, Ernst WE (2008) J Chem Phys 128(15):154320
4. Auböck G, Nagl J, Callegari C, Ernst WE (2008) J Chem Phys 129(11):114501
5. Hauser AW, Auböck G, Callegari C, Ernst WE (2010) J Chem Phys 132(16):164310
6. Hauser AW, Callegari C, Soldán P, Ernst WE (2008) J Chem Phys 129:044307
7. Hauser AW, Callegari C, Soldán P, Ernst WE (2010) Chem Phys 375(1): 73-84
8. (2001) Special issue on Helium nanodroplets: a novel medium for chemistry and physics. J Chem Phys 115(22)
9. Lim IS, Schwerdtfeger P, Metz B, Stoll H (2005) J Chem Phys 122:104103
10. Müller W, Flesch J, Meyer W (1984) J Chem Phys 80(7):3297
11. Watts JD, Gauss J, Bartlett RJ (1993) J Chem Phys 98(11):8718
12. Urban M, Neogrády P, Hubač I (1997) In Bartlett RJ (ed) Recent advances in coupled cluster methods. Recent advances in computational chemistry, vol 3. World Scientific, Singapore, pp 275–306
13. Xie F, Sovkov VB, Lyyra AM, Li D, Ingram S, Bai J, Ivanov VS, Magnier S, Li L (2009) J Chem Phys 130(5):051102
14. Li D, Xie F, Li L, Magnier S, Sovkov V, Ivanov V (2007) Chem Phys Lett 441(1–3):39

15. Guerout R, Soldan P, Aymar M, Deiglmayr J, Dulieu O (2009) Int J Quant Chem 109(14, Sp. Iss. SI):3387
16. Thompson TC, Izmirlian GJ, Lemon SJ, Truhlar DG (1985) J Chem Phys 82:5597
17. Cocchini F, Upton TH, Andreoni W (1988) J Chem Phys 88:6068
18. Werner HJ, Meyer W (1980) J Chem Phys 73(5):2342
19. Werner HJ, Meyer W (1981) J Chem Phys 74(10):5794
20. Werner HJ, Knowles PJ (1985) J Chem Phys 82(11):5053
21. Knowles PJ, Werner HJ (1985) Chem Phys Lett 115(3):259
22. Celani P, Werner HJ (2000) J Chem Phys 112:5546
23. Werner HJ, Knowles PJ, Lindh R, Manby FR, Schütz M, Celani P, Korona T, Rauhut G, Amos RD, Bernhardsson A, Berning A, Cooper DL, Deegan MJO, Dobbyn AJ, Eckert F, Hampel C, Hetzer G, Lloyd AW, McNicholas SJ, Meyer W, Mura ME, Nicklass A, Palmieri P, Pitzer R, Schumann U, Stoll H, Stone AJ, Tarroni R, Thorsteinsson T (2006) Molpro, version 2006.1, a package of ab initio programs. See http://www.molpro.net
24. Roos B, Andersson K (1995) Chem Phys Lett 245:215
25. Bünermann O, Mudrich M, Weidemüller M, Stienkemeier F (2004) J Chem Phys 121(18):8880
26. Ernst WE, Huber R, Jiang S, Beuc R, Movre M, Pichler G (2006) J Chem Phys 124(2):024313
27. Hauser AW, Callegari C, Ernst WE (2009) In Lipscomb WN, Prigogine I, Piecuch P, Maruani J, Delgado-Barrio G, Wilson S (eds) Advances in the theory of atomic and molecular systems. Progress in theoretical chemistry and physics, vol 20. Springer, Netherlands, pp 201–215
28. Barckholtz TA, Miller TA (1998) Int Rev Phys Chem 17(4):435
29. Schön J, Köppel H (1998) J Chem Phys 108(4):1503
30. Koizumi H, Sugano S (1995) J Chem Phys 102:4472
31. Domcke W, Mishra S, Poluyanov LV (2006) Chem Phys 322:405
32. Poluyanov LV, Domcke W (2008) Chem Phys 352:125
33. Meiswinkel R, Köppel H (1989) Chem Phys 129(3):463
34. García-Fernández P, Bersuker I, Aramburu J, Barriuso M, Moreno M (2005) Phys Rev B 71:184117
35. Bersuker IB (2006) The Jahn-Teller effect. University Press, Cambridge

# Chapter 17
# Jahn-Teller Transitions in the Fe(II)Fe(III) Bimetallic Oxalates

**R.S. Fishman**

**Abstract** Because the orbital angular momentum $L_z^{cf}$ on the Fe(II) sites of the Fe(II)Fe(III) bimetallic oxalates is incompletely quenched by the crystal field, the spin-orbit coupling competes with the Jahn-Teller (JT) distortion energy. The value of $L_z^{cf}$ depends on the cation between the bimetallic layers. When $L_z^{cf}$ is sufficiently small, the open honeycomb lattice of each bimetallic layer is distorted at all temperatures below the JT transition temperature. But in a range of $L_z^{cf}$, the lattice is only distorted between lower and upper JT transition temperatures, $T_{JT}^{(l)}$ and $T_{JT}^{(u)}$. For some cations, $L_z^{cf}$ may exceed the threshold required for the cancellation of the moments on the Fe(II) and Fe(III) sublattices at a temperature $T_{comp}$ below the transition temperature $T_c$. Using elastic constants obtained from compounds that exhibit magnetic compensation, we find that $T_{JT}^{(l)}$ always lies between $T_{comp}$ and $T_c$ and that $T_{JT}^{(u)}$ always lies above $T_c$.

## 17.1 Introduction

With their low density, multifaceted functionality, and possible biocompatibility, molecule-based magnets have attracted great attention over the past two decades [1]. Like conventional solid-state magnets, bulk molecule-based magnets are typically crystalline with an extensive number of magnetic ions. The building blocks of molecule-based systems are organic molecules but the magnetic species may be either transition-metal ions or organic radicals. Unlike in conventional magnets, the competing spin-orbit, crystal-field, and exchange energies in a molecule-based

---

R.S. Fishman (✉)
Materials Science and Technology Division, Oak Ridge National Laboratory,
Oak Ridge, TN 37831-6065, USA
e-mail: fishmanrs@ornl.gov

M. Atanasov et al. (eds.), *Vibronic Interactions and the Jahn-Teller Effect: Theory and Applications*, Progress in Theoretical Chemistry and Physics 23, DOI 10.1007/978-94-007-2384-9_17, © Springer Science+Business Media B.V. 2012

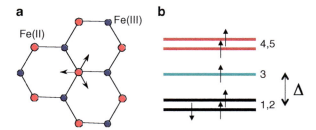

**Fig. 17.1** (a) A portion of the open honeycomb lattice, displaying the three possible displacements of an Fe(II) ion into adjacent hexagons. (b) Level-splitting of the $3d^6$ Fe(II) multiplet in a $C_3$-symmetric crystal-field

magnet can be tuned by the choice of organic cation or by applying pressure. These competing energies produce a remarkable interplay between the structural, chemical, and magnetic properties of molecule-based materials.

In many molecule-based magnets [2], the competing spin-orbit coupling and Jahn-Teller (JT) distortion energies [3] are both important because the orbital angular momentum of a transition-metal ion with a singly-occupied doublet orbital is not quenched by the surrounding crystal field. For a trigonally-distorted crystal field, the low-energy doublet of an Fe(II) ion with filling $3d^6$ is occupied by a single hole, as shown in Fig. 17.1b, and carries an orbital angular momentum $L_z^{cf}$ [4] that can take any value between 0 and 2.

Bimetallic oxalates A[M(II)M'(III)(ox)$_3$] are a versatile class of layered molecule-based magnets with transition metals M(II) and M'(III) arranged in an open honeycomb lattice (sketched in Fig. 17.1a) and coupled by the oxalate bridge ox $= (C_2O_4)^{-2}$ [5, 6]. In Fe(II)Fe(III) bimetallic oxalates, every $S = 2$ Fe(II) spin is antiferromagnetically coupled to the three neighboring $S' = 5/2$ Fe(III) spins within each bimetallic layer. While the choice of organic cation A separating the layers does not affect the sign of this exchange interaction, it does shift the orbital-angular momentum $L_z^{cf}$ on the Fe(II) sites. If $L_z^{cf}$ lies between $l_c \approx 0.28$ and 1, the magnetic moments of the Fe(II) and Fe(III) sublattices exactly cancel [7] at a magnetic compensation (MC) temperature $T_{comp} \approx 30$ K [8, 9] below the transition temperature $T_c \approx 45$ K. When $T < T_{comp}$, these compounds exhibit a metastable state with negative magnetization in small positive fields.

The competition between the spin-orbit and JT energies can produce [10] both upper and lower JT transitions with the open honeycomb lattice distorted by Fe(II) displacements between $T_{JT}^{(l)}$ and $T_{JT}^{(u)}$. This paper studies the general relationship between the magnetic and JT transitions in the Fe(II)Fe(III) bimetallic oxalates. For weak enough elastic energies, $T_c$ can lock onto $T_{JT}^{(l)}$ over a range of $L_z^{cf}$. For stronger elastic energies, $T_c$ may become double valued with magnetic order suppressed below $T_{JT}^{(u)}$. In compounds that exhibit MC, $T_{comp} < T_{JT}^{(l)} < T_c$ and $T_c < T_{JT}^{(u)}$.

## 17.2 Model

The crystal field at each Fe site is produced by the six surrounding oxygen atoms, which form two parallel, equilateral triangles rotated by $48°$ with respect to one another [11]. Because one triangle is bigger than the other, only $C_3$ symmetry is respected by the trigonally-distorted crystal-field potential. While the $L' = 0$ Fe(III) multiplet is unaffected by the crystal-field potential, the $L = 2$ Fe(II) multiplet splits into two doublets and a singlet [7], as shown in Fig. 17.1b, with the doublet $\psi_{1,2}$ lying an energy $\Delta$ below the singlet $\psi_3$.

Ab initio calculations [10] indicate that $\Delta \gg J_c SS'$ and $\Delta \gg |\lambda|S$, where $\lambda \approx -12.65$ meV is the spin-orbit coupling constant for Fe(II). Consequently, Fe(II) ions carry the non-quenched orbital angular momentum $L_z = \pm L_z^{cf}$, where $0 \le L_z^{cf} \le 2$. The choice of cation can alter $L_z^{cf}$ by slightly shifting the position of the Fe(II) atom within the potential landscape produced by the surrounding oxygen atoms.

If the crystal field respected octahedral symmetry, then $L_z^{cf} = 1$ and $\Delta = 0$. The degenerate $\psi_{1,2,3}$ levels would then form the familiar $t_{2g}$ triplet with $L_z = 0$ or $\pm 1$ and the orbital angular momentum $\mathbf{L}$ could be treated as an $L = 1$ operator. But restricted to the orbital doublet $\psi_{1,2}$, the matrix elements $\langle \psi_i | L_\pm | \psi_j \rangle$ ($i, j = 1$ or 2) of the raising and lowering operators $L_\pm = L_x \pm iL_y$ vanish so that the spin-orbit coupling $\lambda L_z S_z = -|\lambda| L_z S_z$ contains no transverse components. Therefore, the orbital angular momentum of the doublet acts like a spin-1/2 Ising variable.

In the absence of spin-orbit coupling, the $S' = 5/2$ Fe(III) moments would always dominate over the $S = 2$ Fe(II) moments. But when the spin-orbit coupling $-|\lambda| L_z S_z$ on the Fe(II) sites is sufficiently strong, the Fe(II) moments will order more rapidly below $T_c$ than the Fe(III) moments. At $T = 0$, the Fe(III) moment again dominates if $S' > S + L_z^{cf}/2$ or $L_z^{cf} < 1$. So MC occurs in a range of $L_z^{cf}$ between $l_c$ and 1.

The Hamiltonian of a non-distorted bimetallic layer is given by

$$H = J_c \sum_{\langle i,j \rangle} \mathbf{S}_i \cdot \mathbf{S}'_j - |\lambda| \sum_i L_i^z S_i^z, \qquad (17.1)$$

where $L_i^z = \pm L_z^{cf}$, the $\langle i, j \rangle$ sum runs over all nearest-neighbors, and the $i$ sum runs over all Fe(II) sites. For antiferromagnetic exchange, $J_c > 0$. Based on Eq. 17.1, the appearance of MC in a range of $L_z^{cf}$ between $l_c$ and 1 was verified using hybrid Monte-Carlo simulations [12] with the spins $\mathbf{S}_i$ and $\mathbf{S}'_j$ treated quantum-mechanically and the orbital angular momenta $L_i^z$ treated classically.

Within mean-field (MF) theory, the eigenvalues of $\psi_{1,2;\sigma}$ on the Fe(II) sites are $\varepsilon_{1\sigma} = \varepsilon_{0\sigma} - |\lambda| L_z^{cf} \sigma$ and $\varepsilon_{2\sigma} = \varepsilon_{0\sigma} + |\lambda| L_z^{cf} \sigma$, where $\varepsilon_{0\sigma} = 3J_c M_{S'}\sigma/2$ and $M' = M_{S'} = 2\langle S'_z \rangle$ is the Fe(III) moment with $\mu_B$ set to 1. It is then straighforward to evaluate the temperature dependence of $M'(T)$ and of the Fe(II) moment $M(T) = M_S(T) + M_L(T)$, where $M_S = 2\langle S_z \rangle$ and $M_L = \langle L_z \rangle$. Since $M$ and $M'$ always have opposite signs in the absence of an external field, we shall take $M < 0$ and $M' > 0$.

Due to the mixing of the eigenstates $\psi_{1\sigma}$ and $\psi_{2\sigma}$, the JT distortion on the Fe(II) sites is described by the matrix

$$\underline{H}_\sigma^{\mathrm{mix}} = \begin{pmatrix} \varepsilon_{1\sigma} & \xi \\ \xi & \varepsilon_{2\sigma} \end{pmatrix}, \tag{17.2}$$

where the distortion energy $\xi$ is proportional to the Fe(II) displacement $Q$. The eigenvalues of $\underline{H}_\sigma^{\mathrm{mix}}$ are $\varepsilon_{a\sigma} = \varepsilon_{0\sigma} + t_\sigma$ and $\varepsilon_{b\sigma} = \varepsilon_{0\sigma} - t_\sigma$ where

$$t_\sigma = -\mathrm{sgn}(\sigma)\sqrt{(\lambda L_z^{cf}\sigma)^2 + \xi^2} \tag{17.3}$$

and we define $t_0 = \xi$. Hence, the doublet splitting $2|t_\sigma|$ is enhanced by the JT effect (strictly speaking, the pseudo-JT effect when $\varepsilon_{1\sigma} \neq \varepsilon_{2\sigma}$). With the convention that $M' = M_{S'} > 0$, the lowest eigenvalue of $\underline{H}_\sigma^{\mathrm{mix}}$ is $\varepsilon_{b,-2}$ with $\sigma = -2$.

Including the JT distortion energy $\xi$, the MF free energy can be written

$$\frac{F}{N} = -T\log\left\{Z_{\mathrm{II}}Z_{\mathrm{III}}e^{3J_cM_SM_{S'}/4T}\right\}$$
$$+ \alpha|\lambda|\left\{\left(\frac{\xi}{|\lambda|}\right)^2 + \gamma_3\left(\frac{\xi}{|\lambda|}\right)^3 + \gamma_4\left(\frac{\xi}{|\lambda|}\right)^4\right\}, \tag{17.4}$$

where

$$Z_{\mathrm{II}} = 2\sum_\sigma e^{-3J_cM_{S'}\sigma/2T}\cosh(t_\sigma/T), \tag{17.5}$$

$$Z_{\mathrm{III}} = 2\sum_{\sigma'} e^{-3J_cM_S\sigma'/2T} \tag{17.6}$$

are the partition functions on the Fe(II) and Fe(III) sites, respectively, containing sums over $\sigma = 0, \pm 1, \pm 2$ and $\sigma' = \pm 1/2, \pm 3/2, \pm 5/2$. The second line in Eq. 17.4 corresponds to the elastic energy proportional to the dimensionless parameter $\alpha$.

We restrict consideration to displacements $Q \sim \xi$ of the Fe(II) ions either directly into one of the open hexagons or towards one of the neighboring Fe(III) ions, with the former displacements sketched in Fig. 17.1a. The different energy costs for these two types of displacements is reflected in the anharmonic $\gamma_3(\xi/|\lambda|)^3$ term in the elastic energy. Fluctuations between the distorted atomic configurations are assumed to be slow compared with the electronic time scales. The JT distortion is not washed out by an average over the three possible displacements of the Fe(II) ion sketched in Fig. 17.1a because the interactions with other Fe(II) ions will break the threefold symmetry about any site.

It is simple to obtain the equilibrium values for $M_S$, $M_{S'}$, and $\xi$ from the extremal conditions $\partial F/\partial M_S = \partial F/\partial M_{S'} = \partial F/\partial \xi = 0$. The rather complex expression for the average orbital angular momentum $M_L = \langle L_z \rangle$ on the Fe(II) sites was given in Ref. [10](b).

## 17.3 Results

Based on the data plotted in Fig. 17.2, Tang et al. [14] attributed two "compensation" temperatures to an Fe(II)Fe(III) compound with cation A = N(n-C$_4$H$_9$)$_4$. However, their experimental data is very well described by our model [10] using the parameters $J_c/|\lambda| = 0.037$, $\alpha = 3.7$, $\gamma_3 = -1.9$, $\gamma_4 = 1.1$, and $L_z^{cf} = 0.3$, which is just above the threshold $l_c \approx 0.28$ for MC. Clearly, the upper "compensation" point assigned in Ref.[14] is actually the lower JT transition marked by a jump in the magnetization. The predicted upper JT transition, not shown in Fig. 17.2, should occur at $T_{JT}^{(u)} \approx 0.41|\lambda|$ or about 60 K. Due to the anharmonic term $\sim \xi^3$ in the elastic energy, both predicted lower and upper JT transitions are first order with $\xi \approx 0.8|\lambda|$, corresponding to an atomic displacement of roughly $Q \approx 0.075$ Å.

While the lower JT transition at $T_{JT}^{(l)} \approx 42$ K $< T_c$ is easy to observe due to the jump in the magnetization, the predicted upper JT transition has not yet been seen. But x-ray measurements [15] on a different sample with the same cation N(n-C$_4$H$_9$)$_4$ indicate that the hexagonal symmetry present at room temperature is absent in the monoclinic lattice at 60 K, implying that 60 K $< T_{JT}^{(u)} <$ 290 K.

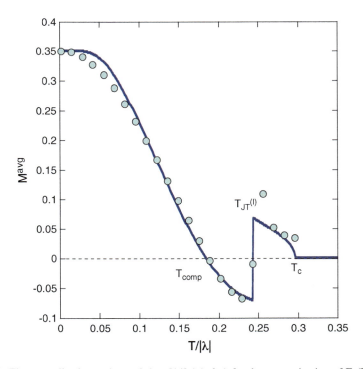

**Fig. 17.2** The normalized experimental data [14] (*circles*) for the magnetization of Fe(II)Fe(III) bimetallic oxalate with cation N(n-C$_4$H$_9$)$_4$ and the theoretical prediction [10] (*solid curve*) for the average magnetization with $J_c/|\lambda| = 0.037$, $\alpha = 3.7$, $\gamma_3 = -1.9$, $\gamma_4 = 1.1$, and $L_z^{cf} = 0.3$

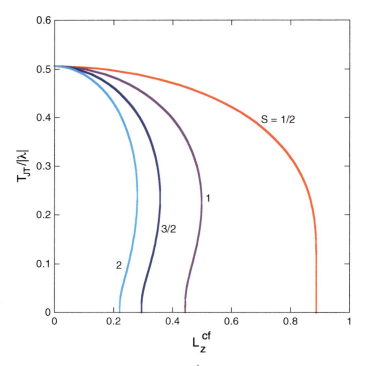

**Fig. 17.3** The JT transition temperature versus $L_z^{cf}$ for $\alpha = 4$, $\gamma_3 = -1.9$, $\gamma_4 = 1.1$, $J_c = 0$, and different values of the spin $S$. $T_{JT}$ can be double valued for ranges of $L_z^{cf}$ when $S > 1/2$

To understand the physical origin of the inverse JT distortion in the Fe(II)Fe(III) bimetallic oxalates, we consider the general problem of a spin-$S$ ion containing a singly-occupied doublet orbital with non-quenched orbital angular momentum $L_z^{cf}$ in the absence of any magnetic coupling ($J_c = 0$). In Fig. 17.3, we plot the JT transition temperature $T_{JT}$ versus $L_z^{cf}$ for $\alpha = 4$, $\gamma_3 = -1.9$, and $\gamma_4 = 1.1$. Notice that $T_{JT}$ depends only on the elastic constants and is independent of $S$ as $L_z^{cf} \to 0$. Both upper and lower JT transitions are found in narrow ranges of $L_z^{cf}$ for $S > 1/2$. This is easy to understand: at low temperatures, the spin states with larger $|S_z|$ and stronger spin-orbit coupling $-|\lambda||L_z S_z|$ dominate the free energy and quench the JT distortion.

Although magnetic order is not responsible for the inverse JT transition, it does compete with the JT distortion in interesting ways. Along with the magnetic temperatures $T_c$ and $T_{comp}$, the JT temperatures $T_{JT}^{(u)}$ and $T_{JT}^{(l)}$ are plotted versus $L_z^{cf}$ in Fig. 17.4 for four different values of the elastic coefficient $\alpha$. Of course, $T_c$ does not depend on $\alpha$ when $T_c > T_{JT}^{(u)}$ or when the JT transition is absent.

For $\alpha = 2$ or 3, $T_{JT}^{(u)}$ always exceeds $T_c$, which locks onto $T_{JT}^{(l)}$ over a range of $L_z^{cf}$, as shown in the insets to Fig. 17.4a,b. Over these ranges of $L_z^{cf}$, the JT distortion is quenched with the development of long-range magnetic order below $T_c$.

# 17 Jahn-Teller Transitions in the Fe(II)Fe(III) Bimetallic Oxalates

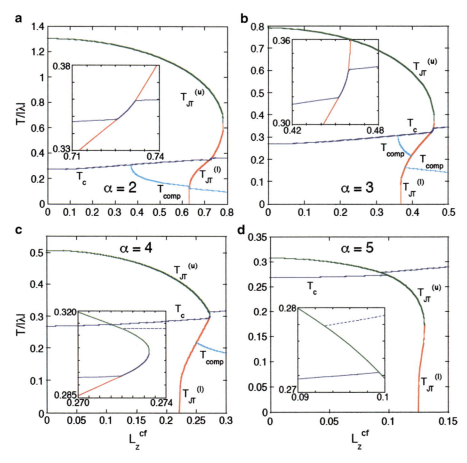

**Fig. 17.4** The transition temperature $T_c$, compensation temperature $T_{comp}$, and upper and lower JT transition temperatures $T_{JT}^{(l)}$ and $T_{JT}^{(u)}$ versus $L_z^{cf}$ for $\gamma_3 = -1.9$, $\gamma_4 = 1.1$, $J_c/|\lambda| = 0.037$, and $\alpha =$ (a) 2, (b) 3, (c) 4, or (d) 5

For $\alpha = 4$, close to the value of 3.7 used to model the experimental data of Ref.[14], $T_c$ again locks in with $T_{JT}^{(l)}$ up to the point where $T_{JT}^{(u)}$ and $T_{JT}^{(l)}$ meet. But as shown in the inset to Fig. 17.4c, $T_c$ is discontinuous and double valued in the range $0.2721 < L_z^{cf} < 0.2733$. Consequently, magnetic order appears just below the dashed line for the upper $T_c$, disappears below $T_{JT}^{(u)}$, and then reappears again below the solid curve for the lower $T_c$, which coincides with $T_{JT}^{(l)}$.

The JT distortion energy $\xi$ and the average magnetization per site $M_{av} = (M + M')/2$ are plotted versus temperature in Fig. 17.5a,b for $\alpha = 4$. A negative value for $M_{av}(T)$ means that the Fe(II) moment $M(T) < 0$ has a larger magnitude than the Fe(III) moment $M'(T) > 0$. As explained above, the JT transition is first order due to the anharmonic term in the elastic energy. For $L_z^{cf} = 0.273, 0.2726$, and

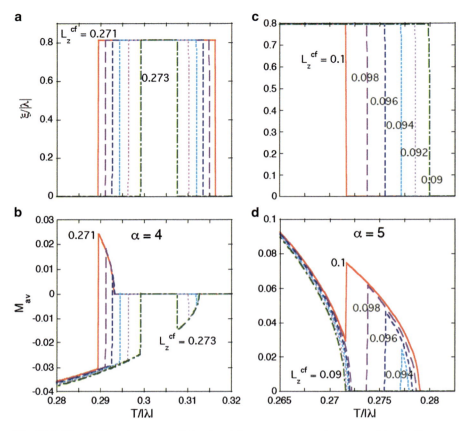

**Fig. 17.5** The JT distortion energy $\xi$ (**a, c**) and the average magnetic moment $M_{av}$ (**b, d**) versus temperature for $\alpha = 4$ (**a, b**) or 5 (**c, d**) and several values of $L_z^{cf}$. Other parameters as in Fig. 17.4

0.2722, the magnetization vanishes when $T_{JT}^{(l)} < T < T_{JT}^{(u)}$ and the lower $T_c$ coincides with $T_{JT}^{(l)}$. For $L_z^{cf} = 0.2718, 0.2714$, and 0.271, the magnetization is positive just below $T_c$ because the magnitude $|M(T)|$ of the Fe(II) moment is suppressed by the JT distortion $\xi \approx 0.81|\lambda|$. Below $T_{JT}^{(l)}$, the average moment changes sign due to the sudden increase in $|M(T)|$. This sign change explains the higher "compensation" point assigned by Tang et al. [14] to the data plotted in Fig. 17.2.

Similar features are seen for $\alpha = 5$ in Fig. 17.4d. Although $T_c$ always exceeds $T_{JT}^{(l)}$, it becomes double valued as it crosses $T_{JT}^{(u)}$. For $0.0932 < L_z^{cf} < 0.0995$, magnetic order appears below the dashed curve for the upper $T_c$ in the inset to Fig. 17.4d, then disappears below $T_{JT}^{(u)}$, and reappears in a first-order transition at the solid curve for the lower $T_c$. This series of transitions is exhibited in Fig. 17.5c,d. For $L_z^{cf} = 0.09$ and 0.092, the upper JT transition occurs at a higher temperature than the magnetic transition. But for $L_z^{cf} = 0.094, 0.096$, and 0.098, the magnetization first develops at the upper $T_c$, is quenched at $T_{JT}^{(u)}$, and then reappears below the lower $T_c$. For $L_z^{cf} = 0.1$, the average magnetization is suppressed but not quenched at $T_{JT}^{(u)}$.

Since the magnetization can change sign but can not vanish at $T_{JT}^{(l)}$, there is only one compensation temperature for most values of $L_z^{cf}$. The relation between $T_{comp}$ and $T_{JT}^{(l)}$ is shown in Fig. 17.4. Generally, $M_{av} < 0$ when $T_{comp} < T < T_c$ and $M_{av} > 0$ when $T < T_{comp}$. For $\alpha = 5$, MC only occurs when $L_z^{cf} > 0.234$ so $T_{comp}$ does not appear in Fig. 17.4d. For $\alpha = 4$, $T_{comp} = T_{JT}^{(l)}$ at $L_z^{cf} = 0.251$ and $T_{comp} < T_{JT}^{(l)}$ for $0.251 < L_z^{cf} < 0.273$, above which the JT transition is absent.

But for $\alpha = 2$ and 3, $T_{comp}$ drops when it crosses $T_{JT}^{(l)}$ with increasing $L_z^{cf}$ due to the jump in the magnitude $|M(T)|$ of the Fe(II) moment. As clearly shown in Fig. 17.4b for $\alpha = 3$, two compensation points may lie on either side of an intervening JT transition. In the range $0.38 < L_z^{cf} < 0.397$, $M_{av}$ is negative below $T_c$, positive between the upper $T_{comp}$ and $T_{JT}^{(l)}$, negative between $T_{JT}^{(l)}$ and the lower $T_{comp}$, and then positive again at low temperatures.

## 17.4 Conclusion

The remarkable agreement between experiment and theory in Fig. 17.2 suggests that Fe(II)Fe(III) bimetallic oxalates have an elastic constant $\alpha \approx 3.7$, with qualitatively the same behavior shown in Fig. 17.4c for $\alpha = 4$. For compounds that exhibit MC, Fig. 17.4c implies that $T_{comp} < T_{JT}^{(l)} < T_c$ and $T_c < T_{JT}^{(u)}$. This prediction is substantiated by Nuttall and Day [9], who consistently observed a small jump in the magnetization between $T_{comp}$ and $T_c$ for several Fe(II)Fe(III) compounds with different cations A. The magnetization always jumps to a higher value with increasing temperature because the JT distortion suppresses the magnitude of the Fe(II) moment above $T_{JT}^{(l)}$. Possibly due to a mixture of different stacking patterns for the bimetallic layers [16], the magnetization jumps observed by Nuttall and Day [9] are much smaller than the one observed by Tang et al. [14]. For compounds that do not exhibit MC, it is likely that $L_z^{cf}$ is small enough that $T_{JT}^{(l)} = 0$ ($L_z^{cf} < 0.222$ for $\alpha = 4$). Based on Fig. 17.4c, however, it is possible that $T_{JT}^{(l)} > 0$ even without MC in a narrow window of $L_z^{cf}$ ($0.222 < L_z^{cf} < 0.251$ for $\alpha = 4$).

Whether the pattern of JT displacements is short- or long-ranged depends on the shape and symmetry of the cation that couples the bimetallic layers. Since the cation $N(n\text{-}C_4H_9)_4$ is *not* $C_3$ symmetric, it may introduce a permanent distortion [10] of the hexagonal lattice at all temperatures. But $N(n\text{-}C_4H_9)_4$ is also just small enough to rotate freely within each unit cell, so it may permit a cooperative, long-ranged JT distortion between $T_{JT}^{(l)}$ and $T_{JT}^{(u)}$. On the other hand, larger cations like $N(n\text{-}C_5H_{11})_4$ will get stuck in different orientations within different unit cells, so that the JT distortion is short-ranged.

To summarize, an inverse JT transition occurs over a narrow range of $L_z^{cf} < 1$ for any ion with spin $S > 1/2$ containing a singly-occupied doublet orbital. In the Fe(II)Fe(III) bimetallic oxalates, magnetic order suppresses the JT distortion and enhances the lower JT transition temperature. Conversely, the JT distortion suppresses and can quench magnetic order below the upper JT transition temperature. A similar range of behavior including first and second-order phase transitions was found by Allen [17] for $UO_2$. We are hopeful that future x-ray measurements will be able to pinpoint the upper JT transition temperature in the Fe(II)Fe(III) bimetallic oxalates.

**Acknowledgments** I would like to acknowledge helpful conversations with Dr. Satoshi Okamoto. This research was sponsored by the U.S. Department of Energy, Office of Basic Energy Sciences, Materials Sciences and Engineering Division.

# References

1. Miller JS, Epstein AJ (1994) Angew Chem internat edit 33:385; Miller JS, Epstein AJ (2000), MRS Bull. November 21; Miller JS (2002) Adv Mat 14:1105
2. Kahn O (1994) Molecular magnetism. VCH, New York
3. See for example, Bersuker IB (2006) The Jahn–Teller effect. Cambridge University Press, Cambridge, and references therein
4. Palii A, Tsukerblat B, Clemente-Juan JM, Coronado E (2010) Int Rev Phys Chem 29:135
5. Tamaki H, Zhong ZJ, Matsumoto N, Kida S, Koikawa M, Achiwa N, Hashimoto Y, Ōkawa H (1992) J Am Chem Soc 114:6974
6. See the review Clément R, Decurtins S, Gruselle M, Train C (2003) Mon für Chem 134:117
7. (a) Fishman RS, Reboredo FA (2007) Phys Rev Lett 99:217203; (b) (2008) Phys Rev B 77:144421
8. (a) Mathonière C, Carling SG, Day P (1994) J Chem Soc Chem Commun 1551; (b) Mathonière C, Nuttall CJ, Carling SG, Day P (1996) Inorg Chem 35:1201
9. Nuttall CJ, Day P (1998) Chem Mat 10:3050
10. (a) Fishman RS, Okamoto S, Reboredo FA (2008) Phys Rev Lett 101:116402; (b) (2009) Polyhedron 28:1740
11. Pellaux R, Schmalle HW, Huber R, Fischer P, Hauss T, Ouladdiaf B, Decurtins S (1997) Inorg Chem 36:2301
12. Henelius P, Fishman RS (2008) Phys Rev B 78:214405
13. (a) Clemente-León M, Coronado E, Galán-Mascarós JR, Gómez-García CJ (1997) Chem Commun 1727; (b) Coronado E, Galán-Mascarós JR, Gómez-García CJ, Ensling J, Gütlich P (2000) Chem Eur J 6:552
14. (a) Tang G, He Y, Liang F, Li S, Huang Y (2007) Physica B 392:337; (b) Bhattacarjee A (2007) Physica B 399:77; (c) Tang G, He Y (2007) Physica B 399:79
15. Watts ID, Carling SG, Day P, Visser D (2005) J Phys Chem Sol 66:932
16. Nuttall CJ, Day P (1999) J Solid State Chem 147:3
17. Allen SJ (1968) Phys Rev B 167:492

# Part V
# Jahn-Teller Effect in Mixed Valence Systems

# Chapter 18
# Coherent Spin Dependent Landau-Zener Tunneling in Mixed Valence Dimers

**Andrew Palii, Boris Tsukerblat, Juan Modesto Clemente-Juan, and Eugenio Coronado**

**Abstract** In this contribution we introduce the concept of single molecule ferroelectric based on the vibronic pseudo Jahn-Teller model of mixed valence dimeric clusters belonging to the Robin and Day class II compounds. We elucidate the main factors controlling the nonadiabatic Landau-Zener tunneling between the low lying vibronic levels induced by a pulse of the electric field. The transition probabilities are shown to be dependent on the both time of the pulse and the total spin of the cluster. A possibility to control the spin-dependent Landau-Zener tunneling by applying a static magnetic field is discussed.

## 18.1  Introduction

Two decades ago it was discovered that the $Mn_{12}$ acetate molecule ($Mn_{12}$) exhibits slow relaxation of magnetization at low temperature due to the presence of the

---

A. Palii (✉)
Institute of Applied Physics, Academy of Sciences of Moldova, Academy Str. 5 Kishinev, Moldova
e-mail: andrew.palii@uv.es

B. Tsukerblat
Chemistry Department, Ben-Gurion University of the Negev, Beer-Sheva 84105, Israel
e-mail: tsuker@bgu.ac.il

J.M. Clemente-Juan
Instituto de Ciencia Molecular, Universidad de Valencia, Polígono de la Coma, s/n 46980 Paterna, Spain

Fundació General de la Universitat de València (FGUV), Valencia, Spain
e-mail: juan.m.clemente@uv.es

E. Coronado
Instituto de Ciencia Molecular, Universidad de Valencia, Polígono de la Coma, s/n 46980 Paterna, Spain
e-mail: eugenio.coronado@uv.es

M. Atanasov et al. (eds.), *Vibronic Interactions and the Jahn-Teller Effect: Theory and Applications*, Progress in Theoretical Chemistry and Physics 23, DOI 10.1007/978-94-007-2384-9_18, © Springer Science+Business Media B.V. 2012

barrier for reversal of magnetization arising from the magnetic anisotropy. Under this condition a single molecule behaves like a tiny magnet and being magnetized by an applied field is able to retain magnetization for a long time. Such kind of systems was called single molecule magnets (SMMs). SMMs behave like classical magnets and at the same time exhibit distinct quantum effects like quantum tunneling of magnetization at low temperature. Presently many classes of SMMs are discovered and the concept of single molecule magnetism is well developed [1, 2]. Nowadays this field remains topical due to a possibility of the design of the memory storage devices of molecular scale, quantum bits (qubits) and study of the fundamental aspects of quantum effects in the nanoscopic objects. Study of SMMs [1–9] revealed a possibility to observe the coherent Landau-Zener (LZ) tunneling [10–12] in this kind of magnetic clusters. Below 0.5 K the hysteresis curves of $Mn_{12}$ system exhibit steps at particular values of the magnetic field. These steps were attributed to the LZ tunneling phenomenon.

In this article we propose the idea of a single molecule ferroelectric (SMF), a molecule that has a large dipole moment and a barrier separating two opposite direction of polarization. If the barrier is large enough this molecule can behave as a tiny ferroelectric and can be referred to as SMF by analogy with SMM. To achieve this goal it is reasonable to employ a mixed valence (MV) clusters, the molecular entities that are composed of the ions in different oxidation degrees and therefore containing an "extra" electron (or hole) that can travel between the metal ions [13–24]. In view of the present topic it is to be mentioned that MV systems have a large internal dipole moment mediated by the itinerant electron. As distinguished from the SMMs in which the barrier for the magnetization reversal is caused by the anisotropic magnetic interactions, the MV dimers exhibit a barrier for the reversal of the electric dipole moment (electron tunneling) that is created by the pseudo Jahn-Teller vibronic interaction. It is notable also that in SMMs the LZ transitions are induced by the time-dependent magnetic field, whereas in MV clusters such transitions are expected to be caused by the time-dependent electric field.

We will consider the situation when the MV system is subjected by the action of the time-dependent electric field which represents the pulse of the Gaussian form added to the static electric field that gives rise to the electric polarization of the system. The static electric field induces a ferroelectric type alignment of the electric dipole moments while the electric field pulse is expected to partially destroy this alignment through the LZ transitions, with the latter effect being dependent on the sweep rate of the electric field at the leading and the trailing edges of the pulse. As distinguished from the conventional studies of MV compounds based on the well elaborated Stark spectroscopy [19–24] that is based on the ability of the applied electric field to significantly change the intervalence band, the consideration of the LZ tunneling effects in MV compounds seems to be an unexplored area and the present article represents the prediction and a first step towards the theoretical description of this phenomenon.

Along with the simplest one-electron MV dimer belonging to the Robin and Day class II [16] we will consider a MV dimer containing localized spins coupled to the itinerant electron through the double exchange [25–27]. It is well known that

18 Coherent Spin Dependent Landau-Zener Tunneling in Mixed Valence Dimers    331

the double exchange tends to stabilize the ferromagnetic state of a MV pair [26]. In MV pairs belonging to the class II, the valence at low temperature is trapped and the double exchange is fully suppressed. It will be shown, however, that in the presence of the electric field pulse the double exchange becomes operative even at zero temperature due to the spin-dependent LZ tunneling that changes the electric dipole moment of a MV cluster. It will be demonstrated that the response of the system to the time-dependent electric field can be controlled by applying a static magnetic field. Such possibility seems to be useful for possible various technological applications, particularly, for the creation of qubits and quantum logical gates [28].

## 18.2 Vibronic Model of the Mixed-Valence Dimer and Evaluation of the Tunneling Splitting

Let us first consider as the simplest example a MV dimer comprising one extra electron delocalized over two spinless cores $A$ and $B$. We will use a well known vibronic model proposed by Piepho, Krausz and Schatz (PKS model) [14, 18] which assumes that the instantly localized mobile electron interacts with the full-symmetric ("breathing") modes $Q_A$ and $Q_B$ located on the sites $A$ and $B$. The so called out-of phase mode $Q_- = (Q_A - Q_B)/\sqrt{2} \equiv Q$ is relevant to the vibronic interaction in the whole system while the in-phase mode $Q_+ = (Q_A + Q_B)/\sqrt{2}$ can be eliminated. The total Hamiltonian of the MV dimer with the equivalent sites $A$ and $B$ can be represented as:

$$H = \left( \frac{P^2}{2M} + \frac{1}{2}kQ^2 \right) I + \frac{l}{\sqrt{2}} Q \sigma_Z + B \sigma_X. \tag{18.1}$$

In Eq. 18.1 $P$ is the kinetic moment operator, $B = \langle \varphi_A | \hat{h} | \varphi_B \rangle$ is the electron transfer integral ($\hat{h}$ is the one-electron part of the Hamiltonian), $l$ is the vibronic coupling parameter, $\sigma_Z$ and $\sigma_X$ are the Pauli matrices defined in the basis of the orbitals $\varphi_A(r) = \varphi_A(r - R_A)$, $\varphi_B(r) = \varphi_B(r - R_B)$ and $I$ is the unit matrix.

It is convenient to rewrite this Hamiltonian in terms of the dimensionless variables $q$ and $p$ in which all energies are measured in $\hbar\omega$ units [18]:

$$q = \sqrt{M\omega/\hbar}Q, \quad \lambda = (2\pi M \omega^3)^{-1/2}l,$$

$$p = (M \hbar\omega)^{-1/2} P, \quad \beta = B/(\hbar\omega). \tag{18.2}$$

In Eq. 18.2 $\omega$ is the vibrational frequency, $M$ is the effective mass of the out-of-phase vibration, $\lambda$ and $\beta$ are the dimensionless vibronic coupling and transfer parameters, respectively. Then the Hamiltonian in which all energies are expressed in $\hbar\omega$ units becomes:

$$H = \hbar\omega \left[ \frac{1}{2} \left( q^2 - \frac{\partial^2}{\partial q^2} \right) I + \lambda q \sigma_Z + \beta \sigma_X \right]. \tag{18.3}$$

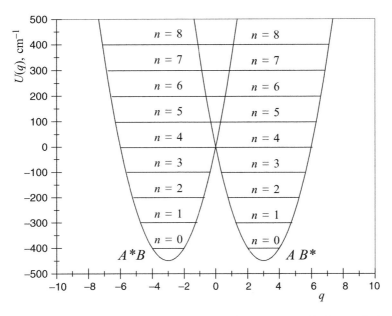

**Fig. 18.1** Adiabatic potential wells and energy pattern of one-electron MV dimer calculated for $\beta = 0$, $\lambda = 3$, $\hbar\omega = 100\,\text{cm}^{-1}$

Applying a shift transformation

$$q' = q + \lambda, \; q'' = q - \lambda \quad (18.4)$$

to the vibrational coordinate, one can present the Hamiltonian as follows:

$$H = \hbar\omega \begin{pmatrix} -\dfrac{\lambda^2}{2} + \dfrac{1}{2}\left(q'^2 - \dfrac{\partial^2}{\partial q'^2}\right) & \beta \\ \beta & -\dfrac{\lambda^2}{2} + \dfrac{1}{2}\left(q''^2 - \dfrac{\partial^2}{\partial q''^2}\right) \end{pmatrix} \quad \begin{matrix} \varphi_A \\ \varphi_B \end{matrix} . \quad (18.5)$$

Let us discuss first the case when $\beta = 0$ (uncoupled sites). In this case the adiabatic potentials of the dimer represent two intersecting parabolas with minima at $q = \pm\lambda$ (Fig. 18.1). The left and right potential wells correspond to the localization of the extra electron on the site $A$ and $B$, respectively. Then taking into account the kinetic energy of the nuclear motion we can present the energy pattern of the dimeric unit as that comprising two equivalent harmonic oscillator patterns corresponding to the left and right potential wells:

$$\varepsilon_A(n) = \varepsilon_B(n) = \hbar\omega\left(-\dfrac{\lambda^2}{2} + n + \dfrac{1}{2}\right). \quad (18.6)$$

The corresponding eigenvectors represent the products of the electronic orbitals $\varphi_A$, $\varphi_B$ and the harmonic oscillator wave-functions with the equilibrium positions shifted to the left and right minima of the adiabatic potential, respectively:

$$\chi_n^A(\mathbf{r},q) = \varphi_A(\mathbf{r})\Phi_n(q+\lambda), \quad \chi_n^B(\mathbf{r},q) = \varphi_B(\mathbf{r})\Phi_n(q-\lambda). \tag{18.7}$$

Now let us proceed to the general case when $\beta \neq 0$. The inclusion of the electron transfer results in the splitting of the adiabatic potential into two sheets defined by the following expressions typical for a pseudo Jahn-Teller molecule with two non-degenerate electronic terms coupled to one low-symmetry mode:

$$U_\pm(q) = \hbar\omega\left(q^2/2 \pm \sqrt{\lambda^2 q^2 + \beta^2}\right). \tag{18.8}$$

Both sheets of the adiabatic potential have an extremum at $q = 0$. For the upper sheet $U_+(q)$ this point is always a minimum. The curvature of the lower sheet $U_-(q)$ depends on the relative value of the parameters $\lambda$ and $\beta$. At weak vibronic coupling, when $\lambda^2 < |\beta|$ we have a single-well sheet $U_-(q)$ and the point $q = 0$ is the minimum. In this case the system proves to be fully delocalized even at low temperatures and belongs to the so-called class III MV compound according to the Robin and Day classification scheme [14, 16, 18]. In the opposite case of strong coupling $\lambda^2 > |\beta|$ $U_-(q)$ represents a double-well sheet with the maximum at $q = 0$ and two minima at

$$q_m^\pm = \pm\sqrt{\lambda^2 - \beta^2/\lambda^2} \tag{18.9}$$

Such a system behaves as localized at low temperatures and delocalized at high temperatures and belongs to the Robin and Day class II compound. Finally for $\lambda^2 \gg |\beta|$ the extra electron is strongly trapped in the deep minimum and the system can be assigned to the fully-localized Robin and Day class I.

In this work we will focus on the situation when the minima of $U_-(q)$ are deep enough, and the system belongs to the class II but it is closer to the class I than to the class III. In this case the electron transfer is expected to produce the shift of the low-lying vibrational levels and a weak tunneling splitting of the low lying levels. We focus on the case of very low temperatures and thus we will only interested in the tunneling splitting of the ground vibrational level with $n = 0$. A simple way of the evaluation of the tunneling splitting $\delta$ is to apply the semiclassical Wentzel, Kramers, and Brillouin (WKB) approach.

The WKB expression for the tunneling splitting energy gap $\delta$ is the following [29]:

$$\delta = 2\hbar\tilde{\omega} \exp\left[-\int_b^c p(q)\,dq\right]. \tag{18.10}$$

In this expression $\tilde{\omega}$ is the frequency of oscillations at the bottom of the well that is in general different from $\omega$, and $p(q)$ is the following linear momentum conjugated with $q$:

$$p(q) = \sqrt{2\left[U_-(q) - \varepsilon_0\right]/\hbar\omega} \qquad (18.11)$$

where $\varepsilon_0$ is the energy of the degenerate level with $n = 0$ that is apparently different from the energy $\varepsilon_A(n = 0) = \varepsilon_B(n = 0)$ of the ground vibrational level in Eq. 18.6 calculated with $\beta = 0$. It follows from Eq. 18.10 that $\delta$ is twice the product of the frequency $\tilde{\omega}$ the system hits the barrier wall and the exponential tunneling factor, which is the probability of tunneling through the barrier at each hit.

In order to calculate $\varepsilon_0$ and $\tilde{\omega}$ we will expend the adiabatic potential $U_-(q)$ in the Taylor series near the minima. For example, expending $U_-(q)$ near the left minimum and retaining the first two nonvanishing terms one obtains:

$$
\begin{aligned}
U_-^{q<0}(q) &\approx U_-(q_m^-) + \frac{1}{2}\left(\frac{d^2 U_-}{dq^2}\right)_{q=q_m^-}(q - q_m^-)^2 \\
&= -\frac{\hbar\omega}{2}\left(\lambda^2 + \frac{\beta^2}{\lambda^2}\right) + \frac{\hbar\omega}{2}\left(1 - \frac{\beta^2}{\lambda^4}\right)(q - q_m^-)^2.
\end{aligned} \qquad (18.12)
$$

Using this result we can define the following effective Hamiltonian for $q < 0$:

$$\hat{H}^{(q<0)} = -\frac{\hbar\omega}{2}\left(\lambda^2 + \frac{\beta^2}{\lambda^2}\right) + \frac{\hbar\omega}{2}\left(1 - \frac{\beta^2}{\lambda^4}\right)(q - q_m^-)^2 - \frac{\hbar\omega}{2}\frac{\partial^2}{\partial(q - q_m^-)^2}. \qquad (18.13)$$

Let us pass to the new coordinate

$$\tilde{q}_- = \gamma\left(q - q_m^-\right), \qquad (18.14)$$

and introduce the effective frequency $\tilde{\omega}$ such that the Hamiltonian takes on the oscillator form

$$
\begin{aligned}
\hat{H}^{(q<0)} &= -\frac{\hbar\omega}{2}\left(\lambda^2 + \frac{\beta^2}{\lambda^2}\right) + \frac{\hbar\tilde{\omega}}{2}\tilde{q}_-^2 - \frac{\hbar\tilde{\omega}}{2}\frac{\partial^2}{\partial\tilde{q}_-^2} \\
&= -\frac{\hbar\omega}{2}\left(\lambda^2 + \frac{\beta^2}{\lambda^2}\right) + \frac{\hbar\tilde{\omega}}{2}\gamma^2\left(q - q_m^-\right)^2 - \frac{\hbar\tilde{\omega}}{2\gamma^2}\frac{\partial^2}{\partial(q - q_m^-)^2}.
\end{aligned} \qquad (18.15)
$$

By comparing Eqs. 18.14 and 18.15 one finds:

$$\tilde{\omega}\gamma^2 = \omega\left(1 - \frac{\beta^2}{\lambda^4}\right), \frac{\tilde{\omega}}{\gamma^2} = \omega. \qquad (18.16)$$

18 Coherent Spin Dependent Landau-Zener Tunneling in Mixed Valence Dimers 335

Solving this system of equations we arrive at the following result:

$$\tilde{\omega} = \omega\sqrt{1 - \frac{\beta^2}{\lambda^4}}, \gamma = \left(1 - \frac{\beta^2}{\lambda^4}\right)^{1/4}. \qquad (18.17)$$

Then the eigenvalues of the Hamiltonian, Eq. 18.15, are the following:

$$\varepsilon_n^{(q<0)} = -\frac{\hbar\omega}{2}\left(\lambda^2 + \frac{\beta^2}{\lambda^2}\right) + \hbar\tilde{\omega}\left(n + \frac{1}{2}\right) = -\frac{\hbar\omega}{2}\left(\lambda^2 + \frac{\beta^2}{\lambda^2}\right)$$

$$+ \hbar\omega\sqrt{1 - \frac{\beta^2}{\lambda^4}}\left(n + \frac{1}{2}\right), \qquad (18.18)$$

and the same expression is evidently true for $\varepsilon_n^{(q>0)}$. Therefore the energy of the ground level is given by

$$\varepsilon_0 = \frac{\hbar\omega}{2}\left(\sqrt{1 - \frac{\beta^2}{\lambda^4}} - \lambda^2 - \frac{\beta^2}{\lambda^2}\right). \qquad (18.19)$$

Finally, the classical turning points are found by solving the equation

$$U_-(q) = \varepsilon_0 \qquad (18.20)$$

Substituting Eqs. 18.17, 18.19 into Eq. 18.10 together with the values $b$ and $c$ found by solving Eq. 18.20 and performing the integration in Eq. 18.10 we obtain the value of $\delta$ for an arbitrary set of the parameters $\lambda$, $\beta$ and $\hbar\omega$. Let us consider the following two examples of such calculations: (i) $\lambda = 3$, $\beta = 1$, $\hbar\omega = 100\,\text{cm}^{-1}$; (ii) $\lambda = 3$, $\beta = 0.5$, $\hbar\omega = 100\,\text{cm}^{-1}$. Performing calculations we obtain in the case (i):

$$b \approx -1.975, c \approx 1.975, \tilde{\omega} \approx 0.994\,\omega, \varepsilon_0 = -405.865\,\text{cm}^{-1}, \delta \approx 0.353\,\text{cm}^{-1} \text{ (i)},$$

$$b = -1.994, c = 1.994, \tilde{\omega} \approx 0.998\,\omega, \varepsilon_0 = -401.466\,\text{cm}^{-1}, \delta \approx 0.27\,\text{cm}^{-1} \text{ (ii)}.$$

$$(18.21)$$

By comparing these two cases one can see that the tunneling splitting increases with the increase of the parameter $\beta$ and/or with the decrease of $\lambda$. Figure 18.2 illustrates the WKB approach showing the adiabatic potential $U_-(q)$, the unsplit ground level $\varepsilon_0$, and the classical turning points $b$ and $c$ in the case (i). At the same time the tunneling splitting is evidently too small to be shown in this scale.

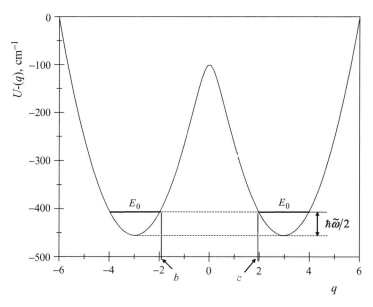

**Fig. 18.2** Illustration for the WKB approach: Lower sheet of the adiabatic potential, the unsplit ground vibrational level, and the classical turning points calculated with $\lambda = 3$, $\beta = 1$, $\hbar\omega = 100\,\mathrm{cm}^{-1}$

## 18.3 Effects of the Time-Dependent Electric Field and Landau-Zener Tunneling in One-Electron Mixed Valence Dimer

Let us first consider the MV dimer in an external static electric field that is directed along the $Z$-axis as shown in Fig. 18.3. The Hamiltonian describing the interaction of MV dimer with the electric field is the following:

$$\hat{H}_d = -\hat{d}_Z E_Z, \qquad (18.22)$$

where $\hat{d}_Z$ is the electric dipole moment operator defined by the following matrix:

$$\hat{d}_Z = \begin{pmatrix} \varphi_A & \varphi_B \\ d_0 & 0 \\ 0 & -d_0 \end{pmatrix}, \qquad (18.23)$$

where $d_0 = eR/2$. In the presence of the electric field the effective two-level Hamiltonian involving only the two low-lying vibrational levels with $n = 0$ is presented by the following $2 \times 2$ matrix:

$$\hat{H}_{\mathit{eff}} = \begin{pmatrix} \chi_0^A & \chi_0^B \\ \varepsilon_0^A & \delta/2 \\ \delta/2 & \varepsilon_0^B \end{pmatrix}, \qquad (18.24)$$

**Fig. 18.3** MV dimer in an applied electric field

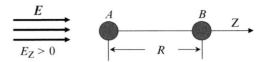

where
$$\varepsilon_0^A = \varepsilon_0 - d_0 E_Z, \quad \varepsilon_0^B = \varepsilon_0 + d_0 E_Z \qquad (18.25)$$
are the energies of the diabatic levels with $A$ and $B$ localization positions, and

$$\chi_0^A = \varphi_A \Phi_0(\tilde{q}_-) = \varphi_A \Phi_0\left[\gamma(q - q_m^-)\right], \quad \chi_0^B = \varphi_A \Phi_0(\tilde{q}_+)$$
$$= \varphi_B \Phi_0\left[\gamma(q - q_m^+)\right]. \qquad (18.26)$$

Solving this matrix we obtain the following expressions for the electronic densities on the sites $A$ and $B$:

$$\rho_A = \frac{1}{2}\left[1 + \frac{d_0 E_Z}{\sqrt{\delta^2 + (d_0 E_Z)^2}}\right], \quad \rho_B = \frac{1}{2}\left[1 - \frac{d_0 E_Z}{\sqrt{\delta^2 + (d_0 E_Z)^2}}\right]. \qquad (18.27)$$

It follows from these formulas that the electric field of chosen direction tends to decrease the electronic density on the site $B$ and to increase it on the site $A$, thus producing the trapping effect as shown in Fig. 18.4. It is seen that the trapping effect is stronger (for the same value of the electric field) for smaller tunneling splitting $\delta$.

Let us consider the situation when in addition to the static electric field the system is subjected by the action of the short symmetric pulse of the electric field that is assumed to be of the bell-shape Gaussian form. The total electric field including both time-independent and time-dependent contributions can be defined as follows:

$$E_Z(t) = E_0 - p E_0 \exp\left(-\frac{t^2}{2\tau^2}\right), \qquad (18.28)$$

where $\tau$ is the characteristic time of the pulse. Time dependences of the applied electric field for $p = 2$ are shown in Fig. 18.5 for two pulses of different durations: $\tau = 1\,s$ and $\tau = 0.5\,s$. Let us note that in order to observe the LZ transitions one needs to have $p > 1$.

In the presence of the time-dependent electric field given by Eq. 18.28 the effective two-level Hamiltonian takes on the following form:

$$\hat{H}_{eff} = \begin{pmatrix} \Delta/2 & \delta/2 \\ \delta/2 & -\Delta/2 \end{pmatrix} \begin{matrix} \chi_0^A \\ \chi_0^B \end{matrix}, \qquad (18.29)$$

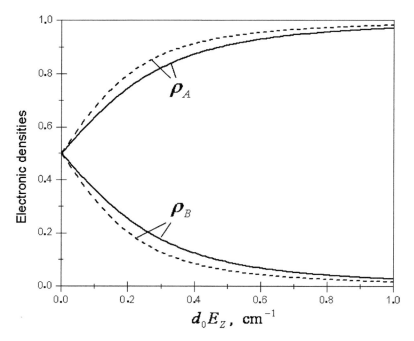

**Fig. 18.4** Electronic densities on the sites $A$ and $B$ calculated as functions of $d_0 E_Z$ at $\delta = 0.353\,\mathrm{cm}^{-1}$ (*solid lines*) and $\delta = 0.27\,\mathrm{cm}^{-1}$ (*dashed lines*)

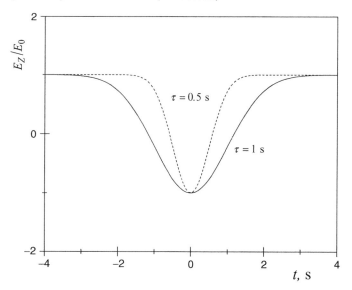

**Fig. 18.5** Time dependences of the applied electric field (p = 2) for $\tau = 1\,s$ (*solid line*), and $\tau = 0.5\,s$ (*dotted line*)

In Eq. 18.29 the value

$$\Delta = \varepsilon_0^A - \varepsilon_0^B = -2d_0E_0\left[1 - pe^{-t^2/(2\tau^2)}\right] \tag{18.30}$$

is the energy gap between the diabatic $(\delta = 0)$ levels. The diabatic levels crossing $(\Delta = 0)$ occurs at times

$$t_T = -t_L = \tau\sqrt{2\ln(p)}, \tag{18.31}$$

where the times $t_T$ and $t_L$ relate to the trailing $(T)$ and the leading $(L)$ edges of the pulse, respectively.

The diagonalization of the $2 \times 2$ matrix in Eq. 18.29 leads to the following expression for the adiabatic energies:

$$\varepsilon_\pm = \pm\frac{1}{2}\sqrt{\Delta^2 + \delta^2} = \pm\sqrt{(d_0E_0)^2\left[1 - p\exp\left(-t^2/2\tau^2\right)\right]^2 + \delta^2/4}. \tag{18.32}$$

The time-dependence of the adiabatic levels are shown in Fig. 18.6 for two different times of the pulse $\tau = 1\,\mathrm{s}$ (upper figure) and $\tau = 0.5\,\mathrm{s}$ (lower figure) assuming $d_0E_0 = 2\,\mathrm{cm}^{-1}$, $\delta = 0.353\,\mathrm{cm}^{-1}$, and $p = 2$. The adiabatic levels pass through the anticrossing points at times $t = t_L$ and $t = t_T$ for which the diabatic levels cross. The adiabatic approximation breaks down in the vicinity of these two anticrossing areas, and the system undergoes the intensive transitions between the ground and excited states via the LZ mechanism [10–12].

The time-dependent sweep rate of the applied electric field $c(t) = \frac{d\Delta}{dt}$ is calculated as follows:

$$c(t) = -\frac{2pd_0E_0t}{\tau^2}e^{-t^2/2\tau^2}. \tag{18.33}$$

At the crossing points the sweep rate takes the values:

$$c(t_L) = -c(t_T) \equiv c = \frac{2d_0E_0}{\tau}\sqrt{2\ln(p)}. \tag{18.34}$$

Using the fact that the time interval for the LZ transitions is narrow, one can assume the linear time dependence of the gap between the diabatic levels near the crossing points. One can thus expand the function $\Delta(t)$ in the Taylor series near the points $t_L$ and $t_T$ and retain only the first nonvanishing terms of this expansion, that is

$$\Delta(t) \approx c(t_L)(t - t_L)\,\mathrm{near}\,t_L;$$

$$\Delta(t) \approx c(t_T)(t - t_T)\,\mathrm{near}\,t_T. \tag{18.35}$$

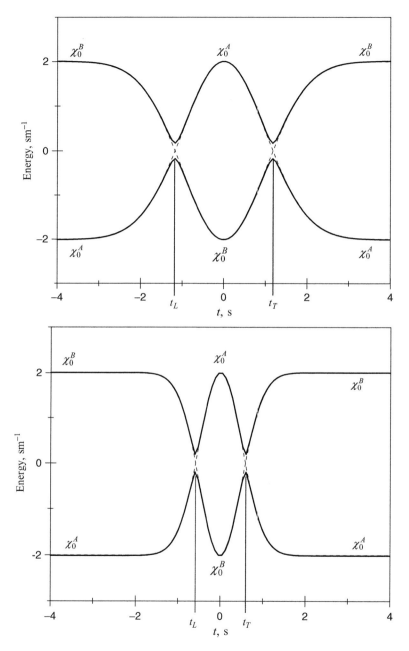

**Fig. 18.6** Time-dependence of the adiabatic (*solid lines*) and diabatic (*dashed lines*) levels calculated with: $d_0 E_0 = 2 \text{cm}^{-1}$, $\delta = 0.353 \text{cm}^{-1}$, $p = 2$ and $\tau = 1\text{s}$ (upper figure); and $\tau = 0.5\text{s}$ (lower figure)

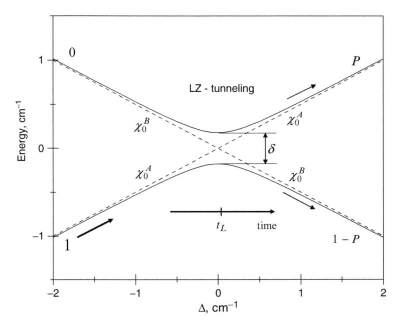

**Fig. 18.7** Illustration for the Landau-Zener tunneling at the leading edge of the pulse: Diabatic (*dashed lines*) and adiabatic (*solid lines*) levels calculated with $\delta = 0.353\,\text{cm}^{-1}$ are plotted as functions of $\Delta$ near the point $\Delta(t_L) = 0$

Under these conditions the LZ theory leads to the following formula for the probability of the transition between the ground and excited states at the leading edge of the pulse [10–12]:

$$P = \exp[-\pi\delta^2/(2\hbar c)]. \tag{18.36}$$

The Landau-Zener tunneling at the leading edge of the pulse is schematically shown in Fig. 18.7, where the diabatic and adiabatic levels are plotted as functions of $\Delta$ near the point $\Delta(t_L) = 0$, provided that $\delta = 0.353\,\text{cm}^{-1}$. It is seen from Fig. 18.7 that the transition from the ground state to the excited one is not accompanied by the change of the localization site of the extra electron, so, the probability in Eq. 18.36 is the probability for system to conserve the $A$-localization of the extra electron, that is $P = P_A^L$. On the contrary, if the system remains in the ground state the extra electron changes its localization, and the probability of this event is

$$P_{A \to B}^L = 1 - P. \tag{18.37}$$

In the adiabatic limit ($c \to 0$) the change of the localization is complete, i.e., $P = 0$, $P_{A \to B}^L = 1$ (all MV dimers remain in the ground state changing the localization) and for very fast sweeps ($c \to \infty$) all dimers are scattered from the ground state to the excited one with the extra electron in all of them being localized at the same site $A(P = 1, P^L_{A \to B} = 0)$. Finally, for the intermediate values of the sweep rate we are dealing with the situation when the change of the localization is partial, that is in some of dimers the extra electron has changed its localization position and in other dimers not $(0 < P < 1, 0 < P^L_{A \to B} < 1)$.

Since the pulse of the electric field is symmetric it is reasonable to assume that the transition probability $P^T_i$ $(i = A, B)$ between the ground and excited states at the trailing edge of the pulse $(t \sim t_T)$ coincides with the probability $P^L_i$ at the leading edge of the pulse $(t \sim t_L)$, that is

$$P^T_i = P^L_i = P, \tag{18.38}$$

where $P$ is given by Eq. 18.36. The probability of the transition from the ground state to the excited one during the total time of the pulse $(t \gg t_T)$ can be obtained with the aid of the following arguments [30]. Let us assume that the ground $\to$ excited transition took place at the leading edge of the pulse with the probability $P$. Then the system will stay in the state $\chi^+_0$ at $t \gg t_T$ if no back excited $\to$ ground transition can occur at the trailing edge of the pulse. The probability of the absence of such back transition is $1 - P$ (the probabilities of the ground $\to$ excited and excited $\to$ ground transitions are apparently equal). The total probability $P_1$ of these two consecutive events (the first event is the ground $\to$ excited transition, and the second event is the absence of the back transition) represents the following product of the probabilities of each of these events:

$$P_1 = P(1 - P). \tag{18.39}$$

However there is another scenario when no transition takes place at the leading edge of the pulse (the probability of this event is $1 - P$), and the system undergoes the ground $\to$ excited transition with the probability $P$ at the trailing edge of the pulse. The probability of this process is given by:

$$P_2 = (1 - P)P \equiv P_1. \tag{18.40}$$

The total probability of the transition from the ground state to the excited one during the time of the pulse $(t \gg t_T)$ is obtained as a sum of the probabilities $P_1$ and $P_2$. It is seen from Fig. 18.6 that this probability is that of changing the $A$-localization to $B$-localization, so we can write down

$$P_{A \to B} = P_1 + P_2 = 2P(1 - P)$$

$$= 2 \exp\left[-\frac{\pi \delta^2 \tau}{4 \hbar d_0 E_0 \sqrt{2 \ln(p)}}\right] \left\{1 - \exp\left[-\frac{\pi \delta^2 \tau}{4 \hbar d_0 E_0 \sqrt{2 \ln(p)}}\right]\right\}. \tag{18.41}$$

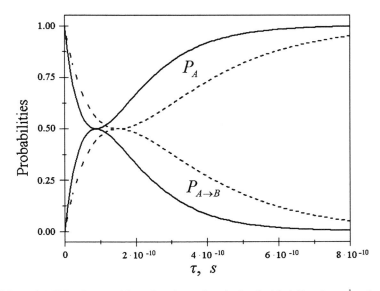

**Fig. 18.8** Probabilities $P_{A \to B}$ and $P_A$ as functions of $\tau$ calculated with $d_0 E_0 = 2\,\text{cm}^{-1}$ and $p = 2$ for the $d^1$–$d^0$ dimer with $\delta = 0.353\,\text{cm}^{-1}$ (*solid lines*) and $\delta = 0.27\,\text{cm}^{-1}$ (*dashed lines*). Alternatively, these curves show $P^S_{A \to B}(\tau)$ and $P^S_A(\tau)$ functions calculated at $\beta = 1$, $\lambda = 3$ and $\hbar\omega = 100\,\text{cm}^{-1}$ for $S = 3/2$ (*solid lines*) and $S = 1/2$ (*dashed lines*) states of the $d^2$–$d^1$ dimer

Finally, the probability for system to remain in the ground state with retaining the $A$-localization of the extra electron is given by

$$P_A = 1 - P_{A \to B}. \tag{18.42}$$

If the time-dependent wave-function $\psi(t)$ was initially $\psi(t \ll t_L) = \chi_0^A$ then, at times $t \gg t_T$, it becomes the following superposition of the states:

$$\psi(t \gg t_T) = c_A \chi_0^A + c_B \chi_0^B, \tag{18.43}$$

where

$$|c_A|^2 = P_A, \ |c_B|^2 = P_{A \to B}. \tag{18.44}$$

The probabilities to conserve the localization of the extra electron ($P_A$) and to change this localization ($P_{A \to B}$) plotted as functions of the time of the pulse $\tau$ are shown in Fig. 18.8 for tunneling splitting values $\delta = 0.353\,\text{cm}^{-1}$ and $\delta = 0.27\,\text{cm}^{-1}$. It is seen that $P_{A \to B}$ is close to 0 and $P_A$ is close to 1 for short and long pulses. At the time point

$$\tau_m = 4\ln(2)\hbar d_0 E_0 (\pi^{-1}\delta^{-2})\sqrt{2\ln(p)} \tag{18.45}$$

the curve $P_{A \to B}(\tau)$ reaches its maximum, and $P_A(\tau)$ becomes minimal with

$$(P_{A \to B})_{\max} = (P_A)_{\min} = 1/2. \tag{18.46}$$

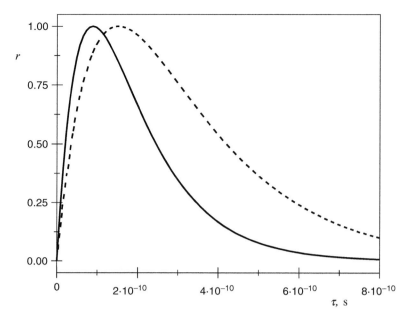

**Fig. 18.9** The $r$ vs. $\tau$ plot calculated with $d_0 E_0 = 2\,\text{cm}^{-1}$ and $p = 2$ for the $d^1$–$d^0$ dimer with $\delta = 0.353\,\text{cm}^{-1}$ (*solid line*) and $\delta = 0.27\,\text{cm}^{-1}$ (*dashed line*). Alternatively, these curves show $r^S$ vs. $\tau$ plots calculated with $\beta = 1$, $\lambda = 3$ and $\hbar\omega = 100\,\text{cm}^{-1}$ for the $S = 3/2$ (*solid line*) and $S = 1/2$ (*dotted line*) states of the $d^2$–$d^1$ dimer

This means that at $\tau = \tau_m$ in a half of MV dimers the extra electron changes the site of localization giving rise to a flip of their electric dipole moments, while for all other values of $\tau$ the number of such dimers is less than half of their total number.

An alternative way to characterize the nonadiabatic LZ transition is to plot the change of the expectation value of the electric dipole moment as a function of $\tau$. This expectation value at $t \gg t_T$ is calculated with the aid of Eqs. 18.23, 18.25, 18.43 and 18.44. We thus obtain that $\bar{d}_Z$ after the pulse is given by:

$$\bar{d}_Z(t \gg t_T) = P_A d_0 - P_{A \to B} d_0 = d_0 \left(1 - 2 P_{A \to B}\right). \tag{18.47}$$

On the other hand the initial (before the pulse) $\bar{d}_Z$ value is

$$\bar{d}_Z(t \ll t_L) = d_0. \tag{18.48}$$

Then the LZ transition will be characterized by the dimensionless value

$$r = \frac{\bar{d}_Z(t \ll t_L) - \bar{d}_Z(t \gg t_T)}{d_0} = 2 P_{A \to B}, \tag{18.49}$$

that characterizes the response of the system to the pulse of the electric field. Figure 18.9 shows the plot of this value as function of $\tau$ calculated for $\delta =$

0.353 cm$^{-1}$ and $\delta = 0.27$ cm$^{-1}$. It follows from Eq. 18.49 that $r \to 0$ (no flipping of the electric dipole moments) if $\tau \to 0$ or $\tau \to \infty$. On the other hand at $\tau = \tau_m$ we find that $\bar{d}_Z(t \gg t_T) = 0$ (the numbers of the electric dipole moments parallel and antiparallel to $Z$ axis are equal and hence the initial electric polarization induced by the time-independent electric field is fully destroyed by the pulse), and the function $r(\tau)$ reaches its maximum value $r_{\max} = 1$ at $\tau = \tau_m$.

## 18.4 Effect of the Double Exchange on the Landau-Zener Transitions

In a more general case the itinerant electron travels over localized spins (the so called spin cores) and the double exchange coupling becomes operative. The interaction of the localized and delocalized spins through the double exchange [24] results in the ferromagnetic properties of MV systems. In this case the transfer integral becomes spin-dependent (it will be denoted as $\beta_S$) as it was discovered for the double exchange mechanism [25–27], with the spin-dependence being the following [26]:

$$\beta_S = \frac{\beta(S+1/2)}{(2s_0+1)}, \tag{18.50}$$

where $S$ is the total spin of the dimer, and $s_0$ is the spin of the spin-core.

We will consider the simplest many-electron MV dimer of $d^2 - d^1$-type. In this case $s_0 = 1/2$, the total spin $S = 1/2, 3/2$, and we find from Eq. (18.50) that $\beta_{1/2} = \beta/2$, $\beta_{3/2} = \beta$. As a result the tunneling splitting $\delta$ proves to be also spin-dependent. Denoting this spin-dependent splitting as $\delta_S$ and keeping in mind that the vibronic parameter $\lambda$ is independent of $S$ one arrives at the conclusion that $\delta_{3/2} > \delta_{1/2}$. Particularly, for $\beta = 1$ one finds $\beta_{1/2} = 0.5$, $\beta_{3/2} = 1$, and the parameters $\delta_{3/2}$ and $\delta_{1/2}$ for this case have been already calculated in Sect. 18.2 (Eq. 18.21):

$$\delta_{3/2} \approx 0.353 \,\text{cm}^{-1}, \delta_{1/2} \approx 0.27 \,\text{cm}^{-1}. \tag{18.51}$$

It is seen from Eqs. 18.41, 18.42 and 18.51 that the probabilities $P_{A \to B}$, $P_A$ and the response $r$ also depend on $S$. These spin-dependences are shown in Figs. 18.8 and 18.9. Thus, the curves shown by solid lines describe the LZ transitions for the state with $S = 3/2$, and those shown by dotted lines relate to the state with $S = 1/2$.

It follows from Eq. 18.45 that $\tau_m \sim \delta^{-2}$, so the position of the extremum in the $P_{A \to B}^S(\tau)$, $P_A^S(\tau)$ and $r^S(\tau)$ dependences is shifted to the area of smaller $\tau$ when we pass from the state with $S = 1/2$ to the state with $S = 3/2$.

Along with the double exchange the Heisenberg-Dirac-Van-Vleck exchange interaction is also operative in the many-electron MV clusters. This interaction acts within each electron localization and is defined by the spin Hamiltonian

$$H_{ex} = -2J \mathbf{s}_A \mathbf{s}_B \tag{18.52}$$

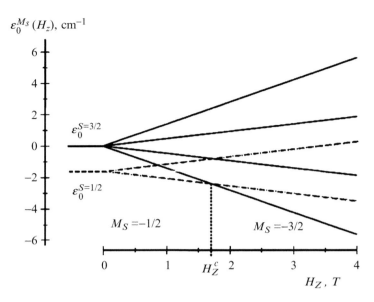

**Fig. 18.10** Crossing of the diabatic levels under the action of the applied static magnetic field at $J = -2\,\text{cm}^{-1}$, $\lambda = 3$, $\beta = 1$, $\hbar\omega = 100\,\text{cm}^{-1}$. The level $\varepsilon_0^{S=3/2}$ at $H_Z = 0$ is chosen as a reference point

where $J$ is the exchange parameter. In most cases of transition metal complexes the exchange interaction is antiferromagnetic ($J < 0$) thus tending to stabilize the energy level with $S = 1/2$ with respect to that with $S = 3/2$. Let us consider the case of relatively weak exchange interaction ($J = -2\,\text{cm}^{-1}$). Then providing $\lambda = 3$, $\beta = 1$, $\hbar\omega = 100\,\text{cm}^{-1}$ we obtain the following values of the spin-dependent diabatic energies $\varepsilon_0^S$ at $E_Z = 0$:

$$\varepsilon_0^{S=1/2} = 2J + \frac{\hbar\omega}{2}\left(\sqrt{1 - \frac{\beta^2}{4\lambda^4}} - \lambda^2 - \frac{\beta^2}{4\lambda^2}\right) \approx -405.466\,\text{cm}^{-1},$$

$$\varepsilon_0^{S=3/2} = -J + \frac{\hbar\omega}{2}\left(\sqrt{1 - \frac{\beta^2}{\lambda^4}} - \lambda^2 - \frac{\beta^2}{\lambda^2}\right) \approx -403.865\,\text{cm}^{-1}, \qquad (18.53)$$

It is seen that the ground diabatic level at $E_Z = 0$ is that with $S = 1/2$. The same is evidently true when $E_Z \neq 0$ and hence in the low temperature limit when the excited levels are not populated, only the $S = 1/2$ component is affected by the pulse of the electric field and participates in the LZ transitions. The LZ tunneling regime can be however changed by applying the static magnetic field. In fact, if this field is strong enough the states with $S = 3/2$ can become active in Landau-Zener transitions. This is illustrated by the right side of Fig. 18.10. When the magnetic field is smaller than the critical value

$$H_Z^c = \frac{1}{g_e \mu_B} \left[ -3J + \frac{\hbar\omega}{2} \left( -\frac{3\beta^2}{4\lambda^2} + \sqrt{1 - \frac{\beta^2}{\lambda^4}} - \sqrt{1 - \frac{\beta^2}{4\lambda^4}} \right) \right] \approx 1.714\,\text{T} \quad (18.54)$$

the Zeeman sublevel with $M_S = -1/2$ arising from the multiplet with $S = 1/2$ is the ground state. The increase of the magnetic field results in the crossing of the Zeeman sublevels, and for $H_Z > H_Z^c$ the state with $M_S = -3/2$ originating from the multiplet with $S = 3/2$ becomes the ground one. Since the electron transfer does not change the total spin projection of the dimer, the Landau-Zener tunneling between the ground Zeeman sublevels of different localizations with $M_S = -3/2$ is described exactly in the same way as the tunneling occurring between the full $S = 3/2$ levels. Figure 18.11 shows the change of the adiabatic levels of the $d^2$–$d^1$ dimer under the action of the static magnetic field. In the absence of the magnetic field the ground adiabatic level possesses the spin $S = 1/2$ (Fig. 18.11a) and this is the only state participating in the LZ transition with the response $r^{1/2}$. At the applied field $H_Z = 3\,\text{T} > H_Z^c$ (Fig. 18.11b) the adiabatic level with $M_S = 3/2$ arising from the state with $S = 3/2$ becomes the ground state and the electric response to the electric field pulse is $r^{3/2}$. The change of the responses $\Delta r = r^{M_S=3/2} - r^{M_S=1/2}$ when passing from $H_Z < H_Z^c$ to $H_Z > H_Z^c$ regime non-monotonically depends on the time of the pulse $\tau$ as shown in Fig. 18.12. At short pulses $r^{M_S=3/2}$ exceeds $r^{M_S=1/2}$ while for long pulses $r^{M_S=3/2} < r^{M_S=1/2}$.

Therefore we arrive at the conclusion that the response of the many-electron MV systems to the electric field pulse can be controlled by the both width of the electric field pulse and static magnetic field.

## 18.5 Concluding Remarks

To summarize the following issues should be mentioned:

1. The electric field pulse has been shown to induce the LZ transitions in MV clusters belonging to the Robin and Day class II. It is shown that the probability to change the site of the localization of the excess electron and thus to reverse the electric dipole moment upon the action of the pulse $(t \gg t_T)$ increases with the increase of the time $\tau$ of the pulse, reaches the maximal value 1/2 and then decreases. On the contrary, the probability to conserve the localization position is decreased when $\tau$ is increased, reaches the minimum that is equal to 1/2 (half of dimers are overturned) and then is increased to 1 (no reversal of the electric dipole moment) at long pulses. This provides a possible way to create controllable superposition of the states with different electron localizations that can be regarded as a proposal for the realization of qubits based on MV dimers;

2. An important peculiarity of the LZ tunneling in many-electron MV dimers is the spin-dependent character of the phenomenon. This opens a way to control the response of the system to the time-dependent electric field by applying the static magnetic field, which can stabilize different spin-states. Such possibility is of potential importance for applications of these systems as molecular electronic devices;

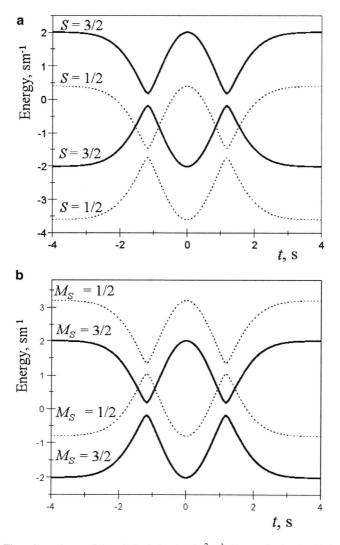

**Fig. 18.11** Time-dependence of the adiabatic levels of $d^2$–$d^1$ dimer calculated with $J = -2\,\text{cm}^{-1}$, $\lambda = 3, \beta = 1, \hbar\omega = 100\,\text{cm}^{-1}$ at $H_Z = 0$ (**a**) and at $H_Z = 3\,\text{T}$ (**b**). The levels with $S = 3/2$ and $S = 1/2$ are shown by *solid* and *dotted lines* respectively. The anticrossing point for the levels with $S = 3/2$ is chosen as a reference point

3. The theory reported here can be regarded only as a first step in the study of the nonadiabatic LZ tunneling in MV compounds. It is applicable only in the low-temperature limit and neglects the decoherence effects arising from the electric dipole-dipole interactions between MV dimers and from the interactions of the electric dipoles with phonons. Therefore the development of a more comprehensive theory of incoherent LZ tunneling in MV clusters accompanied

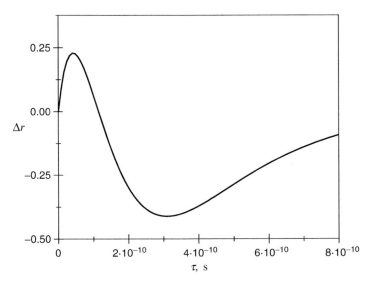

**Fig. 18.12** Change of the responses $\Delta r = r^{M_S=3/2} - r^{M_S=1/2}$ when passing from $H_Z < H_Z^c$ to $H_Z > H_Z^c$ as function of $\tau$

by the experimental study of the LZ transitions seems to be the next step towards the better understanding of this fascinating phenomenon and its possible applications in the design of nanoelectronic devices.

**Acknowledgments** A.V. Palii thanks the Paul Scherrer Institute for financial support that made possible his participation in the Jahn-Teller Symposium. A.V.P. also gratefully acknowledges financial support from STCU (project N 5062) and the Supreme Council on Science and Technological Development of Moldova. B.T. thanks the Israel Science Foundation (ISF), for the financial support (grant no. 168/09). J.M.C.J. and E.C. thank Spanish MICINN (CSD2007-00010 CONSOLIDER-INGENIO in Molecular Nanoscience, MAT2007-61584, CTQ-2008-06720 and CTQ-2005-09385), Generalitat Valenciana (PROMETEO program), and the EU (MolSpinQIP project and ERC Advanced Grant SPINMOL) for financial support.

# References

1. Gatteschi D, Sessoli R, Villain J (2006) Molecular nanomagnets. Oxford University Press, Oxford
2. Gatteschi D, Sessoli R (2003) Angew Chem Int Edu 42:268
3. Miyashita S (1995) J Phys Soc Jpn 64:3207
4. Miyashita S (1996) J Phys Soc Jpn 65:2734
5. Dobrovitski VV, Zvendin AK (1997) Europhys Lett 38:377
6. Gunther L (1997) Europhys Lett 39:1
7. De Raedt H, Miyashita S, Saito K, García-Pablos D, García N (1997) Phys Rev B 56:11761
8. Wernsdorfer W, Sessoli RA, Caneschi D, Gatteschi A (2000) Cornia Europhys Lett 50:552
9. Leuenberger MN, Loss D (2000) Phys Rev B 61:12200

10. Landau L (1932) Phys Z Sowjetunion 2:46
11. Zener C (1932) Proc R Soc Lond A 137:696
12. Stückelberg ECG (1932) Helv Phys Acta 5:369
13. Day P, Hush NS, Clark RJH (2008) Philos Trans R Soc A 366:5
14. Borras-Almenar JJ, Clemente-Juan JM, Coronado E, Palii AV, Tsukerblat BS (2001) In: Miller J, Drillon M (eds) From molecules to materials. Villey-VCH, Weinheim, p 155
15. Tsukerblat B, Klokishner S, Palii A (2009) In: Köppel H. Yarkony DR, Barentzen H (eds) The Jahn-Teller effect. Fundamentals and implications for physics and chemistry, Springer series in chemical physics, vol 97. Springer, Heidelberg, pp 555–620
16. Robin MB, Day P (1967) Adv Inorg Chem Radiochem 10:247
17. Hush NS (1967) Prog Inorg Chem 8:391
18. Piepho SB, Krausz ER, Schatz N (1978) J Am Chem Soc 100:2996
19. Reimers JR, Hush NS (1990) In: Prassides K (ed) Proceedings of the NATO advanced workshop on mixed valency compounds: applications in chemistry, physics and biology; Series C. Kluwer, Dordrecht
20. Oh DH, Sano M, Boxer SG (1991) J Am Chem Soc 113:5880
21. Brunschwig BS, Creutz C, Sutin N (1998) Coord Chem Rev 177:61
22. Treynor TP, Boxer SG (2004) J Phys Chem A 108:1764
23. Kanchanawong P, Dahlbom MG, Treynor TP, Reimers JR, Hush NS, Boxer SG (2006) J Phys Chem B 110:18688
24. Silverman LN, Kanchanawong P, Treynor TP, Boxer SG (2008) Philos Trans R Soc A 366:33
25. Zener C (1951) Phys Rev 82:403
26. Anderson PW, Hasegawa H (1955) Phys Rev 100:675
27. De Gennes P-G (1960) Phys Rev 118:141
28. Loss D, Di Vincenzo DP (1998) Phys Rev A 57:120
29. Polinger V (2003) Adv Quan Chem 44:59
30. Kovarsky VA, Perel'man NF, Averbuch YS, Baranov SA. Todirashcu SS (1980) Nonadiabatic Transitions in a Strong Electromagnetic Field. Pub. Stiintsa Academy of Sciences of Moldova, Kishinev (in Russian)

# Chapter 19
# Mixed Valence Iron Dimer in the Generalized Vibronic Model: Optical and Magnetic Properties

**Serghei M. Ostrovsky**

**Abstract** A mixed valence iron dimer $[L^1Fe_2(\mu - OAc)_2](ClO_4)$ is investigated in the framework of the generalized vibronic model which takes into account both the local vibrations on the metal sites (Piepho-Krausz-Schatz model) and the molecular vibrations changing the intermetallic distance (suggested by Piepho). It is shown that the behaviour of the system is determined by a strong competition between three main processes: double exchange interaction and vibronic coupling with both types of vibrations. The optical and magnetic properties of the regarded compound are reported. The influence of the key parameters of the system on these properties is studied in the framework of the presented theoretical model. The degree of delocalization of the itinerant 'extra' electron and probability distribution in configuration (qQ) space are calculated at different values of temperature.

## 19.1 Introduction

Mixed-valence (MV) transition metal clusters contain two or more metal ions in different oxidation states between which an itinerant 'extra' electron can transfer. These kinds of compounds are of great interest because of their unusual spectral and magnetic properties and the possibility to use these properties for studying the intramolecular electron transfer process in many biologically important systems [1, 2]. For instance, valence delocalized $[Fe_2S_2]^+$ pairs where iron is in the formal oxidation state $+2.5$ were found in a variety of iron-sulfur clusters in ferredoxins. Because of this, a growing interest in the characterization, the modelling and in the theoretical description of such compounds exists [3–6].

The basic theory of the double exchange interaction was formulated in [7, 8]. It was shown that due to electron delocalization process each spin level is split

---

S.M. Ostrovsky (✉)
Institute of Applied Physics of the Academy of Sciences of Moldova, Kishinev, Moldova
e-mail: sm_ostrovsky@yahoo.com

M. Atanasov et al. (eds.), *Vibronic Interactions and the Jahn-Teller Effect:*
*Theory and Applications*, Progress in Theoretical Chemistry and Physics 23,
DOI 10.1007/978-94-007-2384-9_19, © Springer Science+Business Media B.V. 2012

351

into two sublevels, the splitting being linear in the total spin value. So, the moving electron produces strong ferromagnetic effect. The energy levels of the dimeric system with orbitally non degenerate ground state in the presence of both double exchange and Heisenberg-type magnetic exchange interactions can be described by the well known equation:

$$E_{\pm}(S) = -JS(S+1) \pm B(S+1/2), \qquad (19.1)$$

where $J$ is the isotropic exchange parameter, $B$ means the double exchange parameter and $S$ is the total spin of the system.

The moving electron strongly perturbs the ligand environment so the vibronic coupling plays a considerable role in MV systems along with isotropic and double exchange interactions. The background for the consideration of the vibronic effects in this kind of systems was suggested by Piepho, Krausz and Schatz and became well-known as the PKS model [9–11]. PKS model operates with the full-symmetric ("breathing") vibrations of the local environment of each ion of the cluster. The vibronic interaction with the out-of-phase breathing mode is relevant to the electron transfer process. Strong PKS interaction produces a localization effect.

During ten years the PKS model was the only possibility to take into account vibronic interactions in MV systems. Later Piepho [12,13] suggested a new vibronic coupling model for this type of compounds. In this model multi-center vibrations are produced by the modulation of the intermetallic distances. The generalized vibronic model was suggested in [14] with the emphasis on the problem of localization and magnetic properties of the many-electron dimers. As distinguished from the PKS-type of vibrations the coupling with the intercenter vibrations was shown to contribute to the electron delocalization effect and thus produces a ferromagnetic effect in the MV dimers.

The dinuclear mixed-valence $[L^1Fe_2(\mu - OAc)_2](ClO_4)$ complex is completely delocalized over the whole temperature range [15–18] and belongs to Class III systems in the Robin and Day classification scheme [19]. The compound exhibits a broad intervalence absorption band. It is impossible to explain a broad intervalence band of the fully delocalized system in the framework of pure PKS model. A possibility of such kind of broad bands in the Class III systems providing relatively strong intercenter vibrations was mentioned in [14]. In this contribution the $[L^1Fe_2(\mu - OAc)_2](ClO_4)$ complex is investigated in the framework of the generalized vibronic model which takes into account both the PKS and the Piepho-type of vibrations. The results have been presented in part in [20].

## 19.2   Model

Let us consider a dimeric MV system with one itinerant 'extra' electron delocalized over two spin cores $S_1 = S_2$. To explain the behaviour of the regarded compound

19  Mixed Valence Iron Dimer in the Generalized Vibronic Model...  353

a Hamiltonian that includes isotropic exchange interaction, double exchange and vibronic coupling with PKS and Piepho type of vibrations is used:

$$\mathbf{H} = \mathbf{H}_{ex} + \mathbf{H}_{de} + \mathbf{H}_{PKS} + \mathbf{H}_{Piepho}. \tag{19.2}$$

All of these interactions cannot mix states of the system belonging to the different values of the total spin $S$. As a consequence, for each value of the total spin $S$ the block of the full Hamiltonian matrix is as follows:

$$\mathbf{H}(S) = \left[ -JS(S+1) + \frac{\omega}{2}\left( q^2 - \frac{\partial^2}{\partial q^2} \right) + \frac{\Omega}{2}\left( Q^2 - \frac{\partial^2}{\partial Q^2} \right) \right] \begin{matrix} \psi_+^{SM} \ \psi_-^{SM} \\ \begin{pmatrix} 1 & 0 \\ 0 & 1 \end{pmatrix} \end{matrix}$$

$$+ \begin{matrix} \psi_+^{SM} \qquad\qquad\qquad \psi_-^{SM} \\ \begin{pmatrix} B(S+1/2) - \lambda(S)Q & \upsilon q \\ \upsilon q & -(B(S+1/2) - \lambda(S)Q) \end{pmatrix} \end{matrix} \tag{19.3}$$

where $\lambda(S) = (S+1/2)\lambda_0$ is the spin-dependent P-coupling parameter possessing the same spin dependence as the effective (multielectron) double exchange [14], $\upsilon$ means the PKS vibronic coupling parameter, $\omega$ and $\Omega$ are the frequencies of the PKS and P-vibrations respectively. Since the investigated dimeric complex is centrosymmetric, no difference in the energy between the localization of the 'extra' electron on sites 1 and 2 is included in the model. The basis set in Eq. 19.3 is:

$$\psi_\pm^{SM}(r) = (\psi_1^{SM}(r) \pm \psi_2^{SM}(r))/\sqrt{2}, \tag{19.4}$$

where $\psi_1^{SM}(r)$ and $\psi_2^{SM}(r)$ are the electronic wave functions of the system with the 'extra' electron localized on centre 1 and 2, respectively, $M$ means the magnetic quantum number.

The eigenfunctions of (19.3) can be found in the following form:

$$\Phi_\nu^{SM}(r,q,Q) = \sum_{n=0}^{\infty} \sum_{m=0}^{\infty} (\psi_+^{SM}(r)C_{\nu nm}^{+(S)}\chi_n(q)\chi_m(Q) + \psi_-^{SM}(r)C_{\nu nm}^{-(S)}\chi_n(q)\chi_m(Q)),$$

$$\tag{19.5}$$

where $\chi_n(q)$ and $\chi_m(Q)$ are harmonic oscillator functions. In practical calculations for moderate vibronic coupling it is enough to take into account the upper limit in the summations equal to 50 without significant loss of accuracy. In this case the dimension of the matrix for each value of the total spin is equal to 5,000 and the problem can be easily solved.

To obtain the profile of the intervalence absorption band, the dipole intensity of each vibronic line is calculated. The corresponding electronic transitions are z polarized [11, 12]. Following the Franck-Condon approximation one can obtain:

$$D(v'SM \rightarrow vSM) = \left( \frac{N_{v'}^{SM} - N_v^{SM}}{N} \right) \delta_{v'v} \left| \left\langle \psi_+^{SM} \left| m_\gamma \right| \psi_-^{SM} \right\rangle^0 \right|^2, \qquad (19.6)$$

where $m_\gamma$ is the $\gamma$-th component of the electric dipole operator and the matrix element $\langle \psi_+^{SM} | m_\gamma | \psi_-^{SM} \rangle^0$ is evaluated at $q = 0$ and $Q = 0$. $N_v^{SM}$ is the thermal population of the corresponding vibronic level and

$$\delta_{v'v} = \sum_{n,m} \left( C_{v'nm}^{+(S)} C_{vnm}^{-(S)} + C_{v'nm}^{-(S)} C_{vnm}^{+(S)} \right). \qquad (19.7)$$

The intervalence absorption band is the totality of transitions $(v'SM \rightarrow vSM)$ with different spin $S$ and magnetic quantum number $M$. Each of these transitions can be replaced by Gaussian and thus the theoretical intervalence band can be calculated.

The temperature dependence of the magnetic susceptibility can be calculated as:

$$\chi_{\gamma\gamma} = N_A k_B T \frac{\partial^2}{\partial H_\gamma^2} (\ln Z(H_\gamma))_{H_\gamma \rightarrow 0}, \qquad (19.8)$$

where $\gamma = x, y, z$ and $Z(H_\gamma)$ is the partition function of the system in the presence of the external magnetic field. The powder-average magnetic susceptibility is calculated as:

$$\chi = \frac{1}{3} (\chi_{XX} + \chi_{YY} + \chi_{ZZ}). \qquad (19.9)$$

Significant information about the localization (delocalization) of the 'extra' electron can be obtained from the probability distribution in configuration $(qQ)$ space. The expression $P_v^{SM}(qQ) = \int |\Phi_v^{SM}|^2 d\tau_{el}$ represents the probability to find the system at configuration $qQ$ in $\Phi_v^{SM}$ state. This probability is independent of the magnetic quantum number $M$ and has the following form:

$$P_v^S(qQ) = \sum_{n,n'} \sum_{m,m'} \chi_n(q) \chi_{n'}(q) \chi_m(Q) \chi_{m'}(Q) \left( C_{vnm}^{+(S)} C_{vn'm'}^{+(S)} + C_{vnm}^{-(S)} C_{vn'm'}^{-(S)} \right). \quad (19.10)$$

For the obtaining of the probability to find the system at configuration $qQ$ in any state, Eq. 19.10 is summarized over all total spin values, its projections and vibronic levels using corresponding Boltzmann distribution.

Another important value for estimation of the extent of valence trapping is a degree of delocalization of the itinerant 'extra' electron. To calculate this value, the procedure suggested in [11] is used. The value $P_v^{SM}(el) = \int \int |\Phi_v^{SM}|^2 dq dQ$ consists of three parts and specifies the relative contributions of $|\psi_1^{SM}|^2$, $|\psi_2^{SM}|^2$ and $\psi_1^{SM} \psi_2^{SM}$. Since $\psi_1^{SM} \psi_2^{SM}$ is a measure of delocalization, the degree of delocalization of the 'extra' electron in state $\Phi_v^{SM}$ is defined by the square of its coefficient [11]. As a result, for the degree of delocalization of the 'extra' electron in the state $\Phi_v^{SM}$ one finds:

$$(\beta_d^S(v))^2 = \left\{\sum_n \sum_m \left( \left(C_{vnm}^{+(S)}\right)^2 - \left(C_{vnm}^{-(S)}\right)^2 \right)\right\}^2 \qquad (19.11)$$

To obtain the degree of delocalization of the 'extra' electron for the whole system $\bar{\beta}_d^2$ one has to summarize Eq. 19.11 over all total spin values, its projections and vibronic levels using corresponding Boltzmann distribution. The degree of delocalization of the itinerant 'extra' electron together with the probability distribution in configuration space provide an important information about the behaviour of the MV dimer at given temperature.

## 19.3 Analysis

The synthesis, crystal structure and Mössbauer data of the mixed-valence iron dimer $[L^1Fe_2(\mu-OAc)_2](ClO_4)$ were described elsewhere [15]. Mössbauer data show that the regarded compound is completely delocalized over the whole temperature range. The previously reported intervalence absorption band of the investigated dimer [15,16] was measured for the compound dissolved in different solvents. To avoid the influence of solvents, for the present analysis the near infrared absorption measured for the regarded complex pressed in potassium bromide was used. The investigated compound demonstrates a broad intervalence absorption band in the near infrared region at about $8,900\,cm^{-1}$. With the temperature increase this band becomes slightly broader. The position of the maximum of the intervalence absorption band is practically independent of the temperature which is typical for Class III compounds. As it was already mentioned in the framework of pure PKS model the fully delocalized system (Class III) exhibits strong narrow intervalence absorption band. However, the investigated complex shows a broad intervalence band and can be explained in the model with two different types of vibrations only [14, 17]. It is assumed that the low symmetry crystal field removes the orbital degeneracy of high spin Fe(II) and the model described above can be applied to the investigated dimeric complex.

The central part of $[L^1Fe_2(\mu-OAc)_2](ClO_4)$ cluster consists of two iron ions and ten atoms from their nearest surrounding and has $(3N-6) = 30$ metal-ligand vibrational degrees of freedom. For the sake of simplicity only core contributions to the vibrational spectrum are taken into account. For four atoms (two irons and two bridging oxygen atoms) there are the six metal-ligand vibrational degrees of freedom and the problem can be easily solved using standard normal coordinate analysis program. With force constant from [5, 6] the following frequencies of vibrations were obtained: $310\,cm^{-1}$ for PKS and $125\,cm^{-1}$ for Piepho type of vibrations (Fig. 19.1). Of course, the model is oversimplified and it is not possible to

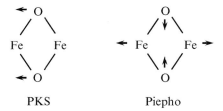

**Fig. 19.1** Normal modes associated with PKS (*left*) and Piepho (*right*) vibrations (the arrows show relative nuclear displacement)

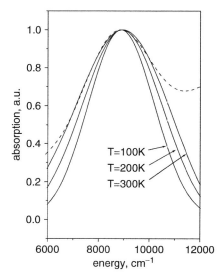

**Fig. 19.2** Intervalence absorption band (*dotted curve*, $T = 300$ K) and theoretical fit (*solid curves*) calculated at $B = 260 \text{cm}^{-1}$, $\omega = 310 \text{cm}^{-1}$, $\upsilon = 3\omega$, $\Omega = 125 \text{cm}^{-1}$, $\lambda_0 = \Omega$

establish a one-to-one correspondence between the real spectrum of the system and these idealized vibrations. However, the obtained values of frequencies look quite reasonable and are used in further calculations.

Figure 19.2 shows the intervalence absorption band for the investigated compound calculated at 100, 200 and 300 K. One can see a good correlation between the shape and the position of the experimental and calculated curves. With the temperature increase the intervalence absorption band becomes broader. The position of the maximum of the theoretical curve is independent of the temperature that is in a good agreement with the experiment.

It should be mentioned that although in the theoretical simulations of the intervalence absorption band there are three different parameters (double exchange, vibronic coupling parameters with PKS and Piepho vibrations) all of them can be more or less accurately obtained from the comparison between the experimental and theoretical data because their influence on the position and the shape of the intervalence absorption band as well as on the delocalization degree of the 'extra' electron is different. The increase of all of these parameters shifts the position of the maximum of the intervalence band to high energy region but the bigger value of the double exchange results in smaller band width while the growth of coupling parameters leads to a broad band contour with a decrease of the degree of delocalization in the case of PKS vibrations and an increase of the degree of delocalization for the Piepho vibrations.

The obtained value of the double exchange parameter $(B = 260\,\text{cm}^{-1})$ looks rather small as compared to the previous considerations [15–17] and needs some explanations. The Piepho vibrations change the intermetallic distances and result thus in the modulation of the transfer integral. Expanding the double exchange parameter in Tailor series to a first approximation one gets [14]:

$$B(R) = B(R_0) - \lambda_0 Q, \tag{19.12}$$

where $R = R_{12}$ is the metal-metal distance, $R_0$ is the metal-metal distance for system consisting of two spin-cores, $Q = R - R_0$ is the vibrational coordinate in the Piepho-model, $\lambda_0 = -(\partial B(R)/\partial R)_{R=R_0}$ is the vibronic coupling parameter. One can see that the obtained value of the double exchange parameter corresponds to the non distorted system $(R = R_0)$. Due to vibronic coupling with Piepho type of vibrations the iron-iron distance changes and the real (effective) value of the double exchange for new nuclear configuration can be calculated using Eq. 19.12.

To explain the magnetic properties of the regarded compound, the zero-field splitting for the ferromagnetic ground state was added to the Hamiltonian (19.3). Only axial component $D_{\text{gr}}$ of zero-field splitting tensor is assumed to be non zero. The experimental magnetic properties of the regarded compound are shown in Fig. 19.3 as well as the theoretical fit (relative error is about 0.4%). The $\chi T$ value at low temperature (but more than 30 K) indicates a ferromagnetic ground state. In the temperature region 50 K–300 K the $\chi T$ product slightly depends on the temperature. The high temperature magnetic behaviour is very sensitive to the isotropic exchange parameter (at given values of the double exchange and vibronic coupling parameters) and allows us to determine this value very accurately. It is illustrated in the inset to Fig. 19.3. The sharp decrease of the effective magnetic moment below 30 K is explained by the zero field splitting of the ferromagnetic ground state. The obtained values of the exchange parameter and the zero field splitting parameter are in a good agreement with the previously reported in [17].

Figure 19.4 shows the probability distribution in the configuration space for the investigated compound at low and at room temperature. At low temperature

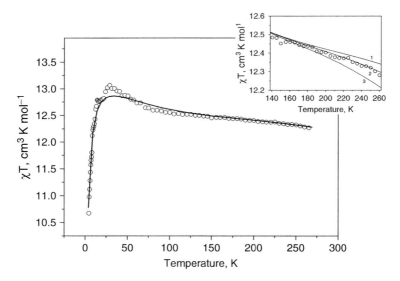

**Fig. 19.3** Temperature dependence of the $\chi T$ product (open circles represent the experimental values, solid curve - best fit obtained with $B = 260\,\text{cm}^{-1}$, $J = -5\,\text{cm}^{-1}$, $\omega = 310\,\text{cm}^{-1}$, $\upsilon = 3\,\omega$, $\Omega = 125\,\text{cm}^{-1}$, $\lambda_0 = \Omega$, $D_{gr} = 4\,\text{cm}^{-1}$, $g_{\parallel} = 1.8$, $g_{\perp} = 2.09$). The *inset* demonstrates the influence of the exchange interaction on high temperature magnetic properties: $J = 5\,\text{cm}^{-1}$ (1), $J = -5\,\text{cm}^{-1}$ (2) and $J = -15\,\text{cm}^{-1}$ (3)

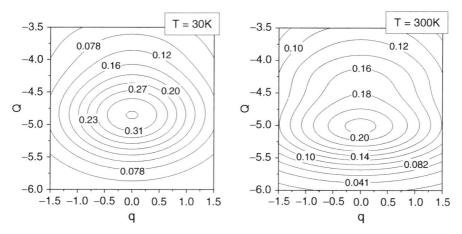

**Fig. 19.4** Contour plots for probability distribution in configuration space: $T = 30\,\text{K}$, $\bar{\beta}_d^2 = 0.97$ (*left*) and $T = 300\,\text{K}$, $\bar{\beta}_d^2 = 0.90$ (*right*)

the probability distribution in configuration space represents a sharp peak. The most probable configuration corresponds to the situation when both subunits are equivalent and have the equal oxidation degree. The 'extra' electron is fully delocalized. With the temperature increase this peak becomes lower and the

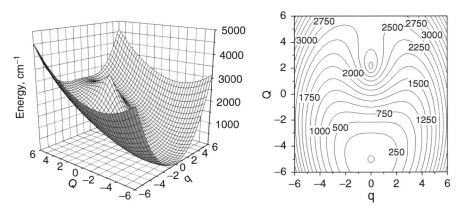

**Fig. 19.5** Adiabatic potential and the corresponding contour plot for $S = 9/2$ ground state

probability to find the system at configuration with smaller values of $Q$ increases. It corresponds to the thermal elongation of Fe-Fe distance. The 'extra' electron remains fully delocalized. The regarded complex belongs to Class III in Robin and Day classification scheme. Respective degrees of delocalization of the 'extra' electron are $\bar{\beta}_d^2 = 0.97$ at $T = 30$ K and $\bar{\beta}_d^2 = 0.90$ at $T = 300$ K.

The energy gaps between $S = 7/2, 5/2, 3/2, 1/2$ states and the ground $S = 9/2$ one are equal to $740 \, \text{cm}^{-1}$, $1{,}220 \, \text{cm}^{-1}$, $1{,}350 \, \text{cm}^{-1}$ and $1{,}390 \, \text{cm}^{-1}$, respectively. Even at room temperature the effect of excited states corresponding to $S = 7/2, 5/2, 3/2$ and $1/2$ is negligible and the behavior of the complex under study is mainly determined by the $S = 9/2$ ground state. The adiabatic potential for the $S = 9/2$ ground state is shown in Fig. 19.5. A single deep minimum is located in $\{0, -\lambda(S)/\Omega\}$ position. The 'extra' electron in this minimum is fully delocalized. The effective value of the double exchange parameter calculated in this minimum (Eq. 19.12) is equal to $885 \, \text{cm}^{-1}$ which is in a good agreement with the previously reported results ($B = 820 \, \text{cm}^{-1}$ in [17] and $B = 896 \, \text{cm}^{-1}$ in [18]).

## 19.4 Summary

A mixed valence iron dimer $[L^1 Fe_2(\mu\text{--}OAc)_2](ClO_4)$ is investigated in the framework of the generalized vibronic model which takes into account both the local vibrations on the metal sites and the molecular vibrations changing the intermetallic distance. The behaviour of the regarded compound is determined by a strong competition between three processes: double exchange interaction, vibronic coupling with Piepho type of vibrations and vibronic coupling with PKS out-of-phase mode. The first and the second interactions lead to delocalization of the itinerant electron while the third one represents a trapping effect that tends to localize the 'extra' electron. It was shown that the properties of the complex under study are mainly determined by the $S = 9/2$ ground state. The corresponding adiabatic surface possesses a deep single minimum with the fully delocalized 'extra' electron. The investigated complex belongs to Class III in Robin and Day classification scheme.

**Acknowledgments** Financial support of STCU (project N5062) is highly appreciated.

# References

1. Bominaar EL, Achim C, Borshch SA, Girerd J-J, Münck E (1997) Inorg Chem 36:3689
2. Achim C, Bominaar EL, Münck E (1998) J Biol Inorg Chem 3:126
3. Papaefthymiou V, Girerd J-J, Moura I, Moura JJG, Münck E (1987) J Am Chem Soc 109:4703
4. Noodleman L, Case DA (1992) Adv Inorg Chem 38:423
5. Gamelin DR, Bominaar EL, Kirk ML, Wieghardt K, Solomon EI (1996) J Am Chem Soc 118:8085
6. Gamelin DR, Bominaar EL, Mathoniere C, Kirk ML, Wieghardt K, Girerd J-J, Solomon EI (1996) Inorg Chem 35:4323
7. Zener C (1951) Phys Rev 82:403
8. Anderson PW, Hasegawa H (1955) Phys Rev 100:675–681
9. Piepho SB, Krausz ER, Schatz PN (1978) J Am Chem Soc 100:2996
10. Schatz PN (1980) In: Brown DB (ed) Mixed-valence compounds. Reidel, Dordrecht, p 115
11. Wong KY, Schatz PN (1981) Prog Inorg Chem 28:369
12. Piepho SB (1988) J Am Chem Soc 110:6319
13. Piepho SB (1990) J Am Chem Soc 112:4197
14. Borras-Almenar JJ, Coronado E, Ostrovsky SM, Palii AV, Tsukerblat BS (1999) Chem Phys 240:149
15. Saal C, Mohanta S, Nag K, Dutta SK, Werner R, Haase W, Duin E, Johnson MK (1996) Ber Buns Phys Chem 100:2086
16. Dutta SK, Ensling J, Werner R, Flörke U, Haase W, Gütlich P, Nag K (1997) Angew Chem 109:107
17. Ostrovsky SM, Werner R, Nag K, Haase W (2000) Chem Phys Lett 320:295
18. Böhm MC, Saal C, Haase W (1999) Inorg Chim Acta 291:82
19. Robin MB, Day P (1967) Adv Inorg Radiochem 10:247
20. Ostrovsky S (2003) In : Haase W, Wrobel S (eds) Relaxation phenomena. Springer, Heidelberg, pp 609–626

# Chapter 20
# Spectroscopic Detection of Hopping Induced Mixed Valence of Ti and Sc in $GdSc_{1-x}Ti_xO_3$ for x Greater than Percolation Threshold of 0.16

**Gerry Lucovsky, Leonardo Miotti, and Karen Paz Bastos**

**Abstract** Only two of the first row transition metal binary oxides are either ferro- or ferri-magnetic. These are $CrO_2$ and $Fe_3O_4$. The electron spin alignment promoting electron spin alignment is associated with a double exchange mechanism requiring mixed valence as well as metallic conductivity. This chapter describes a novel way to realize these two necessary, but not sufficient conditions for double exchange magnetism. These are mixed valence and a hopping conductivity that can promote intra-plane electron spin alignment in a complex oxide host perovskite, $GeSc_{1-x}Ti_xO_3$. This in-plane spin-correlation does necessarily produce for producing spin alignment between alternating $(Sc,Ti)O_2$ atomic planes, especially in distorted perovskite structures. Intra-plane alignment is obtained when the A-atom of a trivalent atom $AB(D)O_3$ peroskite, in this example Gd, is an ordinary metal or a rare earth atom, the B-atom, in this example Sc, is a $d^0$ transition metal, and the dopant atom, D, in this example, $Ti^{3+}$ in $d^1$ state, is a $d^n$ transition with $n \geq 1$, as in $GdSc_{1-x}Ti_xO_3$. This article combines X-ray absorption spectroscopy, multiplet theory, and degeneracy removal by a Jahn-Teller effect mechanisms to demonstrate intra-layer mixed valence for Sc and Ti above a percolation threshold $x \sim 0.16$ at which hopping transport is associated with a metal to insulator transition. This has been observed in epitaxial films, and not in nano-grain nanocrystalline, where the number of Sc atoms in a grain with 2–5 nm dimensions is orders of magnitude too small for observation of an a hopping conductivity that requires a percolation mechanism.

---

G. Lucovsky (✉) • L. Miotti • K.P. Bastos
Department of Physics, North Carolina State University, Raleigh, NC 27695-8202, USA
e-mail: lucovsky@ncsu.edu

M. Atanasov et al. (eds.), *Vibronic Interactions and the Jahn-Teller Effect:*
*Theory and Applications*, Progress in Theoretical Chemistry and Physics 23,
DOI 10.1007/978-94-007-2384-9_20, © Springer Science+Business Media B.V. 2012

## 20.1 Introduction

In the **GeSc$_{1-x}$Ti$_x$O$_3$** distorted orthorhombic perovskite structure, Gd$^{3+}$ is in an f$^7$ high-spin state, and Sc$^{3+}$ is in a d$^0$ state. Substitution of normally tetravalent Ti onto Sc trivalent sites requires that Ti also be trivalent, Ti$^{3+}$, and therefore in a d$^1$ state that maintains charge neutrality. X-ray absorption spectroscopy (XAS) confirms the Ti substitution produces conditions necessary, but not sufficient for double exchange magnetism above a percolation threshold of $\sim$16%, but only in expitaxial thin films. This includes mixed valence for both Ti and Sc, and a metallic conductivity [1, 2]. These are related, and it will be demonstrated that mixed valence comes from in-plane electron hopping electron transport in epitaxial single crystal thin films. Satisfaction of these conditions necessary for double exchange magnetism are demonstrated by interpreting X-ray spectroscopy results with multiplet theory [3–6]. However, the distorted character of the Gd(Sc,Ti)O$_3$ lattice favors an anti-ferromagnetic ordering of alternating (Sc,Ti)O$_2$ planes [7].

Section 20.2 discusses two different and complimentary aspects of theory: (i) charge transfer multiplet(CTM) theory for interpretation of number of dipole-allowed transition features in Sc and Ti L$_{2,3}$ spectra for different initial and final state d$^n$ ion configurations [5, 6], in this instance, n = 0, 1 and 2, and (ii) multiplet theory as applied to occupied d-states, d$^n$, with n > 1, in transition metal oxides [4]. Multiplet theory has been shown in Ref. [8] to apply to O-vacancy defect states through an empirical *equivalent d$^2$* model. The development of this model is addressed in detail in Ref. [8] and the results of the model are used in this chapter to describe strain-induced O-vacancies, and distortions that derive from a cooperative in-plane Jahn-Teller effect (JTE) in the GdSc$_{1-x}$Ti$_x$O$_3$ alloys.

Section 20.3 presents to the spectroscopic results and their interpretation, O K edge and Sc and Ti L$_{2,3}$ spectra. Section 20.4 combines the experimental results of Sect. 20.3 and the theory developed in Sect. 20.2 into a microscopic description of GdSc$_{1-x}$Ti$_x$O$_3$ alloy issues to highlight the differences between long range order in single crystal epitaxial films and intermediate range order on a significantly smaller length scale, $\sim$2–5 nm, in nanocrystalline, nano-grain thin films.

## 20.2 Theory

The first subsection addresses charge CTM coherent transitions important in core level spectroscopy for (i) the O K edge, and (ii) the Ti and Sc L$_{2,3}$ edges [5, 6]. The concept of multiplets is subsequently used in a second subsection to provide the basis for interpretation of O K *pre-edge* spectral features assigned to O-atom vacancy defects introduced by local strain by substituting Ti atoms for Sc atoms in ScO$_2$ planes of the perovskite structure [8–10]. This will be addressed later on in the chapter. At this point it is sufficient to recognize that Sc and Ti trivalent 3+ ions have atomic radii that differ by more than five percent.

**Table 20.1** Splitting of one-electron levels and term symbols into non-spherical irreducible representations that apply in solids, including their internal interfaces

| One electron level | Term symbol | Irreducible representations spanned | |
|---|---|---|---|
| Spherical | Spherical | Octahedral | Tetrahedral |
| s | S | $A_{1g}$ | $A_1$ |
| p | P | $T_{1u}$ | $T_2$ |
| d | D | $E_g + T_{2g}$ | $E + T_2$ |
| f | F | $A_{2g} + T_{1u} + T_{2u}$ | $A_2 + T_1 + T_2$ |
| g | G | $A_{1g} + E_g + T_{1g} + T_{2g}$ | $A_1 + E + T_1 + T_2$ |
| h | H | $E_u + 2T_{1u} + T_{2u}$ | $E + T_1 + 2T_2$ |

## 20.2.1 Charge Transfer Multiplet Theory: Ti and Sc $L_{2,3}$ Transitions

Ti and Sc $L_{2,3}$ transitions are spin conserving transitions from occupied Ti and Sc 2p core levels, $2p_{3/2}$ and $2p_{1/2}$, to empty Ti and Sc 3d states, i.e., virtual bound states [11]. The degeneracy of the $2p_{3/2}$ level is four, and the degeneracy of the $2p_{1/2}$ level is two. CTM transitions for (a) $Ti^{4+} d^0$ and (b) $Ti^{3+} d^1$ ground states, respectively, in spherical symmetry $O_3$ group are represented in Eqs. 20.1a and b:

$$2p^6 3d^0 \rightarrow 2p^5 3d^1 \tag{20.1a}$$

$$2p^6 3d^1 \rightarrow 2p^5 3d^2 \tag{20.1b}$$

The next step in the theoretical approach for interpretation of spectroscopic data is to specify the terms that correspond to the ground and final state configurations [5–7]. These are given, respectively in Eq. 20.2. This aspect of the analysis is important for determining the differences in the number of features in Ti and Sc $L_{2,3}$ spectra in the next section. These samples must be $GdSc_{1-x}Ti_xO_3$ epitaxial films, at compositions exceeding the percolation threshold of ~16.5% Ti. These symbolic term relation equations, corresponding respectively to Eqs. 20.1a and 20.1b are:

$$^1S_0 \rightarrow {}^1F_3 \text{ and} \tag{20.2a}$$

$$^2D_{5/2,3/2} \rightarrow 2G_{9/2,7/2} \tag{20.2b}$$

The final step necessary for interpretation of the relevant Ti and Sc $L_{2,3}$ spectra is to specify the states in the octahedral $O_h$ point group that correspond to the terms in the transition equations expressed in the spherical symmetry $O_3$ group. This is accomplished using the branching ratio Table 20.1 that also includes the tetrahedral Td symmetry. The same terms for tetrahedral symmetry also apply to cubic symmetry. The branching for cubic symmetry with a coordination derived from the $CaF_2$ geometry must include u and g substrates, and an interchange of the 1 and 2 substrates for $T_i$ states. Consider first the CTM theory transition from the empty state, $Ti^{4+}d^0$ to the singly occupied $Ti^{3+}d^1$ state. This is presented in Eq. 20.1a and is repeated here for emphasis: $2p^63d^0 \rightarrow 2p^53d^1$. The ground state term is $^1S_0$ and corresponds to a non-degenerate "s-like state" $^1A_{1g}$ state, and final state term is $^1F_3$ (see Eq. 20.2a). Using Table 20.1, the final states that conserve spin and are orthogonal are triply degenerate $T_{1u}$ states with J values of 1 and 3 [6]. The orthogonality relationship between initial and final states is presented in Eq. 20.3,

$$<^1T_{1u}|^1A_{1g}>=0 \qquad (20.3)$$

This relation insures a non-vanishing dipole moment for radiative transitions. Equation 20.3 applies for all combinations of the $^1A_{1g}$ ground state, and the two triply degenerate $^1T_{1u}$ final states in the $O_h$ octahedral symmetry group. Based on the degenerate character for the final states, six transitions are expected for the $2p^63d^0 \rightarrow 2p^53d^1$ transition. This is observed in cubic $SrTiO_3$, in which the octahedral arrangement is ideal. There are four transitions in the $L_3$, lower energy region of the spectrum, and two transitions in the $L_2$ higher energy region. This counting of transitions applies to cubic $SrTiO_3$, and will be addressed later on in this chapter as well.

The second transition, Eq. 20.1b is $2p^63d^1 \rightarrow 2p^53d^2$. The ground state terms are $^2D_{5/2,3/2}$ "d-like" states, so that the final states must be either "p-like" or "f-like". Based on Table 20.1, the possible spin-conserving final states accessible for the $^2G_{9/2,7/2}$ term symmetry are $^2P_{1/2,3/2}$ and $^2F_{5/2,7/2}$ corresponding, respectively to $L_2$ transitions, and $L_3$ transitions. Based on the triply degenerate character of the final states, twelve transitions are expected for the $2p^63d^0 \rightarrow 2p^53d^1$ transition. This is what is observed in $LaTiO_3$. $L_{2,3}$ spectra will be presented for $d^0$ $SrTiO_3$ and $d^1$ $LaTiO_3$ later on in this chapter.

Proceeding in a similar way, for a $d^2$ ground state, there are also twelve transitions for the $2p^63d^2 \rightarrow 2p^53d^3$ transition. This transition has an $^3F_{4,3,2}$ ground state term symbol, and an $^3H_{6,5,4}$ final state term. There two words four threefold degenerate possible final states, $^3D_{3,2}$ and $^3G_{5,4}$, and therefore 12 spectral features are predicted.

In addition to applying CTM to the Ti and Sc $L_{2,3}$ transitions, the CTM formalism provides a novel quantitative approach for the interpretation of O K edge spectrum that is an improvement over the qualitative discussion in Ref. [5]. The significant aspects of this application of the CTM approach for OK edge spectra are illustrated in Fig. 20.1a, b for $ZrO_2$. The energy ordering of transition energies in the Zr $M_{2,3}$ spectrum, including the 5s satellite features, is displayed in Fig. 20.1a. This includes $3p^64d^0$ to $3p^54d^1$ transitions, as well as $3p^65s^0$ to $3p^55s^1$ *satellite transitions* [5, 6]. The number of states expected in the $M_3$ and $M_2$ energy regimes of the spectra is determined by the number of spin-allowed transitions determined corresponding to term initial and final terms. For idealized $ZrO_2$ in a $CaF_2$ lattice, this corresponds to four dominant states for $L_3$ and $L_2$ portions of the spectrum. These are displayed in Fig. 20.1a [5, 6]. For $ZrO_2$, the average separation between the $M_3$ and $M_2$ equivalent d- and s-state features is equal to $\sim 14\,eV$, and approximately the same spin-orbit splitting of the Zr 3p state, 13.7 eV [12]. It is significant to note that

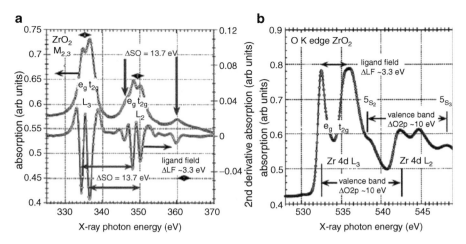

**Fig. 20.1** (a) $M_{2,3}$ XAS and second derivative $M_{2,3}$ spectra for nanocrystalline $ZrO_2$ indicating spin orbit splitting of the Zr 3p core level and ligand field splitting at the Zr site. (b) O K edge spectrum for nanocrystalline monoclinic $ZrO_2$ indicating the energy difference between O2p valence band molecular orbitals, and ligand field splitting at the Zr site

the relatively small 3d–4s atomic state energy difference, ~2.8 eV [13], explains why the satellite 3d to 4s spectral features in Fig. 20.1a *bracket* or *encapsulate* the primary 3p to 4d transitions, rather than occur sequentially as they do in rutile and anatase $TiO_2$ nano-grain films where the 3d to 4s atomic state energy difference is larger, ~5 eV [14, 15].

In the O K edge spectrum for $ZrO_2$, the spitting between the two groups of features corresponding to the $M_3$ and $M_2$ regimes of the Zr $M_{2,3}$ spectrum in Fig. 20.1a is reduced to ~10 eV in Fig. 20.1b. This reduction is associated with a difference in energy between O 2p σ-bonding and 2p π-bonding, or 2p π-bonding and non-bonding valence band O 2p state differences. It derives from a *coherent process* in which the core hole is filed by electrons from O-atom 2p states from the valence band of $ZrO_2$. This converts the O K transitions to O 2p to Zr 4d transitions, with O 2p to Zr 5s satellite features that *encapsulate* the d-state terms.

## 20.2.2 Multiplet Theory of O-Vacancy Defects in $GdSc_{1-x}Ti_xO_3$ Alloys

The removal of a neutral O-atom from a nanocrystalline or single crystal TM elemental or complex oxide results in two electrons being confined within the vacancy site bounded by TM atoms. In the ionic limit, the two electrons donated to the O-atom by the TM atoms to increase the formal charge of O to −2, remain behind when a neutral O-atom is removed. These Hf atoms border on the vacancy site as

**Fig. 20.2** Schematic representation of O-atom vacancy defect in t-HfO$_2$, including ideal tetrahedral coordination, and a Jahn Teller distortion that minimizes repulsions between the two singly occupied Hf orbitals

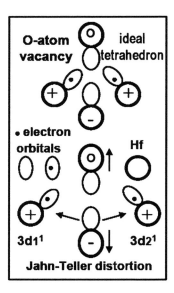

in Fig. 20.2. This model is first applied to idealized HfO$_2$ and ZrO$_2$ in a CaF$_2$ high-temperature phase. This CaF$_2$ geometry for the vacancy is introduced to indicate the nonequivalent nature of the TM atoms bordering the vacancy after a cooperative JTE distortion, and an additional relaxation to minimize electron-electron repulsive interactions. The two electrons are then associated with two of the four Hf atoms bordering the O-vacancy site.

Previously reported density functional theory (DFT) calculations assumed the same two electrons were shared equally among the four TM d-states bordering an O-atom vacancy in ZrO$_2$ or HfO$_2$, and represented could be represented by a doubly occupied one-electron theory band state [8, 9]. This distributed electron configuration results in fractional d-orbital occupancies inconsistent with Hund's rules as applied to strongly correlated d-electron systems [1, 10].

The new model for the O-vacancy addressed in this chapter continues using rutile (r-) TiO$_2$ as an example. The two electrons *in the vacancy site* reside on 3d orbitals of two of the three Ti atoms that border on the O-vacancy site. There is a cooperative JTE distortion that breaks the symmetry of the equilateral triangle site for an idealized rutile lattice of sixfold coordinated Ti atoms in an octahedral arrangement. These octahedra are corner-connected through the threefold coordinated O atoms in a planar geometry independent of longer-range order. The rutile lattice of TiO$_2$ is distorted with an extension in z-direction and octahedra are corner connected. There is a distortion in the x–y plane that reduces geometry from a perfect square to a parallelepiped with two acute and two obtuse angles. These symmetry-reducing atomic displacements make the three Ti atoms bordering the vacancy *nonequivalent* with respect to their local site symmetries.

The derivation of the new model is based on an equivalent Hund's rule occupancy of d-states, uses a term symbol (TS) description. The model is developed in the context the Russell-Saunders, L–S, coupling approximation [5, 10]. The term symbol for a two-electron state is $^{2S+1}L_J$, where L is the total atomic angular momentum, $l_1 + l_2$, S is total electron spin, either 1 for 0, and J is the maximum value of L + S. The symmetry designations, L, such as for S, P and D have values of L = 0, 1, and 2, respectively and are analogous to the atom values of s, p, and d, used for one-electron spherical symmetry states [5, 10]. For a given configuration, with each 3d electron and 4d electron on the same atom, the maximum value of L is 4, and the corresponding integral vector sums of the angular momentum are L = 0, 1, 2, 3, and 4, corresponding to S, P, D, F and G terms. The spin vectors of the two electrons have values of S = 0 or 1, corresponding to singlet and triplet spin states with degeneracies, 2S + 1, equal to 1 and 3, respectively. These specify values of J for each term symbol. Term symbols with different values of both L and S, e.g., $^{1}P_1$, and $^{3}P_{0,1,2}$ have different energies, typically of order 1 eV. Term symbols with the same S and L values, but different values of J, as e.g., $^{3}P_{0,1,2}$ have smaller energy differences usually fractions of an eV; e.g., $0.25 \pm 0.05$ eV [16, 17].

For the O-atom vacancy in $TiO_2$, the two electrons must be in the same 3d shell. These are resident on *nonequivalent Ti atoms* bordering the relaxed O-atom vacancy. These are singly occupied Ti 3d orbitals designated, respectively as $d_1$ and $d_2$. With no restrictions on angular atomic momentum vectors, $l_1$ or $l_2$, and spin, the term symbol for this pair of atomic states has a maximum angular momentum L = 4, and spin values of either 1 or 0, and a maximum value of S = 1. The term symmetry is then $^{3}G_5$ [5, 10]. This includes seventeen different terms: five singlets, $^{1}S_0$, $^{1}P_1$, $^{1}D_2$, $^{1}F_3$ and $^{1}G_4$, and four triplets, $^{3}P_{0,1,2}$, $^{3}D_{1,2,3}$, $^{3}F_{2,3,4}$, and $^{3}G_{3,4,5}$, with twelve different energies after degeneracy removal.

Since, the constraints of the distorted rutile lattice result in partial overlap of radial wave functions for the *two nonequivalent 3d-electrons*, Hund's rules for spin, and the Pauli exclusion principle *can be applied* to these two electrons as well. The maximum value of S = 1, corresponds to triplet states, and the minimum value of S = 0, corresponds to singlet states. The atomic values are of L, $l_1$ and $l_2$, are different, but the sum must be maximized so that $l_1 = 2$ and, $l_2 = 1$. This assignment of $l_j'$s renders the two 3d states orthogonal, thereby satisfying the Pauli exclusion principle. The value of L = 3 for the orthogonal pair of 3d states corresponds to an F symmetry. The maximum values of L, S and J combine to make this symmetry term $^{3}F_4$ with J values of 0, 1, 2, 3, and 4. The $^{3}F_4$ term is included in unrestricted $^{3}G_5$ term symmetry, the term symmetry with no restrictions on either spin, or the values of $l_1$ and $l_2$.

These restrictions on spin and $l_j$ reduce the number of terms to nine: $^{1}S_0$, $^{3}P_{0,1,2}$, $^{1}D_2$, $^{3}F_{2,3,4}$ and $^{1}G_4$ [5]. It is important and significant to recognize that these are the same nine terms for a $d^2$ *high – spin configuration*. Based on this equivalence, the new model has been designated is an *equivalent $d^2$ model*. This also provides the basis for using the Tanabe-Sugano (T–S) diagrams for interpretation of XAS spectra for O-vacancy defects. For $^{3}F_4$ term, the ordering of the term symbols in increasing energy in the spherical symmetry notation is $^{3}F_{2,3,4}$, $^{1}D_2$, $^{3}P_{0,1,2}$, $^{1}G_4$ and $^{1}S_0$ [5–7, 10]. The ordering of these free-ion term energies is the same as the ordering of

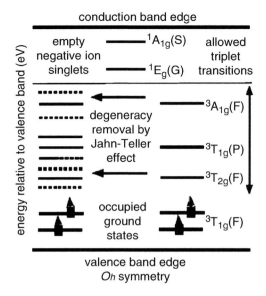

**Fig. 20.3** Schematic energy band diagrams for O-atom vacancy represented by a high-spin *equivalent $d^2$ state* for triplet-to-triplet and negative ion transitions in sixfold coordinated octahedral $O_h$ symmetry

the energy terms on the y-axis of the T–S diagram for a $d^2$ high-spin configuration for octahedral coordination of the Ti atoms [4]. This $d^2$ T–S diagram applies to octahedral coordination the Ti and Sc in the $GdSc_{1-x}Ti_xO_3$ alloy system and tracks the evolution of the states that correspond to the free ion spherical symmetry terms for intermediate values of the ligand field. Figure 20.3 has been constructed from the T–S diagram for a $d^2$ high-spin state in an octahedral ligand field.

In general, O K pre-edge spectra indicate both triplet and singlet final states indicating that the two electrons in the vacancy sites have singlet and triplet ground states that are within a few tenths of an eV of each other. For the transition metal oxides, this is explained by relatively small distortions in the vacancy geometry in which the separation between the respective transition metal atoms is at least 1.6 times the bond-length in an ideal tetrahedral geometry, and may be increased to as much as 2–2.5 times the bond-length of relaxation. This minimizes the exchange energy, and reduces the energy difference between singlet and triplet grounds states to small fractions of an eV. This aspect of O K pre-edge spectra is beyond the scope of this chapter.

## 20.3 Experimental Procedures and Results

Section 20.3.1 presents experimental procedures for sample preparation, and the acquisition of X-ray spectra, and Sect. 20.3.2 presents the experimental results and their interpretations. This includes data reduction by numerical second derivative analysis providing the X-ray photon energy of spectral features that facilitate comparison with (1) the *equivalent* $d^2$ *model* for O-vacancy allowed transitions and negative ion states in the O K pre-edge regime, and (2) CTM theory for the interpretation of the Ti and Sc $L_{2,3}$ spectra for identification of mixed valence and the hopping conductivity mechanism from which this derives. The interpretation of differences in the O K edge and Ti and Sc $L_{2,3}$ spectra for single crystal and nano-grain films, above and below the percolation threshold of $\sim 16.5\%$, Ti provide important information relative to the application of percolation theory approaches to alloy properties.

### 20.3.1 Procedures

Five nm thick $GdSc_{1-x}Ti_xO_3$ films with $x = 0.0$ to $0.25$ were deposited in an ultra-high vacuum system at $300°C$ onto (i) $SrTiO_3$ substrates for epitaxial growth, and (ii) superficially oxidized Si(001) for nano-grain films [14]. X-ray absorption spectroscopy measurements were made at the Stanford Synchrotron Research Lightsource (SSRL) [15, 18, 19]. These spectroscopic studies include the acquisition O K edge, and Sc and Ti $L_{2,3}$ edge spectra. O K edge measurements were made in the pre-edge regime below $\sim 530\,eV$, and the conduction band regime, from $\sim 530$ to $545\,eV$ [19]. Sc and Ti $L_{2,3}$ edge spectra were obtained in the spectral regimes from 390 to $420\,eV$ for Sc, and 440 to $470\,eV$ for Ti [15, 18, 19].

### 20.3.2 Spectroscopic Results and Interpretation

Spectra were taken under conditions that allowed reliable second derivatives to be determined numerically. These included continuous scanning with sampling at 0.02 to $0.03\,eV$ intervals while maintaining signal to noise ratios in excess of at least 3,000. This facilitates three point smoothing of the data, and after the two differentiations, the effective spectral resolution is of order of tenths of an eV. Second derivative analysis of pre-edge XAS spectra reveal $Ti^{3+} d^1$ and O-vacancy features, and Ti and Sc $L_{2,3}$ spectra reveal mixed valence associated with hopping conductivity that involves both Ti and Sc atoms in the $(Sc,Ti)O_2$ planes of the perovskite structure.

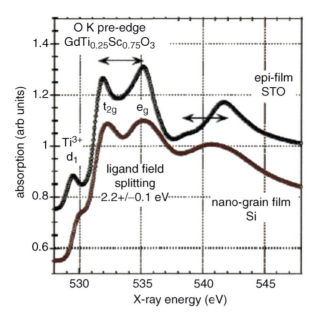

**Fig. 20.4** Comparison of O K edge spectra for a $GdSc_{1-x}Ti_xO_3$ alloy with 25% Ti substituted for Sc for epi-films on STO substrates and nano-grain films on oxidized Si substrates

In the absence of spin-exchange interactions with $Sc^{3+}$ $d^0$ and $f^7$ Gd atoms of the host complex oxide, the $d^1$ state of $Ti^{3+}$ is a singly occupied state above the valence band edge and displays a single excited state below the conduction band edge. The $f^7$ state is spherically symmetric. Studies of OK edge spectra in both epi- and nano-grain alloy films indicate that the pre-edge absorption feature in Fig. 20.4 for the $d^1$ state of the $Ti^{3+}$ feature scales with Ti content. In the epi-films this appears as a discrete feature with discrete side bands, and in the nano-grain films this feature is slightly broader, but otherwise effectively featureless. In addition, the feature in nano-grain films overlaps the onset of transitions to the $t_{2g}$ band edge states of Sc in the nano-grain film in the O K edge spectrum, whilst the $d^1$ absorption of the $Ti^{3+}$ feature in the epi-films is separate and distinct from the Sc $t_{2g}$ state feature.

Figure 20.5a compares the $Ti^{3+}$ second derivative XAS spectra for the $x = 0.25$ alloy for the nano-grain and single crystal epi-films. This composition is above the percolation threshold of $\sim$16.5%, and in the metallic alloy region. The epi-films display distinct features that are assigned to a strain reducing JTE distortion at the Ti alloy atom sites.

The second derivative spectral trace for the nano-grain film is markedly different. The energy of the minimum in the nano-grain film is displaced by +0.5 eV to higher eV, and there is one discrete feature with no significant strong additional side bands. The full-width at half-maximum (FWHM) for the dominant feature in each of these traces is approximately the same, $\sim$0.7 eV. O-atom vacancies at

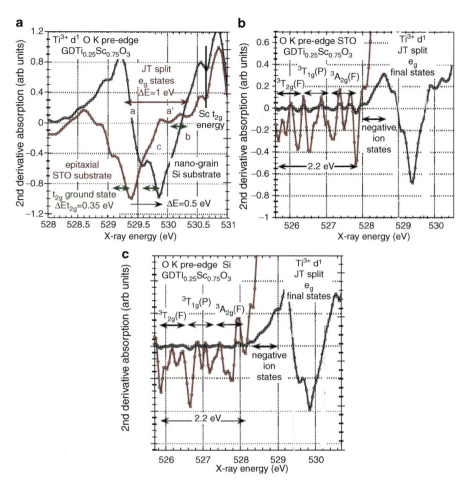

**Fig. 20.5** (a) Comparison of O K pre-edge spectra for a GdSc$_{1-x}$Ti$_x$O$_3$ alloy with 25% Ti substituted for Sc for epi-films in (b) and nano-grain films in (c) in the spectral regime between 525 and 531 eV that includes final state features of Ti$^{3+}$ d$^1$ transitions and O-vacancy transitions and one of two negative ion states

lower eV in Fig. 20.5b, c are for films with x = 0.25. The film in 5(b) is for an epi-film on STO, and the film in 5(c) is for a nano-grain film on Si. The spectral extent of the defect features is the same in each film and is approximately equal to the ligand field splitting extracted from the O K edge spectra in Fig. 20.4. This is ligand field splitting for the Sc atoms in an octahedral field. The equivalent d2 model predicts this relationship between the spectral width of O-vacancy transitions and the ligand field splitting. The vacancy defect features scales with Ti content, and the vacancies are likely strain-induced due to a 9% difference between the radii of sixfold coordinated Ti$^{3+}$ and Sc$^{3+}$. However, the atoms that border the vacancies

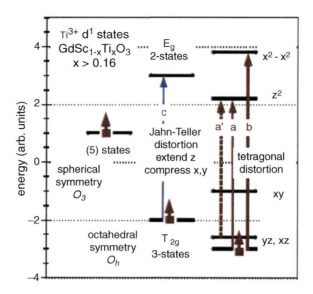

**Fig. 20.6** Schematic representation of changes in Ti 3d-state energies levels in going from a spherically symmetric site with $O_3$ symmetry, to an ideal octahedral site with $O_h$ symmetry, and the vacancy site after a cooperative Jahn Teller distortion, with a distorted tetragonal symmetry

are predominantly Sc atoms. The ionic radius of $Ti^{3+}$ is 0.081 nm, and for $Sc^{3+}$, it is larger and equal to 0.089 nm [20]. The spectra for the epi- and nano-grain films have the same number of features consistent with the *equivalent $d^2$ model*. However, there are differences in these spectra indicative of a higher degree of local disorder in the nano-grain films, and this is expected because the connectivity of the vacancy defects in epi-film benefits from a cooperative JTE, averaging out fluctuations in local order. Equally important percolation theory cannot be applied to the nano-grain films because of the number of Ti sites in a nano-grain with 5 nm dimensions is too small for a statistical approach to apply [21].

The $Ti^{3+}$ feature in the O K pre-edge regime for epi- and nano-grain thin films for a composition of 5% Ti, well below the percolation threshold, display essentially the same derivative features; e.g., a dominant feature at ~530.7 eV, and a weaker feature about 0.35 eV lower. The lower energy feature is tentatively assigned to a $t_{2g}$ ground state transition associated with $Ti^{3+}$ $d^1$ non-interacting sites, and is due to local strain differences determined by the proximity to other near-neighbor $Ti^{3+}$ sites (see energy level diagram Fig. 20.6). This is qualitatively different than the cooperative JTE proposed in the next paragraph to explain the larger splittings in Fig. 20.5b for a composition above the percolation limit.

Figure 20.6 is schematic representation of the changes in the Ti 3d-state energies levels in going from a spherically symmetric site with $O_3$ symmetry, to an ideal octahedral site with $O_h$ symmetry, and finally to a local site that displays a

# 20 Spectroscopic Detection of Hopping Induced Mixed Valence of Ti and Sc...

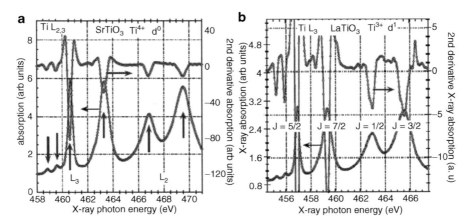

**Fig. 20.7** Ti $L_{2,3}$ and second derivative $L_{2,3}$ spectra for two prevoskites: (**a**) a cubic $d^0$ epi-$SrTiO_3$, and (**b**) a distorted cubic $d^1$ epi-film of $LaTiO_3$

cooperative JTE distortion as in Fig. 20.1. This defect state, like t-$HfO_2$ in Fig. 20.1, has a distorted tetragonal symmetry. The notations, "a b", "a'" and "c" in. Fig. 20.5a, correspond to the transitions marked in Fig. 20.6. There is more structure in Fig. 20.5a than can be explained by the schematic representation in Fig. 20.6, and this is assumed to reflect more distant local environments of Ti $d^1$ states in the (Sc,Ti)$O_2$ plane, or equivalently may be related to multiply connected paths for hopping conductivity.

Figure 20.7a, b compares reference $L_{2,3}$ spectrum and second derivative $L_{2,3}$ spectrum for a $d^0$ $Ti^{4+}$ cubic persovskite, $SrTiO_3$ with ideal octahedral bonding of Ti in Fig. 20.7a, and the $L_{2,3}$ spectrum for $Ti^{3+}$ $d^1$ $LaTiO_3$ with a tetragonally-distorted persovskite structure in which the unit cell is extended in the z-direction in Fig. 20.7b. The undifferentiated $Ti^{4+}$ $d^0$ $L_{2,3}$ and $Ti^{3+}$ $d^1$ $L_{2,3}$ spectra indicate relatively small differences; however, these are amplified in the respective second derivative spectra. It is more significant to count the number of resolved features in the respective second derivative spectra in these spectra than rely on the spectral traces before differentiation as is generally done by de Groot [5–8]. As expected from the analysis in Sect. 1.2, the $Ti^{4+}$ $d^0$ $L_{2,3}$ spectrum displays six features, four in the $L_3$ regime associated with transitions from the $2p_{3/2}$ initial state, and two in the $L_2$ regime associated with features from the $2p_{3/2}$ initial state. In addition to the 12 features expected in $LaTiO_3$, and also expected from the symmetry of the final state, the second derivative spectrum displays weaker features derived from a distortion of the cubic perovskite lattice. Stronger secondary features also are present in $GdScO_3$, which has an even larger distortion from an ideal cubic structure.

There is a significant multiplicity of second derivative features in the $L_3$ portion of the spectrum in an epi-film with 18% Ti in Fig. 20.8, but not in epi-films with 5% and 1% Ti, or in nano-grain films for all Ti concentrations whether they be above or below the percolation threshold. The sixteen features are expected for intra-Ti

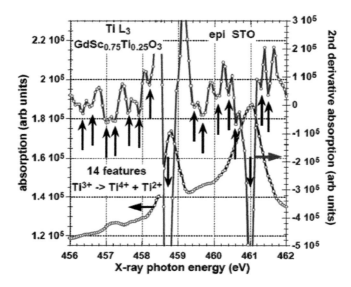

**Fig. 20.8** Ti $L_{2,3}$ and second derivative $L_{2,3}$ spectra for $GdSc_{1-x}Ti_xO_3$ with 25% Ti. The arrows indicate the 16 features predicted for a combination of $Ti^{4+}$, $Ti^{3+}$ and $Ti^{2+}$ initial and final states, respectively for the hopping conductivity transport of an electron $Ti^{3+} + Ti^{3+} \rightarrow Ti^{4+} + Ti^{2+}$

sub-lattice hopping conductivity transport equation in Eq. 20.4 which changes the formal valence of the Ti alloy atoms for a pair of nearest neighbor $Ti^{3+}$ states;

$$Ti^{3+} + Ti^{3+} \rightarrow Ti^{4+} + Ti^{2+}. \qquad (20.4)$$

The total number of features expected in Eq. 20.4 is $Ti^{3+}(6) + Ti^{4+}(4) + Ti^{2+}(6) = 16$, and accounts for the number of features in Fig. 20.8.

There are fourteen features in the $L_2$ portion of epi-film with 18% Ti (spectrum not shown). This is number of features expected from Eq. 20.4. The difference is the number of features for $Ti^{4+}$ is reduced from four in the $L_3$ contribution to two in the $L_2$ contribution.

The comparison of spectral features in the Sc $L_3$ portion of the Sc $L_{2,3}$ spectrum identifies inter-atom $Ti^{3+} d^1$ to $Sc^{3+}d$ electron hopping as well, and is indicated in the comparison between Figs. 20.9a for $GdScO_3$ and Fig. 20.9b for a film with 18% Ti. The inter-atom $Ti^{3+} d^1$ to $Sc^{3+}d$ electron hopping transport is indicated below in Eq. 20.5:

$$Ti^{3+} + Sc^{3+} \rightarrow Ti^{4+} + Sc^{2+}. \qquad (20.5)$$

The total number of features expected in for the Sc $L_3$ portion of the Sc $L_{2,3}$ spectrum is $Sc^{3+}(4) + Sc^{2+}(6+) = 10$ features. This is what is observed in the experimental plot in Fig. 20.9b

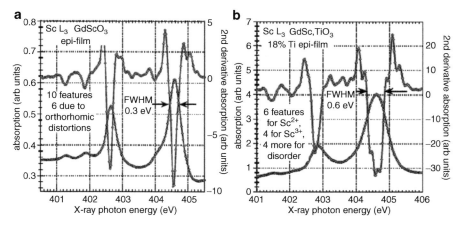

**Fig. 20.9** Sc $L_{2,3}$ and second derivative $L_{2,3}$ spectra for epi-$GdSc_{1-x}Ti_xO_3$: (**a**) for x = 0, $GdScTiO_3$, indicating ten features: six for the octahedral arrangement of O-atoms, and for a tetragonal distortion, and (**b**) for an 18% Ti alloy, x = 0.18, that indicates, fourteen features: ten for hopping conductivity: $Sc^{3+}$ and $Sc^{2+}$, and four for a tetragonal distortion

## 20.4 Summary: Mixed Valence and Electron Transport in $GdSc_{1-x}Ti_xO_3$ Alloys Above the Percolation Threshold

The spectral results above indicate that conditions necessary for double exchange ferro(i)magnetism are met by Ti alloying in epitaxial single crystal thin films above the percolation threshold of ~16.5%, [3]. These are: (i) mixed valance of in-plane Ti atoms as detected in Ti and Sc $L_{2,3}$ spectra in Figs. 20.8 and 20.9b, respectively, (ii) hopping transport and metallic conductivity as determined by sample charging and consistent with spectral detection of mixed valance. This is an intra-atom hopping process, and metallic conductivity is supported by the detection of JTE split final states in the epitaxial films in Fig. 20.5a, and the occurrence of a single band in the nano-grain films where grain size and Ti alloy statistics are too small for a distinct percolation threshold [20]. The sharpness of the features in the epi-film in Fig. 20.5a suggests that the JTE is a cooperative effect that is enabled in large part by the hopping transport [11].

This hopping conductivity is detected by the number of features in $L_{2,3}$ and Sc $L_{2,3}$ spectra above a percolation threshold of ~16.5%, but only in the epi-films that have long range order which is required for the statistical nature of the percolation process [20]. The nano-grains in the films deposited on oxidized Si are too small to provide the concentration-dependant statistics necessary for a percolation process. In point of fact, they must be larger by more than three orders of magnitude, well into the micrometer regime and this cannot be achieved in thin films that are at most 5 nm thick.

**Acknowledgments** The authors acknowledge partial support from the NSF, SRC, AFOSR, DTRA and ARO. XAS spectra were obtained at the Stanford Synchrotron Lightsource (SSRL). We acknowledge helpful discussions with Professor J. L. Whitten in the Department of Chemistry at NC State University, and Professor D. E. Aspnes in the Department of Physics at NC State University. The $GdSc_{1-x}TiO_x$ epitaxial and nanocrystalline thin films were grown by Carolina Adamo and Darrell G. Schlom of the, currently at the Department of Materials Science at Cornell, and University, and previously at Pennsylvania State University.

# References

1. Zener C (1954) Phys Rev 96:1335
2. de Gennes P-G (1960) Phys Rev 118:141
3. Sugano S, Tanabe Y, Kamimura H (1970) Multiplets of transition metal ions. Academic, New York
4. Tanabe Y, Sugano S (1956) J Phys Soc Jpn 11:864, and references therein
5. deGroot F, Kotani A (2008) Core level spectroscopy of solids. CRC Press, Boca Raton
6. DeGroot FMF, Fuggle JC, Thoule BT, Sawatzky GA (1990) Phys Rev B 41:928
7. DeGroot FMF, Fuggle JC, Thoule BT, Sawatzky GA (1990) Phys Rev B 42:5459
8. Xiong K, Robertson J, Gibson MC, Clark SJ (2006) Appl Phys Lett 87:183505
9. Gavartin JL, Muñoz Ramo D, Shlugger AL (2006) Appl Phys Lett 89:082908
10. Cotton FA (1963) Chemical applications of group theory, 2nd edn. Wiley Interscience, New York, Chap 8
11. White RM, Geballe TH (1979) Long range order in solids. Academic, New York, Chap 1
12. Harrison WA (1999) Elementary electronic structure. World Scientific, Singapore
13. Thompson A, Lindau I, Attwood D et al (eds) (2001) X-ray data booklet. Lawrence Berkeley Laboratory, Berkeley, Section 1
14. Lim S-G, Kriventsov S, Jackson TN (2002) J Appl Phys 91:4500
15. Lucovsky G, Seo H, Lee S et al (2007) Jpn J Appl Phys 46:1899
16. Fujimori A, Minami F (1984) Phys Rev B 30:963
17. Bethe H (1929) Ann Physik 3:133
18. Lucovsky G (2007) J Mol Struct 838:187
19. Lucovsky G, Chung KB, Miotti L et al (2009) Solid State Electron 53:1273
20. Huheey JE (1978) Inorganic chemistry, 2nd edn. Harper and Roe, New York, Chap 4
21. Zallen R (1983) The physics of amorphous solids. Wiley, New York, Chap 4

# Part VI
# Cooperative Jahn-Teller Effect and Spin-Coupling Phenomena

# Chapter 21
# Vibronic Approach to the Cooperative Spin Transitions in Crystals Based on Cyano-Bridged Pentanuclear $M_2Fe_3$ (M=Co, Os) Clusters

**Serghei Ostrovsky, Andrew Palii, Sophia Klokishner, Michael Shatruk, Kristen Funck, Catalina Achim, Kim R. Dunbar, and Boris Tsukerblat**

**Abstract** In this article we present a theoretical microscopic approach to the problem of cooperative spin crossover in crystals based on cyano-bridged pentanuclear metal clusters $\{[M(CN)_6]_2[Fe(tmphen)_2]_3\}$ (M = Co, Os) with a trigonal bipyramidal structure. The low-spin to high-spin transition is considered as a cooperative phenomenon that is driven by the interaction of the electronic shells of the Fe ions with the full symmetric deformation of the local surrounding that is extended over the crystal lattice via the acoustic phonon field. Due to the proximity of Fe ions within the metal cluster, the short-range intracluster interactions between these ions via the optic phonon field is included as well. The interrelation between short- and long-range phonon induced interactions between metal ions was shown to determine the type and the temperature of spin transition in cluster systems. The proposed model is applied to the systems with temperature induced spin transition within the single metal ion as well as to the complexes with spin transition caused by the charge transfer between different ions in the cluster. A wide set of the experimental data on the temperature dependence of the magnetic susceptibility

---

S. Ostrovsky • A. Palii • S. Klokishner (✉)
Institute of Applied Physics, Academy of Sciences of Moldova, Kishinev, Moldova
e-mail: sm_ostrovsky@yahoo.com; andrew.palii@uv.es; klokishner@yahoo.com

M. Shatruk
Department of Chemistry and Biochemistry, Florida State University, Tallahassee, FL, USA
e-mail: shatruk@chem.fsu.edu

K. Funck • K.R. Dunbar
Department of Chemistry, Texas A&M University, College Station, TX, USA
e-mail: kchambers@mail.chem.tamu.edu; dunbar@mail.chem.tamu.edu

C. Achim
Department of Chemistry, Carnegie Mellon University, Pittsburgh, PA, USA

B. Tsukerblat (✉)
Chemistry Department, Ben-Gurion University of the Negev, Beer-Sheva, Israel
e-mail: tsuker@bgu.ac.il

---

M. Atanasov et al. (eds.), *Vibronic Interactions and the Jahn-Teller Effect: Theory and Applications*, Progress in Theoretical Chemistry and Physics 23, DOI 10.1007/978-94-007-2384-9_21, © Springer Science+Business Media B.V. 2012

and Mössbauer spectra is interpreted qualitatively and quantitatively. The approach described in this paper represents a theoretical tool for the study of spin-crossover systems based on metal clusters as structural units of the crystal lattice.

## 21.1 Introduction

Cyanide-bridged compounds exhibit a remarkable diversity of structural, magnetic and magnetooptical properties. Recently the interest in studying of cyanide-based compounds has been renewed due to discovered ability of the bridging cyanide ligand to extend the exchange coupling resulting in the high $T_C$ magnets [1–3]. This development has inspired research in the area of polynuclear cyanide clusters with the goal to build up finite systems that demonstrate at the molecular level such fascinating phenomena as single molecule magnet (SMM) behavior [4–11], temperature-induced [12] and light-induced spin crossover and charge transfer [13] induced spin-transitions (CTIST).

Recently a family of pentanuclear clusters of structure $\{[[Fe(tmphen)_2]_3 M(CN)_6]_2\}$ (tmphen $=$ 3,4,7,8-tetramethyl-1,10-phenanthroline, $M = Co, Os$) has been synthesized and reported. A common feature of both types of compounds ($M = Co, Os$) is that they demonstrate spin transitions that can be assigned to a cooperative phenomenon in the corresponding crystals. The common structural unit of these clusters represents a trigonal bipyramid (TBP) as shown in Fig. 21.1. Two M ions in the apical positions of this bipyramid are surrounded by six carbon atoms while three Fe ions in the equatorial plane are coordinated by the nitrogen atoms. Depending on the type of ions in the apical positions the systems display different kinds of behavior. The $[Fe^{II}(tmphen)_2]_3[Co^{III}(CN)_6]_2$ cluster compound demonstrates a reversible temperature induced low spin (*ls*)– high spin (*hs*) transition at the Fe(II) sites while in the $\{[Fe(tmphen)_2]_3[Os(CN)_6]_2\}$ complex a CTIST was found. In the latter cluster compound a reversible transition between *ls* Fe(II)–N $\equiv$ C–*ls* Os(III) and *hs* Fe(III)–N $\equiv$ C–*ls* Os(II) redox pairs with the change of temperature was observed [14]. The occurrence of the *ls-hs* transition was proved by the combination of Mössbauer spectroscopy, magnetic measurements and single crystal X-ray studies.

The nature of spin-crossover transition as a cooperative phenomenon has been understood more than three decades ago in pioneering studies [15]. In line of the theory of the cooperative Jahn-Teller structural ordering [16] it was suggested that the vibronic coupling at the *Fe* sites leads to an effective long range interaction via the acoustic phonon field giving rise to a phase transition accompanied by $ls \leftrightarrow hs$ transformation in Fe ions. The numerous subsequent studies of spin-crossover phenomenon involving magnetic properties and Mössbauer spectra were based on the ideas formulated in Ref. [15] and consequently refer to the systems in which Fe sites in crystals are coupled by long-range interactions. Although the concept of the phonon induced long range metal-metal coupling is universal this approach [15] can not be applied directly to the crystals composed of the

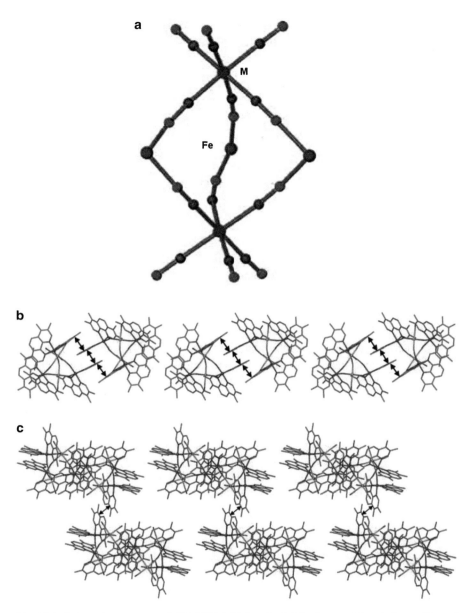

**Fig. 21.1** Schematic structure of $\{[M(CN)_6]_2[Fe(tmphen)_2]_3\}$ clusters (**a**) and the packing of dimers of TBP clusters in the solid state (**b, c**). The π–π interactions are indicated with *arrows*

exchange coupled metal clusters in which along with the long range coupling between Fe ions a short range interaction within the cluster also exists and should be explicitly introduced in the model. Spin-crossover in cluster systems has not been treated theoretically (see review article [17]). One of the tasks to be solved in the present paper is the elucidation of all key interactions governing spin-crossover transitions in cluster systems. To the best of our knowledge a microscopic model of charge transfer induced spin transitions (CTIST) has not been also developed. In the present paper we suggest a unified microscopic approach that allows to treat the cooperative spin transitions in crystals containing $\{[M(CN)_6]_2[Fe(tmphen)_2]_3\}$ clusters as structural units. Both the temperature induced $ls$-$hs$ transition and CTIST are analyzed in the framework of the developed approach. The main mechanisms governing these effects are revealed. Finally, the developed approach is applied to the analysis of the temperature behavior of the observed magnetic properties and Mössbauer spectra of the titled compounds.

## 21.2 The Model

In both studied cluster compounds the metal ions (M = Co, Os) in the axial positions are surrounded by six carbon atoms and are in the $ls$ ground state. The corresponding ground states are the following:

$^2T_2(t_2^5)$ for the $ls$ Os(III),
$^1A_1(t_2^6)$ for the $ls$ Os(II) and $ls$ Co(III).

It should be noted that the phase transformations in crystals accompanied by the change of the metal-ligand distances do not affect the electronic configurations of these ions and therefore it will be assumed that these ground states are retained in the whole temperature range. Each $Fe$ ion in the equatorial position is surrounded by six nitrogen atoms and can be subjected to a $ls$-$hs$ transition. These transitions can be referred to as $ls$ Fe(II) ($^1A_1$ ground term, $S = 0$) to $hs$ Fe(III) ($^6A_1$ ground term, $S = 5/2$) or $hs$ Fe(II) ($^5T_2$ ground term, $S = 2$) in the compounds $Os_2Fe_3$ or $Co_2Fe_3$, respectively. The electronic basis for the $Os_2Fe_3$ cluster compound involves states arising from three configurations: $(ls\text{-}Fe^{II})_3(ls\text{-}Os^{III})_2$ (type **A**), $(ls - Fe^{II})_2$ $(hs - Fe^{III})(ls - Os^{III})(ls - Os^{II})$ (type **B**) and $(ls - Fe^{II})(hs - Fe^{III})_2(ls - Os^{II})_2$ (type **C**).

Let us start from the type **A** configuration. The wave functions corresponding to this configuration carry the quantum numbers of $ls$ Os(III), since $ls$ Fe(II) is diamagnetic. The spin-orbit interaction splits the $^2T_2$ term of $ls$ Os(III) ions into the ground doublet and the excited quadruplet. Since the spin-orbit coupling parameter for the $ls$ Os(III) is about $2,100\,cm^{-1}$, the energy gap between these two groups of levels is more than $3,000\,cm^{-1}$ and the $ls$ Os(III) ions can be regarded as pseudospins 1/2. With regards to the phase transformations considered hereunder this means that the $ls$ Os(III) ions act as a pseudospins 1/2 particles in the whole temperature range. In this approximation each $ls$ Os(III) ion is characterized by the projection of the pseudospin 1/2 and, as a consequence, there are $2 \times 2 = 4$ states of type **A** with different quantum numbers of the two $ls$ Os(III) ions.

The wave functions corresponding to the type **B** configuration are characterized by the quantum numbers of the $ls$ Os(III) and $hs$ Fe(III) and labeled by the positions of $ls$ Os(III) and $hs$ Fe(III) within the TBP. There are three possible locations of *hs* Fe(III) within three equatorial sites and two possible locations of *ls* Os(III) within two apical positions. The *hs* Fe(III) ($^6A_1$ ground term) is characterized by the projection of its spin and has 6 different magnetic quantum numbers while the *ls* Os(III) can be described by two different projections of pseudospin 1/2. As a result, there are $3 \times 2 \times 6 \times 2 = 72$ different states of type **B** configuration.

The wave functions corresponding to the type **C** configuration are characterized by the quantum numbers of two *hs* Fe(III) and by the positions of these *hs* iron ions within the TBP. There are three possible distributions of two *hs* Fe(III) within three equatorial sites. Since each of *hs* Fe(III) is described by 6 different magnetic quantum numbers (spin projections), there are $3 \times 6 \times 6 = 108$ different states of type C configuration. As a consequence, the total basis (type A + type B + type C) includes $4 + 72 + 108 = 184$ cluster states.

As for the $Co_2Fe_3$ cluster compound, for each of three Fe(II) ions there are 15 *hs* states and one *ls* state. Thus, the total basis for this complex includes $16 \times 16 \times 16 = 4{,}096$ states.

For both compounds the total Hamiltonian can be written as follows:

$$H = H_{st} + H_{sr} + H_{intra} + H_{inter} + H_{cr}. \tag{21.1}$$

where the first term represents the interaction of Fe ions with the spontaneous all-round full symmetric lattice strain, the second one is the short-range interactions between Fe ions belonging to the same TBP, the third and the fourth terms are the intra-ion and inter-ion interactions (like spin-orbit coupling, exchange interaction, Zeeman perturbation etc), respectively, and the last part takes into account the energy difference between the different electronic configurations of the cluster under study in the crystal field. Later on these contributions will be given explicitly for the two systems under the study.

We start with the interaction of Fe ions with the spontaneous all-round full symmetric lattice strain. A crystal is considered to be built of $m$ TBPs, in each TBP $n$ Fe molecules can undergo spin transition. In $Co_2Fe_3$ the corresponding transition is *ls* Fe(II) $\rightarrow$ *hs* Fe(II) while in the $Os_2Fe_3$ cluster compound the transition *ls* Fe(II) to *hs* Fe(III) is accompanied by the electron transfer to the Os ion. It is supposed that the mechanism responsible for both spin transitions is the interaction of Fe ions with the spontaneous all-round full symmetric lattice strain [15, 18–20]. In spin crossover compounds the medium between molecules is more easily affected by deformation than that inside of the molecule, so a difference between the molecular and intermolecular space is made. We introduce the internal molecular

$$\varepsilon_1 = \left( \varepsilon_{xx}^1 + \varepsilon_{yy}^1 + \varepsilon_{zz}^1 \right) \Big/ \sqrt{3}$$

and external (intermolecular volume)

$$\varepsilon_2 = \left( \varepsilon_{xx}^2 + \varepsilon_{yy}^2 + \varepsilon_{zz}^2 \right) \Big/ \sqrt{3}$$

full symmetric strains, the corresponding bulk moduli are $c_1$ and $c_2$. In each molecule the interaction of the Fe ion with the $\varepsilon_1$ strain is considered.

The contribution of the uniform strains $\varepsilon_1$ and $\varepsilon_2$ in the crystal Hamiltonian can be written as:

$$H_{st} = \frac{1}{2} nmc_1 \Omega_0 \varepsilon_1^2 + \frac{1}{2} nmc_2 (\Omega - \Omega_0) \varepsilon_2^2 + \upsilon_1 \varepsilon_1 \sum_{k,\alpha} \tau_k^\alpha + \upsilon_2 \varepsilon_1 nm \qquad (21.2)$$

where $\alpha = 1,\ldots,n$ and $k = 1,\ldots,m$; $\Omega_0$ and $\Omega$ are the molecular and unit cell volumes, respectively; $\upsilon_1 = (\upsilon_{hs} - \upsilon_{ls})/2$, $\upsilon_2 = (\upsilon_{hs} + \upsilon_{ls})/2$, where $\upsilon_{hs}$ and $\upsilon_{ls}$ are the constants of interaction of the *hs* iron ions (Fe(II) for $Co_2Fe_3$ or Fe(III) for $Os_2Fe_3$) and *ls* Fe(II) with the full symmetric strain $\varepsilon_1$, respectively. In Eq. 21.2 the first two terms describe the elastic energy of the deformed crystal, the third and the forth ones correspond to the coupling of the *d*-electrons with the $\varepsilon_1$ deformation. The matrix $\tau_k^\alpha$ is diagonal and has the dimension of the whole basis of the problem under study (4096 states for $Co_2Fe_3$ and 184 states for $Os_2Fe_3$ cluster compounds). The *i*-th diagonal matrix element of the $\tau_k^\alpha$ matrix is 1 if in the *i*-th cluster state the $\alpha$-th Fe ion is in the *hs* configuration (*hs* Fe(II) for $Co_2Fe_3$ and *hs* Fe(III) for $Os_2Fe_3$) or -1 if this iron ion is in the *ls* Fe(II) configuration.

For a uniform crystal compression (or extension) the ratio of $\varepsilon_2/\varepsilon_1$ is roughly supposed to be

$$\varepsilon_2 \approx \varepsilon_1 \frac{c_1}{c_2} \qquad (21.3)$$

Minimizing the right side of Eq. 21.2 with respect to the strain $\varepsilon_1$ and taking into account Eq. 21.3 one finds:

$$\varepsilon_1 = -\frac{A}{nm} \upsilon_1 \sum_{k,\alpha} \tau_k^\alpha - A\upsilon_2, \qquad (21.4)$$

where

$$A = \frac{c_2}{c_1 \left[ c_2 \Omega_0 + c_1 (\Omega - \Omega_0) \right]}. \qquad (21.5)$$

Finally, after substitution of Eq. 21.4 back into Eq. 21.2 one obtains:

$$H_{st} = -B \sum_{k,\alpha} \tau_k^\alpha - \frac{J}{2nm} \sum_{k,\alpha} \sum_{k',\alpha'} \tau_k^\alpha \tau_{k'}^{\alpha'} \qquad (21.6)$$

where $B = A\upsilon_1\upsilon_2$ and $J = A\upsilon_1^2$. Thus, one can conclude that the coupling to the strain gives rise to an infinite range interaction between all molecules in the crystal (see Fig. 21.1b, c). The model of the elastic continuum so far introduced satisfactorily describes only the long-wave acoustic vibrations of the lattice [21]. Therefore, the obtained intermolecular interaction corresponds to the exchange via the field of long-wave acoustic phonons.

Possible mechanisms of short-range interactions are the direct interaction of the electric quadrupoles and the interaction via the field of the optic phonons. The effect of the exchange via the optic phonons is larger than that arising from electric

# 21 Vibronic Approach to the Cooperative Spin Transitions in Crystals Based...

quadrupole-quadrupole interactions [16, 22]. The short-range interactions within TBP can be written as:

$$H_{sr} = -J_0 \sum_k \sum_{\alpha < \alpha'} \tau_k^\alpha \tau_k^{\alpha'}, \qquad (21.7)$$

where the parameter $J_0$ describes the interaction of Fe ions within the trigonal bipyramid.

The intra-ion and inter-ion interactions to be included into consideration depend on the compound under study. For the $Co_2Fe_3$ cluster compound the single ion terms can be written as:

$$H_{\text{intra}} = -\lambda \kappa \sum_{k\alpha} s_k^\alpha l_k^\alpha + \Delta \sum_{k\alpha} \left( (l_Z^{k\alpha})^2 - 2/3 \right) + \mu_B \mathbf{H} \sum_{k\alpha} \left( g_0 s_\alpha^k - \kappa l_\alpha^k \right) \qquad (21.8)$$

where the first term is the spin-orbit coupling operating within the cubic $^5T_2(t_2^4 e^2)$ term of the $hs$-Fe(II) ion, the second term describes the axial crystal field splitting of the $^5T_2$ $(l = 1)$ term into an orbital singlet $(m_l = 0)$ and an orbital doublet $(m_l = \pm 1)$ and the third term refers to the Zeeman interaction for the $hs$-Fe(II) ion and contains both the spin and orbital contributions, $\mu_B$ is the Bohr magneton, and $g_0$ is the spin Lande factor. The through space interactions between the equatorial $hs$-Fe(II) ions are expected to be negligible due to large metal-metal distances ($>6$ Å). Since in the $Co_2Fe_3$ complex the apical positions are occupied by diamagnetic $ls$-Co(III) ions, the magnetic exchange in the $ls$-Co(III)–$hs$-Fe(II) pairs is not operative.

For the $Os_2Fe_3$ cluster compound the single ion and inter-ion terms are written as a sum of the exchange interaction and the Zeeman perturbation:

$$H_{\text{inter}} + H_{\text{intra}} = \sum_k H_{ex}^k + \mu_B \mathbf{H} \sum_{k\alpha} \left( g_0 S_\alpha^k + g_1 s_\alpha^k \right) \qquad (21.9)$$

In Eq. 21.9 $S = 5/2$ is the spin of $hs$ Fe(III) and $s = 1/2$ is the pseudospin of $ls$ Os(III), $g_0 = 2.0$, $g_1 = 1.55$ is the Lande factor for the low-lying Kramers doublet of $ls$ Os(III). In Ref. [23] it was shown that the magnetic exchange between $ls$ Os(III) and $hs$ Mn(II) through the cyanide bridge can be described by the Ising type Hamiltonian:

$$H_{ex} = -2J_{ex} S_Z s_Z \qquad (21.10)$$

For the Mn-Os pair the parameter $J_{ex}$ is about $-30\,\text{cm}^{-1}$. Since the electronic configurations of $hs$ Fe(III) and $hs$ Mn(II) are the same $(3d^5)$ it is reasonably to assume that the magnetic exchange between $ls$ Os(III) and $hs$ Fe(III) via the cyanide bridge can be also described by Eq. 21.10 with a similar value of the exchange coupling constant.

The last part in the Hamiltonian, Eq. 21.1, accounting for the energy difference between the different electronic configurations of the cluster under the study can be written as:

$$H_{cr} = \frac{\Delta_0}{2} \sum_{k\alpha} \tau_k^\alpha. \qquad (21.11)$$

For the $Co_2Fe_3$ complex $\Delta_0$ represents the effective energy gap between the $hs$ and $ls$ states of Fe(II) ion in the cubic crystal field while for the $Os_2Fe_3$ complex this parameter is the effective gap between the energies of $ls$ Fe(II)–N≡C–$ls$ Os(III) and $hs$ Fe(III)–N≡C–$ls$ Os(II) redox pairs with allowance only for crystal field splitting and intra- and intercenter Coulomb interaction.

The problem of interaction with the strain $\varepsilon_1$ can be solved in the mean field approximation. In this approximation the Hamiltonian Eq. 21.6 is decomposed into the sum of the single-ion Hamiltonians:

$$H_{st} = -\left(B + J\bar{\tau}\right) \sum_{k,\alpha} \tau_k^\alpha, \tag{21.12}$$

where $\bar{\tau} = Tr(\rho \tau_k^\alpha)$ and $\rho$ is the density operator. As a consequence, the total crystal Hamiltonian Eq. 21.1 can be represented as a sum of single cluster Hamiltonians.

Both magnetic measurements and Mössbauer spectroscopy for hydrated crystals of $Co_2Fe_3$ and $Os_2Fe_3$ demonstrate the presence of some amount of $hs$-Fe(II) and $hs$-Fe(III) ions, respectively, even at very low temperatures. Therefore, hereafter for both compounds we will denote the fraction of Fe ions which are in the $hs$ state at all temperatures by $x$ and the concentration of those ions which undergo the $ls$-$hs$ transition as $(1-x)$. The number $p_i$ of TBP $Co_2Fe_3$ clusters in which $i$ of three Fe(II) ions are in the $hs$ configuration in the whole temperature range is estimated as:

$$p_i = C_i^3 x^i \left(1 - x\right)^{3-i}, \tag{21.13}$$

where $C_i^r = r!/i!(r-i)!$ and $i = 0, 1, 2, 3$. For the $Os_2Fe_3$ compound the spin transition arises from charge transfer from the Fe ion to the Os one. Since each TBP cluster contains two Os ions the cluster will never be in a configuration containing three hs-Fe(III) ions. Therefore, while determining the probabilities $p_i$ in this case we take into account the presence of trigonal bypyramids containing only zero, one or two Fe(III) ions which are in the hs-state in the whole temperature range.

In the mean field approximation the total Hamiltonian of the crystal decomposes into the sum of Hamiltonians $\tilde{H}_k$ which describe clusters with $k$ Fe ions that undergo spin-crossover or CTIST. Therefore the order parameter $\bar{\tau}$ for the $Co_2Fe_3$ system can be calculated from the following equation:

$$\bar{\tau} = p_3 + p_2 \frac{Tr\left(Exp(-\tilde{H}_1/kT)\tau_k^\alpha(1)\right)}{Tr\left(Exp(-\tilde{H}_1/kT)\right)} + p_1 \frac{Tr\left(Exp(-\tilde{H}_2/kT)\tau_k^\alpha(2)\right)}{Tr\left(Exp(-\tilde{H}_2/kT)\right)}$$

$$+ p_0 \frac{Tr\left(Exp(-\tilde{H}_3/kT)\tau_k^\alpha(3)\right)}{Tr\left(Exp(-\tilde{H}_3/kT)\right)} \tag{21.14}$$

here the notations $\tilde{H}_1$, $\tilde{H}_2$, $\tilde{H}_3$ are introduced for the mean field Hamiltonians of individual clusters containing 1, 2, and 3 spin crossover Fe ions, respectively. These Hamiltonians can be easily derived from Eqs. 21.1 and 21.7–21.12. In the case of the $Os_2Fe_3$ system with its specifics described above the first term in the right hand side of Eq. 21.14 should be omitted. In Eq. 21.14 the symbol $i$ labeling the matrices $\tau_k^\alpha(i)$ corresponds to the number of Fe ions that undergo spin transition in the cluster.

## 21.3 Estimation of the Characteristic Parameters

To reduce the number of the fitting parameters we start with the estimation of the value of the parameter $J$ that describes the interaction with the full symmetric lattice strain. The matrix elements $\upsilon_{hs}$ and $\upsilon_{ls}$ that contribute to $\upsilon_1$ can be expressed as the mean values of the derivatives of crystal field energies in the $ls$ and $hs$ states

$$\upsilon_{ls} = \left\langle ls \left| \left( \frac{\partial W(r,R)}{\partial R} \right)_{R=R_{ls}} \right| ls \right\rangle \frac{R_{ls}}{\sqrt{3}},$$

$$\upsilon_{hs} = \left\langle hs \left| \left( \frac{\partial W(r,R)}{\partial R} \right)_{R=R_{hs}} \right| hs \right\rangle \frac{R_{hs}}{\sqrt{3}}. \tag{21.15}$$

Here $R_{hs}, R_{ls}$ are the metal-ligand distances in the corresponding states, $W(r, R)$ is the potential energy (crystal field) acting on the electronic shell of the Fe ion. For an octahedral complex $FeX_6$, the values $\upsilon_{hs}$ and $\upsilon_{ls}$ corresponding to the electronic configurations $t_2^4 e^2$ and $t_2^6$ of Fe(II) ion are proportional to the cubic crystal field parameters $Dq^{hs}$ and $Dq^{ls}$:

$$\upsilon_{ls} = 120Dq^{ls}/\sqrt{3}, \quad \upsilon_{hs} = 20Dq^{hs}/\sqrt{3}. \tag{21.16}$$

As a consequence, the parameter $\upsilon_1$ is represented as

$$\upsilon_1 = 10(Dq^{hs} - 6Dq^{ls})/\sqrt{3}. \tag{21.17}$$

The typical values of the crystal field parameters for the Fe(II) ion are the following: $Dq^{hs} = 1,176 \, \text{cm}^{-1}$ and $Dq^{ls} = 2,055 \, \text{cm}^{-1}$ [17]. Using Eq. 21.17 one obtains the following estimate: $\upsilon_1 = -6.4 \times 10^4 \, \text{cm}^{-1}$. For spin-crossover compounds the main change in the volume occurs for the intermolecular space and, hence one can assume that $c_1 \gg c_2$. In the $Co_2Fe_3$ cluster compound the unit cell volume per one Fe ion is $\Omega = 1026 \, \text{Å}^3$ and $\Omega_0$ is approximately $64 \, \text{Å}^3$. The typical value of $c_2$ is about $10^{11} \, \text{dyne/cm}^2$ and for spin-crossover compounds $c_2 \approx (0.05 \div 0.1)c_1$. Using these values one can find that the parameter $J$ falls into the range of $20 \div 80 \text{cm}^{-1}$. The value of the parameter $B$ in Eq. 21.12 can be estimated in a similar way, however, this parameter will be further incorporated in the energy gap $\Delta_0$ so that this gap becomes $\Delta_{hs-ls} = \Delta_0 - 2B$.

To estimate the parameter $J$ for the $Os_2Fe_3$ cluster compound using the described above procedure one needs to know the parameter $Dq^{hs}$ for the $hs$ Fe(III) ion. To the best of our knowledge this parameter is not available from the literature. Therefore, to obtain the parameter of the interaction of the Fe ion with the full symmetric lattice strain in $Os_2Fe_3$ complex an alternative procedure can be applied. One can take into account that when passing from the $ls$ Fe(II) electronic state to the $hs$ Fe(III) one, the vibrational relaxation occurs in the final electronic state, and the nuclei start to oscillate around the new equilibrium positions:

$$q_{eq} = -\frac{v_{A_1}^{hs} - v_{A_1}^{ls}}{\hbar\omega}. \tag{21.18}$$

Here $v_{A_1}^{hs}$ and $v_{A_1}^{ls}$ are the constants of the interaction of $hs$ Fe(III) and $ls$ Fe(II) with the local full symmetric $A_1$ mode, respectively. The transition from $ls$ Fe(II) to the $hs$ Fe(III) is accompanied by the change $\Delta R$ in the average metal – ligand bond length. Since the nearest surrounding of the Fe ion is a distorted octahedron, this change is connected with the equilibrium positions of the nuclei in the $hs$ Fe(III) state by the relation:

$$\Delta R = \sqrt{\frac{\hbar\omega}{6f}} q_{eq}, \tag{21.19}$$

where $f$ is the mean force constant of the full symmetric mode and $q_{eq}$ is given by Eq. 21.18.

The constant of interaction of the Fe ion with the full symmetric strain $\varepsilon_1$ can be written as:

$$v_{hs(ls)} = v_{A_1}^{hs(ls)} \sqrt{\frac{2f}{\hbar\omega}} R_0^{hs(ls)}, \tag{21.20}$$

where $R_0^{hs(ls)}$ is the average distance between the $hs$ Fe(III) ($ls$ Fe(II)) ion and its nearest neighbors. Finally, the coupling constant $v_1$ can be written as:

$$v_1 = -\sqrt{3} f R_0 \Delta R. \tag{21.21}$$

Here $R_0$ is the average metal – ligand distance for $hs$ Fe(III) and $ls$ Fe(II) ions. The change in the metal – ligand distances in course of the $ls$ Fe(II) to the $hs$ Fe(III) transition is about $0.18\text{Å}$ [14]. Using the mean force constant value $f = 2.6 \times 10^5$ dyne/cm and $R_0 = 2.0\,\text{Å}$ one finds $v_1 = -8.16 \times 10^4 \text{cm}^{-1}$. For the compound under the study the unit cell volume per one iron ion is $\Omega = 830\text{Å}^3$

# 21 Vibronic Approach to the Cooperative Spin Transitions in Crystals Based...

and $\Omega_0$ is approximately $64\,\text{Å}^3$. Under the assumption that the main change in the volume falls at the intermolecular space ($c_1 >> c_2$) and the value of $c_2$ is about $10^{11}\,\text{dyne/cm}^2$ one obtains $J = 160\,\text{cm}^{-1}$.

## 21.4 Results and Discussion

For the calculation of the temperature dependence of the order parameter, the self-consistent procedure was applied. The calculations of magnetic properties were based on the Hamiltonian given in Eqs. 21.1 and 21.7–21.12. The molar magnetization $M_\beta$ and the molar susceptibility $\chi_{\beta\beta}$ were calculated from the following expressions:

$$M_\beta = N_A k_B T \frac{\partial \ln Z(H_\beta)}{\partial H_\beta}; \quad \chi_{\beta\beta} = \frac{M_\beta}{H_\beta} \tag{21.22}$$

where $Z(H_\beta)$ designates the partition function which accounts for contributions from the Fe ions which are in the $hs$-state over the entire temperature range as well as the contribution from clusters with different numbers of spin crossover ions or those participating in the CTIST in the compounds $Co_2Fe_3$ and $Os_2Fe_3$, respectively, $H_\beta$ is the external magnetic field ($\beta = X, Y, Z$) and $N_A$ is the Avagadro's number.

We start with the analysis of the experimental data for the $Co_2Fe_3$ cluster compound. Figure 21.2 shows the experimental data for $Co_2Fe_3$ complex. The result of the best fit procedure developed in the model so far described is represented by the solid curve. In course of the best fit procedure the condition $J < J_0$ is implied in accord with the results reported in Ref. [22] where the optic phonons were found to promote larger interaction constants since the atoms are moving against each other and generate large electric fields. From Fig. 21.2 it is seen that a good agreement with the experimental data is obtained. The best fit parameters are the part of the figure caption. At temperatures below $100\,\text{K}$ the $\chi T$ values show that the Fe(II) ions are mainly in the $ls$-state with a small admixture of $hs$ ions. In the temperature range of $150$–$300\,\text{K}$, the $\chi T$ product gradually increases, thus indicating the $ls$-$hs$ transition at the Fe(II) centers. The parameter $J$ of the long-range cooperative electron-deformational interaction obtained from the best fit procedure falls within the limits estimated above. Relatively small values of the parameters $J$ and $J_0$, as compared to the gap $\Delta_{hs-ls} = \Delta_0 - 2B$, are also in agreement with the observed gradual temperature dependence of $\chi T$ and noticeable increase of $\chi T$ at temperatures higher than $150\,\text{K}$. Finally, the percentage of Fe(II) ions which are always in the $hs$-state is found to be 10% that is very close to that obtained from the Mössbauer spectra [12].

Now let us proceed to the analysis of the magnetic behavior of $Os_2Fe_3$ cluster compound. The result of the best fit procedure is shown in Fig. 21.3 as a solid

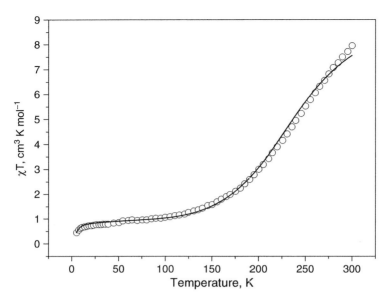

**Fig. 21.2** Temperature dependence of the $\chi T$ product for Co$_2$Fe$_3$ cluster compound. The *solid line* represents a theoretical fit with $\lambda = -103\,\text{cm}^{-1}$, $x = 10\%$, $\Delta_{hs-ls} = 640\,\text{cm}^{-1}$, $\Delta = 180\,\text{cm}^{-1}$, $J = 35\,\text{cm}^{-1}$, $J_0 = 45\,\text{cm}^{-1}$, $\kappa = 1$

line together with the experimental data. The best fit parameters are given in the Figure 21.3 caption. One can observe a good agreement between the experimental data and the theoretical curve at the temperatures above 30 K. The decrease of the experimental values of $\chi T$ at $T < 30$ K is probably caused by the intercluster exchange interaction that was not included in the model. The obtained value of the parameter of the short range interaction is larger than the calculated value of the parameter of long range cooperative interaction that agrees well with the preliminary estimations. The best fit value of $\Delta_0 - 2B$ is negative and, at first glance, it looks like the energies of the cluster configurations with one or two *hs* Fe(III) ions (types **B** and **C**) are lower than those of the configuration with three *ls* Fe(II) (type **A**). This would mean that the *hs* Fe(III) fraction should be significant even at very low temperature that is not true. In fact, a distinctive feature of the cluster compounds is the presence of short range interaction between iron ions belonging to the same cluster. The short range interaction, Eq. 21.7, results in the additional stabilization of the type **A** configuration with respect to that for the types **B** and **C** configurations. Due to this additional stabilization the **A** configuration becomes lower in energy and becomes the ground one. Cluster configurations of the types **B** and **C** are higher in energy by the value $E_B = \Delta_0 - 2B + 4J_0$ and $E_C = 2\Delta_0 - 4B + 4J_0$ with respect to the ground configuration. For the set of the best fit parameters one obtains that $E_B$ is about $600\,\text{cm}^{-1}$ and $E_C$ is about $490\,\text{cm}^{-1}$. From this it follows that at low temperatures the main part of iron ions are in the *ls* Fe(II) state. With the temperature increase up to 150 K practically no change in the *hs* and *ls* fractions is observed. With the further temperature growth, CTIST starts and the amount of *hs* Fe(III) increases.

# 21 Vibronic Approach to the Cooperative Spin Transitions in Crystals Based...

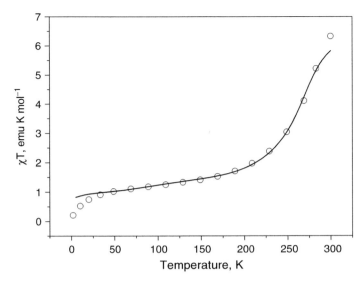

**Fig. 21.3** Temperature dependence of $\chi T$ product for $Os_2Fe_3$ cluster compound. The *solid line* represents a theoretical fit calculated with $\Delta_0 - 2B = -110\,\text{cm}^{-1}$, $J_0 = 178\,\text{cm}^{-1}$ and $x = 6\%$

**Table 21.1** Parameters used for simulation of Mössbauer spectra in $Co_2Fe_3$ compound

| T, K | $\delta$, mm/s | $\Delta E_Q$, mm/s | Fe type |
|---|---|---|---|
| 4.2 | 0.45 | 0.46 | ls-Fe(II) |
|  | 1.12 | 2.85 | hs-Fe(II) |
| 220 | 0.41 | 0.44 | ls-Fe(II) |
|  | 1.04 | 2.6 | hs-Fe(II) |
| 300 | 0.38 | 0.37 | ls-Fe(II) |
|  | 0.99 | 2.3 | hs-Fe(II) |

For both compounds the Mössbauer spectra can be simulated with the aid of the following superposition of the Lorenz curves:

$$F_C(\Omega) = n(hs)\frac{\Gamma}{\Gamma^2 + (\Omega - \delta_{hs} \pm \Delta E_{hs}/2)^2}$$
$$+ n(ls)\frac{\Gamma}{\Gamma^2 + (\Omega - \delta_{ls} \pm \Delta E_{ls}/2)^2} \quad (21.23)$$

The first term in Eq. 21.23 represents the contribution of *hs* Fe ions (Fe(III) or Fe(II) in the case of $Os_2Fe_3$ or $Co_2Fe_3$, respectively) while the second term is the contribution of *ls* Fe(II) ions. The factors $n(hs)$ and $n(ls)$ are the fractions of the *hs* and *ls* iron ions in the crystal at a definite temperature. The values of the parameters used in the calculations of the Mössbauer spectra are listed in Tables 21.1 and 21.2. The calculated Mössbauer spectra are shown in Figs. 21.4 and 21.5 by

**Table 21.2** Parameters used for simulation of Mössbauer spectra in $Os_2Fe_3$ compound

| T, K | $\delta$, mm/s | $\Delta E_Q$, mm/s | Fe type |
|------|------|------|---------|
| 4.2 | 0.4 | 0.36 | ls Fe(II) |
| | 0.55 | 0.55 | hs Fe(III) |
| 220 | 0.42 | 0.34 | ls Fe(II) |
| | 0.6 | 0.6 | hs Fe(III) |
| 300 | 0.42 | 0.3 | ls Fe(II) |
| | 0.6 | 0.6 | hs Fe(III) |

solid lines. One can see a very good agreement between the theoretical curves and the experimental data. The model presented above gives a good explanation of the magnetic properties and the Mössbauer spectra in crystals containing $\{[M(CN)_6]_2[Fe(tmphen)_2]_3\}$ clusters as structural units.

## 21.5 Concluding Remarks

In this article, we have presented a microscopic approach to the problem of cooperative spin transitions in crystals containing metal clusters as structural units. The *ls-hs* transition is considered as a cooperative phenomenon that is driven by the interaction of electronic shells of Fe ions with the fully symmetric deformation of the local coordination environment that is extended over the crystal lattice via the acoustic phonon field.

On the basis of this approach we were able to describe both the temperature induced and charge transfer induced spin transitions in crystals containing the cyano-bridged pentanuclear clusters $\{[M(CN)_6]_2[Fe(tmphen)_2]_3\}$ (M = Co, Os) with a trigonal bipyramidal structure. A random distribution of the spin-crossover ions in clusters was taken into account. The spin transitions were treated within the molecular field approximation. A competition between short- and long-range interactions was shown to determine the type and temperature of the spin transition in cluster systems. For both compounds we were able to interpret qualitatively and quantitatively a wide set of the experimental data on the temperature dependence of magnetic susceptibility and Mössbauer spectra with the unified set of parameters that fall inside the range of reasonable estimations and experimental data for similar entities. Finally, it should be underlined that the model contains a number of adjustable parameters that should be estimated independently from the *ab initio* calculations.

These results unambiguously demonstrate the cooperative nature of both temperature induced and charge transfer induced spin transitions in cluster compounds $\{[M(CN)_6]_2[Fe(tmphen)_2]_3\}$ (M = Co, Os). These transitions lead to a crossing of the *ls* and *hs* levels due to a structural phase transformation induced by the ordering of the local deformations through the field of acoustic phonons.

**Fig. 21.4** Mössbauer spectra of the $Co_2Fe_3$ cluster compound at $T = 4.2$ K, 220 K and 300 K. Points – experiment, thick *solid lines* – theoretical curves calculated with the set of the best fit parameters. Contributions from *ls*- and *hs*-Fe(II) ions are shown with *dashed* and *dotted lines*, respectively. The half-width of the individual lines: $\Gamma = 0.16 \,\text{cm}^{-1}$ (4.2 K), $0.18 \,\text{cm}^{-1}$ (220 K), $0.24 \,\text{cm}^{-1}$ (300 K)

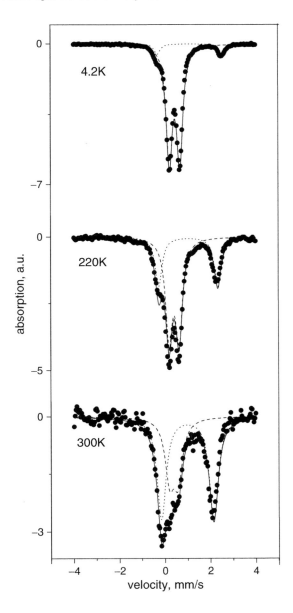

**Fig. 21.5** Mössbauer spectra for the Os$_2$Fe$_3$ cluster compound at T = 4.2 K, 220 K and 300 K. Points – experiment, thick *solid lines* – theoretical curves calculated at the set of best fit parameters. Contributions from *ls* Fe(II) and *hs* Fe(III) are shown in *dot* and *dashed lines*, respectively. The half-width of the individual lines is 0.16 mm/s (4.2 K) and 0.23 mm/s (220 and 300 K)

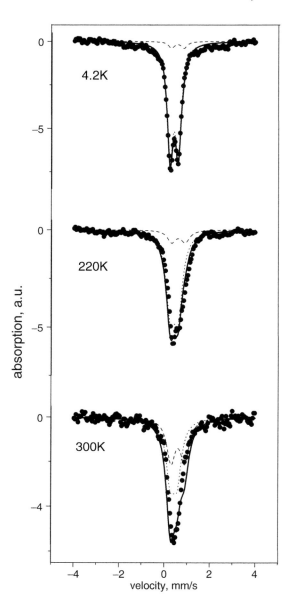

**Acknowledgments** S. O., A. P., S. K. thank STCU (project N5062) for the financial support. Financial support of the Supreme Council for Science and Technological Development of Moldova is also appreciated. B. T. and K. R. D. gratefully acknowledge financial support of the Binational US-Israel Science Foundation (BSF grant no. 2006498). B. T. thanks the Israel Science Foundation for additional financial support (ISF grant no. 168/09). KRD gratefully acknowledges support from the Department of Energy (DE-FG02-02ER45999) for the synthetic portion of this work.

# References

1. Holmes SM, Girolami GS (1999) J Am Chem Soc 121:5593
2. Entley WR, Girolami GS (1995) Science 268:397
3. Mallah T, Thiebaut S, Verdaguer M, Veillet P (1993) Science 262:1554
4. Scuiller A, Mallah T, Verdaguer M, Nivorozkhin A, Tholence JL, Veillet P (1996) New J Chem 20:1
5. Van Langenberg K, Batten SR, Berry KJ, Hockless DCR, Moubaraki B, Murray KS (1997) Inorg Chem 36:5006
6. Beltran LMC, Long JR (2005) Acc Chem Res 38:325
7. Song Y, Zhang P, Ren X-M, Shen X-F, Li Y-Z, You X-Z (2005) J Am Chem Soc 127:3708
8. Li DF, Parkin S, Wang GB, Yee GT, Prosvirin AV, Holmes SM (2005) Inorg Chem 44:4903
9. Berlinguette CP, Vaughn D, Cañada-Vilalta C, Galán-Mascarós JR, Dunbar KR (2003) Angew Chem Int Ed 42:1523
10. Wang C-F, Zuo J-L, Bartlett BM, Song Y, Long JR, You X-Z (2006) J Am Chem Soc 128:7162
11. Choi HJ, Sokol JJ, Long JR (2004) Inorg Chem 43:1606
12. Shatruk M, Dragulescu-Andrasi A, Chambers KE, Stolan SA, Bominaar EL, Achim C, Dunbar KR (2007) J Am Chem Soc 129:6104
13. Berlinguette CP, Dragulescu-Andrasi A, Sieber A, Güdel H-U, Achim C, Dunbar KR (2005) J Am Chem Soc 127:6766
14. Hilfiger MG, Chen M, Brinzari TV, Nocera TM, Shatruk M, Petasis DT, Musfeldt JL, Achim C, Dunbar KR (2010) Angew Chem Int Ed 49:1410
15. Kambara T (1979) J Chem Phys 70:4199
16. Kaplan MD, Vekhter BG (1995) Cooperative phenomena in Jahn-Teller crystals Plenum, New York
17. Gütlich P, Hauser A, Spiering H (1994) Angew Chem Int Ed Engl 33:2024
18. Klokishner SI, Varret F, Linares J (2000) Chem Phys 255:317
19. Klokishner SI, Linares J (2007) Phys Chem C 111:10644
20. Klokishner SI, Linares J, Varret F (2001) J Phys Condens Matter 13:595
21. Gehring GA, Gehring KA (1975) Rep Prog Phys 38:1
22. Baker JM (1971) Rep Prog Phys 341:109
23. Mironov VS (2007) Dokl Phys Chem 415:199

# Chapter 22
# On the Interplay of Jahn–Teller Physics and Mott Physics in the Mechanism of High $T_c$ Superconductivity

**H. Ushio, S. Matsuno, and H. Kamimura**

**Abstract** Based on the model proposed by Kamimura and Suwa which bears important characteristics born from the interplay of Jahn–Teller Physics and Mott Physics, it is shown that the feature of Fermi surfaces is the Fermi pockets constructed by doped holes under the coexistence of a metallic state and of the local antiferromagnetic order. Then it is discussed that the phonon-involved mechanism based on the Kamimura–Suwa model leads to the d-wave superconductivity. Further it is shown that $T_c$ is higher in the cuprates with $CuO_5$ pyramid than those with $CuO_6$ octahedron. Finally a new phase diagram for underdoped cuprates is proposed.

## 22.1 Introduction

In 1986 Bednorz and Müller discovered high temperature superconductivity in copper oxides by doping hole carriers into $La_2CuO_4$ [1]. Their motivation was that higher $T_c$ could be achieved for copper oxide materials by combining Jahn–Teller (JT) active Cu ions with the structural complexity of layer-type perovskite oxides.

---

H. Ushio (✉)
Tokyo National College of Technology, 1220-2 Kunugida-chou, Hachioji 193-0997, Japan
e-mail: ushio@tokyo-ct.ac.jp

S. Matsuno
Shimizu General Education Center, Tokai University, 3-20-1 Orido Shimizu-ku, Shizuoka 424-8610, Japan
e-mail: smatsuno@scc.u-tokai.ac.jp

H. Kamimura
Research Institute for Science and Technology, Tokyo University of Science, 1-3 Kagurazaka, Shinjuku-ku, Tokyo 162-8601, Japan
e-mail: kamimura@rs.kagu.tus.ac.jp

M. Atanasov et al. (eds.), *Vibronic Interactions and the Jahn-Teller Effect:*
*Theory and Applications*, Progress in Theoretical Chemistry and Physics 23,
DOI 10.1007/978-94-007-2384-9_22, © Springer Science+Business Media B.V. 2012

Undoped copper oxide $La_2CuO_4$ is an antiferromagnetic Mott insulator, in which an electron correlation plays an important role [2]. Thus we may say that undoped cuprates are governed by Mott physics. On the other hand, the $CuO_6$ octahedrons in undoped $La_2CuO_4$ are elongated by the JT effect. In order to investigate the mechanism of high temperature superconductivity, most of models assumed that the doped holes itinerate through the orbitals extended over a $CuO_2$ plane in the systems consisting of an array of $CuO_6$ octahedrons. Those models are called "single-component theory", because the orbitals of hole carriers extend only over a $CuO_2$ plane.

In 1989, Kamimura and his coworkers showed by first-principles calculations that the apical oxygen in the $CuO_6$ octahedrons tend to approach towards $Cu^{2+}$ ions, when $Sr^{2+}$ ions are substituted for $La^{3+}$ ions in $La_2CuO_4$, in order to gain the attractive electrostatic energy in ionic crystals such as cuprates [3,4]. As a result the $CuO_6$ elongated by the JT effect shrink by doping holes. This suppression of the JT distortion is called "anti-Jahn–Teller effect" [5]. By this effect the energy separation between the two kinds of orbital states which have been split originally by the JT effect becomes smaller by doping hole-carriers, where the spatial extension of one kind is parallel to the $CuO_2$ plane while that of the other kind is perpendicular to it.

By taking account of the anti-Jahn–teller effect, Kamimura and Suwa proposed that one must consider these two kinds of orbital states equally in forming the metallic state of cuprates, and they constructed a metallic state coexisting with the local antiferromagnetic (AF) order [6]. This model is nowadays called "Kamimura–Suwa (K–S) model" [7]. Since these two kinds of orbitals extend not only over the $CuO_2$ plane but also along the direction perpendicular to it, the K–S model represents a prototype of "two-component theory", in contrast to the single-component theory.

In this paper we will discuss two important subjects based on the K–S model. The first one is concerned with the shape of Fermi surfaces in underdoped cuprates. The second one is the occurrence of d-wave superconductivity in phonon mechanism. It is caused by the interplay of the strong electron–phonon interactions and local AF order [8, 9].

The organization of the present paper is the following: It consists of seven parts. After Introduction, in Sect. 22.2 we will explain the important role of "Mott Physics" and "Jahn–Teller Physics". Then we will describe the essential features of the K–S model which bears important characteristics born from the interplay of Jahn–Teller Physics and Mott Physics in Sect. 22.3. At Sect. 22.4 we will explain from the calculated many-body-effects included energy bands that the key features of the Fermi surfaces in underdoped cuprates are Fermi pockets, In Sect. 22.5 the mechanism of d-wave superconductivity is explained and in Sect. 22.6 a new phase diagram is described. Section 22.7 is devoted to conclusion and concluding remarks.

## 22.2 The Important Role of "Mott Physics" and "Jahn–Teller Physics"

Figure 22.1 shows the energy-level landscape starting from the orbitally doubly-degenerate $e_g$ and triply-degenerate $t_{2g}$ states of a $Cu^{2+}$ ion in a $CuO_6$ octahedron with octahedral symmetry in $La_2CuO_4$ to the molecular orbital states in the $CuO_6$ octahedron with Jahn–Teller distortion. By the JT effect the Cu $e_g$ orbital state at octahedral symmetry at the second column from the left splits into the $a_{1g}$ and $b_{1g}$ orbital states by the Jahn–Teller effect as shown at the third column from the left. As shown in middle column, they form the antibonding and bonding molecular orbitals of $A_{1g}$ and $B_{1g}$ symmetry with the molecular orbitals constructed from the in-plane oxygen $p_\sigma$ and apical oxygen $p_z$ orbitals in a Jahn–Teller distorted $CuO_6$ octahedron with tetragonal symmetry, respectively. These molecular orbitals are denoted by $a_{1g}^*$, $a_{1g}$, $b_{1g}^*$ and $b_{1g}$, as shown at the middle column, where the asterisk * represents the antibonding orbital. Here the $a_{1g}$ anti-bonding orbital $|a_{1g}\rangle$ is constructed by a Cu $d_{z^2}$ orbital and six surrounding oxygen p orbitals including apical O $p_z$-orbitals, while $b_{1g}$ bonding orbital $|b_{1g}\rangle$ consists of four in-plane O $p_\sigma$ orbitals with a small

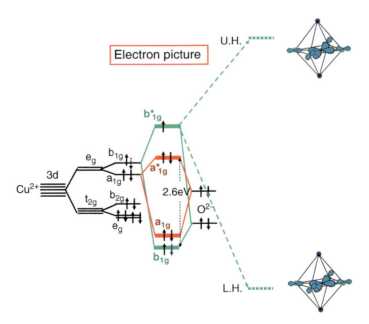

**Fig. 22.1** Energy-level landscape due to the interplay of Jahn–teller physics and Mott physics. The *right column* shows the splitting of the highest half-filled $b_{1g}^*$ state into L.H. and U.H. bands by the Hubbard $U$ interaction. The spins of localized electrons in the L.H. band at the neighboring A and B sites form an AF order. The energy in this figure is taken for electron-energy but not hole-energy. And the spatial distribution of localized electron is shown in this figure schematically

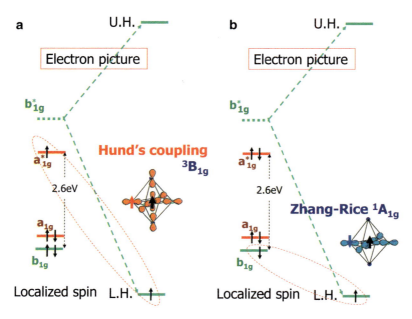

**Fig. 22.2** Energy level structures for the case where an electron is removed. (**a**) An down-spin $a_{1g}^*$ electron is removed. (**b**) An up-spin $b_{1g}$ electron is removed. The interaction of two electrons surrounded by each *red dotted circle* lowers the energy to get multiplet, Hund's coupling $^3B_{1g}$ or Zhang-Rice $^1A_{1g}$

Cu $d_{x^2-y^2}$ component parallel to a CuO$_2$ plane. In an undoped case, seven electrons occupy these molecular orbitals, so that the highest occupied $b_{1g}^*$ state is half filled, resulting in an $S = 1/2$ state, where a $b_{1g}^*$ antibonding orbital has mainly Cu $d_{x^2-y^2}$ character. Following Mott physics, we introduce the Hubbard $U$ interaction ($U = 10$ eV) as a strong electron-correlation effect. Then the half-filled $b_{1g}^*$ state splits into the lower and upper Hubbard bands denoted by L.H. and U.H. at the right column in the figure. An electron in the $b_{1g}^*$ state occupies L.H. band. The localized electrons in the L.H. band give rise to the localized spins around the Cu sites. These localized spins form the antiferromagnetic (AF) order by the superexchange interaction via intervening O$^{2-}$ ions in undoped La$_2$CuO$_4$.

Now let's consider the case where a hole is doped in the system in Fig. 22.1. This means that an electron is removed. If we do not take into account the many body interaction, one may remove an electron from the highest $a_{1g}^*$ orbital which is doubly occupied. However, if we take into account the many body interaction, such as exchange interactions, one may remove an electron from the $a_{1g}^*$ orbital or the $b_{1g}$ orbital. In case the $b_{1g}$ electron with up spin is removed, the remaining down-spin electron gains energy by the spin-singlet exchange interaction with a localized up-spin, resulting in the Zhang-Rice spin-singlet $^1A_{1g}$ [12] as shown in Fig. 22.2b. In case the $a_{1g}^*$ electron with down spin

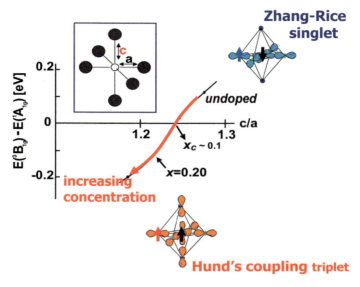

**Fig. 22.3** Calculated energy difference in the lowest state between Hund's coupling triplet and Zhang-Rice singlet as a function of Sr concentration in LSCO, where the distance between apical O and Cu, $c$, decreases with increase of Sr concentration, by Kamimura and Eto [10, 11]. The energy difference between the two multiplets is small

is removed, the remaining up-spin $a^*_{1g}$ electron gains energy by Hund's coupling with a localized up-spin, resulting in the Hund's coupling triplet state $^3B_{1g}$, as shown in Fig. 22.2a. The Zhang-Rice spin-singlet $^1A_{1g}$ and Hund's coupling triplet $^3B_{1g}$ are shown in the figure schematically.

When $Sr^{2+}$ ions are substituted for $La^{3+}$ ions in LSCO, the distance between apical O and Cu decreases, reflecting the anti-Jahn–Teller effect. Kamimura and Eto [10, 11] calculated the lowest state energies of $^1A_{1g}$ and $^3B_{1g}$ multiples by taking account of the anti-Jahn–Teller effect. The calculated energy difference between the lowest energies of $^1A_{1g}$ and $^3B_{1g}$ is shown in Fig. 22.3, as a function of the apical O–Cu distance $c$, where $c$ decreases with increasing the Sr concentration.

## 22.3 Brief Review of Kamimura–Suwa Model (K–S model)

In this section we will describe the electronic structure of a metallic state in underdoped LSCO calculated by Kamimura and Suwa [6], emphasizing the important roles due to the interplay of Jahn–Teller physics and Mott physics.

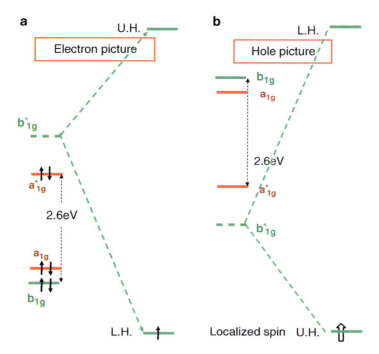

**Fig. 22.4** Energy level structures in the electron picture and the hole picture, where the undoped case is shown. (**a**) is the electron picture and (**b**) is the hole picture

## 22.3.1 Hund's Coupling Triplet and Zhang-Rice Singlet—Results of the First Principles Cluster Calculation

In a previous section, we have explained how the two kinds of multiplets, $^3B_{1g}$ and $^1A_{1g}$, appear when an electron is removed from the undoped La$_2$CuO$_4$. In this section, instead of the electron picture so far adopted as shown in Fig. 22.4a, we adopt a hole picture shown in Fig. 22.4b.

Let us dope a hole into the undoped La$_2$CuO$_4$. Then there appear two kinds of many-particle states called multiplets. One case is that a dopant hole occupies an antibonding $a_{1g}^*$ orbital shown in Fig. 22.5a. In this case its spin becomes parallel to a localized spin in the $b_{1g}^*$ orbital, resulting in a spin-triplet state, because the Hund's coupling makes the spins of holes in $a_{1g}^*$ and $b_{1g}^*$ orbitals with different symmetry parallel. As a result the energy of $a_{1g}^*$ orbital state for a single hole decreases by the Hund's coupling energy of 0.5 eV as shown in Fig. 22.5a. As named in a previous section, this spin-triplet multiplet is called the "Hund's coupling triplet" denoted by $^3B_{1g}$, which is schematically shown in Fig. 22.5a.

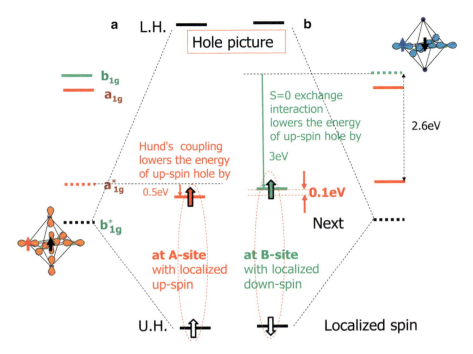

**Fig. 22.5** Energy level structures in the case where a hole is doped. There appear two kinds of many-particle states called multiplets. One is $^3B_{1g}$ (**a**) and another is $^1A_{1g}$ (**b**)

The other case is that a dopant hole occupies a bonding $b_{1g}$ orbital, as shown in Fig. 22.5b, and its spin becomes antiparallel to the localized spin in the antibonding $b_{1g}^*$ orbital, since the dopant and localized holes occupy the orbitals of the same symmetry. As a result the energy of $b_{1g}$ orbital state for a single hole decreases by the spin-singlet exchange interaction of 3.0 eV as shown in Fig. 22.5b. This spin-singlet multiplet may correspond to the "Zhang-Rice singlet" in the t-J model [12], and it is denoted by $^1A_{1g}$. The Zhang-Rice singlet state is schematically shown in Fig. 22.5b.

By the first-principles cluster calculations which take into account the Madelung potential due to all the ions surrounding a $CuO_6$ cluster in LSCO and also the anti-Jahn–Teller effect, Kamimura and Eto showed that the lowest-state energies of these two multiplets are nearly equal as shown in Fig. 22.3 [10, 11], when both the Hund's coupling exchange interaction and the spin-singlet exchange interaction are included. According to their calculated result the energy difference between the highest occupied orbital states $a_{1g}^*$ in $^3B_{1g}$ multiplet and $b_{1g}$ in $^1A_{1g}$ multiplet becomes only 0.1 eV for the optimum doping ($x = 0.15$) in $La_{2-x}Sr_xCuO_4$, where the anti-Jahn–Teller effect is included, as shown in Fig. 22.5.

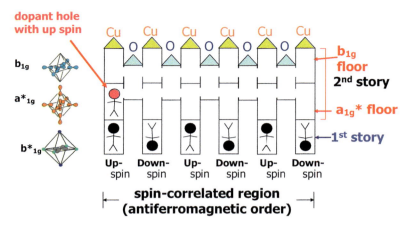

**Fig. 22.6** Two-story house model to explain the coexistence of AF order and metallic state

## 22.3.2 Explanation of the K–S Model by the Two-Story House Model

Based on the results of the first-principles cluster calculations for the lowest state of CuO$_6$ octahedron as a function of hole-concentration $x$ in La$_{2-x}$Sr$_x$CuO$_4$ (LSCO) by Kamimura and Eto and by assuming that the localized spins form an AF order by the superexchange interaction via intervening oxygen ions, Kamimura and Suwa [6] constructed a metallic state of LSCO for its underdoped regime. Since this model has already been explained in detail in refs. [6] and [7], here we explain the key-features of the model in a heuristic way using the picture of a two-story house model shown in Fig. 22.6.

In this figure the first story of a Cu house with (yellow) roof is occupied by the Cu localized spins, which form the AF order in the spin-correlated region by the superexchange interaction. The second story in a Cu house consists of two floors due to the anti-JT effect, a lower $a_{1g}^*$ floor and an upper $b_{1g}$ floor. The second story between neighboring Cu houses are connected by oxygen rooms with (blue-color) roof, reflecting the hybridization of Cu d and O p orbitals. In the second story a hole-carrier with up spin enters into the $a_{1g}^*$ floor at the left-hand Cu house due to Hund's coupling with a Cu localized up-spin in the first story (Hund's coupling triplet). By the transfer interaction the hole is transferred into the $b_{1g}$ floor at the neighboring Cu house (the second from the left) through the oxygen room, where the hole with up spin forms a spin-singlet state with a localized down spin at the second Cu house from the left (Zhang-Rice singlet).

The key feature of the K–S model is that the hole-carriers in the underdoped regime of LSCO form a metallic state, by taking the Hund's coupling triplet and the Zhang-Rice singlet alternately in the presence of the local AF order without destroying the AF order, as shown in the figure. Since the second story consists of the two floors of different symmetry, the two-story house model represents the two-component theory. As seen in Fig. 22.6, the characteristic feature of the K–S model is the coexistence of the AF order (the first story) and a normal, metallic (or a superconducting) state (the second story) in the underdoped regime. This feature in the K–S model (two-component theory) is different from that of the single-component theory. One may call the K–S model to be the three-component theory, if we count the component of hole carriers and localized hole.

## 22.4 Effective Hamiltonian and the Shape of Fermi Surface in K–S Model

Kamimura and Suwa introduced the effective Hamiltonian in order to describe the two-story house model. By using their effective Hamiltonian, Kamimura and Ushio [13, 14] calculated effective energy band in AF order, by treating the effect of the localized spin system as an effective magnetic field acting on the hole carriers. The coexistence of AF order and metallic state is suggested by many experiments [15–22]. As a result they could separate the localized hole-spin system in the AF order and the hole-carrier system from each other.

### 22.4.1 Effective Hamiltonian for the K–S Model

The effective Hamiltonian which has been introduced by Kamimura and Suwa [6] (see also ref. [7]) in order to describe the "Two-story house model" consists of four parts: The one-electron Hamiltonian $H_{\text{sing}}$ for $a_{1g}^*$ and $b_{1g}$ orbital states, the transfer interaction between neighboring $CuO_6$ octahedrons $H_{\text{tr}}$, the superexchange interaction between the Cu $d_{x^2-y^2}$ localized spins $H_{\text{AF}}$, and the exchange interactions between the spins of a dopant hole and of $d_{x^2-y^2}$ localized hole within the same $CuO_6$ octahedron $H_{\text{ex}}$. Thus we have

$$
\begin{aligned}
H = {} & H_{\text{sing}} + H_{\text{tr}} + H_{\text{AF}} + H_{\text{ex}} \\
= {} & \sum_{i,m,\sigma} \varepsilon_m C_{im\sigma}^{\dagger} C_{im\sigma} \\
& + \sum_{\langle i,j \rangle, m,n,\sigma} t_{mn} \left( C_{im\sigma}^{\dagger} C_{jn\sigma} + \text{h.c.} \right) \\
& + J \sum_{\langle i,j \rangle} \mathbf{S}_i \cdot \mathbf{S}_j + \sum_{i,m} K_m \mathbf{s}_{i,m} \cdot \mathbf{S}_i,
\end{aligned}
\tag{22.1}
$$

where $\varepsilon_m$ ($m = a_{1g}^*$ or $b_{1g}$) represents the one-electron energy of the $a_{1g}^*$ and $b_{1g}$ orbital states, $C_{im\sigma}^{\dagger}$ and $C_{im\sigma}$ are the creation and annihilation operators of a dopant hole with spin $\sigma$ in the $i$-th $CuO_6$ octahedron, respectively, $t_{mn}$ the transfer integrals of a dopant hole between $m$-type and $n$-type orbitals of neighboring $CuO_6$ octahedrons, $J$ the superexchange interaction between the spins $S_i$ and $S_j$ of $d_{x^2-y^2}$ localized holes in the $b_{1g}^*$ orbital at the nearest neighbor Cu sites $i$ and $j$ ($J > 0$ for AF interaction), and $K_m$ the exchange integrals for the exchange interactions between the spin of a dopant hole $s_{im}$ and a $d_{x^2-y^2}$ localized spin $S_i$ in the $i$-th $CuO_6$ octahedron. There are two exchange integrals, $K_{a_{1g}^*}$ and $K_{b_{1g}}$, for the Hund's coupling triplet and Zhang-Rice singlet, respectively, where $K_{a_{1g}^*} < 0$ and $K_{b_{1g}} > 0$. The appearance of the two kinds of exchange interactions in the fourth term is due to the interplay of Mott physics and Jahn–Teller physics. This is the key-feature of the K–S model.

### 22.4.2 Energy Band and Fermi Surface

By replacing the localized spins $S_i$'s in $H_{ex}$ by their average value $\langle S \rangle$ in the mean-field sense, Kamimura and Ushio calculated the "one-electron type" energy band for a carrier system assuming a periodic AF order. Here "one-electron type" means the inclusion of many-body-effects in the energy bands. In their calculation they neglected the electron–electron interactions between doped holes in the effective Hamiltonian (22.1), because these are very weak due to two reasons: One is because the concentration of hole-carriers in the underdoped regime is low and the other is because the wave functions of hole-carriers with up and down spins in a $CuO_6$ octahedron occupy $a_{1g}^*$ and $b_{1g}$ orbitals, respectively.

In Fig. 22.7b, the calculated exchange interaction included energy band structure for up-spin (or down-spin) doped holes for LSCO is shown for various values of wave-vector $\mathbf{k}$ and symmetry points in the antiferromagnetic (AF) Brillouin zone, where the AF Brillouin zone is adopted because of the coexistence of a metallic state and the AF order. Here one should note that the energy in this figure is taken for electron-energy but not hole-energy. Further the Hubbard bands for localized $b_{1g}^*$ holes which contribute to the local AF order are separated and do not described in this figure.

In the undoped $La_2CuO_4$, all the energy bands in Fig. 22.7b are occupied fully by electrons so that $La_2CuO_4$ is an antiferromagnetic Mott insulator, consistent with experimental results. In this respect the present effective energy band structure is completely different from the ordinary LDA energy bands [23, 24]. When Sr are doped, holes begin to occupy the top of the highest band in Fig. 22.7b marked by #1 at $\Delta$ point which corresponds to $(\pi/2a, \pi/2a, 0)$ in the AF Brillouin zone. At the onset concentration of superconductivity, the Fermi level is located just below the top of the #1 band at $\Delta$, which is a little higher than that of the $G_1$ point. Here the $G_1$ point in the AF Brillouin zone lies at $(\pi/a, 0, 0)$, and corresponds to a saddle point of the van Hove singularity.

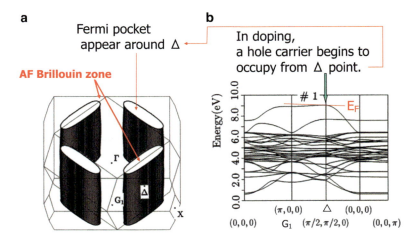

**Fig. 22.7** (a) The Fermi surface of hole-carriers for $x = 0.15$ calculated for the highest energy band. Here two kinds of Brillouin zones are also shown. One at the outermost part is the ordinary Brillouin zone corresponding to an ordinary unit cell consisting of a single $CuO_6$ octahedron and the inner part is the Brillouin zone for the AF unit cell in LSCO. Here the $k_x$ axis is taken along $\overline{\Gamma G_1}$, corresponding to the x-axis (the Cu–O–Cu direction) in a real space. (b) The exchange interaction included band-structure for up-spin (or down spin) dopant holes. The highest occupied band is marked by the #1 band (right). The $\Delta$-point corresponds to $(\pi/2a, \pi/2a, 0)$, while the $G_1$-point corresponds to $(\pi/a, 0, 0)$. It should be noticed that the Hubbard bands for localized $b_{1g}^*$ holes are not described in this figure

Based on the calculated band structure shown in Fig. 22.7b, Kamimura and Ushio [13, 14] calculated the Fermi surfaces for the underdoped regime of LSCO. In Fig. 22.7a the Fermi surface structure of hole-carriers calculated for $x = 0.15$ is shown as an example, where the Fermi surface (FS) consists of four Fermi pockets of extremely flat tubes around $\Delta$ point, $(\pi/2a, \pi/2a, 0)$ and other three equivalent points in the AF Brillouin zone. Recently Meng and his coworkers reported the existence of the Fermi pocket structure in the ARPES measurements of underdoped $Bi_2Sr_{2-x}La_xCuO_{6+\delta}$ (La-Bi2201) [25]. Their results are the clear experimental evidence for our prediction of the Fermi-pocket-structure for underdoped LSCO in 1994 [13].

## 22.5 High $T_c$ Superconductivity—Appearance of d-Wave Superconductivity in K–S Model

In high $T_c$ superconductivity, d-wave superconductivity is observed in $\pi$-junction experiment for hole-doped YBCO by Wollmam et al. [26] and by Kirtley et al. [27]. Based on the K–S model we explain the mechanism of d-wave superconductivity by phonon mechanism, following the theory developed by Kamimura and coworkers [8, 9].

### 22.5.1 Mechanism of d-Wave Superconductivity

In connection with the observed finite size of a spin-correlated region of the AF order [15, 16], Hamada and his coworkers [28] and Kamimura and Hamada [29] determined the ground state of the K–S model in a two-dimensional (2D) square lattice system with 16 ($4 \times 4$) localized spins by the exact diagonalization method. They clarified that, in the presence of hole-carriers, the localized spins in a spin-correlated region tend to form an AF order rather than a random spin-singlet state. Then they clarified that the hole-carriers can lower the kinetic energy of itineration in the spin-correlated region by taking the two kinds of orbitals alternately in the lattice of AF order. This mechanism of lowering the kinetic energy of the hole-carriers has led to the coexistence of a metallic state and the local AF order, which is the key-point of the K–S model.

Then in this section we will explain the following new mechanism of d-wave superconductivity in cuprates based on the K–S model: The interplay among the phonon mechanism, the local AF order and the characteristic feature of the K–S model that carrier holes occupy different orbitals with different symmetry in neighbouring sites leads to the d-wave pairing mechanism. First, we will point out that the phase relation between the wave functions of a hole carrier with up and down spins plays an important role in creating the d-wave superconductivity. As a result the matrix elements of electron–phonon interactions have a spin-dependent property. Let us start showing a sketch of the wavefunctions of a hole carrier with up- and down-spins in the tight binding scheme in Fig. 22.8a,b, respectively. As seen in Fig. 22.8, the wavefunction of a down-spin carrier is displaced from that of an up-spin carrier by the vector $\mathbf{a}$ representing the Cu–O–Cu distance. Thus, the wavefunctions of a hole carrier with up- and down-spins have the following phase relation in K–S model:

$$\psi_{\mathbf{k}\downarrow}(\mathbf{r}) = \exp(i\mathbf{k} \cdot \mathbf{a})\,\psi_{\mathbf{k}\uparrow}(\mathbf{r}) \tag{22.2}$$

Now let us define the matrix elements of electron–phonon interactions for up- and down-spin carriers between hole-states $\mathbf{k}$ and $\mathbf{k}'$ scattered by a phonon with wave vector $\mathbf{q}$, $V_\uparrow(\mathbf{k}, \mathbf{k}', \mathbf{q})$ and $V_\downarrow(\mathbf{k}, \mathbf{k}', \mathbf{q})$, respectively. By using the phase relation (22.2), these matrix elements have the following spin-dependent property:

$$V_\uparrow(\mathbf{k}, \mathbf{k}', \mathbf{q}) = \exp(i\mathbf{K} \cdot \mathbf{a})V_\downarrow(\mathbf{k}, \mathbf{k}', \mathbf{q}) \tag{22.3}$$

where $\mathbf{K} = \mathbf{k} - \mathbf{k}' - \mathbf{q}$ is a reciprocal lattice vector in the AF Brillouin zone and $\mathbf{a}$ is a Cu–O–Cu distance. In ordinary BCS case $\exp(i\mathbf{K} \cdot \mathbf{a})$ is always $+1$, and thus the spectral function is always positive leading to the s-wave superconductivity. In K–S model, however, $\exp(i\mathbf{K} \cdot \mathbf{a})$ takes the value of $+1$ or $-1$, because of $\mathbf{K} = (n\pi/a, m\pi/a, 0)$ with $n + m =$ even. This is a characteristic feature of the K–S model. By using the relation Eq. 22.3, the effective interactions of a pair of holes from $(\mathbf{k}\uparrow, -\mathbf{k}\downarrow)$ to $(\mathbf{k}'\uparrow, -\mathbf{k}'\downarrow)$ via a phonon of wave vector $\mathbf{q}$ is expressed as

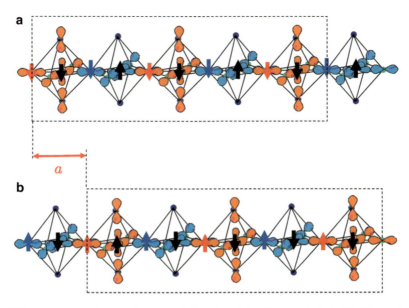

**Fig. 22.8** A wave function for down-spin hole (**a**) and that for up-spin hole (**b**). They coincide with each other, if one of them is displaced by vector **a** (a is Cu–O–Cu distance)

$$V_\uparrow(\mathbf{k},\mathbf{k}',\mathbf{q}) \cdot V_\downarrow(\mathbf{k},\mathbf{k}',\mathbf{q}) = \exp(i\mathbf{K}\cdot\mathbf{a})|V_\uparrow(\mathbf{k},\mathbf{k}',\mathbf{q})|^2$$

$$= \begin{cases} +1 & \text{for } n = \text{even(normal-scattering)} \\ -1 & \text{for } n = \text{odd(umklapp-scattering)}. \end{cases}$$

(22.4)

In the K–S model one should notice that the process of scattering from **k** to **k**′ by a phonon of wave vector **q** includes the two sub-processes of normal- and umklapp-scatterings. Suppose that a hole occupies the state A in one sections of Fermi pockets in Fig. 22.7. Then a hole is scattered by phonon with wavevector **q** from A to state C′ on a Fermi pocket. Since state C′ is equivalent to state B by the translation of a reciprocal lattice vector $(-\mathbf{Q}_1) - \mathbf{Q}_2$, the effective interaction from A to B is attractive according to equation (22.4) with $n = 2$. On the other hand, when a hole is scattered from state A to C on Fermi pockets in the shaded area in Fig. 22.9, we notice that state C is equivalent to state B by the translation of a reciprocal lattice vector $(-\mathbf{Q}_2)$. Thus the effective interaction is repulsive by equation (22.4) with $n = m = 1$. Since state C′ is inside the AF Brillouin zone which can be placed in the first Brillouin zone by the translation of AF-reciprocal lattice vector $\mathbf{K} = n\mathbf{Q}_1 + m\mathbf{Q}_2$ with $n + m$ being an even number while state C is outside the AF Brillouin zone, one may say that a scattering process from A to B via C′ is a normal scattering while a scattering process from A to B via

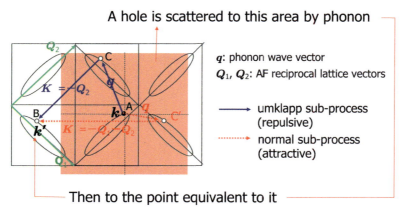

**Fig. 22.9** The process of scattering from **k** to **k**′. It includes normal subprocess and umklapp subprocess

C is umklapp scattering. Thus the total effective interactions between hole-states **k** and **k**′ is expressed by the sum of normal term and umklapp term as follows,

$$[V_\uparrow(\mathbf{k},\mathbf{k}',\mathbf{q}) \cdot V_\downarrow(\mathbf{k},\mathbf{k}',\mathbf{q})]_{\text{total}}$$
$$= V_\uparrow(\mathbf{k},\mathbf{k}',\mathbf{q})_{\text{normal}} \cdot V_\downarrow(\mathbf{k},\mathbf{k}',\mathbf{q})_{\text{normal}}$$
$$+ V_\uparrow(\mathbf{k},\mathbf{k}',\mathbf{q})_{\text{umklapp}} \cdot V_\downarrow(\mathbf{k},\mathbf{k}',\mathbf{q})_{\text{umklapp}} \qquad (22.5)$$

Normal and umklapp scatterings have different sign, so that if the normal term is larger, the effective interaction is attractive while if the umklapp term is larger, the effective interaction is repulsive.

As an example of calculated results, in Fig. 22.10 we show the results of the **k** and **k**′ dependence of the electron–phonon spectral function $\alpha^2 F_{\uparrow\downarrow}(\Omega,\mathbf{k},\mathbf{k}')$ with spin-singlet calculated by Kamimura and coworkers for one of out-of-plane modes, $A_{1g}$ mode (Fig. 22.10a) in LSCO with tetragonal symmetry [8], where $F_{\uparrow\downarrow}(\Omega,\mathbf{k},\mathbf{k}')$ is the momentum-dependent density of phonon states in energy and $\alpha^2$ is the square of the electron–phonon coupling constant.

In the right column of Fig. 22.10 we can see that the momentum-dependent spectral function varies by taking values with + and − sign in each section of ① to ④ in Fig. 22.10c, when the wave vector **k**′ changes from the section ① of Fermi pocket to the section ④ while **k** is fixed in Fig. 22.10b. This is clearly a d-wave behavior. By using this spectral function, we can obtain the **k**′ dependence of a gap function. The obtained d-component gap function $\Delta(\mathbf{k})$ varies as a function of $(\cos(k_x a) - \cos(k_y a))$ and the wavefunction of a Cooper pair has a spatial extension of $d_{x^2-y^2}$ symmetry.

The calculated result for d-wave spectral function is shown as a function of phonon energy $\Omega$ in Fig. 22.11a. Out-of-plain phonon modes, an example of which

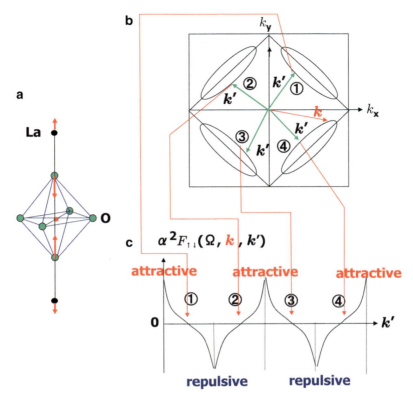

**Fig. 22.10** An example of phonon modes which contribute to the formation of Cooper pairs (**a**). Changing **k**′ as shown in (**b**), the spectral function varies as shown in (**c**)

is shown in Fig. 22.11b, contribute to the positive part of spectral function. The positive part of spectral function contributes to superconductivity. In-plain phonon modes, an example of which is shown in Fig. 22.11c, contributes to the negative part of spectral function. The negative part of spectral function does not contribute to superconductivity.

## 22.5.2 Finite Size Effects on Superconductivity

Before calculating $T_c$, we would like to point out that a mean free path $\ell_0$ in a metallic region where K–S model stands up all right is much expanded from the spin-correlated region $\lambda_s$ by the spin fluctuation effect in the localized spin system. The spin-correlated region $\lambda_s$ is shown by a red circle in Fig. 22.12. On a boundary, localized spins are frustrated. The spin-flip time in a frustration region, $\tau_s = \hbar/J$,

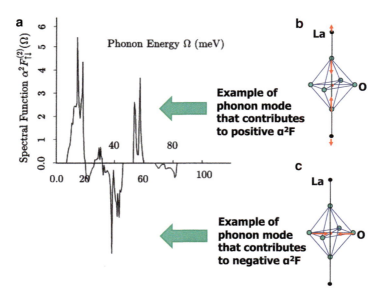

**Fig. 22.11** The calculated d-wave component of electron–phonon spectral function, for LSCO as a function of phonon frequency $\Omega$

is $10^{-14}$ s while the traveling time of a hole over $\lambda_s$, $\tau_s = \lambda_s/v_F$, is $10^{-13}$ s, where $v_F$ is the Fermi velocity of a hole carrier. Thus the spin-flip time is shorter than the traveling time of a hole carrier over $\lambda_s$ (Fig. 22.12)

Before a hole moves rightward out of spin-correlated region, several spin-flips occur and spin-correlated region moves rightward. So that, a hole can move coherently beyond spin-correlated region.

### 22.5.3 The Hole-Concentration Dependence of $T_c$ and of the Isotope Effect Calculated by Kamimura et al

In Fig. 22.13 we show the hole-concentration dependences of $T_c$ and the isotope effect $\alpha$ for LSCO calculated by Kamimura et al. [9]. In this figure the calculated results are compared with experimental results. In doing so we introduced one disposal parameter, that is the mean free path of a hole carrier. Kamimura et al. determined it so as to reproduce the highest value of $T_c$ at the optimum doping in LSCO. The mean free path thus determined for the optimum doping of LSCO is 300 Å. Then we notice good agreement between the theoretical results by the K–S model and experiments in Fig. 22.13a. In this figure the sharp increase of DOS at the onset of superconductivity gives rise to a sharp increase of $T_c$. The breakdown of Cooper pairs by the finite size gives rise to a sharp decrease of $T_c$.

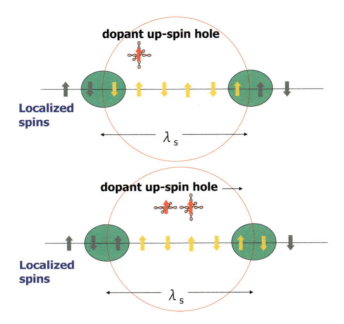

**Fig. 22.12** A spin-correlated region and a dopant hole. On a boundary, localized spins are frustrated, so that a metallic region expanded from the spin-correlated region

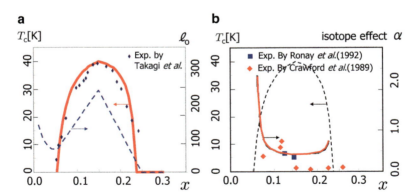

**Fig. 22.13** The calculated results on the hole-concentration dependence of $T_c(x)$ (**a**) and isotope effect (**b**) for LSCO

The isotope effect on $T_c$ in LSCO depends on the hole concentration critically. The present calculated result in Fig. 22.13b shows that it is remarkably large near the onset concentration of superconductivity. This is consistent with the experimental result reported by Bishop and coworkers [30].

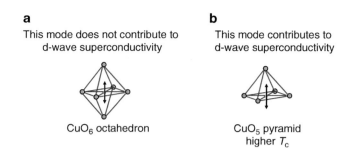

**Fig. 22.14** The phonon modes which contribute to superconductivity in CuO₅ pyramid but do not contribute in CuO₆ octahedron

## 22.5.4 Why $T_c$ is Higher in a Cuprate with $CuO_5$ Pyramid than that with $CuO_6$ Octahedron

We will comment on why $T_c$ is higher in a cuprate with CuO₅ pyramid than that with CuO₆ octahedron. It is because the number of out-of plane modes which contribute to d-wave superconductivity is larger in a cuprate with CuO₅ pyramid than that with CuO₆ octahedron. For example, let's consider the $\mathbf{q} = 0$ case. The Cu out of plane mode does not contribute to d-wave superconductivity in CuO₆ octahedron because CuO₆ octahedron has a symmetry with respect to CuO₂ plane, while the Cu out of plane mode contributes to d-wave superconductivity in CuO₅ pyramid, as shown in Fig. 22.14b.

Then we would like to respond to the question why a transition temperature is so high in cuprates. There are three reasons. One is that the crystal structures of cuprate superconductors are complicated, so that the number of phonon modes which contribute to superconductivity is large. Second is that the electron–phonon interaction is large in these Jahn–Teller materials. In particular, the change of the on-site energies due to the displacement of atoms in each phonon contributes siginificantly to the electron–phonon interactions, in addition to the change of transfer interaction from the tight-binding view [8]. The last one is that the density of states (DOS) is large because cuprates have a character of two-dimensionality and thus a peak in DOS may appear at the point of van Hove singularity.

## 22.6 New Phase Diagram Without the Pseudogap Hypothesis

In the recent decade considerable attention has been paid to the phenomenological idea of a pseudogap. When a portion of the large Fermi surface in cuprates were not seen in the ARPES experiments, an idea of pseudogap was proposed as a

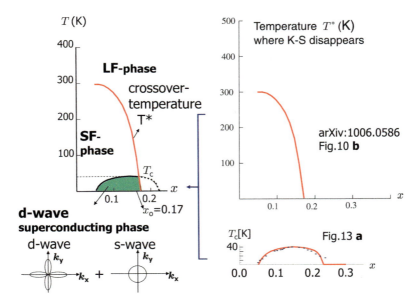

**Fig. 22.15** A new phase diagram proposed by Kamimura and Ushio [33, 34]

kind of gap to truncate the Fermi surface in the single particle spectrum [31, 32]. According to it, a pseudogap phase develops below a temperature called $T^*$ which depends on the hole concentration $x$ in the underdoped regime of cuprates. In the K–S model, however, the pseudogap hypothesis is not necessary because Fermi surface does not exist in the outside of Fermi pockets in the antinodal region. Recently, without the pseudogap hypothesis Kamimura and Ushio proposed a new phase diagram shown in Fig. 22.15. They have shown that the $T^*(x)$ represents a crossover from the a phase of Fermi pockets to a phase of large Fermi surface [34]. Hereafter we designate the former and latter phases the SF- and LF-phases, respectively.

Combining the result of hole concentration dependence of $T^*$ and the result of hole concentration dependence of $T_c$ shown in Fig. 22.13a, Kamimura and Ushio suggested that a green area surrounded by the $T_c$ and $T^*$ curves in the phase diagram represents a d-wave superconducting phase. However, as we already mentioned in Sect. 22.5.2 the finite size effect of a metallic region may mix a superconducting phase of other symmetry than the d-wave such as an s-wave phase. Thus we think that a superconducting phase in the green area may not show pure d-wave superconductivity, for example, a mixed phase of d-wave + s-wave superconductivity. This conjecture is consistent with the experimental result reported by Müller [35].

## 22.7 Conclusion and Concluding Remarks

In this paper we have described on the basis of the K–S model how the interplay of Mott physics and Jahn–Teller physics plays an important role in determining the superconducting as well as the metallic state of underdoped cuprates. It was pointed out for underdoped cuprates that Mott physics gives rises to the existence of local antiferromagnetic order due to the localized spins while that the anti-Jahn–Teller effect as a central issue of Jahn–Teller physics creates the existence of two kinds of orbitals parallel and perpendicular to a $CuO_2$ plane which are energetically nearby. The K–S model which bears important characteristics born from the interplay of Jahn–Teller Physics and Mott Physics has led to the coexistence of the local AF order and a metallic states above $T_c$. Further the coexistence of Hund's coupling spin-triplet exchange and Zhng-Rice spin-singlet exchange is also the important feature of the interplay. This coexistence has led to the appearance of Fermi pockets. Further below $T_c$ the superconductivity and antiferromagnetism coexist, leading to the appearance of d-wave superconductivity even in the phonon-involved mechanism. Two points are important to realize d-wave superconductivity. One is the coexistence of AF order and metallic state. The other is the fact that an up-spin hole carrier occupies $a_{1g}^*$ orbital in A-site where localized spin is up and $b_{1g}$ orbital in B-site where localized spin is down.

## References

1. Bednorz JG, Müller KA (1986) Z Phys B 64:189
2. Anderson PW (1987) Science 235:1196
3. Shima N, Shiraishi K, Nakayama T, Oshiyama A, Kamimura H (1989) In: Sugano T et al (eds) Proceedings of the JSAP-MRS international conference on electronic materials. Materials research society, Pittsburg, p 51
4. Oshiyama A, Shima N, Nakayama T, Shiraishi K, Kamimura H (1989) In: Kamimura H, Oshiyama A (eds) Mechanism of high temperature superconductivity. Springer series in materials science, vol 11. Springer, Berlin, p 111
5. Kamimura H, Ushio H, Matsuno S, Hamada T (2005) Theory of copper oxide superconductors. Springer, Berlin
6. Kamimura H, Suwa Y (1993) J Phys Soc Jpn 62:3368–3371
7. Kamimura H, Hamada T, Ushio H (2002) Phys Rev B 66:054504
8. Kamimura H, Matsuno S, Suwa Y, Ushio H (1996) Phys Rev Lett 77:723
9. Kamimura H, Hamada H, Matsuno S, Ushio H (2002) J Supercond 15:379
10. Kamimura H, Eto M (1990) J Phys Soc Jpn 59:3053
11. Eto M, Kamimura H (1991) J Phys Soc Jpn 60:2311
12. Zhang FC, Rice TM (1988) Phys Rev B 37:3759
13. Kamimura H, Ushio H (1994) Solid State Commun 91:97
14. Ushio H, Kamimura H (1995) J Phys Soc Jpn 64:2585
15. Mason T, Schroder A, Aeppli G, Mook HA, Haydon SM (1996) Phys Rev Lett 77:1604
16. Yamada K, Lee CH, Wada J, Kurahashi K, Kimura H, Endoh Y, Hosoya S, Shirane G, Birgeneau RJ, Kastner MA (1997) J Supercond 10:343
17. Yamada K et al (1998) Phys Rev B 57:6165

## 22 On the Interplay of Jahn–Teller Physics and Mott Physics...

18. Kao Y-J, Si Q, Levin K (2000) Phys Rev B 61:R11898
19. Christensen NB et al (2004) Phys Rev Lett 93:147002
20. Tranquada JM (2004) Nature 429:534
21. Haydon SM, Mook HA, Dai P, Perring TG, Dogan F (2004) Nature 429:531
22. Mukuda H et al (2006) Phys Rev Lett 96:087001
23. See, for example, Mattheiss LF (1987) Phys Rev Lett 58:1028
24. See, also, Yu J, Freeman AJ, Xu J-H (1987) Phys Rev Lett 58:1035
25. Meng J et al (2009) Nature 462:335
26. Wollmam DA et al (1993) Phys Rev Lett 71:2134
27. Kirtley JR et al (1995) Nature 373:225
28. Hamada T, Ishida K, Kamimura H, Suwa Y (2001) J Phys Soc Jpn 70:2033
29. Kamimura H, Hamada T (2003) In: Srivastava JK, Rao SM (eds) Models and method of high-$T_c$ superconductivity: some frontal aspects, vol 2, Chap 2. Nova Science, New York
30. Bishop AR, Bussmann-Holder A, Cardona M, Dolgov OV, Furrer A, Kamimura H, Keller H, Khasanov R, Kremer RK, Manske D, Mueller KA, Simon A (2007) J Supercond Novel Magn 20:393
31. Marshall DS et al (1996) Phys Rev Lett 76:4841
32. Norman MR et al (1998) Nature 392:157
33. Kamimura H, Ushio H (2008) J Phys Conf Ser 108:012030
34. Kamimura H, Ushio H (2010) arXiv:1006.0586 cond-mat.supr-con. 3 June 2010
35. Müller KA (2002) Phil Mag Lett 82:279

# Chapter 23
# Vibronic Theory Approach to Ising Model Formalism and Pseudospin-Phonon Coupling in H-Bonded Materials

**S.P. Dolin, Alexander A. Levin, and T.Yu. Mikhailova**

**Abstract** The investigation of the lowest adiabatic potential energy surface (APES) of the $AO_4$ structural unit in KDP ($KH_2PO_4$)-family materials as a function of its proton surrounding is reported. This approach results in the Ising form of the crystal pseudospin Hamiltonian. Besides the electronic structure characteristics of the $AO_4$ tetrahedra, the Ising model interaction parameters depend also on the vibronic constants. Numerical estimations of scale for vibronic contributions into Ising parameters (about a quarter of the total value of the maximal Ising parameter V) was obtained by using diffraction data for $KH_2PO_4/KD_2PO_4$ crystal structure change under ferroelectric transition together with the results of quantum-chemical calculation of $PO_4^{3-}$ force constants. The estimates show the really tangible role of the vibronic terms in the Ising parameters formation, though the physical reasons of these terms remained somewhat unclear. The situation is clarified by revealing the relationship between the above-mentioned approach and the theory of pseudospin-phonon (proton-lattice) coupling which leads to the vibronic contributions to the Ising parameters.

## 23.1 Introduction: Induced Vibronic Effects

As is well known, besides the spontaneous distortions of symmetric polyatomic systems due to Jahn-Teller effect or strong pseudo Jahn-Teller effect, there exist also the distortions induced by atomic substitutions. The familiar example is the deformation of the coordination polyhedron of the central A atom in the closed-shell complexes [1] and molecules $AL_n$ when the substitution of one or several ligands

---

S.P. Dolin • A.A. Levin (✉) • T.Yu. Mikhailova
NS Kurnakov Institute of General and Inorganic Chemistry, RAS, 119991 Leninskii prospect, 31, Moscow, Russia
e-mail: Levin@igic.ras.ru

M. Atanasov et al. (eds.), *Vibronic Interactions and the Jahn-Teller Effect: Theory and Applications*, Progress in Theoretical Chemistry and Physics 23, DOI 10.1007/978-94-007-2384-9_23, © Springer Science+Business Media B.V. 2012

lead to the change of distances A-L also for the other, non-substituted ligands (so-called mutual influence of ligands [2]). The origin of such deformations [2] is different from the origin of strong pseudo Jahn-Teller effect thus in this case one might talk about specific induced vibronic effect.

The distortions of the molecular structural units in the crystals with intermolecular H/D bonds caused by the change of protons or deuterons sites at these bonds [2–4] may be considered as another manifestation of the same effect. The typical example of such distortions is the displacement of P or As atoms along local z-axis with relation to the centers of $PO_4/AsO_4$ tetrahedral anions upon the structural phase transition from proton disordered paraelectric phase into ordered ferroelectric phase in the KDP-like ferroelectric materials [5]. The present paper considers the role of this induced vibronic effect in the Ising model interaction parameters formation for the case of KDP-like crystals. It is find out that the appearance of vibronic terms in the Ising parameters is due to the pseudospin-phonon coupling account.

## 23.2 The Substitution Operator and Basic Equations

The central point of induced vibronic phenomena treatment is the concept of the "substitution operator", that actually has been used earlier in the "molecular orbitals perturbation theory" [6,7] for studying the structure, reactivity and reactions mechanisms for organic molecules. Concerning the heteroligand molecular systems this operator describes how the change of the ligands acts on the eigen functions of the electronic Hamiltonian of unperturbed homoligand system $AL_n$. Then the lowest APES of a heteroligand system may be obtained by means of the second-order perturbation theory approach where the perturbation operator is the sum of the vibronic interaction operator for $AL_n$ and the substitution operator [2].

Proceeding from the many-electron operators to the one-electron ones, it is not difficult to demonstrate [2] (for the case of the closed-shell systems which we are interested in) that the corresponding expression for the lowest APES has the form:

$$E(Q) = S_0 + \frac{1}{2}\sum_{v} K_v Q_v^2 - 2\sum_{n,m} \frac{S_{nm}^2}{\omega_{nm}} - 4\sum_{v}\sum_{n,m} \frac{S_{nm} A_{nm}^v}{\omega_{nm}} Q_v. \qquad (23.1)$$

Here $Q = \{Q_v\}$ is the set of normal (or symmetry) coordinates for the $AL_n$ system, $K_v$ (that are considered positive) and $A_{nm}^v$ are its force constants and orbital vibronic constants, respectively [8], indices n and m relating to the occupied ($\psi_n$) and vacant ($\psi_m$) molecular orbitals (MOs) respectively. The $\omega_{nm}$ in (23.1) denote the energy gaps between these MOs and $S_{nm}$ are the matrix elements $S_{nm} = < \psi_n|H_s|\psi_m >$ of one-electron substitution operator $H_s$ that describes the ligands substitution influence on the molecular orbitals of $AL_n$. The first item $S_0$ in the rhs of (23.1) accounts in the first order of perturbation theory the amendment of the $AL_n$ ground state energy and doesn't take into account the change of MOs and the geometry of the system upon substitution [2].

Relation (23.1) leads immediately to the following expressions for two characteristic parameters of the APES considering – its minimum coordinates $Q_v^f$ and the depth $E(Q_v^f)$ of the minimum with regard to the minimum of the $AL_n$ lowest APES:

$$Q_v^f = \frac{4}{K_v} \sum_{n,m} \frac{S_{nm} A_{nm}^v}{\omega_{nm}}. \tag{23.2}$$

$$E(Q_v^f) = S_0 - 2 \sum_{n,m} \frac{S_{nm}^2}{\omega_{nm}} - 8 \sum_v \frac{1}{K_v} \left( \sum_{n,m} \frac{S_{nm} A_{nm}^v}{\omega_{nm}} \right)^2. \tag{23.3}$$

Finally, let's slightly elucidate the applying modeling of the $H_s$ operator [2]. One may assume that the ligands substitution in the $AL_n$ system may be described by the replacing of energies of the atomic orbitals (AOs) of the substituted atoms by the corresponding energies for the AOs of the substituents. Then we obtain

$$S_{nm} = \sum_i c_{in} c_{im} \Delta \alpha_i. \tag{23.4}$$

Here $c_{in}$, $c_{im}$ are the coefficient of the AO $\chi_i$ in the MOs $\psi_n$, $\psi_m$, and $\Delta \alpha_i$ is the orbital energy (Coulomb integral) change of this AO due to substitution.

## 23.3 The Proton Redistribution as "Ligands Substitution"

The possibility of the use of the substitution operator and the induced vibronic effect concepts to establish the mechanism of the ferroelectric properties formation in KDP family materials becomes readily clear from the spatial structure of these materials (Fig. 23.1). Its structural units are the alkali metal cations and the closed-shell tetrahedral molecular ions $AO_4$ ($A = P$ or $As$) with formal charge $q = -3$; according to the ESR measurements there are no unpaired electrons in the crystals at hand besides X-ray or $\gamma$-irradiated species [9]. The tetrahedra are connected by vertexes with strong short double-well hydrogen bonds [5,9].

Thus each hydrogen may occupy two equivalent positions: O–H...O or O...H–O. Hence each oxygen in the crystal may be found in the state $O(-)$ or $O(...)$, in other words, may be connected to the next hydrogen either by covalent (O–H) or by hydrogen (O...H) bond. Therefore any redistribution of protons over equilibrium sites at H-bonds causes the transformation of several or even all oxygen atoms from one state to another, that may be formally treated as somewhat specific mutual substitution of two chemically different kinds of O ligands at the vertexes of the $AO_4$ tetrahedra. Surely the equivalent description is also possible, where the actual proton distribution is considered with regard to the hypothetic "protophase"

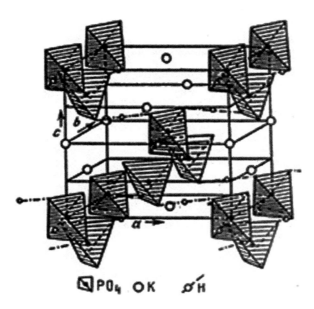

**Fig. 23.1** Crystal structure of KDP and its analogues

of the material, where the hydrogen atoms are placed in the centers of the H-bonds and all oxygen atoms are in the same state. Then the protons transfer to their actual equilibrium positions leads again to the appearance of two different types of the oxygens instead of previously equivalent oxygen atoms. Let us mention that any description yields in the same cooperative dependence between the states of the O atoms in different $AO_4$ tetrahedra, because in the real material two oxygen atoms at the opposite ends of the each H-bond are always in different states.

From such "chemical" viewpoint the formation and redistribution of two types of oxygens over the vertexes of $AO_4$ tetrahedra is the essential process, while the proton redistribution boils down to the trigger role for mutual transmutations of oxygen sorts. Thus one may assume that the component of the crystal potential energy $H$ sensitive to the hydrogen distribution is determined by the sum

$$H = \sum_i E_i \tag{23.5}$$

of the "inner" energies $E_i$ for all $AO_4$ tetrahedra where each summand is defined by relation (23.3). Indeed all the other interactions, i.e. the interactions between $AO_4$ tetrahedra (mostly of the $AO(-)_2O(\ldots)_2$ constitution), between K ions and also $K - AO_4$ interactions come to the lattice's sum alike the Madelung energy for the lattice comprised by cations $K^{q+}$ and anions $[H(-)_2AO(-)_2O(\ldots)_2]^{q-}$ that depends slightly on proton redistribution.

## 23.4 The Pseudospin Formalism and Ising Model

To clarify the relationship between pseudospin-phonon coupling and formula (23.5) one should revert to common formulation of the microscopic theory for H-bonded ferroelectrics of order-disorder type in terms of proton distribution over H-bonds. In the simplest version of the theory neglecting proton tunneling this distribution is considered [9, 10] in terms of the set of dichotomic pseudospin variables $\sigma_i^z = +1, -1$. Here two values of the i-th variable indicate two possible equilibrium sites of proton at the i-th H-bond according to two symmetrical minima of proton adiabatic potential.

Then one can obtain that the crystal potential energy $H$ has the Ising form:

$$H = -(1/2)\Sigma_{(i \neq j)} J_{ij} \sigma_i^z \sigma_j^z, \tag{23.6}$$

where $J_{ij}$ are Ising model interaction parameters, describing an effective two-particle coupling between protons in the crystal. As it was demonstrated in [3] this relation for $H$ follows directly from (23.5) as the valence state of each O atom in any tetrahedron is definitely related with the type of the neighboring hydrogen: H(-) or else H(...), i.e. with one of two values of the correspondent pseudospin variable. In fact it is easy to see also that the summands $S_0$ from (23.3) vanish in (23.5) and hence in (23.6) (as well as in all the other related formulae) because of their mutual compensation by summation over the lattice. Thus expressing all remaining matrix elements $S_{nm}$ in linear dependence on the pseudospin variables we come just to the formula (23.6) where all Ising parameters are expressed in terms of the electronic structure and the vibronic properties of the structural units for the material at hand.

## 23.5 Ising Parameters

As a representative example let us give the form of two Ising parameters used as a rule:

$$U = J_{\parallel} = c_a^2 \Delta \alpha^2 \left( \frac{l_b^2}{4\omega_{\perp}} - \frac{l_e^2}{2\omega_{\parallel}} + \frac{l_b^2 A_{\perp}^2}{\omega_{\perp}^2 K_{\perp}} - \frac{2l_e^2 A_{\parallel}^2}{\omega_{\parallel}^2 K_{\parallel}} \right) \tag{23.7}$$

$$V = J_{\perp} = c_a^2 \Delta \alpha^2 \left( \frac{l_b^2}{4\omega_{\perp}} + \frac{l_b^2 A_{\perp}^2}{\omega_{\perp}^2 K_{\perp}} \right) \tag{23.8}$$

Here the parameter U corresponds to the coupling of the "upper" (1,2) or "lower" (3,4) protons and V corresponds to the analogous coupling between "upper" and "lower" protons, Fig. 23.2. Symbols $\omega_{\parallel}$ and $\omega_{\perp}$ denote respectively the energy gaps $e-a^*$ and $b-a^*$ between the e, b MO's (HOMO) and $a^*$ MO (LUMO) for the

**Fig. 23.2** Numeration of O and H atoms in PO$_4$ tetrahedron

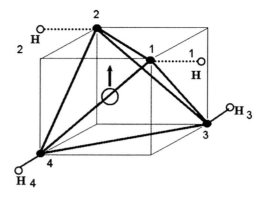

**Fig. 23.3** The MO diagram for PO$_4$ tetrahedron; σ-MO's approximation

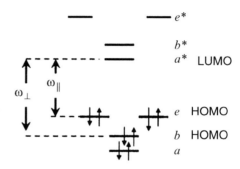

AO$_4$-tetrahedron with actual S$_4$ symmetry in the KDP/DKDP protophase (Fig. 23.3). The coefficients of the symmetrized atomic σ – AOs of oxygens in these MO's are denoted as $l_e, l_b, c_a$. The symbols A$_\|$, A$_\perp$, K$_\|$ and K$_\perp$ denote respectively the orbital vibronic constants and force constants for the AO$_4$-tetrahedron distortions that are accompanied by the central atom displacements (with respect to the oxygen atoms) in the xy plane or along z axis in the protophase. Finally, the $|\Delta\alpha|$ value means the change in magnitude of the Coulomb integral for the oxygen's localized σ –orbital, that accompanies oxygen atom transition from its state in the protophase to the state O(-)/O(...).

## 23.6 Electronic and Vibronic Contributions

The formulae (23.7) and (23.8) show the significant property of the Ising parameters in the considered model. It is evident that each of both parameters may be represented as a sum of two contributions:

$$U = U_{el} + U_{vib}, V = V_{el} + V_{vib} \qquad (23.9)$$

Here first contribution includes the values that characterize the electronic structure of unperturbed $AO_4$-tetrdahedra in the protophase and the matrix elements of the substitution operator. It doesn't contain the vibronic constants. At the same time the second contribution depends on vibronic constants.

## 23.7 Vibronic Terms Estimates

Formulae like (23.7) and (23.8) allow in principle to calculate the "electronic" and "vibronic" terms in the Ising parameters values by means of quantum-chemical *ab initio* procedures thus demonstrating the role of vibronic constants in Ising parameters formation. It is enough however to begin with the much easier approach in order to estimate the significance of these constants.

Using (23.2) for the $AO_4$-tetrdahedron in the frames of FMO e,b, a* approximation mentioned above it is easy to obtain the relation that attributes the $V_{vib}$ value to this tetrahedron distortion in the ferroelectric phase and with the force constant of the correspondent normal mode:

$$V_{vib} = (1/16)K_\perp (Q_\perp^f)^2 \qquad (23.10)$$

The form of the $Q_\perp^f$ mode in (23.10) may be found from neutron diffraction data [5] obtained for KDP and DKDP for the temperatures below as well as above ferroelectric phase transition. These data are presented in Table 23.1 where atom labeling is shown at Fig. 23.2. The $K_\perp$ force constants for this mode may be estimated for example from quantum-chemical calculations for the isolated ions $PO_4^{3-}$, $H_2PO_4^-$, $H_4PO_4^+$. The simplest way to estimate $V_{vib}$ summand in the parameter V is the use of the constant $K_6$, taken from the $PO_4^{3-}$ calculation (DFT, MP2-MP4) for stretching mode $Q_6$ of the symmetry $t_2$ ($K_6 = 8$–$9\,mdyn/A$) neglecting the contributions from the bending mode $Q_9$ of the same symmetry ($K_9 \approx 2.5\,mdyn/A$); widely known forms of these modes for tetrahedral systems see in [1] or [2]. It yields in the $V_{vib}$ values of 120 K (KDP) and 190 K (DKDP). According to our previous calculations [11] the estimating of V neglecting vibronic contribution

**Table 23.1** Interatomic distances (in Å) and bond angles (in degrees) for KDP/DKDP according to [5]

|  | KDP 293 K | KDP $T_c + 5K$ | KDP $T_c - 20K$ | DKDP 293 K | DKDP $T_c + 5K$ | DKDP $T_c - 10K$ |
|---|---|---|---|---|---|---|
| Distances |  |  |  |  |  |  |
| P–O (3,4) | 1.5403 | 1.5429 | 1.5719 | 1.5412 | 1.5427 | 1.5785 |
| P–O (1,2) |  |  | 1.5158 |  |  | 1.5091 |
| Angles |  |  |  |  |  |  |
| O3–P–O4 | 110.52 | 110.58 | 106.65 | 110.50 | 110.56 | 105.74 |
| O1–P–O2 |  |  | 114.80 |  |  | 115.73 |

gives 340–400 K for KDP and 480–560 K for DKDP. Thus the account of this term (in other words the account of pseudospin-phonon coupling, see Sect. 23.8) results in the V parameter growth approximately by a third.

## 23.8 Pseudospin-Phonon Coupling

Finally, let us note the relation of vibronic theory approach with the theory of pseudospin-phonon coupling.

Summation of expressions (23.1) for the lowest APES'es of all $AO_4$-tetrdahedra over the KDP-like crystal's lattice yields in

$$H_1 = \sum_t \left\{ -2 \sum_{nm} \frac{S_{tnm}^2}{\omega_{tnm}} + \frac{1}{2} \sum_v K_{tv} Q_{tv}^2 - 4 \sum_{vnm} Q_{tv} \frac{S_{tnm} A_{tnm}^v}{\omega_{tnm}} \right\} \qquad (23.11)$$

Here an additional index $t$ denotes the $t$-th tetrahedron; remind that summation over the lattice eliminates the terms $S_0$ in (23.1). As it follows from (23.4) (see also Sect. 23.4), any matrix element $S_{tnm}$ may be expressed linearly with the $\sigma_{t\alpha}^z = +1, -1 (\alpha = 1, 2, 3, 4)$ variables determining the protons positions on the four next neighboring H-bonds for $t$-th $AO_4$-tetrahedron. Substituting then $S_{tnm}$ in (23.11) one has

$$H_1 = -\frac{1}{2} \sum_{t, \alpha \neq \beta} J_{t\alpha\beta}^{el} \sigma_{t\alpha}^z \sigma_{t\beta}^z + \frac{1}{2} \sum_{tv} K_{tv} Q_{tv}^2 - \sum_{tv\alpha} K_{t\alpha v} Q_{tv} \sigma_{t\alpha}^z. \qquad (23.12)$$

The last relation may be considered as somewhat modified version of the local representation [12, 13] for the total Hamiltonian of the coupled proton-phonon system in the case of several modes $Q_{tv}$ for each $AO_4$-tetrahedron; that responds to more detailed description of the lattice geometry upon ferroelectric transition in KDP and DKDP. It is readily seen that (23.12) allows to express $H_1$ in the Ising form, basing on the dependence of the tetrahedra geometry on the nearest proton positions. By using relation (23.2) for $Q_{tv} = Q_{tv}^f$ and the dependence of the $S_{tnm}$ on $\sigma_{t\alpha}^z$, we come straightforwardly to the Ising-type (23.6) expression for the sum of all three terms in the rhs of (23.12). In this expression the Ising parameters contain vibronic contributions, alike the U,V parameters, but the reason of the arising of these contributions becomes clearer and lies in the fact that the Hamiltonian (23.12) contains in explicit form the terms which describe the pseudospin-phonon coupling.

**Acknowledgments** The authors are deeply indebted to Prof. I.B. Bersuker for stimulating discussion. This work is supported by RFBR, grant 08-03-00195.

# References

1. Bersuker IB (2010) Electronic structure and properties of transition metal compounds. Introduction to the theory, 2nd edn. Wiley, New York
2. Levin AA, D'yachkov PN (2002) Heteroligand molecular systems. Bonding, shapes and isomer stabilities. Taylor & Francis, London/New York
3. Levin AA, Dolin SP (1996) J Phys Chem 100:6258
4. Levin AA, Dolin SP (1998) Russ J Coord Chem 24:270
5. Nelmes RJ, Tun Z, Kuhs WF (1987) Ferroelectrics 71:125
6. Dewar MJS (1969) The molecular orbital theory of organic chemistry. McGraw Hill, New York
7. Dewar MJS, Dougherty RC (1975) The PMO theory of organic chemistry. Plenum Press, New York
8. Bersuker IB (1984) The Jahn-Teller effect and vibronic interactions in modern chemistry. Plenum Press, New York
9. Blinc R, Žekš B (1974) Soft modes in ferroelectrics and antiferroelectrics. North-Holland Inc. Oxford/New York
10. Strukov BA, Levanyuk AP (1998) Ferroelectric phenomena in crystals. Springer, Berlin
11. Dolin SP, Mikhailova TY, Solin MV, Breslavskaya NN, Levin AA (2010) Int J Quant Chem 110:77
12. Blinc R (1982) Croatica Chem Acta 55:7
13. Blinc R, Žekš B (1982) J Phys C Solid State Phys 15:4661

# Chapter 24
# Virtual Phonon Exchange Influence on Magnetic Properties of Jahn-Teller Crystals: Triple Electronic Degeneracy

**Michael D. Kaplan and George O. Zimmerman**

**Abstract** Mutual influence of magnetic interactions and competing two types of virtual phonon exchange interactions is considered in the framework of the cooperative Jahn-Teller effect theory. It is shown that this competition in the presence of magnetic interactions can result in a new type of structural tetragonal –tetragonal transition. The corresponding virtual phonon exchange interactions drastically affect the magnetic crystal properties. Comparison with the experimental data is presented.

## 24.1 Introduction

Properties of crystals with structural phase transitions are significantly modified not only due to (and near) the phase transitions itself, but, first of all, because of the orbital degeneracy of the ground electronic states of the ions forming the crystal lattice or one of its sublattices [1–4]. This electronic degeneracy leads to a specific coupling between the electronic and lattice subsystems of the crystal, and that in its turn results in the cooperative Jahn-Teller effect (CJTE). The last one is the most important microscopic mechanism of the structural phase transitions. Because the virtual phonon exchange (VPE) between the crystal site electrons is the central part of the CJTE, it is not unexpected that all properties of the JT crystals are significantly influenced by the VPE. This, in part, is related to the magnetic properties of crystals with electronic orbital degeneracy of ions.

---

M.D. Kaplan (✉)
Department of Chemistry and Physics, Simmons College, 300 The Fenway, Boston,
MA 02115, USA
e-mail: michael.kaplan@simmons.edu

G.O. Zimmerman
Department of Physics, Boston University, 590 Commonwealth Ave., Boston,
MA 02215, USA

M. Atanasov et al. (eds.), *Vibronic Interactions and the Jahn-Teller Effect:*
*Theory and Applications*, Progress in Theoretical Chemistry and Physics 23,
DOI 10.1007/978-94-007-2384-9_24, © Springer Science+Business Media B.V. 2012

The interrelation of magnetic and structural orderings is part of the JT field subject which had started with reviews by Reinen and Friebel [5] and Khomsky and Kugel [6]. However in this presentation we are focusing mainly on the role of the virtual phonon exchange interaction and do not discuss the consequence of the superexchange at orbital degeneracy for the existence of different types of magnetic structures in crystals.

Actually, even the subject of magnetic properties in crystals with CJTE is also not quite new. The symmetrical aspects of it were considered by Vekhter and Kaplan [4] for the case of the tetragonal crystals with correspondingly small multiplicity of orbital degeneracy of the JT ions. However, some new situations arise in the case of the cubic symmetry crystals with high local symmetry of the JT ion surrounding, when triple degenerate electronic states are allowed. This leads to some new interesting results which could possibly explain some experimental data. This is of particular importance since the interest in experiments with the JT crystals containing the triplet ground state ions is growing [7–10].

The physics of this class of materials is much less understood because of many additional complications that take place for these systems in comparison with low symmetry crystals with, for example, tetragonal zircon structure. One of the complications is the result of the degeneracy of the vibrational mode active in the Txe JT-problem. In CJTE this leads to many competing structural orderings in the crystal. Experiments show ferroelastic (we'll call it ZZ-type) ordering and the antiferrodistortive (in ab-plane) XY-type of ordering, among others, as the most typical and important.

Because of that, we are considering a CJTE model with competition of these two orderings caused by corresponding two types of VPE. As it will be shown below, the magnetic properties of JT crystals are especially sensitive to this type of VPE interaction competition.

## 24.2  Physics of VPE Competition and Crystal Hamiltonians

The mechanism of the VPE competition leading to the ZZ- or XY- orderings is relatively simple. It could be understood in the following terms. At ZZ-ordering all structural units (for example, octahedrons) are elongated along z –direction. The result is the overall not-zero tetragonal strain $U_\theta = 2U_{zz} - U_{xx} - U_{yy}$ with the elongation of the crystal cell along Z. At the XY-ordering the crystal is divided in two sublattices. Each of them is elongated in the XY-plane (along X- or Y-direction correspondingly), so that the crystal cell dimensions remain equal in the X- and Y- directions. However each of the sublattices is now equally compressed along the Z- direction [11]. So that overall the crystal cell is again tetragonally distorted. However the sign of the distortion is opposite to the previous one. Naturally these two orderings are competing, and the strongest interaction defines the total crystal's distortion at the structural transition. The magnetic interactions – the external magnetic field Zeeman interaction or orbital exchange

24 Virtual Phonon Exchange Influence on Magnetic Properties...

or dipole-dipole interaction – could significantly influence or be influenced by this VPE competition. As a result of that mutual influence the unusual magnetic properties of the crystals take place. Analysis of these effects is at the center of our discussion.

The Hamiltonian under discussion could be described as

$$
H_1 = \sum_{\kappa} \hbar\omega_{\kappa} \left( b_{\kappa}^+ b_{\kappa} + \frac{1}{2} \right) - \sum_{m\kappa} [V_{m\kappa}^{\vartheta}(3L_{zm}^2 - 2) + V_{m\kappa}^{\varepsilon}(L_{xm}^2 - L_{ym}^2)](b_{\kappa} + b_{-\kappa}^+)
$$
$$
- g\beta H_z \sum_m L_{zm} \tag{24.1}
$$

In the above equation, the first term describes the free phonons, the second one is related to the electron-phonon interaction with the phonons creating the local $E_g$-distortions (these distortions become active in the JT effect for high local symmetry only). The last term in (24.1) corresponds to the magnetic interactions; here $H_z$ could represent the external magnetic field or a combination of the external field and the magnetic molecular field caused by magnetic exchange, superexchange, or dipole-dipole interactions. Of course, the total JT crystal Hamiltonian H in addition to $H_1$ contains terms $H_{add}$ that describe the crystal elastic energy and the interaction of the electrons with the homogeneous $e_g$-symmetry strain $U_\theta$ and $U_\varepsilon$

$$
H_{add} = \frac{1}{2} C_0 \Omega (U_\theta^2 + U_\varepsilon^2) - g_0 E \sqrt{\frac{C_0 E \Omega}{N}} \sum_m [U_\theta(3L_{zm}^2 - 2) + U_\varepsilon(L_{xm}^2 - L_{ym}^2)] \tag{24.2}
$$

In Eqs. 24.1 and 24.2, $b_\kappa, b_\kappa^+$ are the phonon operators, $C_0 = 1/2(C_{11} - C_{12})$ is the symmetrized module of elasticity, $U_\varepsilon = U_{xx} - U_{yy}$ is the crystal orthorhombic strain, and the $L_{x,y,z}$ are the orbital moment operators. The electronic operators in the vibronic term of the Hamiltonian are represented on the triplet basis as

$$
3L_z^2 - 2 = \begin{pmatrix} 1 & 0 & 0 \\ 0 & 1 & 0 \\ 0 & 0 & -2 \end{pmatrix}, \quad L_x^2 - L_y^2 = \begin{pmatrix} -1 & 0 & 0 \\ 0 & 1 & 0 \\ 0 & 0 & 0 \end{pmatrix} \tag{24.3}
$$

The advantage of using this crystal model with the triple degenerate electronic state for studying the ZZ- and XY-ordering competition is that it is possible to get the corresponding virtual phonon exchange interactions exactly, without any approximations, by a unitary transformation

$$
\tilde{H} = e(iR)He(-iR) \tag{24.4}
$$

$$
R = \sum_m [g_m^\theta(3L_{zm}^2 - 2) + g_m^\varepsilon(L_{xm}^2 - L_{ym}^2)] \tag{24.5}
$$

$$g_m^{\theta,\varepsilon} = i\sum_\kappa \left( \frac{V_{m\kappa}^{\theta,\varepsilon*}}{\hbar\omega_\kappa} b_{-\kappa}^+ - \frac{V_{m\kappa}^{\theta,\varepsilon*}}{\hbar\omega_\kappa} b_\kappa \right) \tag{24.6}$$

As a result of this transformation (24.4) the virtual phonon exchange and the Zeeman interactions could be written as

$$H_{\text{int}} = -\sum_{mn\kappa,m\neq n} \left[ \frac{V_{m\kappa}^{\theta*}V_{n\kappa}^\theta}{\hbar\omega_\kappa}(3L_{zm}^2 - 2)(3L_{zn}^2 - 2) + \frac{V_{m\kappa}^{\varepsilon*}V_{n\kappa}^\varepsilon}{\hbar\omega_\kappa}(L_{xm}^2 - L_{ym}^2)(L_{xn}^2 - L_{yn}^2) \right.$$
$$\left. +2\frac{V_{m\kappa}^{\theta*}V_{n\kappa}^\varepsilon}{\hbar\omega_\kappa}(3L_{zm}^2 - 2)(L_{xn}^2 - L_{yn}^2) \right] \tag{24.7}$$

$$\tilde{H}_{Zeem} = -g\beta H_z \sum_m [\cos 2g_m^\varepsilon L_{zm} - \sin 2g_m^\varepsilon \tau_m], \tag{24.8}$$

where the electronic operator in the last term of (24.8) is $\tau_m = \begin{pmatrix} 0 & 1 & 0 \\ 1 & 0 & 0 \\ 0 & 0 & 0 \end{pmatrix}$.

It is important to mention that the interaction operators (24.7) and (24.8) are the exact result of the transformation. In the electron interaction (24.7) only the two first terms contribute to the molecular field JT fields, while the last term does not – this term is responsible for the dynamic processes that are beyond the scope in this article.

The interaction (24.7) can lead to a variety of structurally and correspondingly electronically ordered crystal states. This multiplicity of the ordered phases is directly related to the degeneracy of the vibronic $e_g$-mode active in the CJTE under discussion [12]. In part, the Hamiltonian (24.7) could lead to the ZZ-ferroordering, when at each JT site the average of the electronic operator $\overline{3L_{zI}^2 - 2} = \overline{3L_{zII}^2 - 2} \neq 0$, or to the XY-ordering, when $\overline{3L_{xI}^2 - 2} = \overline{3L_{yII}^2 - 2} \neq 0$ (the roman superscripts I, II correspond to the crystal sublattices).

The electronic operators $3L_x^2 - 2$, $3L_y^2 - 2$ can be evidently represented by the operators in (24.7) using the formulae:

$$3L_x^2 - 2 = -\frac{1}{2}(3L_z^2 - 2) + \frac{3}{2}(L_x^2 - L_y^2)$$

$$3L_y^2 - 2 = -\frac{1}{2}(3L_z^2 - 2) - \frac{3}{2}(L_x^2 - L_y^2) \tag{24.9}$$

Of course, the real crystal phase ordering described by the Hamiltonian (24.7) depends upon the concrete crystal structure. Below we are going to discuss the competition of the ZZ- and XY- orderings in the molecular field approximation only. Correspondingly only terms of the Hamiltonian contributing to the thermodynamics of crystals with this type of ordering are taken into account. In

# 24 Virtual Phonon Exchange Influence on Magnetic Properties...

the molecular field approximation the two sublattice crystal Hamiltonian under discussion is

$$H_m^I = -A^{I,II}(0)\overline{(3L_{xII}^2 - 2)}(3L_{ymI}^2 - 2) - B^{I,II}(0)\overline{(3L_{zII}^2 - 2)}(3L_{zmI}^2 - 2)$$
$$-h_z^I \gamma L_{zmI} - P_z(3L_{zmI}^2 - 2)$$
$$H_m^{II} = -A^{II,I}(0)\overline{(3L_{yI}^2 - 2)}(3L_{xmII}^2 - 2) - B^{II,I}(0)\overline{(3L_{zI}^2 - 2)}(3L_{zmII}^2 - 2)$$
$$-h_z^{II} \gamma L_{zmII} - P_z(3L_{zmII}^2 - 2) \qquad (24.10)$$

where I, II are the sublattice indices, $A^{I,II}(0)$ are the JT molecular field constants, the third term in $H_m^{I,II}$ is related to the magnetic interaction with the magnetic (superexchange or dipole-dipole) molecular field constant $J(0)$

$$h_z^{I,II} \equiv g_z \beta \gamma H_z + J(0) \gamma^2 L_{zI,II} \qquad (24.11)$$

and the external uniaxial pressure is represented by the last term, where

$$P_z \equiv g_0^E \frac{p_z}{\sqrt{C_{0E} \Omega N}} \qquad (24.12)$$

with the $g_0^E$ abbreviation for the electron-strain interaction constant.

The interplay of the structural ZZ- and XY- orderings and the role of the magnetic interactions in that competition could be easily understood basing on one of the Hamiltonians from (24.10) (let's accept for simplicity that P = 0). As it is shown on the Fig. 24.1, The triple degenerate in the structural paraphrase ground electronic state could be split at the temperatures below the critical temperature $T_s$ of the structural transition by the JT molecular field. However the ground electronic state formed by this splitting depends upon the type of the structural ordering. As is shown on Fig. 24.1 at $T < T_s$ the ZZ-ordering could lead to a singlet ground state and an excited doublet state. Instead of that the molecular field stabilizing the XY-type of ordering creates doublet degenerate electronic state and a singlet excited state.

The $h_z$ magnetic interaction behaves differently in these two situations. At the ZZ-ordering $h_z$ linearly splits the doublet components of the excited state and does not have any influence on the ground state. Meanwhile at the XY-type of ordering the magnetic interaction is increasing the gap between one of the components of the ground doublet and the excited singlet. As a result of that in the case of the ZZ-ordering magnetic field can drastically change the structural ordering in the crystal. At a big enough critical magnetic interaction value ($h_z = h_{zcrit}$) one of the excited doublet components becomes the ground electronic state which corresponds to the sign change of the tetragonal crystal distortion. However in the case of the XY-ordering the magnetic interaction only stabilizes the existing structural ordering.

In crystals with the CJTE, structural ordering is always accompanied by ordering of the electronic states (orbital ordering) and all their properties are significantly dependent upon the magnetic interactions and their magnitude relatively to the JT molecular fields.

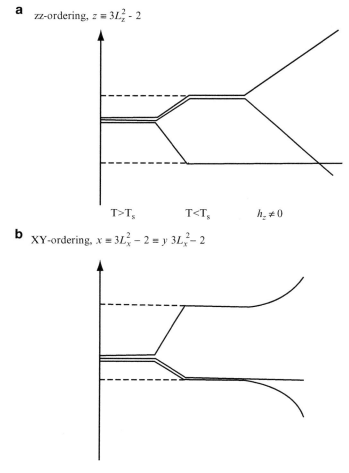

**Fig. 24.1** Magnetic field influence on the energy levels and structural ordering. (**a**) ZZ-ordering, $z \equiv \overline{3L_z^2 - 2}$. (**b**) XY-ordering, $x \equiv \overline{3L_x^2 - 2} = y \equiv \overline{3L_y^2 - 2}$

## 24.3 Homogeneous Crystal Strain and Magnetic Moment

From the Hamiltonians (24.2) and (24.8) it follows that the averages of the electronic operators $3L_z^2 - 2$, $3L_x^2 - 2$, $3L_y^2 - 2$, and $L_z$ define the homogeneous strains of the crystal sublattices (or entire crystal) and its magnetic moment. Because at either structural transitions the crystal acquires tetragonal symmetry, and the magnetic interaction $h_z$ does not change this fact, the both crystal sublattices are always equivalent and x = y (see Fig. 24.1).

The ordering parameters of the phase transitions under discussion $-x$, $z$, and $\overline{L_z}$ are (here one should be careful to distinguish the order parameters x, y, and z from the traditional meaning of the letters denoting the Cartesian axes,) described by the following equations:

$$x = \frac{\frac{9Ax}{D}\sinh\frac{D}{2kT} - \cosh\frac{D}{2kT} + \exp\left(\frac{3[Ax-2(Bz+P)]}{2kT}\right)}{2\cosh\frac{D}{2kT} + \exp\left(\frac{3[Ax-2(Bz+P)]}{2kT}\right)} \tag{24.13}$$

$$z = 2\frac{\cosh\frac{D}{2kT} - \exp\left(\frac{3[Ax-2(Bz+P)]}{2kT}\right)}{2\cosh\frac{D}{2kT} + \exp\left(\frac{3[Ax-2(Bz+P)]}{2kT}\right)} \tag{24.14}$$

$$\overline{L_z} = 4\frac{\frac{h_z}{D}\sinh\frac{D}{2kT}}{2\cosh\frac{D}{2kT} + \exp\left(\frac{3[Ax-2(Bz+P)]}{2kT}\right)} \tag{24.15}$$

In the Eqs. 24.13–24.15 $D \equiv \sqrt{9A^2x^2 + 4h_z^2}$ and $h_z$ is defined in Eq. 24.11. Analysis of the Eqs. 24.13 and 24.14 easily allows proving that they describe the crystal tetragonal distortion of the different signs – cubic crystal compression and expansion along z-direction.

When one of the three interfering interactions (two types of VPE interactions and one of a magnetic type) is dominant ($A \gg B$, J; or $B \gg A$, J; or $J \gg A$, B) the corresponding order parameter equations have a standard form

$$x = 2\frac{\exp(\frac{3Ax}{2kT}) - \exp(-\frac{3Ax}{2kT})}{2\exp(\frac{3Ax}{2kT}) + \exp(-\frac{3Ax}{2kT})} \tag{24.16}$$

$$z = 2\frac{1 - \exp(\frac{3Bz}{2kT})}{2 + \exp(-\frac{3Bz}{2kT})} \tag{24.17}$$

$$\overline{L_z} = 2\frac{\sinh\frac{h_z}{kT}}{2\cosh\frac{h_z}{kT} + 1} \tag{24.18}$$

where it is assumed that the uniaxial pressure $P_z = 0$.

It is important to mention that while the VPE interactions in the Eqs. 24.16 and 24.17 stabilize Z-tetragonal distortions of different signs, the magnetic interaction due to the magnetostriction effect always supports positive z-strain described below

$$z = 2\frac{\cosh\frac{h_z}{kT} - 1}{2\cosh\frac{h_z}{kT} + 1} \tag{24.19}$$

In part that means that even in case of the antiferromagnetic ordering, when the sublattice magnetic molecular fields have the opposite signs, the elastic magnetostrictive deformation is the same for each of the sublattices.

## 24.4 Numerical Results

Let's consider first the results of the magnetic field influence on the ZZ-type of the structural ordering in the crystals. In this case the JT molecular field constant B is bigger than the VPE constant A corresponding to the XY ordering.

The temperature dependence of the tetragonal distortion along the X(Y)-direction in a crystal sublattice is shown on the Fig. 24.2 for $B = 1$ and $A = 0.5$ (on this and other figures all parameters are given in arbitrary energy units so that only their relative magnitude is important). It is clear that at the zero magnetic fields a regular first order structural transition takes place. The external magnetic field is reducing the x-tetragonal crystal strain and at a critical value of $h_{zcrit} = 1.0$ sharply changes the sign of the distortion to the opposite value. After that increasing the magnitude of the magnetic interaction $h_z > h_{zcrit}$ leads to a sharper first order transition at lower temperatures. Above the critical temperature of the structural phase transition a regular paraelastic temperature change of the order parameter takes place. When the magnetic interactions become big enough ($h_z \geq 1.6$) the magnetostriction effect overcomes the spontaneous crystal strain caused by ZZ-type of VPE and there is no structural transition.

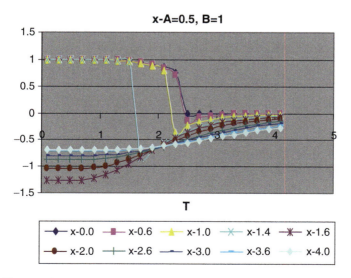

**Fig. 24.2** The x-strain temperature dependence in crystals with the structural ZZ-ordering for different magnitudes of the magnetic field

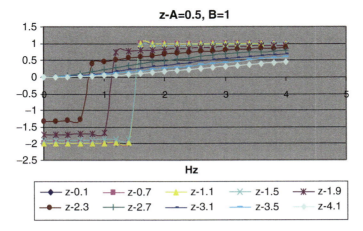

**Fig. 24.3** External magnetic field dependence of the tetragonal z-strain at the different temperatures. The magnitudes of the temperature are shown in the column on the *right*

The influence influence of the magnetic field on the crystal structural ordering is more clearly shown in the calculations of the tetragonal z-strain as a function of the external magnetic field at different temperatures.

The field dependence of magnetostriction is shown in Fig. 24.3 for the same pair of the A and B constants corresponding to the ZZ-ordering. A sharp change of the sign of the crystal z-strain – an overturn of the total crystal deformation – takes place at low temperatures. It is a field induced tetragonal ($z < 0$) – tetragonal ($z > 0$) phase transition. The increase of the temperature decreases the ZZ-molecular field, thus allowing the overturn of the z-strain in a smaller magnetic field. At the temperatures $T \geq 2.7$ there is no structural transition and spontaneous strain. At these temperatures, instead of a field induced first order structural transition, an ordinary paramagnetic paraelastic magnetostriction develops. At small magnetic fields, the magnetostriction is approximately quadratic in $H_z$.

It is easy to see that the results of the calculations for $z(H_z)$ shown in Fig. 24.3 are in complete agreement with the calculations of x(T) on the Fig. 24.2.

The tetragonal-tetragonal structural transition dramatically influences the magnetic moment of the crystal. At the ZZ-ordering as a result of the VPE interaction, as it is shown on the Fig. 24.1, the ground electronic state is changing at the critical magnetic fields.

The initial ground electronic state is characterized by a zero average magnetic moment (Fig. 24.4). However at the critical magnetic fields, when one of the components of the excited doublet crosses the ground electronic state, the situation is changing. This corresponds to a sharp magnetic increase with the external magnetic field. The situation is similar to the behavior of the Ising antiferromagnet [13] in a longitudinal external magnetic field and even more to a metamagnetoelasticity effect [14] in a JT antiferroelastic. But here it is accompanied by a new type of

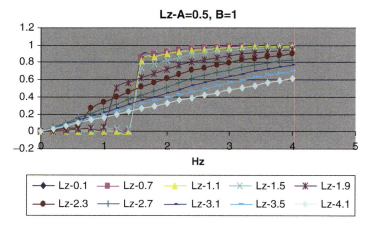

**Fig. 24.4** Field dependence of the magnetic moment at different temperatures (for each curve the T-magnitude is shown in the column on the *right*)

tetragonal – tetragonal structural transition. Once again it should be noted that this Ising type behaviour of the magnetic moment is related to a change in a structural ordering, not in a magnetic ordering. The jump of the magnetic moment is related to the overturn of the total crystal strain, not to an overturn of one of the crystal sublattices.

Naturally, the interplay of the VPE interactions and the magnetic interaction leads to significant changes in the magnetic nonlinear crystal properties. In case of the ZZ-ordering this nonlinearity is evident from the graphs in Fig. 24.3. If the structural ordering is of XY-type, the magnetic interactions support the ordered phase and smear out the phase transition anomalies. However the specific nonlinearity induced by VPE is still there and it becomes more pronounced when the temperature is closer to the critical temperatures of the transition in the absence of magnetic interactions [11] in Fig. 24.5 we present the magnetic nonlinearities for comparing the cases of the ZZ- and XY-orderings at the different molecular field constants A and B.

## 24.5 Conclusions

A microscopic theory of the mutual influence of magnetic interaction and two types of competing VPE interactions potentially leading to different types of structural ordering in crystals is developed. We present the analysis of the most typical for high symmetry crystals ferroelastic ZZ-ordering and antiferroelastic XY-ordering in the framework of the CJTE theory.

It is shown that in case of closeness of the VPE molecular field constants the role of the magnetic interactions could become crucial. The magnetic interactions can

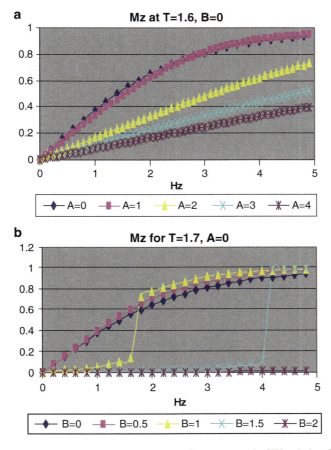

**Fig. 24.5** Nonlinear magnetic behavior of the magnetic moment at the XY-ordering (B = 0, A ≠ 0) and ZZ-ordering (A = 0, B ≠ 0)

change the structurally ordered crystal phase from tetragonal with z > 0 to another tetragonal phase with z < 0.

Of course, at the new type of the tetragonal-tetragonal transition all magnetic properties of the crystal are anomalous and unusual.

Different numerical calculations of magnetic and magnetoelastic properties partially presented in this article give the opportunity to compare the developed theoretical approach with the experimental data. This comparison is done for $MnV_2O_4$ crystals [10]. The experimental data obtained for this compound support the conclusion that in $MnV_2O_4$ crystals a situation with constant B > A, J is realized. In this compound a ferroelastic transition takes place at high temperatures. At lower temperatures a magnetic antiferroordering occurs. This structural ordering is accompanied by the XY structural ordering in the crystals. The possibility of such a situation is discussed in detail in this article.

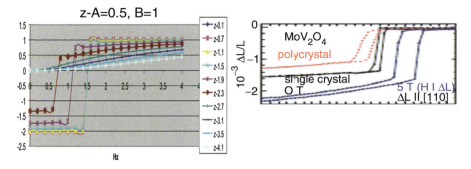

**Fig. 24.6** Comparison of the theoretical calculations of magnetostriction at different temperatures (on the *left*) with the experimental data for the MnV$_2$O$_4$ crystal obtained by Suzuki et al. [10] (on the *right*)

The magnetic and magnetoelastic measurements for MnV$_2$O$_4$ crystals are in agreement with the calculations presented in Figs. 24.2–24.4, and 24.6. On the Fig. 24.6 the results of the calculations or the magnetic field dependence of the static magnetostriction (left part of the Fig. 24.6) are compared with the measurement data obtained for the MnV$_2$O$_4$ crystal in [10]. It is clear that the data are in qualitative agreement. Of course, metastability, polycrystallinity of the samples, and smaller interactions in the crystals between the spins, electrons, and phonons should be additionally analyzed for improving the theory and its agreement with the experiment.

Taking into account the different approximations accepted at the development of the theoretical approach considered in this article, the agreement is satisfactory.

# References

1. Englman R (1972) The Jahn-Teller effect in molecules and crystals. Wiley, New York/London
2. Gehring GA, Gehring KA (1975) Rep Prog Phys 38:1
3. Bersuker IB, Polinger VZ (1989) Vibronic interaction in molecules and crystals. Springer, Berlin/New York
4. Kaplan MD, Vekhter BG (1995) Cooperative phenomena in Jahn-Teller crystals. Plenum Press, New York/London
5. Reinen D, Friebel C (1979) Struct Bonding 37:1
6. Kugel KI, Khomskii DI (1977) Usp Fiz Nauk 136:621
7. Plumier R, Sougi M (1987) Solid State Commun 64:53
8. Plumier R, Sougi M (1989) Physica (Amsterdam) 155B:315
9. Adachi K (2005) Phys Rev Lett 95:197202
10. Suzuki T, Katsumura M, Taniguchi K, Arima T, Katsufuji T (2007) Phys Rev Lett 98:127203
11. Kaplan MD, Zimmerman GO (2008) Phys Rev B 77:104426
12. Hock K-H, Schroder G, Thomas H (1978) Z Phys B 30:403
13. Mattis DC (1965) The theory of magnetism. Harper & Row Publishers, New York/Evanston/London
14. Kaplan MD (2009) In: Koppel H, Yarkony D, Barentzen H (eds) The Jahn-Teller effect. Fundamentals and implications for physics and chemistry. Springer Series of Chemical Physics vol 97, Springer, Berlin, p 653

# Index

## A
Ab-initio, viii, 179
Ab-initio theory, 281–298
ADF. *See* Adiabatic to diabatic transformation
Adiabatic potential energy surface (APES), 7–11, 13, 17–19, 30–31, 34–35, 63, 146, 147, 151–155, 157–159, 220–222, 232, 237, 239, 241, 242, 247, 281, 309, 420, 421, 426
Adiabatic to diabatic transformation (ADF), 281–298
$AlH_2$, 11
Alkali trimers, 301–315
Anisotropic E-type spectrum, 87, 91
Antiferromagnetic (AF) order, 398, 400, 416, 436
Anti-Jahn–Teller effect, 398, 401, 403
Antisymmetric spin state, 40
$AO_4$ structural unit, 425
APES. *See* Adiabatic potential energy surface
Asymmetric radial wave function, 367

## B
$B_{80}$, viii, 265–278
Barrier and Tunneling splitting, 4, 11–14, 22, 89, 96, 100–102, 106, 107, 121, 129–131, 133, 135–138, 144, 331, 337, 343, 345
$BaTiO_3$, 16–19
Benzene cation, 25–36
$BH_2$, 11
Bimetallic oxalates, viii, 317–326
Bipolaron in the $CuO_2$, 20, 21
B3LYP functional, 246, 258–260, 267
Boron
    buckyball, viii, 265–278
    fullerenes, 266–268

## C
$C_{60}$, 215–228, 231, 232, 237, 239, 242, 246, 254, 255, 257, 259–261, 266–271, 274, 276, 278
CASPT2. *See* Complete active space with second-order perturbation theory
CASSCF. *See* Complete active space self-consistent field
Cesium, 303, 305
Character table, 61, 62
Charge transfer multiplet (CTM), 206, 362–365, 369
CJTE. *See* Cooperative Jahn–Teller effect
Clebsch–Gordan coefficients, 46, 234, 248, 262
$CO_3$, 7–9, 26–27
Complete active space self-consistent field (CASSCF), 8, 75–77, 132, 254, 255, 258–259, 291, 304, 306
Complete active space with second-order perturbation theory (CASPT2), 101, 132–134, 304
Complex oxide, 195, 205, 365, 370
Conduction band edge, 202, 203, 205, 206, 209, 370
Cooperative electron-deformational interaction, 389
Cooperative Jahn–Teller effect (CJTE), ix, 4, 14–15, 22, 202, 372–373, 429, 430, 432, 433, 438
Covalence bonding, 17
Critical temperature, 232, 432, 436, 438
Crystal
    pseudospin Hamiltonian, 386
    strain, 87, 89, 94, 102, 434–436, 438
$Cs_3$, viii, 302–307, 309–315
$CsCuCl_3$, 14

M. Atanasov et al. (eds.), *Vibronic Interactions and the Jahn-Teller Effect: Theory and Applications*, Progress in Theoretical Chemistry and Physics 23, DOI 10.1007/978-94-007-2384-9, © Springer Science+Business Media B.V. 2012

CTM. *See* Charge transfer multiplet
$Cu^{2+}$ doped MgO, viii, 106, 130
$CuO_6$ octahedron, 10, 398, 399, 404–407, 414
$CuO_5$ pyramid, 414
Curl Condition, viii, 281, 282, 286

## D

$d^9$ and $d^7$ impurities in cubic lattices, 106, 107
Degeneracy, vii, 5–7, 26, 40, 59–60, 81, 86, 106, 109, 113, 117, 119, 122, 125, 164, 198–204, 217, 232, 236–237, 242, 251, 282, 306–308, 355, 363, 429–440
    removal, 65, 66, 198, 200–203, 367
    of vibrational levels, 42
Density functional theory (DFT), viii, 8, 10, 25–36, 48–59, 71, 72, 78, 79, 194, 196, 216, 217, 219, 221, 227, 254, 255, 258, 259, 267, 366, 425
DFT. *See* Density functional theory
Diabatic potential energy surface matrix, 290–291, 294, 296
Displacive ferroelectrics, 5, 15–20, 22
Distorted perovskite structure, 362, 373
Distortional axes, 232, 239–242
Double exchange, 51, 330, 331, 345–347, 351–353, 357–359, 362, 375
    magnetism, 362, 375
$d^0$ transition metal atom, 362–364, 370, 373
$d^1$ transition metal atom, 362–364, 371–374
d-wave superconductivity, 398, 407–411, 414–416
Dynamic, 11, 12, 33, 39–56, 63, 105–140, 148–149, 217, 220, 224–226, 432
    JTE, viii, 85–102

## E

$Eg \otimes e$, 4, 11–14, 31, 41, 86, 89, 94, 97, 99, 102, 106–123, 144, 146–154, 181–183, 185, 188, 189, 225, 297, 302, 306–310, 313
Elastic moduli, 148, 149
Electric field pulse, ix, 330, 331, 337, 342, 344, 346, 347
Electron
    delocalization, ix, 51, 351–352, 354–355, 357–359
    spin, 89, 196–198, 206, 209, 345, 367, 399–401
Electronic transition, viii, 134, 164, 165, 169, 181, 182, 190, 305, 353

Electron paramagnetic resonance (EPR), viii, 4, 11–13, 16, 19, 86–89, 92, 96, 97, 100–102, 106–112, 122–123, 125–131, 134, 136–138, 153, 157, 158
Electron spin resonance (ESR), 16, 206, 421
Epikernel principle, 60, 65, 70, 74, 76, 79–81
Epitaxial film, 362, 363, 375
EPR. *See* Electron paramagnetic resonance
Equivalent $d^2$ model, 194–199, 206, 207, 209, 362, 367, 369, 372
ESR. *See* Electron spin resonance
Exchange energy, 197
Extended polaron, 20–22

## F

Fermion, 197, 198
Fermi pockets, 398, 407, 409, 415
Ferroelectricity, 4, 15, 18, 22
Ferroelectric transition, ix, vii, 3, 5, 15–20, 425, 426
Finite size effects, 411–412
Force constant, 9, 11, 17, 144, 147, 157–159, 174, 175, 309, 356, 388, 420, 424, 425
Formation energy, 276, 277, 303
Fullerene, viii, 231, 245–262, 266–268
    anions, 215–217, 232
Fulleride, vii, viii, 228

## G

Gd, 362, 370
$GeSc_{1-x}Ti_xO_3$, 362
Ground state potential energy surface, 32, 102
Group-theoretical calculations, 42–45, 59–81
$g$-Tensor values, 135

## H

Harmonic frequencies, 28, 305, 310, 312, 314
HBF, 11
H-bonded materials, 419–426
HCO, 11
$^4$He droplets, 164, 165, 174–176
$HfO_2$, viii, 193–210, 366, 373
Highest occupied molecular orbital (HOMO), 216–219, 221, 232, 258, 271, 272, 274–277, 423–424
High-spin states, 209, 300, 362, 368
High-temperature superconductivity (HTSC), vii, 4, 5, 20, 22, 397, 398
Hilbert space, 282–286, 297

Index 443

HNN, 11
HOMO. *See* Highest occupied molecular orbital
Hopping conducitivity, 369, 373–375
HTSC. *See* High-temperature superconductivity
Hückel theory, 216, 217, 219, 220, 227
Hund's coupling triplet state $^3B_{1g}$, 401–404
Hyperfine tensor values, 89, 91, 95, 102, 134

**I**

Ideal tetrahedral geometry, 368
IDP. *See* Intrinsic distortion path
Influence of random strains, 125–128
Intervalence
  absorption band, 352–357
  optical bands, 40, 55
Intrinsic distortion path (IDP), 27, 31–36
Inverse Jahn–Teller (JT)
  distortion, 322
  transition, 322, 326
Ionic radius, 372
Ising model, 419–426
Isotope effect, 412–414
Isotropic spectrum, 87, 101

**J**

Jahn–Teller (JT), vii–ix, 3–22, 25–36, 39–56, 59–81, 85–102, 105–140, 143–160, 163–176, 179–190, 193–210, 281–298, 301–315, 317–326, 329–349, 351–375, 379–394, 397–416, 419–426, 429–440
  active coordinate, 64, 65, 68–70, 74, 76, 77, 79–81
  active molecules, 26, 31, 34, 35, 72
  active vibration, 34, 41, 72, 144
  coupling, viii–ix, 42, 54, 87, 92, 100, 126, 220, 221, 224, 227, 306–309, 313–315
  symmetry, 66–69, 71, 80
Jahn–Teller effect (JTE), vii, viii, ix, 4–11, 25–36, 60, 67–71, 76, 85–102, 105–139, 143–160, 165, 179, 182, 215–228, 271, 308, 362, 419
Jahn–Teller molecular field, 431–433, 436
JT. *See* Jahn–Teller
JT bipolaron in HTSC cuprates, 20
JTE. *See* Jahn–Teller effect
JT polarons, 20

**K**

$K_3$, viii, 302–307, 310–314
Kamimura-Suwa (K-S) model, 401–405
$KCuF_3$, 15
Kernel, 62, 63, 65, 74, 76, 77, 79

**L**

$La_2CuO_4$, 397–400, 402, 406
Landau–Zener tunneling, ix, 329–349
$La_{0.5}Sr_{1.5}MnO_4$, 15
$LaTiO_3$, 364, 373
Lattice anharmonicity, 18
$La_{2-x}Sr_xCuO_4$ (LSCO), 20, 401, 403, 404, 406, 407, 410, 412–413
$LiCuO_2$, 8, 10
Lowest unoccupied molecular orbital (LUMO), 216–219, 232, 233, 254, 258–260, 272, 274–277, 423–424
$L_{2,3}$ spectrum, 201–202, 373, 374
LUMO. *See* Lowest unoccupied molecular orbital
$Lu_2O_3$, 199, 203–205, 209

**M**

Magnetic
  moment, 318, 324, 434–436, 438, 439
  susceptibility, 354, 392
Magnetic compensation (MC), 318, 319, 321, 325
Magnetization, 318, 321–325, 329–330, 389
MC. *See* Magnetic compensation
Metallic conductivity, 362, 375
MgO
  $Cu^{2+}$, vii, 4, 11–14, 106, 111, 112, 123, 130–138
Mixed valence (MV), 40, 41, 50–56, 361–375
  dimmers, ix, 329–349, 351–360
  trimmers, 50–55
Molecular orbitals (MOs), viii, 29, 32, 61, 199–202, 216–219, 223, 228, 268, 271, 274, 364, 365, 399, 400, 420–421
MOs. *See* Molecular orbitals
Mössbauer spectra, 378–380, 382, 386, 389, 390–394
Mott physics, ix, 397–416
Multideterminental-DFT, 26, 29, 31–35
Multimode
  JTE, 25–36
  problem, 13, 26, 33, 35

Multiplet theory, viii, ix, 193–210, 362–368
Multivibronic operators, 46, 50

**N**
$Na_3$, 119, 283, 286–291, 296, 301
NAC. *See* Non-adiabatic coupling
Nanocrystalline, viii, 193–210, 365
   transition metal (TM) oxides, 194, 197,
     199–204, 362
Nano-grain, 196–197, 199–204, 362, 365,
     369–373, 375
Negative ion states, 201–209, 369, 371
Neutron diffraction, 425
$NH_2$, 11
NMR, 16, 19
Non-adiabatic coupling (NAC), 281–286,
     288–293, 296
Non-crystalline oxides, viii, 194, 204–210
Normal modes, 26, 27, 30–32, 34–36, 71, 113,
     182, 234, 246, 272, 286, 287, 293, 356,
     425
Normal-scattering, 409–410
Nuclear dynamics, viii, 119, 281–298

**O**
O-atom vacancies, 194–196, 202, 205, 207,
     366–368, 370–371
O K edge, 198–204, 206, 209–210, 362, 364,
     365, 369–371
O K pre-edge, 199–204, 206, 209, 362, 368,
     369, 371, 372
Optical spectra, 133–134, 170, 179, 180, 187,
     190, 195
Orbitals
   angular momentum, 110, 318–320, 322
   Cu $d_{x2-y2}$, 400
   Cu $d_{z2}$, 15
   O $p_\sigma$, 268, 269, 399
   O $p_z$, 399
   ordering, 4, 14–15, 433
Order-disorder phase transitions, 15–20
Ordinary paramagnetic paraelastic
     megnetostriction, 437
Orthogonalization, 49, 50
O-vacancy, 194–199, 201, 205, 209, 362,
     365–369, 371
   transitions, 198, 371

**P**
$p^3 \otimes h$-Jahn-Teller coupling, viii, 231–242
Paraelectric temperature change, 436

Parameters, viii, 6, 26–29, 33, 34, 41, 53–55,
     87–91, 93, 94, 98–100, 129, 131–133,
     136, 143–160, 164, 165, 174, 180, 181,
     187–190, 201, 216, 217, 224, 227, 284,
     288, 302, 303, 307–315, 320, 321, 323,
     331, 333, 335, 345, 346, 352, 353, 357,
     358, 382, 385–394, 412, 420, 421,
     423–425, 435, 436
Percolation threshold, 361–375
Perovskite, vii, 4, 15, 17, 18, 22, 362, 369, 373,
     397
PES. *See* Potential energy surface
Phase
   diagram, ix, 398, 414–415
   transitions, vii, 5, 14–20, 22, 326, 380, 420,
     425, 429, 435–438
   velocity, 146–148, 150, 154–157
Phonons, ix, 14, 16, 101, 130, 144, 164–166,
     169–171, 173–176, 179–190, 346, 380,
     384–385, 389, 392, 398, 407–411, 431,
     440
Photoelectron spectrum, 246, 249–253,
     255–258, 260, 261
Piepho–Krausz–Schatz (PKS) model, 51–53,
     55, 56, 331, 352–353, 355–357, 359
PJTE. *See* Pseudo JTE
Polaronic stripes, 20, 22
Population of the high-spin state, 302, 368
Potassium (K), viii, 106–108, 133, 135, 136,
     302–308, 310–314, 355
Potential energy surface (PES)
   extrema, 65, 66
   extremal (extremum) points, 63, 66
   minima, 63, 66, 71
   saddle points, 63
Pseudo-JT activity, vii–ix, 5–11, 35, 40–42,
     50–55, 59–81, 113, 119, 151, 164, 165,
     169, 182, 266, 267, 272–274, 288, 302,
     305, 320, 419–420
Pseudo JTE (PJTE), 4–11, 15–18, 20–22,
     60, 67–71, 73, 74, 77, 79–81, 151,
     167–169, 179–182, 184, 185, 189, 190,
     267–273, 276–278
Pseudospin-phonon coupling, 419–426

**Q**
Quadratic coupling, 12, 167, 221, 237–240,
     242, 258, 309, 313
Quartet states, 303, 305, 306

**R**
Random crystal strain, 87, 89, 94
$Rb_3$, viii, 302–307, 310–314

Reaction coordinates, 27
Relativistic, vi, vii, 302, 307–310
Relaxation time, 135, 144, 146, 148–149, 151–153, 157, 158, 160
Renner–Teller effect (RTE), 9, 60
Representation
   antisymmetric, 64, 74, 81
   degenerate, 70
   direct product, 47, 64, 68
   full-symmetric, 64, 68, 69
   irreducible (IR), 61–62, 64–69
   non-degenerate, 52, 63, 69
   one-dimensional, 61, 65, 68, 69, 164
   symmetric, 61, 64, 68, 69
Resonant Raman scattering (RRS), viii, 163–176
Results of ab initio calculations, 253–254, 258–260
Robin and Day classification of mixed valence compounds, 333, 352, 358, 360
RRS. *See* Resonant Raman scattering
RS2C, 305, 305, 307
RTE. *See* Renner–Teller effect
Rubidium (Rb), viii, 302–307, 310–314

**S**
SA-CASSCF method. *See* State-averaged complete active space self consistent field method
Sc, ix, 361–375
$Sc^{2+}$, 123, 374, 375
$Sc^{3+}$, 362, 370–372, 374, 375
Scanning tunnelling microscopy (STM), viii, 215–228, 242
Sc $L_{2,3}$ spectrum, 363, 369, 374, 375
(Sc,Ti)O2 plane, 362, 369, 373
Second order
   JTE, 7, 271
   perturbation theory, 6–7, 22, 132, 249, 250, 255–257, 260
$Si_3$, 8, 10
Singlet, 52, 54, 74–76, 79, 89, 94–97, 99, 100, 106, 117, 119–122, 126–130, 137, 139, 193, 194, 196, 198, 199, 201–204, 207, 209, 222, 319, 367, 368, 385, 400–404, 406, 408, 410, 416, 433
   ground state, 106, 121, 196, 222, 433
$Si_3N_4$ Si oxynitride, 194, 199, 204–210
$SiO_2$, viii, 193–210
Soft mode, 16, 20
Soft phonon dynamics, 164, 165, 169–174, 176

Spin
   crossover, 7, 8, 40, 380–382, 386–388, 392
   momentum, 197, 198, 322
Spin-orbit coupling (SOC), vii, viii, 301–315, 318, 319, 322, 382, 383, 385
$SrTiO_3$, 364, 369, 373
Stabilization energy, 14, 27, 30, 34, 106, 134, 144, 153, 157, 159–160, 247, 256, 258, 310
State
   degenerate, viii, 15, 26, 29, 31, 181, 190, 202
   electron (electronic), vii–ix, 5–8, 17, 18, 28, 32, 33, 54, 60, 61, 65–81, 86, 90, 91, 96, 106, 107, 112, 123, 157, 164–168, 170, 171, 173, 179–181, 190, 205, 209, 232, 233, 246–248, 250, 254, 261, 262, 281–283, 302, 311, 388, 429, 430, 433, 437
   excited, 6–10, 16, 26, 45, 63, 70, 80, 101, 117, 122, 124, 139, 143, 164–174, 176, 180, 182, 196, 222, 226, 281–299, 302, 305, 341, 342, 358, 372, 433
   split, 32, 72, 307, 400
State-averaged complete active space self consistent field (SA-CASSCF) method, 8, 255, 293
Static, vi, viii, 11–12, 63, 86, 94–97, 105–140, 151–153, 217, 220–223, 226, 258, 302–304, 330, 331, 336, 337, 344, 346, 347, 398, 440
Static and dynamic Jahn–Teller effect, 107–112
STM. *See* Scanning tunnelling microscopy
Strain, 86, 87, 89, 90, 92–102, 107, 108, 123–131, 136–139, 144, 149, 266, 273, 276, 277, 362, 370–372, 383–384, 386–388, 430, 431, 433–438
Strain-induced mixed valence, 129, 362, 371
Structural tetragonal-tetragonal transition, 437–438
Substitutional JT effect, 273
Superconductivity, vii, ix, 4, 20–22, 228, 232, 242, 397–416
Superexchange interaction, 14, 400, 404, 406
Surface interaction, 215–228
Symmetric
   radial wavefunction, 193, 367
   spin state, 7, 41, 63, 193, 205, 322, 347
Symmetry
   B1g, 20, 134, 399–406, 416
   $C_{2h}$, 221, 237–239, 241
   $D_{2h}$, 33–34, 221, 237, 238, 241
   in the dynamic Jahn-Teller problem, 39–56

Symmetry *(cont.)*
  $A_{1g}$, 20, 32–34, 36, 41, 132, 134, 201, 207, 364, 399–403, 410
Symmetry dapted basis, 45–50
Synchrotron X-ray spectroscopy multiplet theory, 362

## T

Tanabe-Sugano (T-S), 201
  diagram, 194, 196–199, 206, 209, 367, 368
Tetragonal distortion, 96–97, 109–110, 148, 151–152, 157, 196, 375, 435, 436
Tetragonally elongated geometries, 99, 110
Theorem of instability, 6
Ti, ix, 16–18, 197, 198, 200, 209, 210, 361–375
$Ti^{2+}$, 374
$Ti^{3+}$, 198–199, 362–364, 369–374
$Ti^{4+}$, 19, 363–364, 373, 374
Ti $L_{2,3}$ spectrum, 199–200, 362, 369, 373, 374
Time-dependent electric field, 330, 331, 336–345, 347
$TiO_2$, 193, 196, 197, 199–201, 204–206, 209, 365–367
TJE. *See* Jahn–Teller effect
Total energy gradient, 259, 260
Totally symmetric normal modes, 26, 31, 34–35, 71
Transition metal (TM) oxides, 194, 196–206, 209, 210, 362
Trianion, $C_{60}^{3-}$, 231–242
Tripet, 197
Triplet ground state, 175, 193, 197, 198, 209, 368, 430
T-S. *See* Tanabe-Sugano
Tunneling splitting, 4, 11–14, 22, 89, 96–97, 100–102, 106, 107, 121, 129–131, 133, 135–138, 144, 331–337, 343, 345
Tunnelling current, 216, 228
Two-band theory, 16

Two-component theory, 398, 405
Two-story house model, 404–405

## U

Ultrasonic attenuation, 144, 147, 149, 158
Ultrasound, 144, 154, 160
  attenuation, 147–148, 157
Umklapp-scattering, 409
Unitary transformation, 254, 283, 289, 431

## V

van der Waals molecules, 302
Vibronic
  interaction, 41, 50, 163–165, 167, 169, 180–182, 330, 352, 420
  state of E symmetry, 167
  states, 55, 89, 91–92, 96, 100, 121, 131, 132, 167, 180, 184, 185, 248–253, 257
  theory, ix, 5, 16, 18, 19, 419–426
Vibronic coupling, vii–ix, 4–7, 11–12, 14–17, 26, 30, 34, 40–42, 50, 51, 53, 55–56, 72, 86, 114, 147, 164–167, 169–170, 175–176, 180, 181, 184, 185, 188, 273, 313, 331, 352–353, 357, 359, 380
  constant, 12, 17, 72, 144, 146, 150, 151, 153, 155–158, 245–263, 310
Virtual phonon exchange, ix, 429–440

## X

X-ray absorption fine structure (XAFS), 16, 19
X-ray absorption spectroscopy (XAS), 194, 198, 201, 205, 206, 208, 209, 362, 365, 367, 369, 370
X-ray scattering, 16

## Z

Zhang-Rice-singlet $^1A_{1g}$, 402–404, 406
$ZrO_2$, 194, 200, 201, 203–205, 207, 364–366